Bacteria from Fish and Other Aquatic Animals

A Practical Identification Manual

Bacteria from Fish and Other Aquatic Animals

A Practical Identification Manual

Nicky B. Buller

Senior Microbiologist
Department of Agriculture
South Perth
Western Australia

CABI Publishing

CABI Publishing is a division of CAB International

CABI Publishing
CAB International
Wallingford
Oxfordshire OX10 8DE
UK

CABI Publishing
875 Massachusetts Avenue
7th Floor
Cambridge, MA 02139
USA

Tel: +44 (0)1491 832111
Fax: +44 (0)1491 833508
E-mail: cabi@cabi.org
Web site: www.cabi-publishing.org

Tel: +1 617 395 4056
Fax: +1 617 354 6875
E-mail: cabi-nao@cabi.org

A catalogue record for this book is available from the British Library, London, UK.

Library of Congress Cataloging-in-Publication Data
Buller, Nicky B.
    Bacteria from fish and other aquatic animals : a practical identification manual / Nicky B. Buller.
        p. cm.
Includes bibliographical references and index.
    ISBN 0-85199-738-4
  1. Aquatic animals--Microbiology.   I. Title.
    QR106.B85 2004
    579.3'176--dc21   2003009624

ISBN 0 85199 738 4

Typeset by AMA DataSet, UK.
Printed and bound in the UK by Biddles Ltd, King's Lynn.

Contents

List of Tables and Figures

Tables

Figures

Photographic section after p. xiv

Foreword

While aquatic animal diseases have been a concern for centuries, a profusion of marine and fresh-water aquaculture and environmental concerns across the globe in the last 75 years has prompted increased interest in these diseases, particularly those caused by bacteria. As aquaculture continues to become more intensive and expands into new frontiers, fish health problems are likely to become more significant. No group of marine or freshwater animals, including mammals, fish, birds, molluscs, shellfish, reptiles and amphibians, have escaped the effects of bacterial diseases. Hundreds of bacterial species can be either pathogenic to wild and cultured aquatic animals or pose a potential disease threat under favourable conditions. Furthermore, the costs incurred by governments, private aquaculture and the public, due to bacterial-related diseases and attempts to control them, totals millions of dollars annually as a result of lost aquatic resources. In order to successfully cope with these disease-producing microbes in a cost-effective manner, prompt and accurate identification is essential.

Until now there has been no single source available for use in identifying bacterial microbes from so many diverse marine and freshwater animals. However, *Bacteria from Fish and Other Aquatic Animals: a Practical Identification Manual* now provides just such a source with global application. This practical, user-friendly identification manual will be of great value to inexperienced and experienced bacteriologists, microbiology teachers and/or students, aquatic animal health researchers or diagnosticians, as well as to workers in public health facilities or medical laboratories who work with marine and freshwater fish, birds, mammals, molluscs, shellfish, reptiles or amphibians. Aided by numerous tables and colour figures the author discusses conventional bacterial identification procedures, commercially available data-based identification kits, molecularly based PCR and 16S rDNA sequencing, thus providing utility to a broad scientific sector. In this single volume one can find biochemical, biophysical and molecular characteristics of nearly 400 species of aquatic bacteria, media on which they are cultured and a brief discussion of many diseases with which they are associated. Assembling this manual was a monumental task and its author, Nicky B. Buller, is to be highly commended for providing this invaluable addition to aquatic microbiology.

John A. Plumb
Department of Fisheries and Allied Aquacultures
Auburn University
Alabama, USA

Acknowledgements

I would like to acknowledge Dr Jeremy Carson (Department of Primary Industries, Water and Environment, Tasmania), and Nick Gudkovs (Australian Animal Health Laboratories, Geelong, Victoria) for allowing me to visit their respective laboratories, particularly in the early part of my career, for discussions on fish bacterial diseases and isolation techniques. In addition, to Dr Jeremy Carson and his laboratory who, over the years, have confirmed the identification of some of our *Vibrio* and *Flavobacterium* strains isolated from diagnostic cases. In particular, for confirmation of the identity of Animal Health Laboratory Department of Agriculture (AHLDA) diagnostic strains mentioned in this manual (*Flavobacterium columnare*, *Listonella anguillarum*, *Vibrio agarivorans*, *Vibrio halioticoli*, and *Vibrio mediterranei*). To Dr Annette Thomas (Department of Primary Industries, Queensland), thank you for the gift of cultures of *Vibrio alginolyticus*, *Vibrio (carchariae) harveyi*, and *Streptococcus iniae* and for our many discussions over the years on veterinary and fish pathogens. Thank you to Dr Bruno Gomez-Gil (CIAD/Mazatlán Unit for Aquaculture, Mexico) for test results on *V. rotiferianus* not listed in the literature, and for supplying the paper on *Vibrio pacinii*, before publication, for inclusion in this manual. To Dr Fabiano Thompson (Laboratory for Microbiology, Ghent University, Belgium) for providing further API 20E reactions for *Vibrio brasiliensis*, *Vibrio neptunius*, *Vibrio xuii* not listed in the journal article. To the fish pathologists who I have worked with, in particular Dr Brian Jones and in memorandum Dr Jeremy Langdon, who have helped me to a better understanding of fish diseases. I am also grateful to the Fisheries Research and Development Council for funding to enable me to attend workshops on fish diseases held at the University of Tasmania in 1996 and attendance at the fish bacteriologists' workshops held at the Australian Animal Health Laboratories, Geelong, in 2000 and 2001.

The photograph of *Renibacterium salmoninarum* was obtained from the Animal Health Laboratory (AHL) archives. There was no documentation as to the source of this photograph and thus I am unable to acknowledge the origin. The rest of the photographs were taken by the author and were of cultures obtained as diagnostic submissions at AHL, type strains, or from Dr A. Thomas. The photographs of *Flavobacterium columnare* adhering to gill tissue are courtesy of Dr Brian Jones.

A great many microbiologists and other scientists have been responsible over the years for developing and refining the media and techniques used for isolation and identification of bacteria. The media quoted from the literature in this book are referenced to include the originator and refiners of those media. My apologies if I am remiss in leaving anybody out.

Due to the expense of producing the photographic plates, assistance towards their cost was achieved through sponsorship from the following organizations. Their assistance is gratefully acknowledged.

Principal sponsor:
Agriculture, Fisheries and Forestry – Australia (AFFA), Aquatic Animal Health Subprogram, funded through Fisheries Research and Development Corporation (FRDC).

Other sponsors:
Department of Fisheries, Western Australia
Oxoid Australia
Animal Health Laboratories, Department of Agriculture, Western Australia.

The contribution by the Animal Health Laboratories, Department of Agriculture, Western Australia, is also gratefully acknowledged for the use of digital photographic equipment, culture media and bacterial cultures used for the photographic section.

I would also like to acknowledge Mr Tim Hardwick at CAB International, for his assistance in bringing this manuscript into publication.

Introduction

This manual attempts to provide a source that enables the identification of bacteria that may be found in animals that inhabit the aquatic environment. The emphasis is on bacteria from farmed aquatic animals. In the words of Louis Pasteur, 'chance favours the prepared mind'; therefore, an informed microbiologist will have a better chance of identifying those bacteria.

Our knowledge about the isolation and identification of bacteria from aquatic animals and the aquatic environment is expanding at a rapid rate. New organisms, be they pathogens, environmental, normal flora or potential probiotics, are being described and reported each month. This has happened due to an increase in aquaculture research, an increase in intensive fish farming systems, an increase in the international trade of live aquatic animals and products, and the emergence of new diseases. More and more laboratories are becoming involved in the isolation and identification of these bacteria in either a diagnostic or research capacity. In this manual there is an emphasis on bacteria of interest to the aquaculture industry either as pathogens, normal flora or strains that may be used as probiotics. Some bacteria that have been isolated from diverse habitats are also included. This manual attempts to provide these laboratories with an up-to-date and standardized database of methods and biochemical identification tables that can be used to isolate and identify bacteria from aquatic sources. Molecular diagnostics is becoming more routine in many laboratories and a section on molecular identification of bacteria using the PCR and 16S rDNA sequencing is also included in this manual.

Many laboratories receive samples not only from veterinary sources, but also samples from zoo animals such as penguins, seals, seabirds, and aquatic mammals both captive and wild. From other aquatic sources, samples for analysis may come from fish, both wild and cultured, freshwater and marine, aquarium fish, tropical fish, and cultured aquatic animals such as abalone, pearl oysters, seahorses, lobster, crayfish, yabbies, marron and prawns. All these hosts have their own microflora and potential bacterial pathogens and are found in a diverse range of habitats from tropical to cool temperate climates. This manual includes as many of those bacteria as possible that may be found during examination of samples from this diverse range of host and habitat. Not only pathogens, but also environmental and saprophytic organisms, are included to aid in the understanding of the microflora that may be found in such samples. Many bacteria from the more extreme environments have also been included as, with the increase in aquaculture throughout the world, and the increasing knowledge about the microflora of such habitats, these organisms may find their way into some laboratories via the samples submitted. Therefore, those isolates that are capable of growing on the isolation media recommended for aquatic organisms such as ZoBell's or Marine agar 2216 (Difco) are included in this manual. In addition, I have included some of the Antarctic organisms that have been suggested as a low-cost food source for some marine finfish because they are rich in omega-3 polyunsaturated fatty acids significant in the diet (Nicols *et al.*, 1996) and thus they may be cultured from samples that come into a laboratory.

Medical laboratories are also required to identify an increasing number of bacteria from aquatic habitats that may be involved in clinical infections. This book may also assist in the identification of such bacteria that are not normally listed in the commercial databases such as API (bioMérieux).

Results of phenotypic test results reported in the literature can be confusing. It is important to perform tests by the methods that have been used by the reporting literature. In this book, the majority of biochemical tests have been performed according to West and Colwell (1984) and Cowan and Steel (1970), and these methods are listed. Results are also included from the commercial identification kits available, namely, API 20E, API 50CH and API-ZYM from bioMérieux. Results from these tests are listed in the appropriate tables. Phenotypic tests that may produce different results between biochemical tube media and commercial identification kits include citrate reaction, decarboxylases, indole, and some carbohydrates. Where different strains have been used in the literature and different phenotypic results reported, the results of these organisms have been listed separately. This is an attempt to enable microbiologists to provide the best possible identification of an organism isolated from a diagnostic or research sample.

A clearly defined set of biochemical tests is used as much as possible in this manual and from this the majority of pathogens and non-pathogens encountered from aquatic sources can be identified to genus level and, and for the more commonly known bacteria, to species level. The aim of having a defined set is so that laboratories that make in-house media can prepare media that will enable as many bacteria as possible to be cultured and identified in the routine laboratory without having to prepare an excessive number of test and growth media. Some of the problems with variations in biochemical reactions reported in the literature are due to different methods being used. This manual provides a standardized set of biochemical identification methods for aquatic organisms and the reactions reported here are based on this defined set.

Bacteria that require specialized media and identification tests are included here to assist laboratories in identification of these organisms. In the case of *Brucella*, *Mycoplasma* and *Mycobacteria*, these methods are intended only as a guide, as these organisms should always be sent to a laboratory that specializes in their identification. Other media, such as alternative methods for the detection of carbohydrate fermentation by *Flavobacterium* species for example, are also included.

The terms fermentation and utilization are often confused in the literature, and in some cases it has been difficult to assess by which method the test was done. It is important to distinguish between fermentation and utilization as they are separate methodologies and a bacterium may show a positive test for fermentation of a carbohydrate, yet negative for utilization of that same carbohydrate when tested as a sole carbon growth source. Basically, fermentation refers to the fermentation or breakdown of a carbohydrate, commonly called a 'sugar'. The breakdown products are detected by a pH change in the medium indicated by a colour change in the pH indicator, usually phenol red. Utilization refers to a test where a carbon source is assessed as a sole growth source for that bacterium. There are no other nutrients in the medium and growth is observed macroscopically, seen as an increase in the turbidity of the test medium. There are no pH indicators in utilization tests. The exception to this is citrate utilization using the Simmons method.

Cryptococcus (a yeast) is also included in this manual, because although it is not a bacterium, it is a zoonotic hazard to fish pathologists, veterinary pathologists, microbiologists and other laboratory personnel who may be dealing with samples from aquatic mammals that are susceptible to this organism. It is therefore included so as to alert staff to the biological hazards of some samples. Other zoonotic organisms include *Brucella*, *Mycobacteria* and *Nocardia*. Many of the bacteria from aquatic sources may cause infections in humans and these are listed in Table 1.1.

Layout of the Manual

This manual is divided into sections basically according to the steps taken to isolate and identify an unknown bacterium. Experienced microbiologists may find some sections basic, but these are included for the benefit of students and newly graduated microbiologists or for researchers who are unfamiliar with the techniques of bacteriology, particularly those used in a diagnostic laboratory.

Thus the sections encompass the host and the microbe, isolation techniques, phenotypic (biochemical) identification techniques, molecular identification and a media section.

In the phenotypic identification section is a flow chart (see Fig. 4.1, p. 138) that directs the microbiologist to the appropriate biochemical identification table required for the identification of the unknown organism. The biochemical identification tables are named according to genus such as *Aeromonas* and *Vibrio*, or to Gram reaction and cell shape or oxidase reaction.

In the conventional identification tables, the organisms are listed in alphabetical order (with the exception of the *Vibrio* tables, 4.21 and 4.22) under headings of Pathogen or Environmental. The Pathogen and Environmental differences refer to the pathogenicity for fish and aquatic animals, and generally not to humans or terrestrial animals. The inclusion of biochemical reactions for saprophytic and other species is intended to assist with the identification and ensure that the correct identification is made where species have similar results. In the *Vibrio* tables, the organisms are listed according to their groupings based on ODC, LDC and ADH reactions. The intention is to use these groupings as the starting point for identification, similar to a flow chart. Tables for the API databases have the organisms listed alphabetically.

Significance

Fish and other aquatic animals (farmed and wild) are prone to bacterial infections in the same way as land animals, especially when they are stressed. Disease may occur systemically or be confined to external surfaces such as the skin or gills. In many instances, the pathogenic bacteria are ubiquitous in the environment, or may form part of the normal internal bacterial flora of an aquatic animal. One study suggested that up to 28 different *Vibrio* species may be found in the hepatopancreas (10^4 CFU/g), intestine and stomach (10^6 CFU/g) of healthy shrimp. The *Vibrio* species identified included *V. alginolyticus*, *V. parahaemolyticus*, *V. cholerae* and *P. damselae*. In diseased states only one or two *Vibrio* species are found (Gomez-Gil *et al.*, 1998). Therefore, many factors need to be considered in making a disease diagnosis, such as clinical signs and symptoms, pathology, amount of growth of the cultured bacteria, the numbers of different species cultured, the tissue site from which it was isolated, and sterility of the specimen collection (Lightner and Redman, 1998).

Bacterial microflora on the surface of fish are heterogeneous in their salt requirement for growth. This requirement for salt (halophilic) is usually retained after serial subculture. Likewise the gut of fish can be composed almost exclusively of halophilic vibrios (Liston, 1957; Simidu and Hasuo, 1968). Therefore, when attempting culture for pathogens, their salt requirement needs to be taken into account. This also applies to the biochemical identification tests.

Bacterial diseases affecting aquatic animals are detailed in texts such as Austin and Austin (1999), and Woo and Bruno (1999) (see Chapter 8 'Further Reading and Other Information Sources'). In this manual, the diseases and the bacterial cause are presented as a quick reference tabular format only.

Photographs of Culture and Microscopic Appearance of Organisms

----

The microscopic and cultural appearance of 31 species of bacteria and some of the biochemical test results are detailed here. Obviously not all bacteria can be presented, however, genera and species have been photographed on commonly used media to show their cultural appearance on that medium. Some species such as *Vibrio mimicus* and *Vibrio cholerae* are photographed to indicate how similar they can be by cultural appearance and how similar they may appear to motile *Aeromonas* species. Likewise the motile *Aeromonas* species all appear similar on blood agar media, whereas the non-motile *Aeromonas salmonicida* has a distinctive colony appearance, slow-growing, with pigment production after a few days incubation.

Reactions of some biochemical tests are included for those not familiar with these reactions.

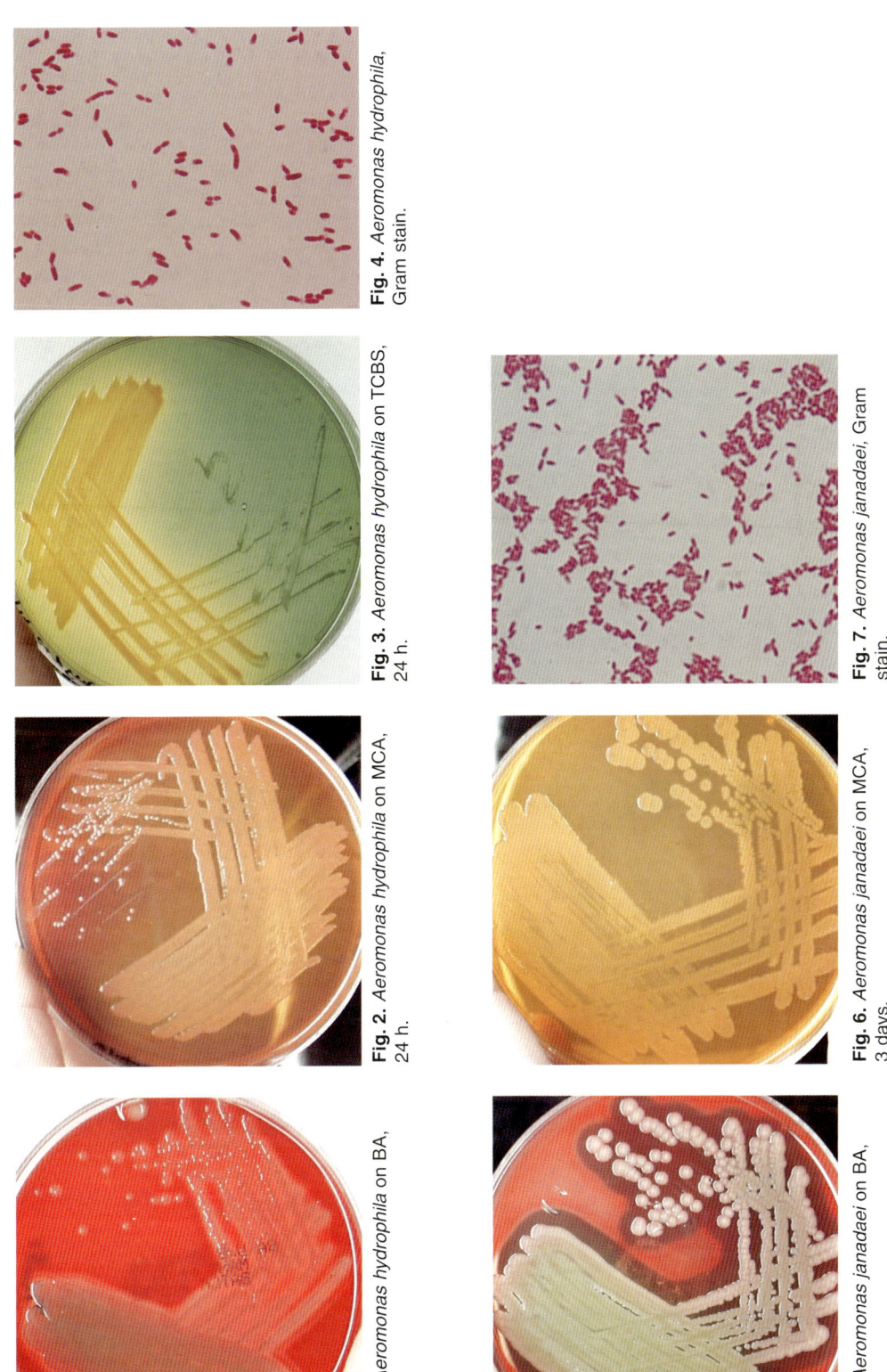

Fig. 4. *Aeromonas hydrophila*, Gram stain.

Fig. 3. *Aeromonas hydrophila* on TCBS, 24 h.

Fig. 7. *Aeromonas janadaei*, Gram stain.

Fig. 2. *Aeromonas hydrophila* on MCA, 24 h.

Fig. 6. *Aeromonas janadaei* on MCA, 3 days.

Fig. 1. *Aeromonas hydrophila* on BA, 24 h.

Fig. 5. *Aeromonas janadaei* on BA, 48 h.

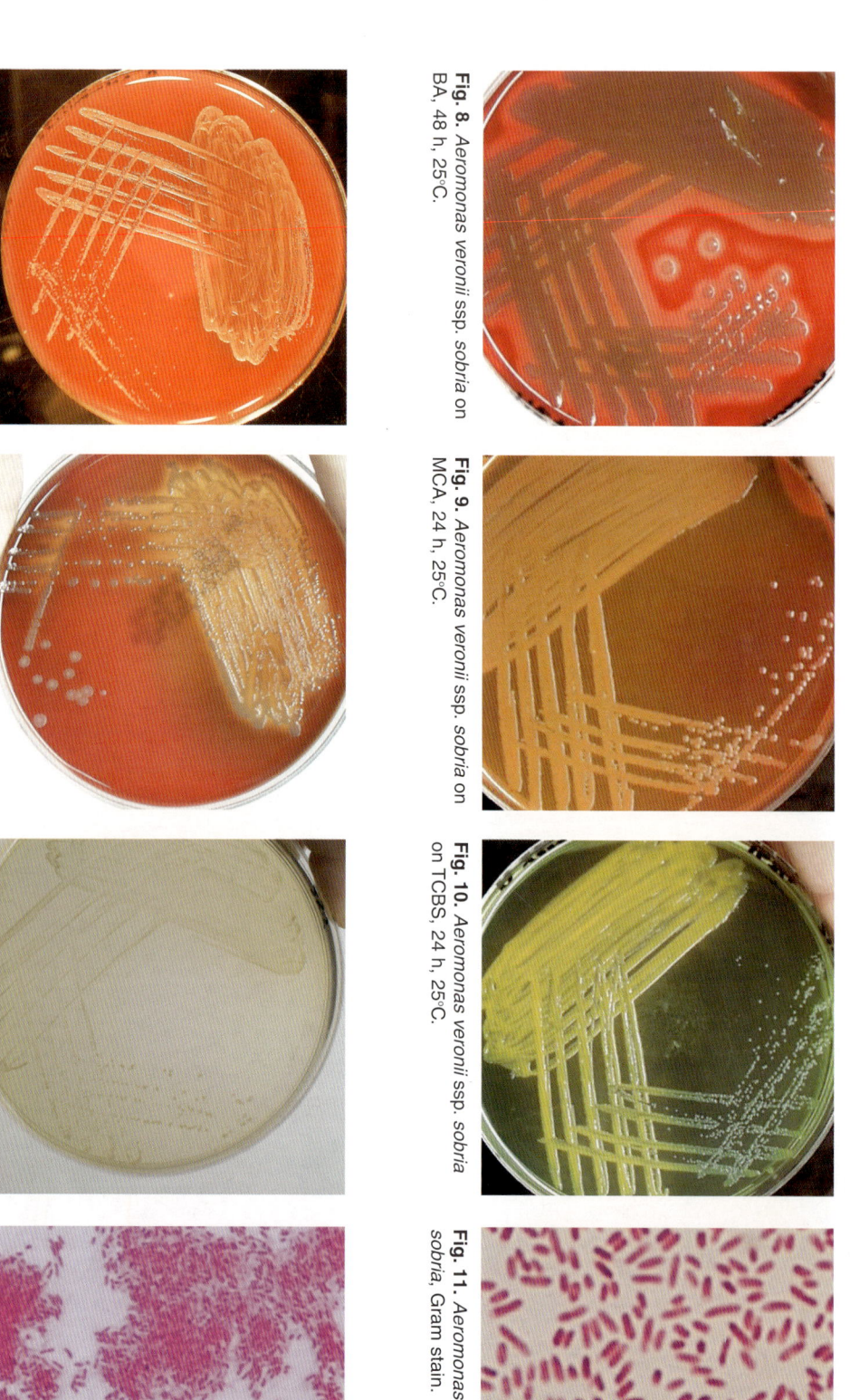

Fig. 8. *Aeromonas veronii* ssp. *sobria* on BA, 48 h, 25°C.

Fig. 9. *Aeromonas veronii* ssp. *sobria* on MCA, 24 h, 25°C.

Fig. 10. *Aeromonas veronii* ssp. *sobria* on TCBS, 24 h, 25°C.

Fig. 11. *Aeromonas veronii* ssp. *sobria*, Gram stain.

Fig. 12. A typical *Aeromonas salmonicida* (Australian strain) on BA, 3 days.

Fig. 13. A typical *Aeromonas salmonicida* (Australian strain) on BA, 7 days.

Fig. 14. A typical *Aeromonas salmonicida* (Australian strain) showing pigment on NB agar.

Fig. 15. A typical *Aeromonas salmonicida* (Australian strain), Gram stain.

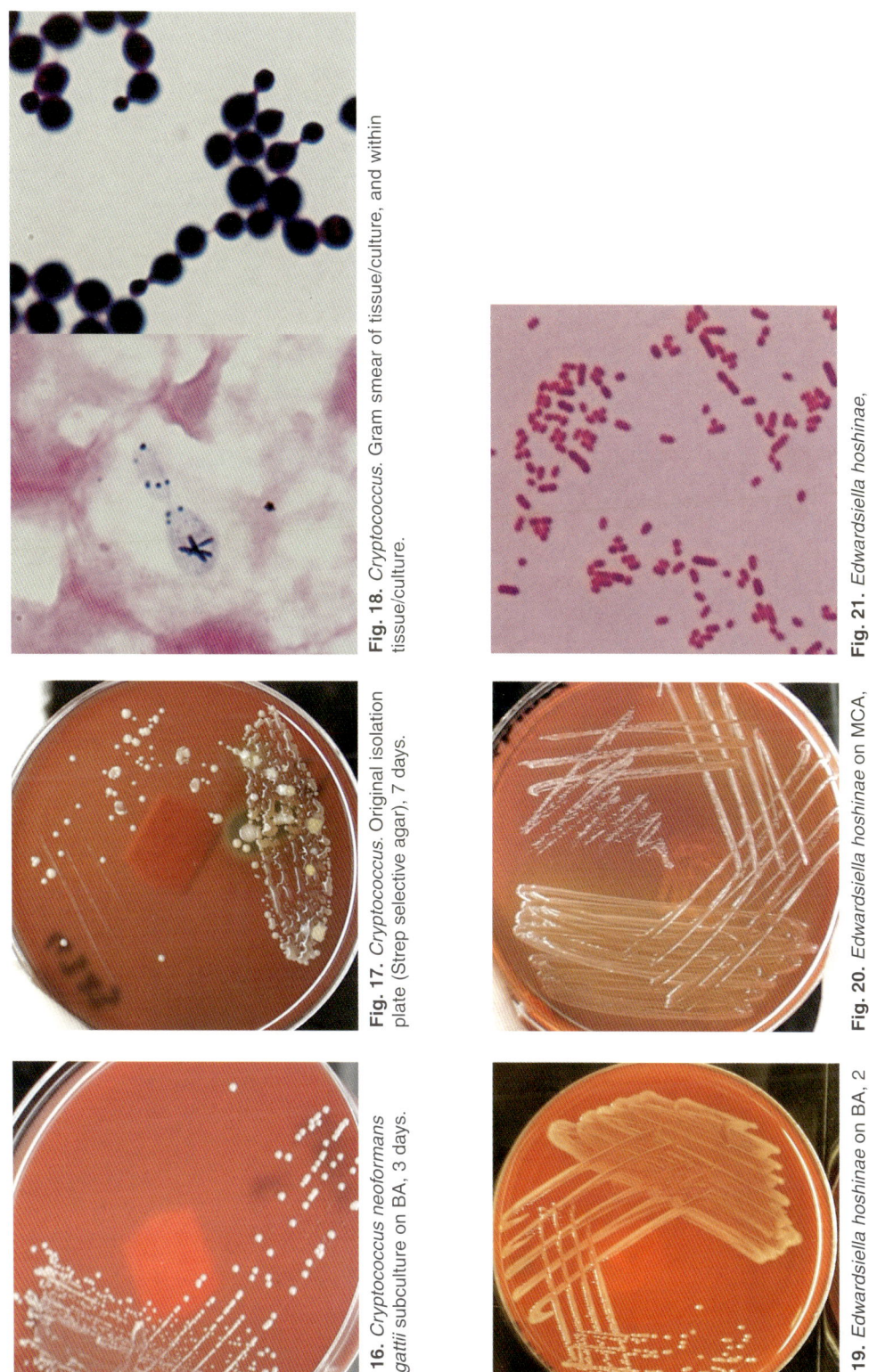

Fig. 16. *Cryptococcus neoformans* var. *gattii* subculture on BA, 3 days.

Fig. 17. *Cryptococcus*. Original isolation plate (Strep selective agar), 7 days.

Fig. 18. *Cryptococcus*. Gram smear of tissue/culture, and within tissue/culture.

Fig. 19. *Edwardsiella hoshinae* on BA, 2 days, 25°C.

Fig. 20. *Edwardsiella hoshinae* on MCA, 24 h, 25°C.

Fig. 21. *Edwardsiella hoshinae*, Gram stain.

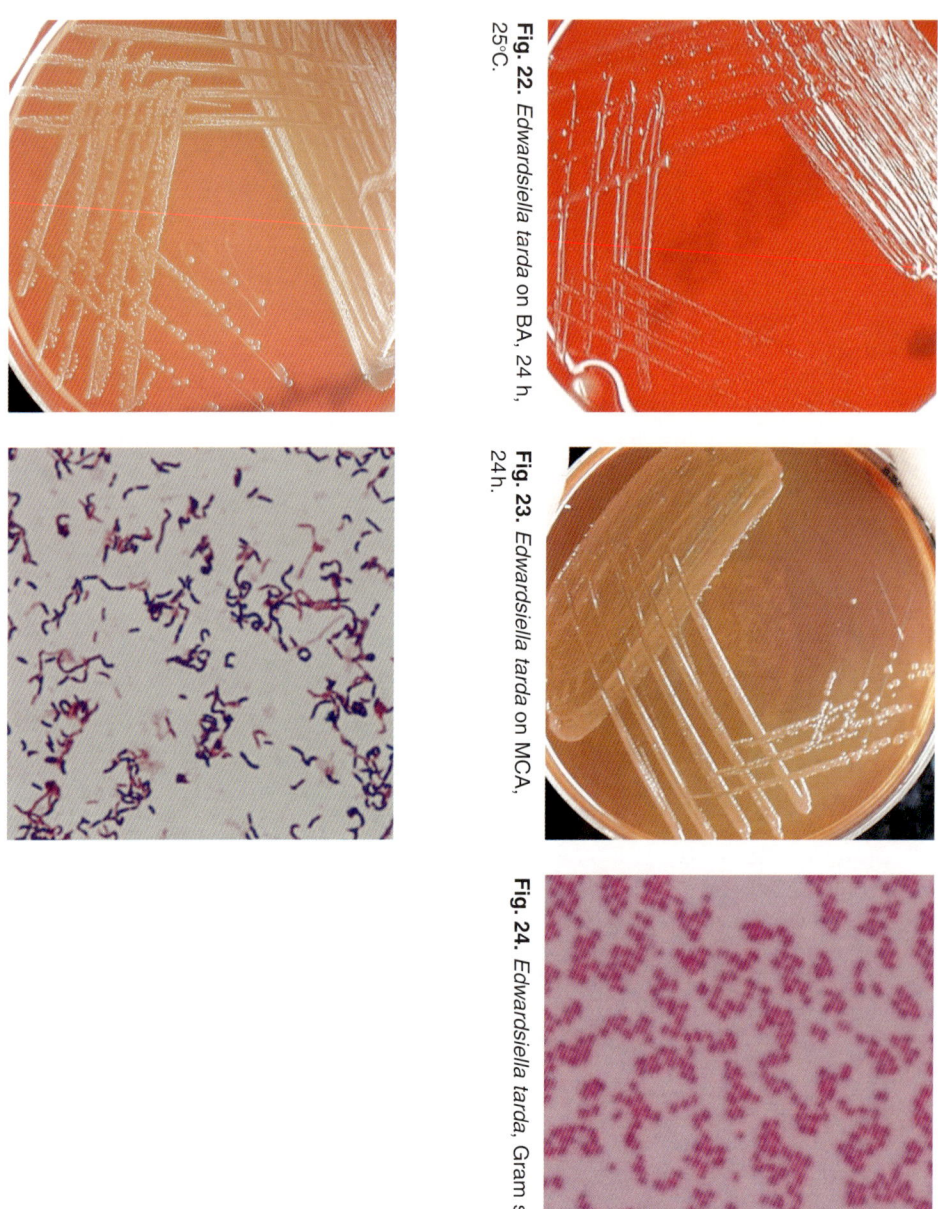

Fig. 22. *Edwardsiella tarda* on BA, 24 h, 25°C.

Fig. 23. *Edwardsiella tarda* on MCA, 24 h.

Fig. 24. *Edwardsiella tarda*, Gram stain.

Fig. 25. *Erysipelothrix rhusiopathiae* on BA, 24 h, 25°C.

Fig. 26. *Erysipelothrix rhusiopathiae*, Gram stain.

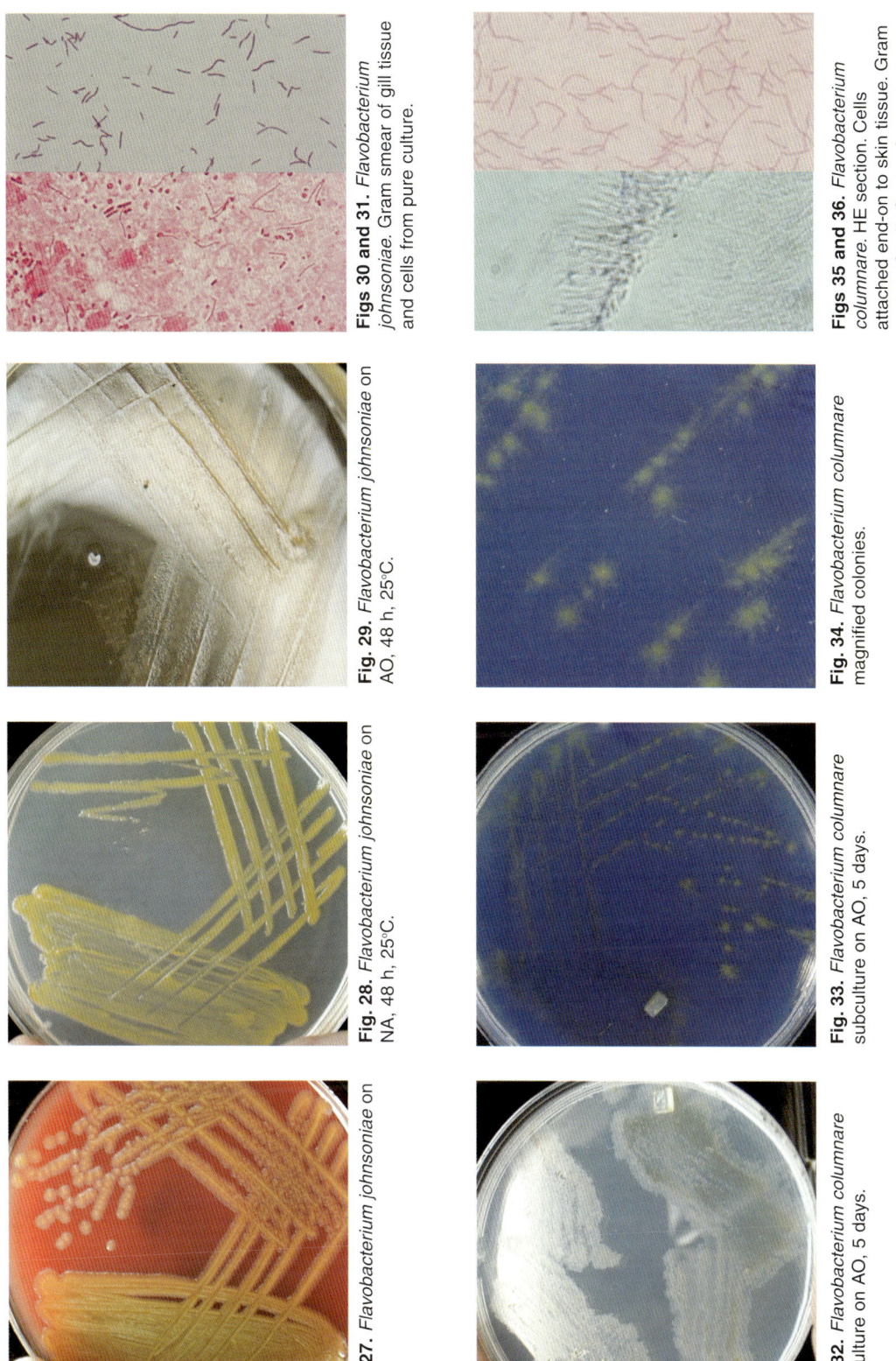

Fig. 27. *Flavobacterium johnsoniae* on BA.

Fig. 28. *Flavobacterium johnsoniae* on NA, 48 h, 25°C.

Fig. 29. *Flavobacterium johnsoniae* on AO, 48 h, 25°C.

Figs 30 and 31. *Flavobacterium johnsoniae*. Gram smear of gill tissue and cells from pure culture.

Fig. 32. *Flavobacterium columnare* subculture on AO, 5 days.

Fig. 33. *Flavobacterium columnare* subculture on AO, 5 days.

Fig. 34. *Flavobacterium columnare* magnified colonies.

Figs 35 and 36. *Flavobacterium columnare*. HE section. Cells attached end-on to skin tissue. Gram smear of cells from culture.

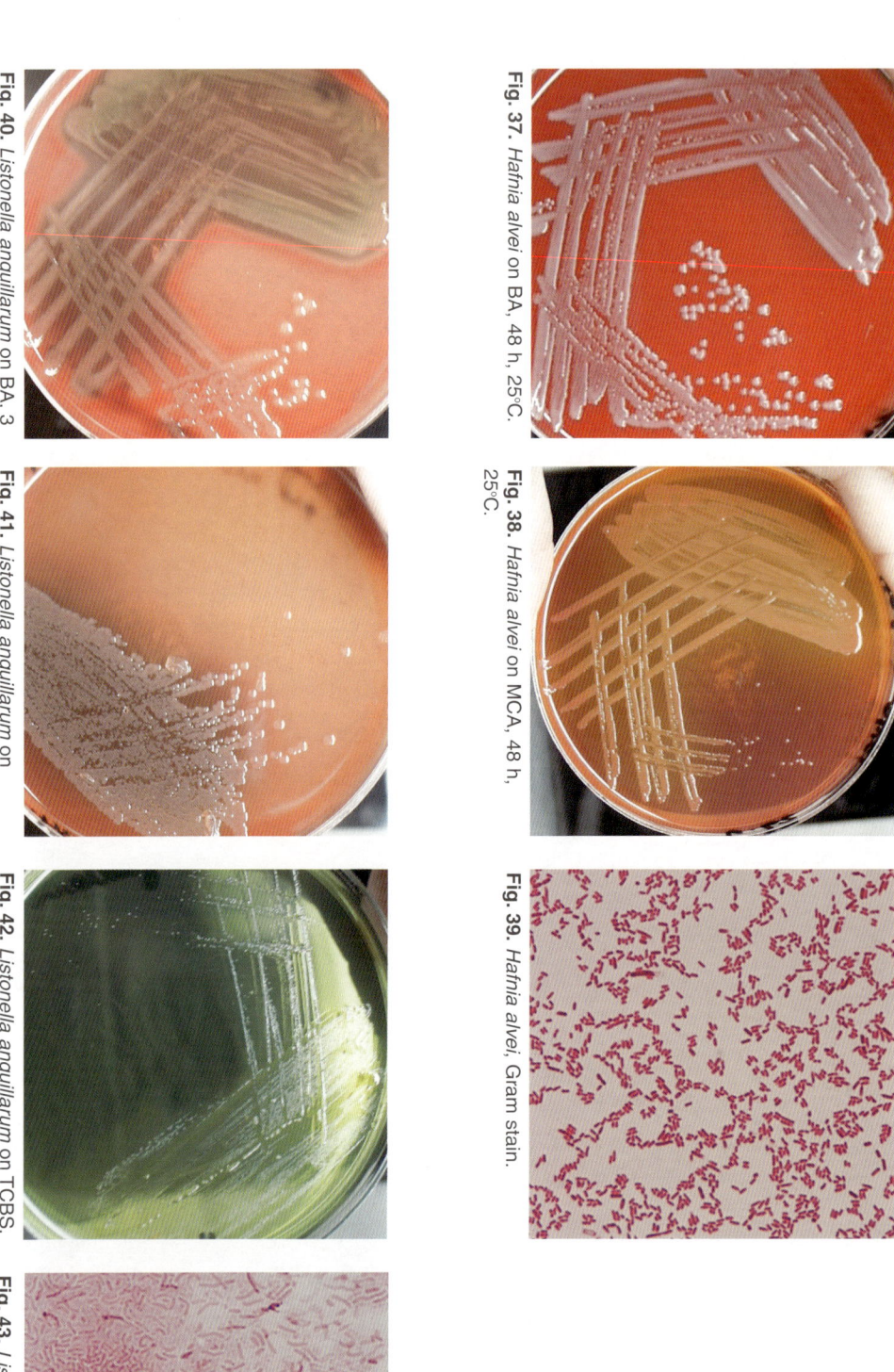

Fig. 37. *Hafnia alvei* on BA, 48 h, 25°C.

Fig. 38. *Hafnia alvei* on MCA, 48 h, 25°C.

Fig. 39. *Hafnia alvei*; Gram stain.

Fig. 40. *Listonella anguillarum* on BA, 3 days, 25°C.

Fig. 41. *Listonella anguillarum* on MSA-B, 24 h, 25°C.

Fig. 42. *Listonella anguillarum* on TCBS, 48 h, 25°C.

Fig. 43. *Listonella anguillarum*, Gram stain.

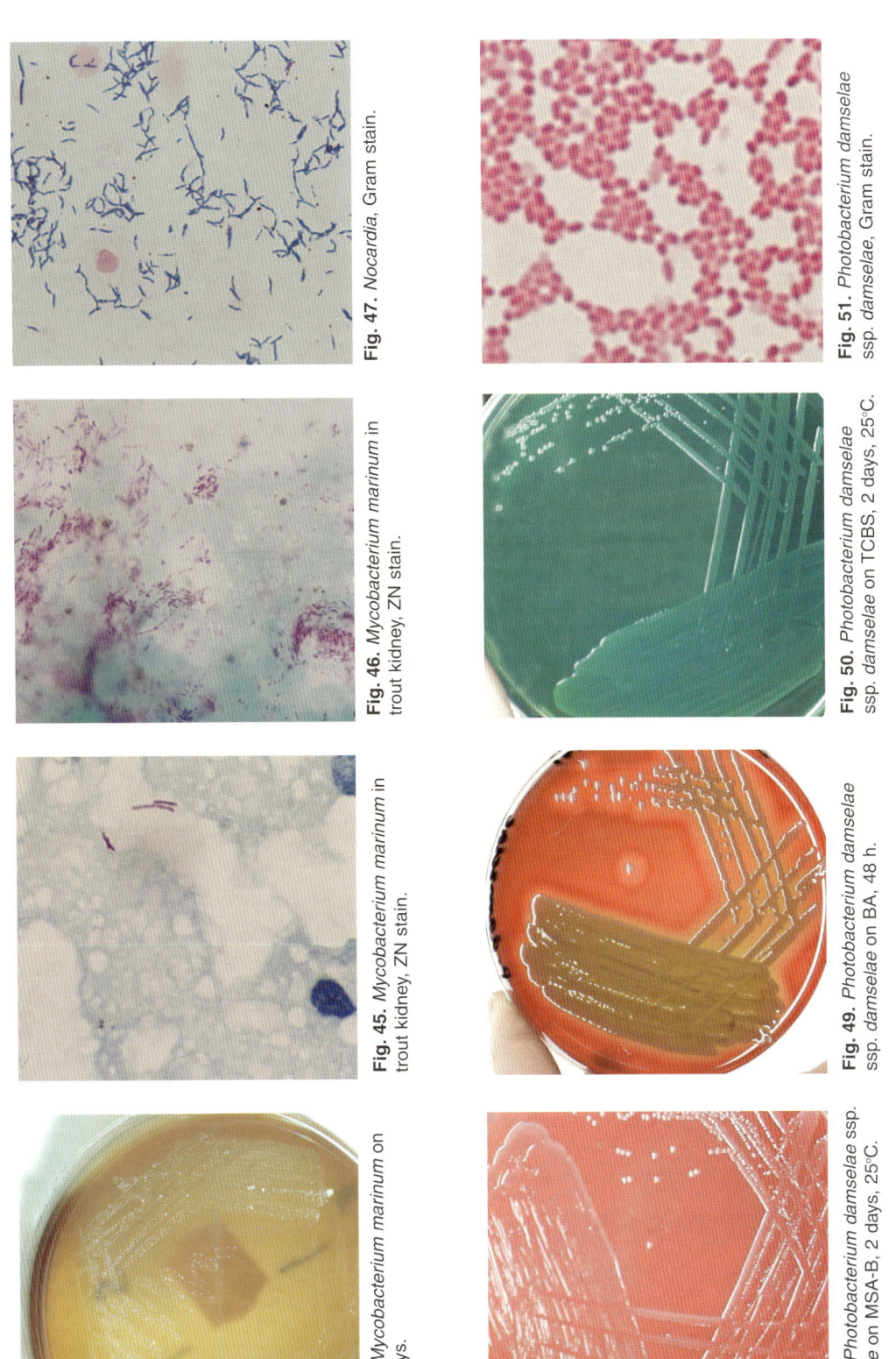

Fig. 47. *Nocardia*, Gram stain.

Fig. 46. *Mycobacterium marinum* in trout kidney, ZN stain.

Fig. 45. *Mycobacterium marinum* in trout kidney, ZN stain.

Fig. 44. *Mycobacterium marinum* on BA, 5 days.

Fig. 51. *Photobacterium damselae* ssp. *damselae*, Gram stain.

Fig. 50. *Photobacterium damselae* ssp. *damselae* on TCBS, 2 days, 25°C.

Fig. 49. *Photobacterium damselae* ssp. *damselae* on BA, 48 h.

Fig. 48. *Photobacterium damselae* ssp. *damselae* on MSA-B, 2 days, 25°C.

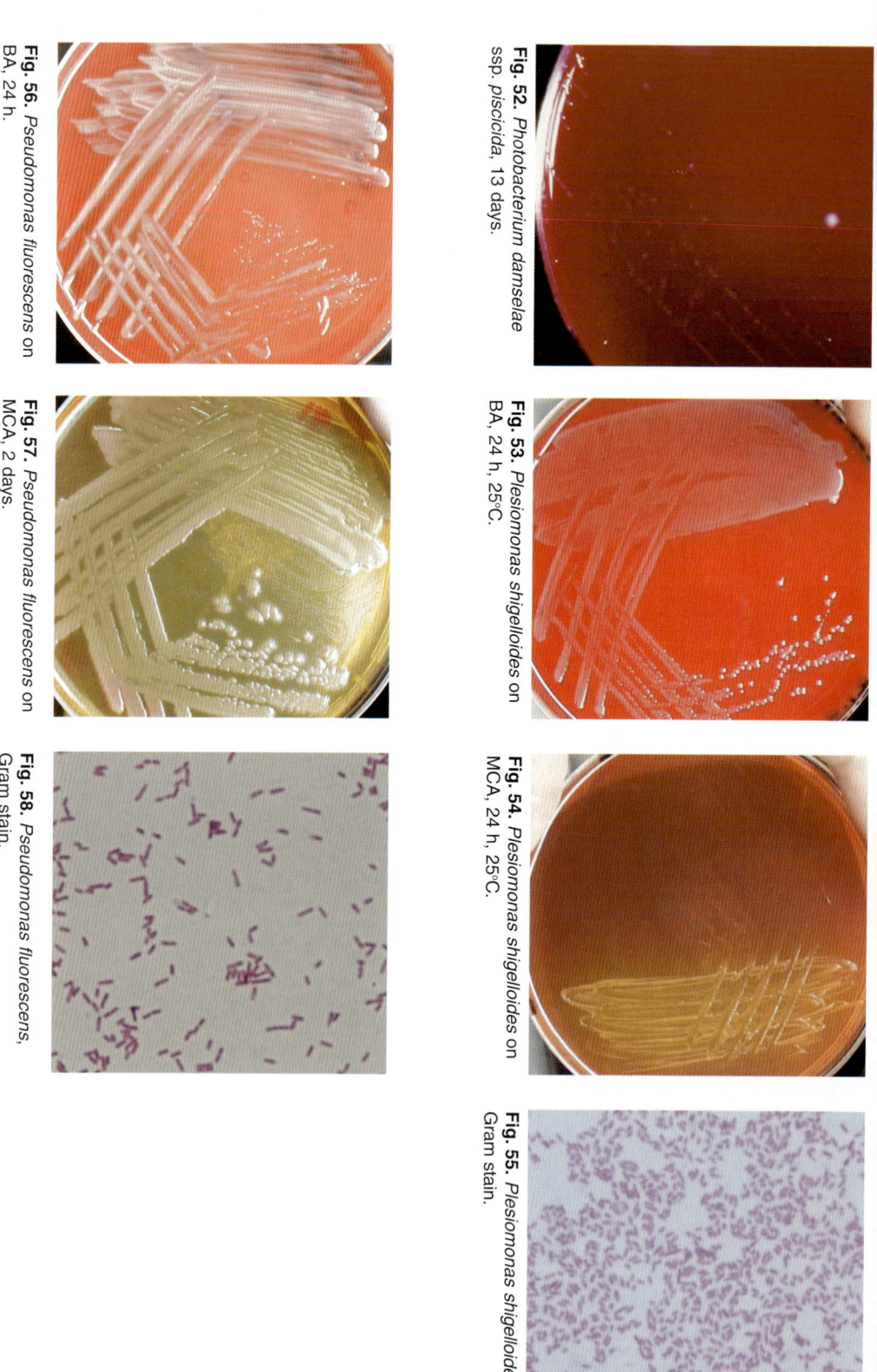

Fig. 52. *Photobacterium damselae* ssp. *piscicida*, 13 days.

Fig. 53. *Plesiomonas shigelloides* on BA, 24 h, 25°C.

Fig. 54. *Plesiomonas shigelloides* on MCA, 24 h, 25°C.

Fig. 55. *Plesiomonas shigelloides*, Gram stain.

Fig. 56. *Pseudomonas fluorescens* on BA, 24 h.

Fig. 57. *Pseudomonas fluorescens* on MCA, 2 days.

Fig. 58. *Pseudomonas fluorescens*, Gram stain.

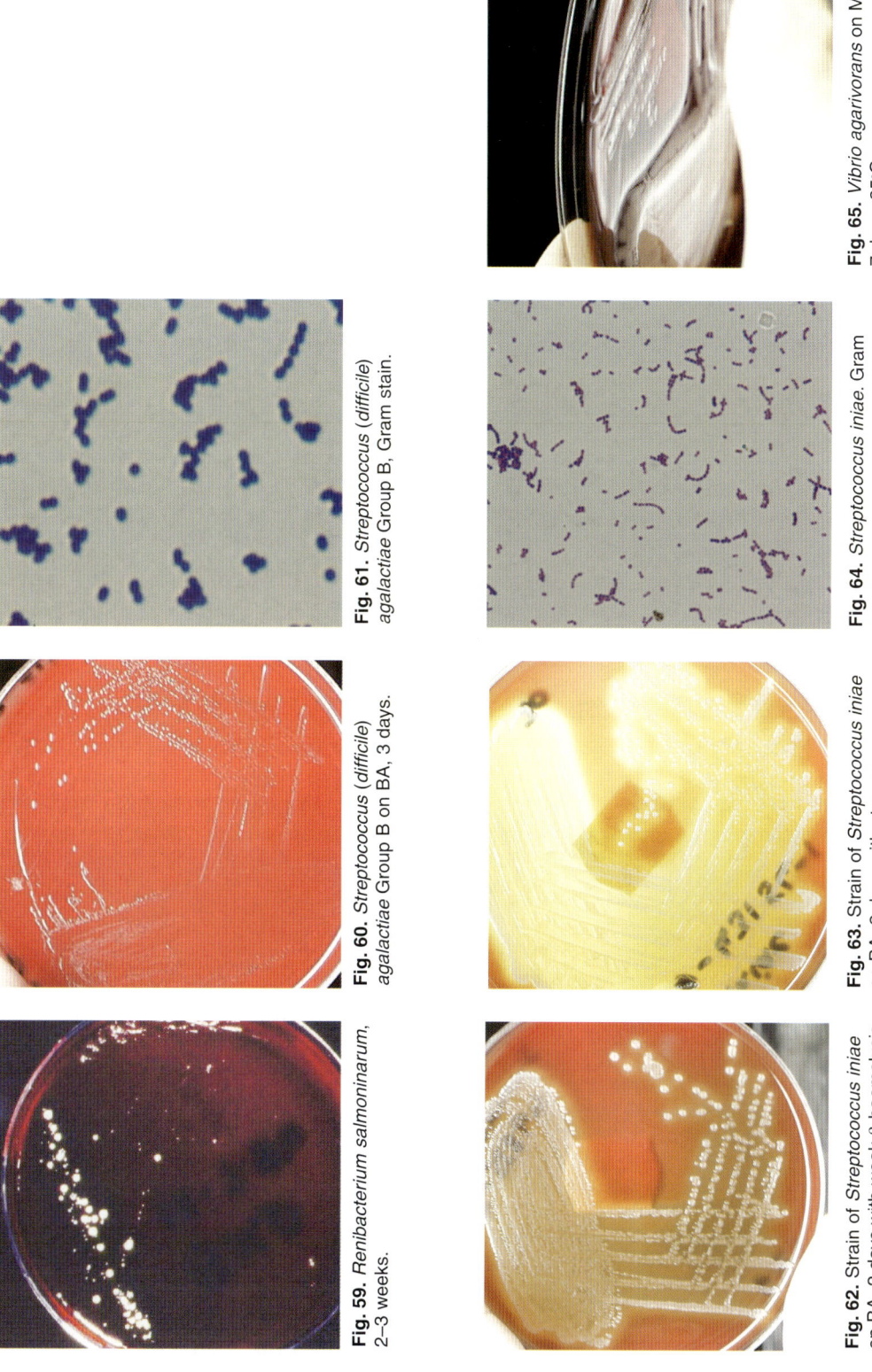

Fig. 59. *Renibacterium salmoninarum*, 2–3 weeks.

Fig. 60. *Streptococcus (difficile) agalactiae* Group B on BA, 3 days.

Fig. 61. *Streptococcus (difficile) agalactiae* Group B, Gram stain.

Fig. 62. Strain of *Streptococcus iniae* on BA, 2 days with weak β-haemolysis.

Fig. 63. Strain of *Streptococcus iniae* on BA, 2 days with stronger β-haemolysis.

Fig. 64. *Streptococcus iniae.* Gram smear.

Fig. 65. *Vibrio agarivorans* on MSA-B, 7 days, 25°C.

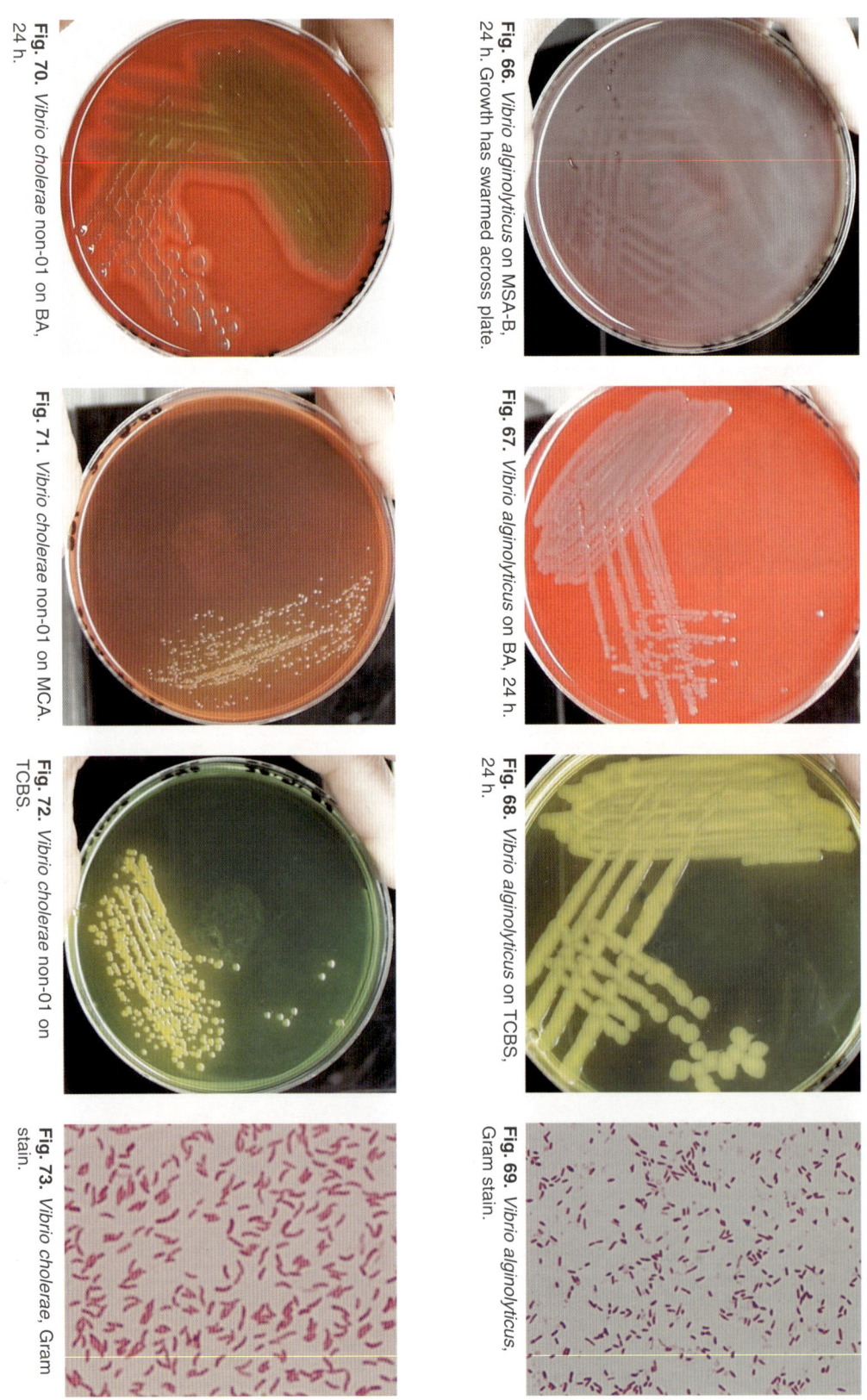

Fig. 66. *Vibrio alginolyticus* on MSA-B, 24 h. Growth has swarmed across plate.

Fig. 67. *Vibrio alginolyticus* on BA, 24 h.

Fig. 68. *Vibrio alginolyticus* on TCBS, 24 h.

Fig. 69. *Vibrio alginolyticus*, Gram stain.

Fig. 70. *Vibrio cholerae* non-01 on BA, 24 h.

Fig. 71. *Vibrio cholerae* non-01 on MCA.

Fig. 72. *Vibrio cholerae* non-01 on TCBS.

Fig. 73. *Vibrio cholerae*, Gram stain.

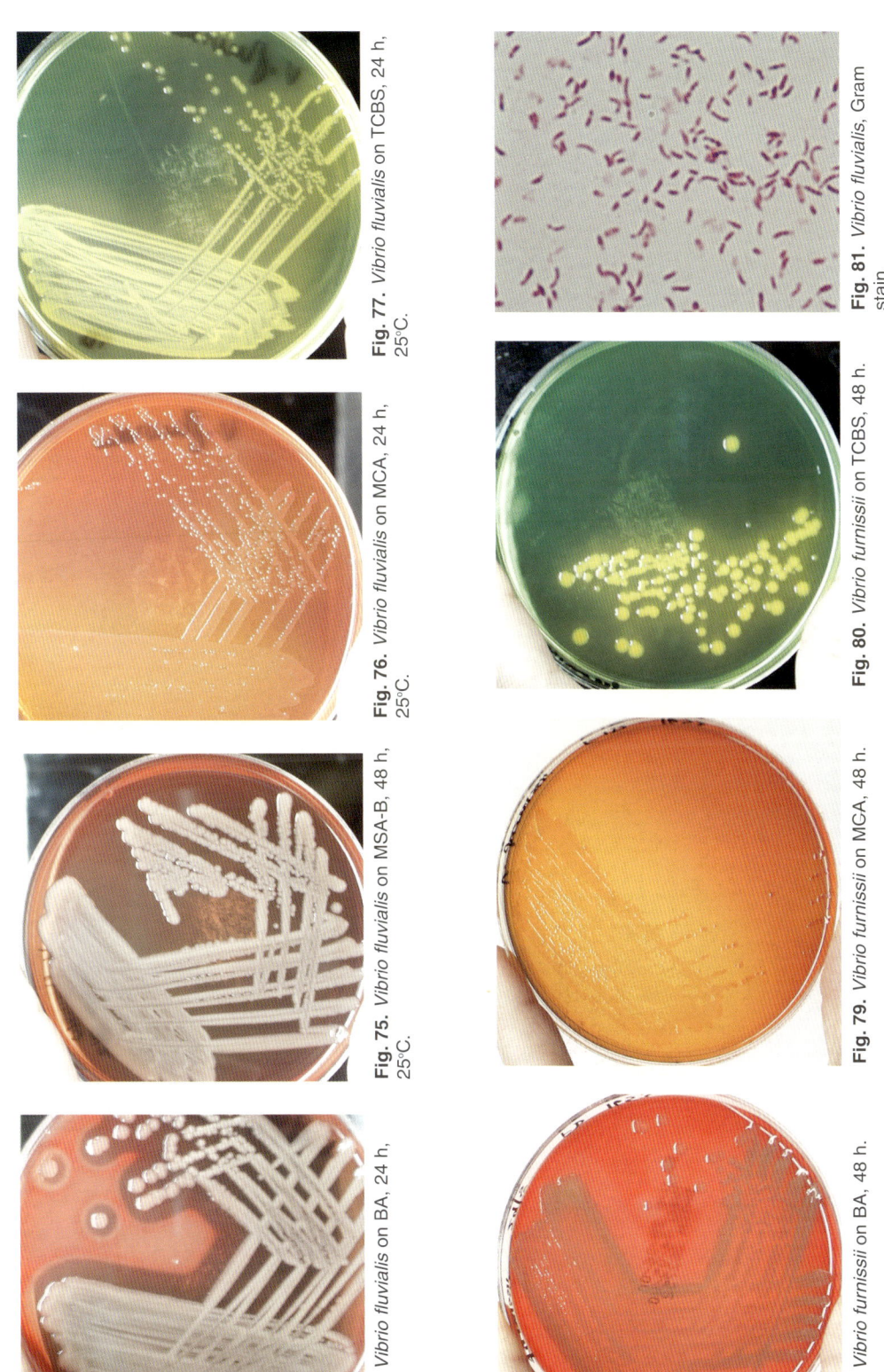

Fig. 77. *Vibrio fluvialis* on TCBS, 24 h, 25°C.

Fig. 76. *Vibrio fluvialis* on MCA, 24 h, 25°C.

Fig. 75. *Vibrio fluvialis* on MSA-B, 48 h, 25°C.

Fig. 74. *Vibrio fluvialis* on BA, 24 h, 25°C.

Fig. 81. *Vibrio fluvialis*, Gram stain.

Fig. 80. *Vibrio furnissii* on TCBS, 48 h.

Fig. 79. *Vibrio furnissii* on MCA, 48 h.

Fig. 78. *Vibrio furnissii* on BA, 48 h.

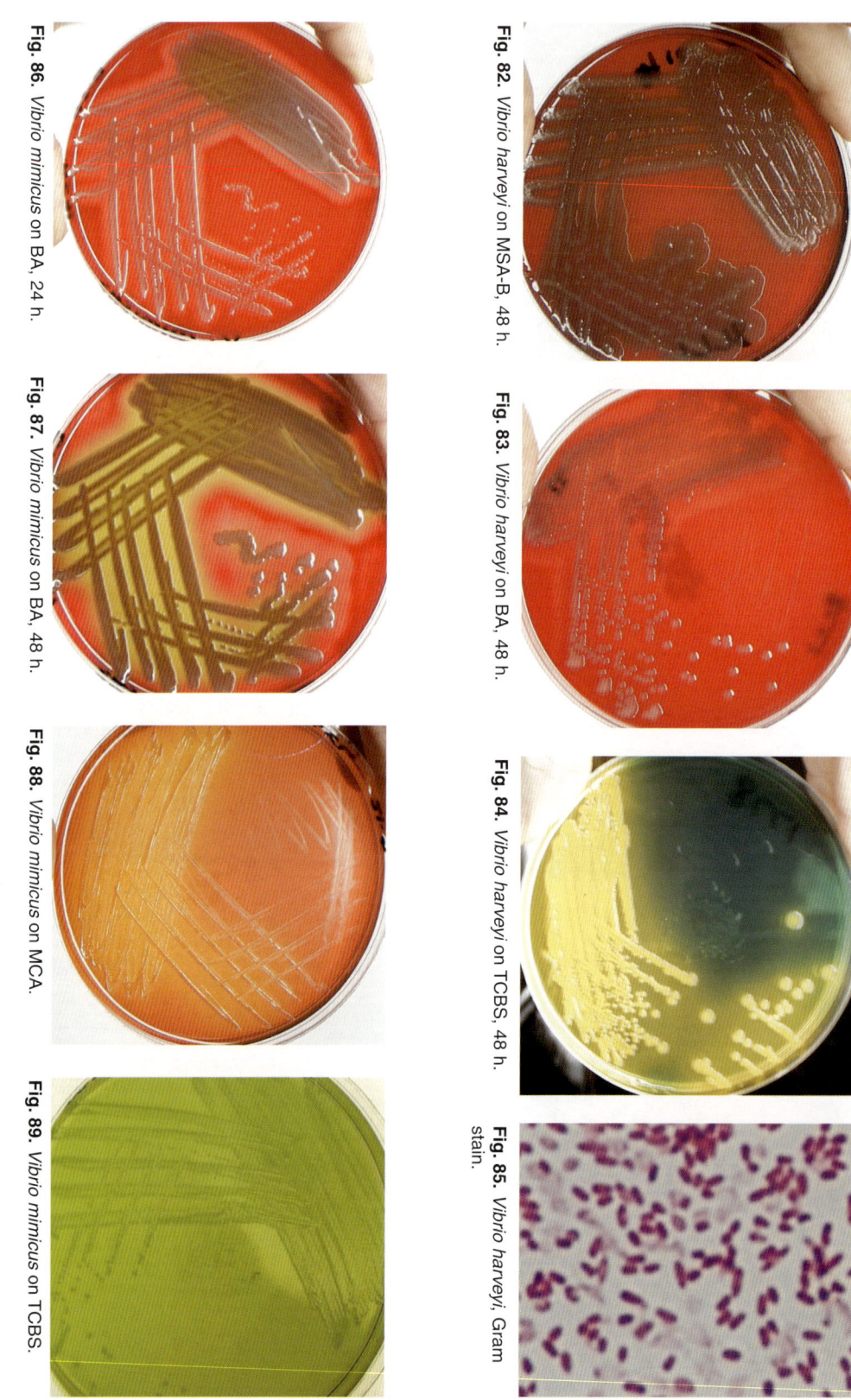

Fig. 82. *Vibrio harveyi* on MSA-B, 48 h.

Fig. 83. *Vibrio harveyi* on BA, 48 h.

Fig. 84. *Vibrio harveyi* on TCBS, 48 h.

Fig. 85. *Vibrio harveyi*, Gram stain.

Fig. 86. *Vibrio mimicus* on BA, 24 h.

Fig. 87. *Vibrio mimicus* on BA, 48 h.

Fig. 88. *Vibrio mimicus* on MCA.

Fig. 89. *Vibrio mimicus* on TCBS.

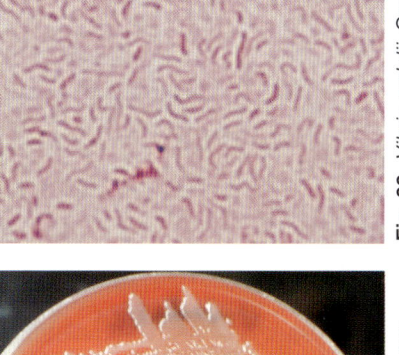

Fig. 93. *Vibrio ordalii*, Gram stain.

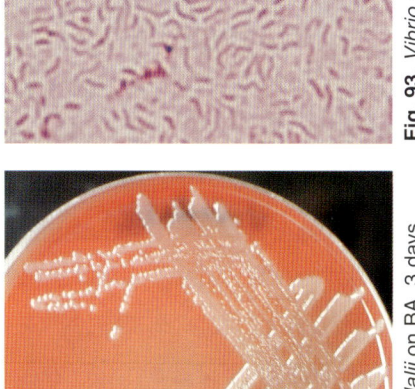

Fig. 92. *Vibrio ordalii* on BA, 3 days, 25°C. No growth on TCBS.

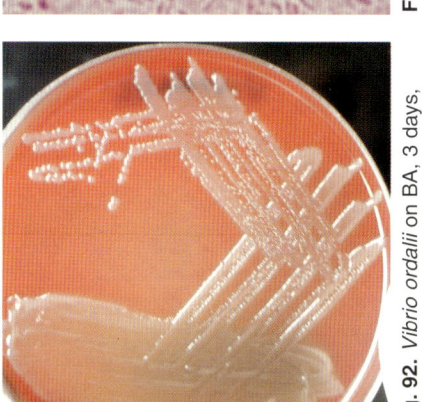

Fig. 91. *Vibrio ordalii* on MSA-B, 2 days.

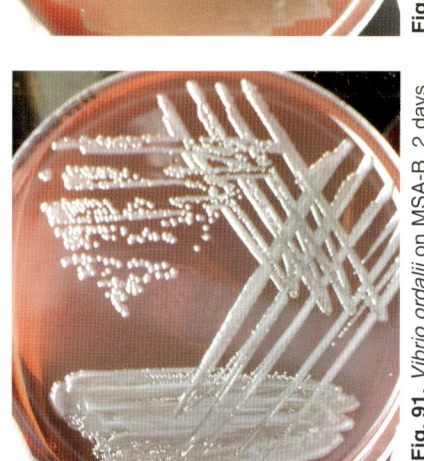

Fig. 90. *Vibrio mimicus*, Gram stain.

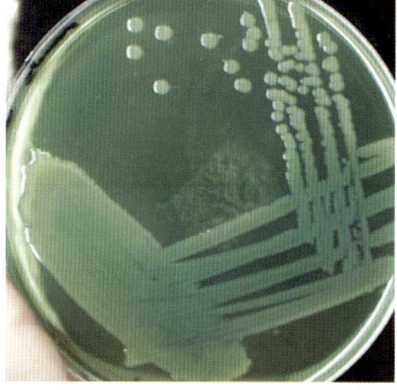

Fig. 97. *Vibrio parahaemolyticus.* Gram smear.

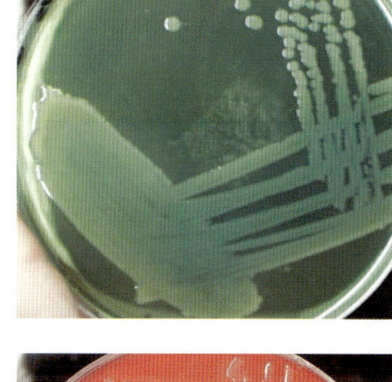

Fig. 96. *Vibrio parahaemolyticus* on TCBS, 24 h.

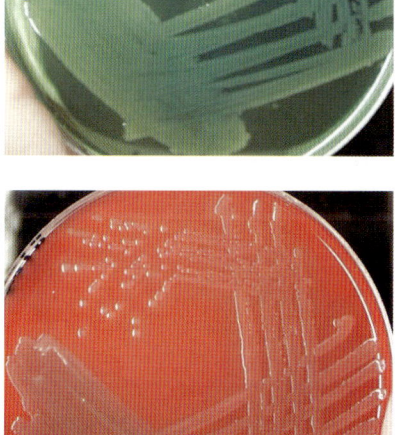

Fig. 95. *Vibrio parahaemolyticus* on BA, 24 h, 25°C.

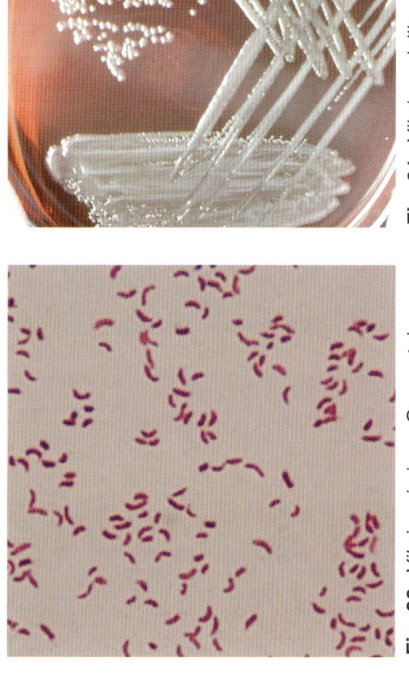

Fig. 94. *Vibrio parahaemolyticus* on MSA-B, 48 h. Growth has swarmed across plate.

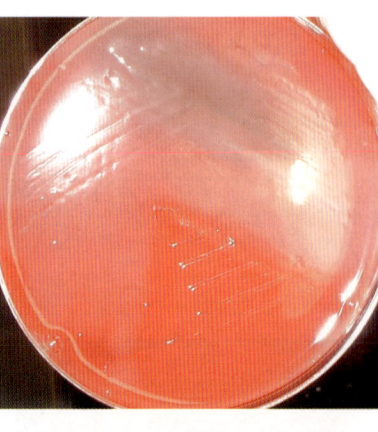

Fig. 98. *Vibrio proteolyticus* on BA, 24 h, 25°C. Showing swarming colonies.

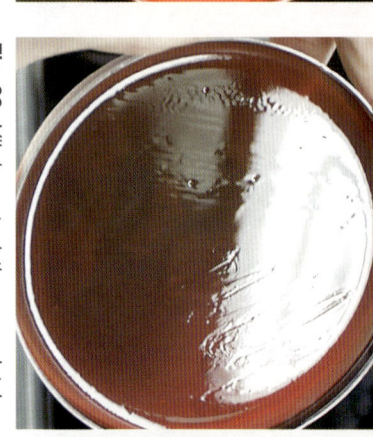

Fig. 99. *Vibrio proteolyticus* completely covers on MSA-B, 24 h.

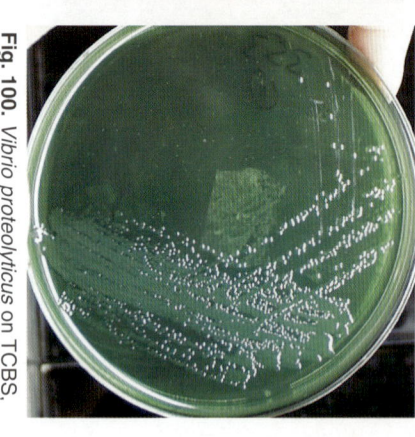

Fig. 100. *Vibrio proteolyticus* on TCBS, 24 h.

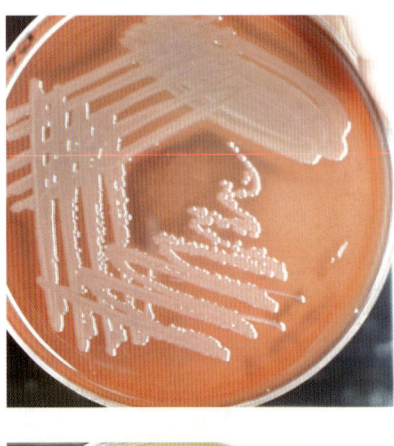

Fig. 101. *Vibrio tubiashii* on MSA-B, 2 days, 25°C.

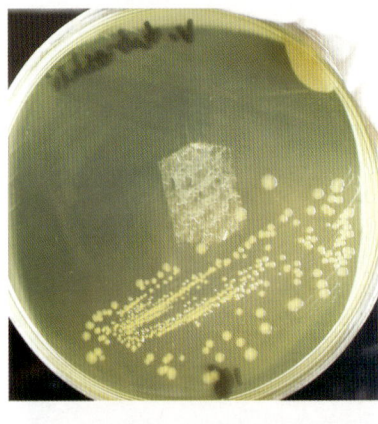

Fig. 102. *Vibrio tubiashii* on TCBS, 3 days, 25°C.

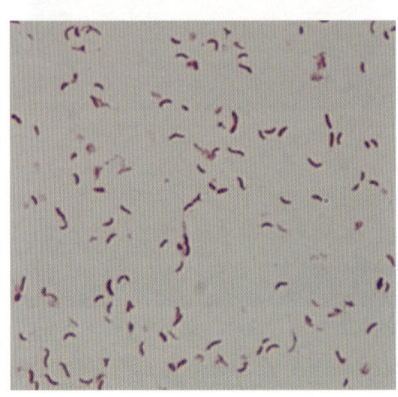

Fig. 103. *Vibrio tubiashii*, Gram stain.

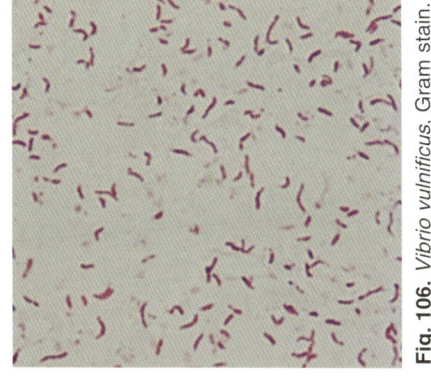

Fig. 104. *Vibrio vulnificus* on MSA-B, 2 days, 25°C.

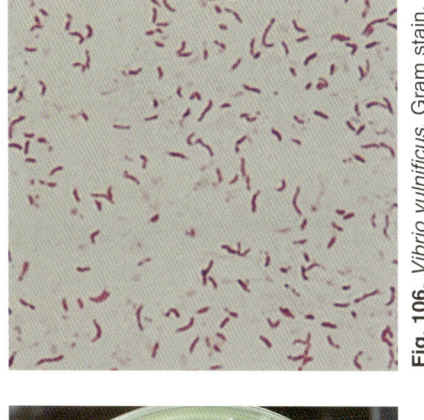

Fig. 105. *Vibrio vulnificus* on TCBS, 2 days, 25°C.

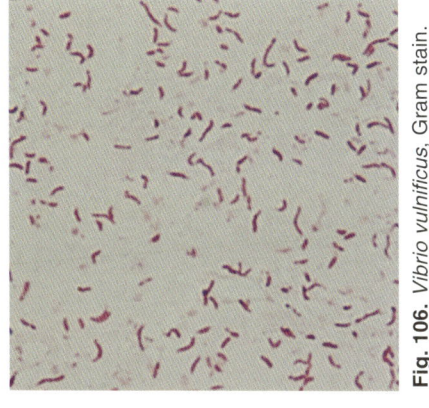

Fig. 106. *Vibrio vulnificus*, Gram stain.

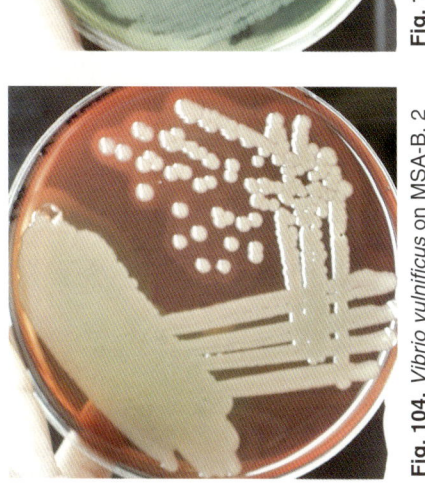

Fig. 107. *Yersinia ruckeri* on BA, 24 h, 25°C.

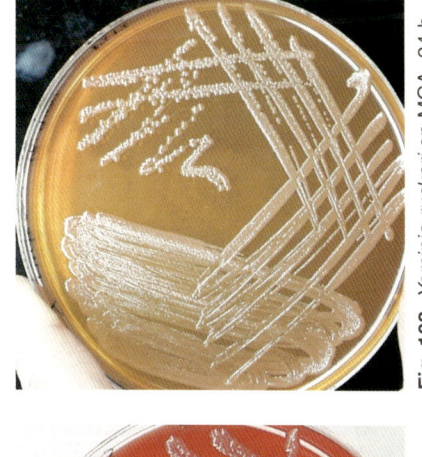

Fig. 108. *Yersinia ruckeri* on MCA, 24 h, 25°C.

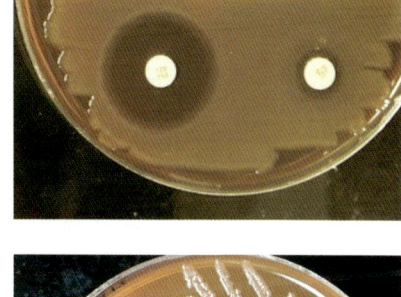

Fig. 109. *Vibrio* identification discs. Top = 150 μg disc (sensitive). Bottom = 10 μg disc (resistant).

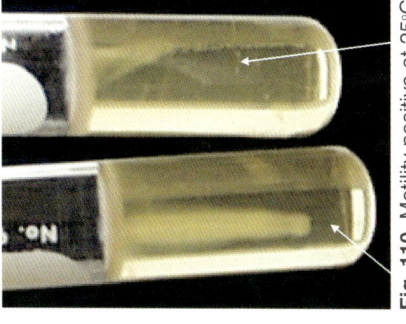

Fig. 110. Motility positive at 25°C (left), but negative at 37°C (right) for *Yersinia ruckeri*.

Fig. 111. Carbohydrate fermentation reaction. Sucrose-positive (yellow) and sucrose-negative (red), 24 h, 25°C.

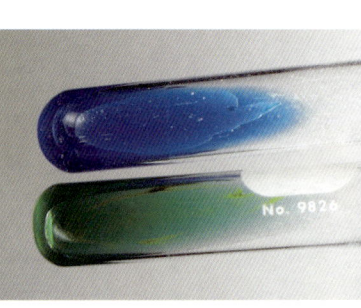

Fig. 112. Citrate test. *Yersinia ruckeri*, citrate-positive (blue) at 24°C, but citrate-negative (green) at 37°C.

Fig. 113. Decarboxylase reactions. Left to right: arginine dihydrolase (negative); lysine decarboxylase (positive); ornithine decarboxylase (positive); and control tube (negative).

Fig. 114. Decarboxylase reactions. Left to right: arginine dihydrolase (positive); lysine decarboxylase (positive); ornithine decarboxylase (negative); and control tube (negative).

Fig. 115. DNase reaction. Negative reaction on left and positive reaction on right.

Fig. 116. Indole reaction. Negative on the left and positive on the right. *Vibrio parahaemolyticus* with 0.85% NaCl, and 2% NaCl, respectively.

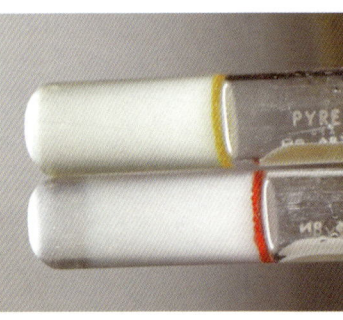
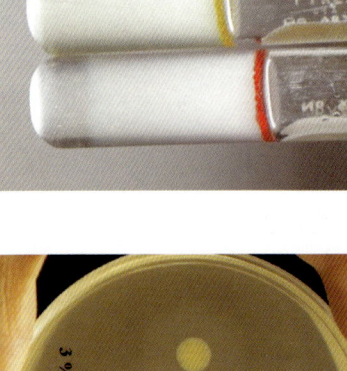

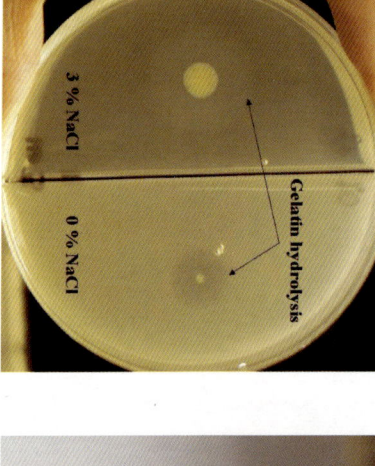

Fig. 117. Plate showing gelatin hydrolysis and growth on 3% and 0% NaCl.

3% NaCl

0% NaCl

Gelatin hydrolysis

Fig. 118. Methyl Red reaction. Positive.

1

Aquatic Animal Species and Organism Relationship

1.1 Host Species, Bacteria and Disease

This chapter deals with the relationship between the host species and the bacterial flora that may be either part of the normal flora of that host, or pathogenic for that host. This information is presented in two formats. Table 1.1 lists the aquatic animal hosts in alphabetical order under their common name. The scientific name is in parentheses. Some hosts are grouped under their Family name, which is in capitals. For example, trout and salmon are listed under SALMONIDS; dolphin, porpoises, seals and whales are listed under SEA MAMMALS; aquarium fish are listed under ORNAMENTALS. The adjacent columns in the table list the bacteria that have been reported to be either pathogens of the host or that are considered part of the normal flora, the tissue site of infection, or the pathology presented and the disease state. Some organisms are considered to be opportunistic pathogens and in a healthy host may be part of the normal flora. In a stressed animal, these same bacteria may overcome host defence mechanisms and cause morbidity or infection in the animal. There are some organisms that have been identified and isolated from a host, but the pathogenicity of the organism is unknown, as virulence studies were not carried out.

In the second format, in Table 1.2, the information is presented by listing the bacteria in alphabetical order, with adjacent columns listing the name of the disease, the tissue site where the organism may be found, the aquatic animals where the organism has been reported, and the geographical location of the disease.

1.2 Bacterial Diseases

The following section provides more detail than in the tables on some of the more commonly recognized fish bacterial diseases. The diseases are described in a brief form, as there are other texts available that provide more comprehensive detail on diseases of fish. See recommended texts in Chapter 8, 'Further Reading and Other Information Sources'.

Bacillary necrosis of *Pangasius*

This is a recently reported disease found in freshwater catfish (*Pangasius hypophthalmus* Sauvage) in Thailand. The causative organism has been identified as *Edwardsiella ictaluri*, which in catfish in America causes the disease known as enteric septicaemia of catfish. The disease in *Pangasius* presents as multifocal, white lesions, irregular in size and found in internal organs, predominantly the kidney, liver and spleen. In histology the lesions appear as areas of necrosis and pyogranulomatous inflammation. The causative bacterium was identified as *E. ictaluri* biochemically; however, when examined microscopically the bacterial cells showed a greater pleomorphism in length and size than normally seen with other strains of *E. ictaluri* (Crumlish *et al.*, 2002).

Table 1.1. Host species and organism relationship.

Host	Pathogen	Normal flora	Tissue site	Disease status	Ref
Abalone		*Pseudoalteromonas* spp., *Shewanella* spp. *Vibrio agarivorans* (pathogenicity not determined) *Vibrio mediterranei*	Tissue	Mortality, lesions	135
Haliotis discus hannai		*Vibrio halioticoli*	Gut	Normal microflora	678
Red abalone (*Haliotis rufescens*)	*Vibrio alginolyticus*		Sick larvae are unable to swim and remain at the bottom of the pond	Mortality in larvae	30
Haliotis tuberculata	*Vibrio (carchariae) harveyi*		White pustules on foot	Mortality	576
Japanese abalone (*Sulculus diversicolor supratexta*)	*Vibrio harveyi* (strain was non-luminescent, ODC-negative, urease-negative)		White spots on foot, tissue lesions, necrotic degeneration in muscle fibres	Mass mortality, lose ability to adhere	581
Small abalone (*Haliotis diversicolor supertexta*)	*Vibrio parahaemolyticus*		Organism in haemolymph	Withering syndrome, mass mortality	499
Alga					
Marine alga (*Ulva lactuca*)		*Pseudoalteromonas ulvae*		Has anti-fouling properties	231
Red alga (*Delesseria sanguinea*)		*Zobellia galactanovorans*			61
Alligator					
Alligator mississippiensis	1. *Edwardsiella tarda* 2. *Mycoplasma alligatoris* 3. *Pasteurella multocida* 4. *Staphylococcus*	5. *Edwardsiella tarda, Salmonella miami, S. java, S. hartford*	1, 3. Large intestine, cloaca. Congestion in kidneys with necrotic foci, peritonitis, ulcers in gastric mucosa 2. Pulmonary oedema, interstitial pneumonia, pericarditis, myocarditis, meningitis, synovitis 3, 4. Lungs	1, 5. Nephritis. Pathogenicity not conclusive 2. Acute multisystemic inflammatory disease 3, 4. Pneumonia	128 129 520 804 823
Amberjack. Japanese amberjack. See Yellowtail					
Anchovy (*Engraulis mordax*)	*Tenacibaculum maritimum*		Haemorrhagic lesions on snout, eye and midsection of the body	Disease	154
Arctic Charr (*Salvelinus alpinus* Linnaeus). See under SALMONIDS					

Species	Organism	Signs/Pathology	Disease/Condition	Page
Artemia spp. See under shrimp – brine shrimp				
Ayu (*Plecoglossus altivelis* Temminck and Schlegel)	1. *Flavobacterium psychrophilum* 2. *Listonella anguillarum* 01 and 02 (European designation) 3. *Pseudomonas anguilliseptica* 4. *Pseudomonas plecoglossicida* 5. *Renibacterium salmoninarum* 6. *Streptococcus iniae* 7. *Vibrio cholera* non-01 (negative for ornithine decarboxylase)	4. Haemorrhagic ascites 5. White nodules in kidney, abdomen swollen with fluid, exophthalmia	1. Cold water disease 2. Vibriosis 3. Disease 4. Mortality, bacterial haemorrhagic ascites (BHA) 5. Bacterial kidney disease (BKD) 6. Mortality, streptococcosis 7. Mass mortality	434 442 561 564 568 582 712 722 757 803
Baitfish. American baitfish (*Pimephales promelas*)	See Minnows			
Barramundi (*Lates calcarifer* Bloch)	See Bass			
Bass				
European seabass (*Dicentrarchus labrax* Linnaeus)	1. *Aeromonas hydrophila* 2. *Photobacterium damselae* ssp. *piscicida* 3. *Pseudomonas anguilliseptica* 4. *Mycobacterium marinum* and *Mycobacterium* spp. 5. *Streptococcus iniae* 6. *Tenacibaculum maritimum*	1. Enlargement of spleen, erythema and swelling of anus 2. No obvious gross pathology, enlarged spleen with white nodules 3. Organism isolated from head–kidney and spleen 4. Degeneration in eye, exophthalmia, skin lesions, necrotic areas in gills 4, 5. Organisms in heart and spleen, extreme splenomegaly 6. Pale skin zones with yellow edges, leading to necrotic lesions on body around fins, oral cavity, eyes and gills	1. Mortalities 2. Fish pasteurellosis 3. Haemorrhagic septicaemia 4. Mycobacteriosis 5. Exudative meningitis and panophthalmitis 6. Dermal necrosis. Fingerlings very susceptible. Stress-related in older fish	60 91 96 140 183 209 227
Largemouth bass (*Micropterus salmoides* Lacepède)	*Edwardsiella tarda*	*Edwardsiella ictaluri* Intestine, liver, spleen	Internal organs pale and anaemic, blood watery, haemorrhagic nodules in stomach wall and serosa of intestine	627 823

continued

Table 1.1. *Continued.*

Host	Pathogen	Normal flora	Tissue site	Disease status	Ref
Sea bass (*Lates calcarifer* Bloch) Also known as barramundi or barramundi perch	1. *Flavobacterium johnsoniae* 2. *Photobacterium damselae* ssp. *damselae* 3. *Photobacterium damselae* ssp. *piscicida* 4. *Streptococcus iniae* 5. *Vibrio harveyi*		1. Superficial skin erosion on posterior flanks, pectoral fins and occasionally the lower jaw 3. Exophthalmia, haemorrhagic skin lesions, septicaemia 4. Organism isolated from brain	1. Disease in juveniles 2, 3, 4, 5. Mortalities	127 135 145 734
Sea bass (*Lateolabrax japonicus* Cuvier), Japanese seaperch	*Nocardia seriolae*		White-yellow nodules in gill, heart, kidney, liver, spleen	Nocardiosis	155
Sea bass (*Puntazzo puntazzo* Cuvier)	*Aeromonas hydrophila*		Enlargement of spleen, erythema and swelling of anus	Mortalities	227
Striped bass (*Morone saxatilis* Walbaum), (*Roccus saxatilis*)	1. *Carnobacterium piscicola* 2. *Corynebacterium aquaticum* 3. *Mycobacterium marinum* and *Mycobacterium* spp. 4. *Photobacterium damselae* ssp. *piscicida* 5. *Serratia marcescens* 6. *Streptococcus iniae*	7. *Vibrio cholerae* non-01 and *Vibrio mimicus* Did not cause deaths in a virulence assay; however, may be opportunistic pathogen in stressed fish	1. Hyperaemia and haemorrhage in liver, kidney, spleen and brain 2. Organism in brain tissue, exophthalmia 3. Nodular lesions in all organs 4. Organism isolated from internal organs, enlarged kidney and spleen. Abnormal skin pigmentation 5. Necrosis of muscular tissues	1. Mortality 2. Disease 3. Mortality 4. Fish pasteurellosis 5. Mortality in fingerlings 6. Streptococcosis 7. Present in apparently healthy wild fish. May cause mortality in stressed farmed fish	73 75 76 235 333 337 339 474 507 708
White sea bass (*Atractoscion nobilis* Ayres). Also known as White weakfish. Family *Serranidae*	*Tenacibaculum maritimum*		Lesions in eye, operculum. Lesions on body ranging from scale loss to deep ulcers in musculature	Disease in juveniles	154
Grouper Orange-spotted grouper (*Epinephelus coioides* Hamilton, and *E. tauvina* Forsskål). Family *Serranidae*	1. *Streptococcus iniae* 2. *Vibrio harveyi*		1. Susceptible to infection	1. Streptococcosis	847 848
Blenny, Viviparous blenny (*Zoarces viviparus* Linnaeus)	Atypical *Aeromonas salmonicida*		Red ulcers with white margins. Organism also isolated from internal organs	Skin ulcers, septicaemia	832

	Organism			Ref.
Bream				
Black sea bream (*Acanthopagrus latus* Houttuyn). Known as Yellowfin seabream	1. *Pseudomonas anguilliseptica*		1. Haemorrhage and ulcerative lesions	Not susceptible to infection with *Streptococcus iniae* 1. Associated with 'winter disease'. Septicaemia	225 569 848
Common bream, Carp bream (*Abramis brama* Linnaeus)	*Aeromonas salmonicida* ssp. *achromogenes*		Large open lesions, areas of descalation. Typical 'furuncles' not seen	Mortalities	534
One-spot sea bream (*Diplodus sargus kotschyi* Steindachner)				Not susceptible to infection with *Streptococcus iniae*	848
Red sea bream, Japanese seabream (*Pagrus major* Temminck and Schlegel). Family *Sparidae*	1. *Edwardsiella tarda* 2. *Listonella anguillarum* 3. *Tenacibaculum maritimum*		1. Septicaemia, focal suppurative or granulomatous lesions, cutaneous ulcerations	1. Edwardsiellosis 2, 3. Infection	
Sea bream (*Sparus auratus* Linnaeus). Also known as Gilt-head sea bream. Family *Sparidae*	1. *Listonella anguillarum* 2. *Photobacterium damselae* ssp. *damselae* 3. *Photobacterium damselae* ssp. *piscicida* 4. *Pseudomonas anguilliseptica* 5. *Streptococcus agalactiae* 6. *Streptococcus iniae* 7. *Vibrio alginolyticus* 8. *Vibrio harveyi* 9. *Vibrio splendidus*		2. Lethargy, distended abdomen, haemorrhages on fins and tail, pale liver 3. No external changes except anus red and protruded, abdomen distended, fluid in abdominal cavity, congested spleen, petechial haemorrhages on liver, granulomatous lesions in spleen and kidney, multifocal tissue necrosis 4. Erratic swimming at water surface, sink to the bottom of the cage and die. Ascites, renal haemorrhage 5. Haemorrhagic areas on body, mouth, eye, operculum and fins 1, 7, 8, 9. Ulcers, haemorrhages, exophthalmia	1, 2. Mortality 3. Fish pasteurellosis 4. Associated with 'winter disease', haemorrhagic septicaemia associated with keratitis 5. Streptococcosis – epizootic with 100% mortality 6. Exudative meningitis and panophthalmitis 1, 3, 7, 8. 9. Primary pathogens in virulence studies	57 58 60 96 225 242 751 786 853
Silver bream, White bream (*Blicca bjoerkna* Linnaeus)	*Aeromonas salmonicida* ssp. *achromogenes*		Large open skin lesions surrounded by areas of descalation. Typical 'furuncles' not seen	Furunculosis, mortality	534
Carp				
Bighead carp (*Aristichthys nobilis*)		*Edwardsiella ictaluri*			627

continued

Table 1.1. *Continued.*

Host	Pathogen	Normal flora	Tissue site	Disease status	Ref
Common carp, Koi carp (*Cyprinus carpio carpio* Linnaeus)	1. *Aeromonas bestiarum* 2. Atypical *Aeromonas salmonicida* 3. *A. veronii* ssp. *veronii* 4. *Citrobacter freundii* 5. *Flavobacterium columnare* 6. *Yersinia ruckeri*	7. *Aeromonas eucrenophila* 8. *A. sobria* were non-pathogenic to carp in virulence studies (452) 9. *Streptococcus iniae* and *S. agalactiae* (*S. difficile*) were non-pathogenic in virulence studies (234)	1, 3. Haemorrhage, necrosis, ulcers 2. Ulcers, lesions 3. Also bloody ascites fluid 4. In fingerlings 5. Gills 6. Can be infected with or without clinical signs 7. Ascites	1, 3. Pathogenic on virulence studies 2. Ulcerating dermal lesions 4. Heavy mortalities, septicaemia, opportunistic infection 5. Disease 6. Yersiniosis 7. Non-pathogenic	135 209 234 271 379 425 452 473 760
Caucasian carp, Crucian carp (*Carassius carassius* Linnaeus)	*Escherichia vulneris*		Haemorrhage in eyes, swollen, dark coloured abdomen, yellow liver, yellow fluid in intestine	Mortality	51
Iberian toothcarp, Spanish toothcarp (*Aphanius iberus* Valenciennes)	*Vibrio parahaemolyticus*		External haemorrhage, tail rot	Mortality	10
Silver carp (*Hypophthalmichthys molitrix* Valenciennes)	1. *Providencia* (*Proteus*) *rettgeri* 2. *Staphylococcus aureus*		1. Large red ulcerative lesions on the abdomen, base of the pectoral fin and on the head. Organism isolated from internal organs. Pond had been fertilized with poultry faeces from which *P. rettgeri* was isolated 2. Red cornea becoming opaque Degeneration of eye tissue	1. Mass mortality 2. Eye disease	79 688
Catfish					
Catfish species (*Ictalurus* spp. Rafinesque)	*Flavobacterium columnare*		Organism in kidney		88
Black bullhead (*Ameiurus melas* – valid name), (*Ictalurus melas* Rafinesque – scientific name)	1. *Edwardsiella ictaluri* 2. *Flavobacterium columnare*		2. Yellow-edged skin erosion on head	1. Enteric septicaemia of catfish (ESC), edwardsiellosis 2. Columnaris disease	88
Blue catfish (*Ictalurus furcatus* Valenciennes)	*Edwardsiella ictaluri*			ESC	334

Animal species	Organism	Clinical signs / location	Disease / condition	Ref.
Brown bullhead (*Ictalurus nebulosus*)	1. *Edwardsiella ictaluri* 2. *Edwardsiella tarda*	1. Infection in the brain, systemic dissemination and localization of the organism in the visceral organs and musculature and cutaneous ulcers 2. Septicaemia, focal suppurative or granulomatous lesions, cutaneous ulcerations	1. ESC, edwardsiellosis 2. Edwardsiellosis, opportunistic infection	334
Channel catfish (*Ictalurus punctatus* Rafinesque)	1. *Aeromonas hydrophila* 2. *Bacillus mycoides* 3. *Carnobacterium piscicola* 4. *Edwardsiella ictaluri*, anaerobic strains also isolated 5. *Edwardsiella tarda* 6. *Yersinia ruckeri*	2. Ulcerative skin lesions and focal necrosis of epaxial muscle 3. Hyperaemia and slight haemorrhage in liver, kidney, spleen and brain 5. Bacteria isolated from lesions on skin and superficial muscle, necrosis in organs 6. Haemorrhagic rings around the eyes and frontal foramens	1, 2, 3. Infection and mortality 4. ESC 5. Edwardsiellosis, enteric septicaemia, opportunistic infection Yersiniosis	73 203 307 334 547 783
Freshwater catfish, Sutchi catfish (*Pangasius hypophthalmus* Sauvage)	*Edwardsiella ictaluri*	Mortality, multifocal irregular, white lesions on internal organs. Necrosis and pyogranulomatous inflammation	Bacillary necrosis of *Pangasius*	194
Walking catfish (*Clarias batrachus* Linnaeus) (*Clarias gariepinus*)	1. *Aeromonas hydrophila* 2. *Edwardsiella ictaluri*		1. Ulcerative disease, mortality 2. ESC	29 426
White catfish (*Ameiurus catus* Linnaeus)	*Edwardsiella ictaluri*		ESC	334
Chub, European chub (*Leuciscus cephalus* Linnaeus)	Atypical *Aeromonas salmonicida*	Skin ulcers and fin rot	Mortality	837
Coalfish	*Photobacterium iliopiscarium*	Intestine	Non-pathogenic	599 767
Cod, Atlantic cod (*Gadus morhua* Linnaeus)	1. Atypical *Aeromonas salmonicida* 2. *Listonella anguillarum* serotype 02 3. *Vibrio salmonicida* 4. *Carnobacterium* spp. 5. *Photobacterium iliopiscarium*	2. Lesions 4, 5. Intestine	1. Skin ulcers 2. Infection, vibriosis 4. Non-pathogenic	186 232 599 712 767

continued

Table 1.1. *Continued.*

Host	Pathogen	Normal flora	Tissue site	Disease status	Ref
Coral					
Oculina patagonica	*Vibrio shilonii* (said to be a later subjective synonym of *V. mediterranei*)		Binds to coral	Coral bleaching	59, 458, 742
Pocillopora damicornis	*Vibrio corallilyticus*		White spots seen at 3–5 days and complete tissue destruction after 2 weeks	Tissue lysis and death	83, 84
Crab					
Blue crab (*Callinectes sapidus*)		*Vibrio cholerae*-like (2% of crabs), *Vibrio parahaemolyticus* (23% of crabs), *Vibrio vulnificus* (7% of crabs)	Haemolymph, digestive tract	Present in apparently healthy crabs	212
Swimming crab (*Portunus trituberculatus*)	*Vibrio harveyi* (initially called *Vibrio sp. zoea*)			Mass mortality in zoeal larvae	390
Crawfish					
American freshwater crayfish	*Vibrio cholerae*			Mortalities	507
Red swamp crawfish (*Procambarus clarkii*)	*Vibrio mimicus*				
Crayfish					
Australian freshwater crayfish Yabby (*Cherax albidus*)	*Vibrio mimicus*		Haemolymph	Mortality, vibriosis	135, 230, 840
Red claw (*Cherax quadricarinatus*)					
Crevalle, Trevally, Jack crevalle (*Caranx hippos* Linnaeus)	*Vibrio harveyi*		Dermal lesions	Infection	453
Crocodile					
		Edwardsiella tarda		Pathogenicity not known	
Crocodilus acutus	*Erysipelothrix rhusiopathiae*		Blackened plaques on scales	Cutaneous lesions	408
Caiman crocodile (*Caiman crocodilus*)	*Erysipelothrix rhusiopathiae*		Necrotic foci under the scales of the abdomen and the jaw	Septicaemia	408
Crocodylus niloticus	*Mycoplasma crocodyli*		Swollen joints. Organism also found in lungs	Exudative polyarthritis	441

Species	Organism		Signs	Outcome	Page
Dab (*Limanda limanda* Linnaeus)	Atypical *Aeromonas salmonicida*		Round, red ulcers with white margin of necrotic tissue	Skin ulcers	832
Dace, Common dace (*Leuciscus leuciscus* Linnaeus)	1. *Aeromonas salmonicida* ssp. *salmonicida* 2. Atypical *Aeromonas salmonicida*		1, 2. Skin ulcers	1. Furunculosis 2. Infection	323 352
Damselfish (Family *Pomacentridae*)					
Blacksmith (*Chromis punctipinnis* Cooper)	*Photobacterium damselae* ssp. *damselae*		Skin ulcers	Fatal infection due to production of a cytolysin	504
Staghorn damselfish, Yellowtail clownfish (*Amblyglyphidodon curacao* Bloch, *Amphiprion clarkii* Bennett)	*Pseudoalteromonas piscicida*	*Listonella anguillarum* and *Vibrio parahaemolyticus* did not appear to be involved in disease in experimental infections	Diseased eggs, seen as white brain and spinal cord, sunken irregular eyes, heart receded into yolk sac	Infection, mortality	572
Danio	See under ORNAMENTAL				
Discus fish	See under ORNAMENTAL				
Dolphin	See under SEA MAMMALS				
Eel					
American eel (*Anguilla rostrata* Lesueur)	1. *Aeromonas salmonicida*		1. Focal lesions progress to large de-pigmented necrotic patches then detach at dermo-epidermal junction to form large ulcers that expose underlying muscle	1. Ulcerative skin disease, morbidity	584
European eel (*Anguilla anguilla* Linnaeus)	1. Atypical *Aeromonas salmonicida* 2. *Listonella anguillarum* serotype 05 3. *Pseudomonas anguilliseptica* 4. *Vibrio furnissii* 5. *Vibrio vulnificus* serovar 04 6. *Yersinia ruckeri*	7. *Aeromonas encheleia*	1. Skin lesions 2. Lesions 3. Subcutaneous haemorrhages. Abdominal distension, organism in internal organs 4. Haemorrhages in intestinal tract 5. Organism cultured from gills, intestinal contents, kidney, spleen 6. Can be infected with or without clinical signs	1. Infection 2. Vibriosis 3. Septicaemia 4. Mortality. Virulent in pathogenicity studies 5. Disease 6. Yersiniosis 7. Non-pathogenic	201 209 240 241 271 323 356 541 712

continued

Table 1.1. *Continued.*

Host	Pathogen	Normal flora	Tissue site	Disease status	Ref
Japanese eel (*Anguilla japonica* Temminck and Schlegel)	1. Atypical *Aeromonas salmonicida* 2. *Edwardsiella tarda* 3. *Flavobacterium columnare* 4. *Pseudomonas anguilliseptica* 5. *Vibrio vulnificus* biogroup 2, serovar E contains virulent and avirulent strains	Other biotypes of *V. vulnificus* are non-pathogenic for eels	2. Septicaemia, focal suppurative or granulomatous lesions, cutaneous ulcerations, abscessed or ulcerative lesions in kidney or liver 4. Haemorrhagic and ulcerative lesions. Lesions in mouth, operculum, brain, liver and kidney 5. Lesions	2. Edwardsiellosis, 'paracolon disease' 4. 'Sekiten-byo' (red spot disease) 5. Vibriosis	26 506 746 757 799 800 830
Eel (*Anguilla reinhardtii*)	*Photobacterium damselae* ssp. *damselae*				429
Eel	*Lactococcus garvieae*				
Eel (*Hyperoplus lanceolatus* Le Sauvege)	Atypical *Aeromonas salmonicida*		Haemorrhages on snout, caudal fin and tail. Typical umbonate furuncles on flank	Skin ulcers	197
Elvers	*Aeromonas allosaccharophila*			Disease	527
Green moray eel (*Gymnothorax funebris*) Spotted moray eel (*G. moringa*)	*Mycobacterium-triplex*-like		Florid skin nodules around the head and trunk that are soft grey, gelatinous and tan-coloured	Proliferative skin disease	345
Sand eel (*Ammodytes lancea* Cuvier)	Atypical *Aeromonas salmonicida*		Haemorrhages on snout, caudal fin and tail. Typical umbonate furuncles on flank	Skin ulcers	197
Flounder					
Flounder (*Platichthys flesus* Linnaeus)	Atypical *Aeromonas salmonicida* and oxidase-negative strains		Epidermal ulcerations – dark, red wounds usually round but some irregular forms seen	Skin ulcer disease	323 829 831 832
Greenback flounder (*Rhombosolea tapirina* Günther)	Atypical *Aeromonas salmonicida*			Ulcerative dermal lesions	826

Species	Organism	Description	Condition	Page
Japanese flounder, also known as Olive flounder (*Paralichthys olivaceus* Temminck and Schlegel)	1. Atypical *Aeromonas salmonicida* 2. *Edwardsiella tarda* 3. *Lactococcus garvieae* 4. *Nocardia seriolae* 5. *Photobacterium damselae* ssp. *piscicida* 6. *Streptococcus iniae* 7. *Tenacibaculum maritimum* 8. *Vibrio ichthyoenteri* 9. *Weissella hellenica* strain DS-12	2. Septicaemia, focal suppurative or granulomatous lesions, cutaneous ulcerations 4. Tubercules in kidney, gills and spleen. Abscesses on epidermis 5. Haemorrhagic septicaemia and white areas of granuloma in the kidney, liver, spleen 8. Intestinal necrosis, opaque intestines 9. Present in intestine	1. Infection 2. Edwardsiellosis 3. Streptococcosis 4. Nocardiosis 5. Fish pasteurellosis 6, 7, 8. Mortality 9. Probiotic potential	139 273 385 389 442 455 567 570
Summer flounder (*Paralichthys dentatus* Linnaeus)	1. *Mycobacterium* spp. with homology to *M. marinum* and *M. ulcerans* 2. *Vibrio (carchariae) harveyi*	1. Granulomas in the kidney, large red-tan coloured multilobulated mass in kidney 2. Reddening around anal area, distended abdomen filled with fluid, enteritis and necrosis	1. Mycobacteriosis 2. Flounder necrotizing enteritis (FINE)	373 710
Grayling (*Thymallus thymallus* Linnaeus)	1. *Aeromonas salmonicida* ssp. *salmonicida* 2. Atypical *Aeromonas salmonicida* 3. *Yersinia ruckeri*	1, 2. Skin ulcers	1. Furunculosis 2, 3. Infection	209 323 352
Green knifefish (*Eigemannia virescens* Valenciennes)	*Edwardsiella ictaluri*		Enteric septicaemia	426
Greenling – marine fish (*Hexagrammos otakii*)	Atypical *Aeromonas salmonicida*		Infection	385
Grouper	See Sea bass			
Haddock (*Melanogrammus aeglefinus* Linnaeus). Family Gadidae – cod and haddock	Atypical *Aeromonas salmonicida*	Skin ulcers	Infection	323
Halibut				
Atlantic halibut (*Hippoglossus hippoglossus* Linnaeus)	1. Atypical *Aeromonas salmonicida* 2. *Tenacibaculum ovolyticum*	1. Skin ulcers 2. Dissolves chorion and zona radiata of the egg shells	1. Infection 2. Opportunistic pathogen of eggs and larvae	323 324
Greenland halibut (*Reinhardtius hippoglossoides* Walbaum)	*Arthrobacter rhombi*	Organism isolated from spleen and internal organs	Pathogenicity not determined	600

continued

Table 1.1. *Continued.*

Host	Pathogen	Normal flora	Tissue site	Disease status	Ref
Herring Baltic herring (*Clupea harengus membras* Linnaeus)	1. *Pseudomonas anguilliseptica*	2. *Photobacterium iliopiscarium*	1. Haemorrhages in the eye 2. Intestine	1. Disease 2. Non-pathogenic	503 599 767
Human (*Homo sapiens*)	1. *A. allosaccharophila* 2. *Aeromonas caviae* 3. *Aeromonas hydrophila* ssp. *dhakensis* 4. *Aeromonas hydrophila* 5. *A. janadaei* 6. *A. schubertii* 7. *A. sobria* 8. *A. trota* 9. *A. veronii* ssp. *sobria* 10. *Aeromonas veronii* ssp. *veronii* 11. *Brucella abortus* 12. *Brucella* species 13. *Burkholderia* (*Pseudomonas*) *pseudomallei* 14. *Chromobacterium violaceum* 15. *Cryptococcus neoformans* 16. *Edwardsiella tarda* 17. *Erysipelothrix rhusiopathiae* 18. *Granulicatella elegans* 19. *Halomonas venusta* 20. *Mycobacterium marinum* 21. *Photobacterium damselae* ssp. *damselae* 22. *Rahnella aquatilis* 23. *Raoultella planticola* 24. *Shewanella algae* 25. *Streptococcus iniae* 26. *Vagococcus fluvialis* 27. *Vibrio cholerae* 01 & 0139 28. *Vibrio cholerae* non-01		1, 2. Faeces 3, 4. Faeces 5. Wound infection, stool, blood 6. Wound, abscess, blood, pleural fluid 7. 9, 10. Faeces 11. Respiratory infection, abortion 12. Headaches, lassitude, sinusitis 13. Wound infections, pneumonia and septicaemia 14. Wound infections 15. Cerebral spinal fluid 16. Intestine 17. Suppurating skin lesions 18. Endocarditis 19. Wound with watery discharge following fish bite 20. Skin lesions 22. Contaminated intravenous fluid 24. Leg ulcers, septicaemia, otitis media, faeces 25. Wounds following handling fish 26. Bite-wound, blood culture isolates and from peritoneal fluid 27. Faeces 28. Blood culture, faeces 29. Blood and spinal fluid 30, 31, 32, 33. Faeces 34. Wounds, faeces	1. Diarrhoea, renal failure in infant 2, 3, 4, 8. Diarrhoea 5. Wound infection, diarrhoea 6, 8. Infection 7. Acute renal failure in an infant 9, 10. Diarrhoea 11, 12. Brucellosis 13. Melioidosis 14. Infection 15. Zoonosis, mortality 17. Erysipeloid 18. Endocarditis 19. Wound infection 20. Infection 21. Wound infection 22. Sepsis. Also infection in immunocompromised hosts 23. Septicaemia in newborns 24. Infection 25. Cellulitis 26. Clinical infections 27. Cholera 28. Cellulitis, meningoencephalitis, septicaemia, diarrhoea 29. Meningitis 30, 31, 32. Diarrhoea 33. Gastroenteritis, food poisoning 34. Infection, food poisoning	120 123 125 142 143 151 177 196 255 259 276 310 346 347 348 378 383 392 482 527 555 588 653 732 792 816 822

Common name (species)	Organism	Description	Disease	Ref.
	29. *Vibrio cincinnatiensis* 30. *Vibrio fluvialis* 31. *Vibrio furnissii* 32. *Vibrio hollisae* 33. *Vibrio parahaemolyticus* 34. *Vibrio vulnificus*			
Japanese medaka (*Oryzias latipes* Temminck and Schlegel)	*Mycobacterium abscessus*		Mycobacteriosis	736
Lamprey	*Aeromonas salmonicida* ssp. *salmonicida*	May be carriers of the disease. Organism found in kidney	Succumb to infection following stress in stress-test experiments	352
Lobster. American lobster (*Homarus americanus, H. gammarus* L)	1. *Aerococcus viridans* var. *homari* 2. *Listonella anguillarum*	1. Organism in haemolymph	1. Gaffkemia, high mortality 2. Vibriosis	114 299 827
Mackerel				
Mackerel (*Scomber scombrus* Linnaeus)	*Mycobacterium* spp.	Grey-white nodules in spleen and visceral organs	Mycobacteriosis	515
Japanese jack mackerel (*Trachurus japonicus* Temminck and Schlegel)	1. *Lactococcus garvieae* 2. *Listonella anguillarum* 3. *Vibrio trachuri* (latest information suggests this is a junior synonym of *V. harveyi*)	3. Haemorrhages in organs and exophthalmia	3. Disease	400 743
Mahi-mahi (*Coryphaena hippurus* Linnaeus) also known as Dolphin fish	*Pantoea* (*Enterobacter*) *agglomerans*	Haemorrhages in eye, and dorsal and lateral musculature	Mortality, opportunistic pathogen	325
Makonbu (*Laminaria japonica*)	1. *Pseudoalteromonas bacteriolytica* 2. *Pseudoalteromonas elyakovii*	1. Produces a red pigment on the *Laminaria* and causes damage to the seed supply 2. Degrades fronds	1. Red spot disease 2. Spot disease	677 679
Milkfish (*Chanos chanos* Forsskål)	*Vibrio harveyi*	Eye lesions, exophthalmia, opaqueness, haemorrhages in eyes	Eye disease and mortality	390

continued

Table 1.1. Continued.

Host	Pathogen	Normal flora	Tissue site	Disease status	Ref
Minnow					
Baitfish, American baitfish, Fathead minnow (*Pimephales promelas* Rafinesque)	*Yersinia ruckeri*			Yersiniosis	
Bullminnows (*Fundulus grandis* Baird)	*Streptococcus* spp. Non-haemolytic group B		Exophthalmia, petechial haemorrhage in abdomen, oedema in liver	Mortality, streptococcosis	637
Eurasian minnow (*Phoxinus phoxinus* Linnaeus)	Atypical *Aeromonas salmonicida*		Extensive haemorrhagic skin lesions	Mass mortality	331
Minnow	*Listonella anguillarum*			Fish kill	135
Menhaden (*Brevoortia patronus* Goode)	*Streptococcus agalactiae*			Streptococcosis	
MOLLUSCS					
Bivalve (*Nodipecten nodosus*)		*Vibrio brasiliensis*, *V. neptunius*	*V. neptunius* isolated from diseased and healthy animals	Pathogenicity not known	740
Clam					
Manila clam (*Ruditapes philippinarum*) and (*R. decussatus*)	*Vibrio tapetis*		Invasion of pallial (mantle) cavity and adherence to the periostracal lamina at the growing edge of the shell. Main characteristic feature is a brown conchiolin deposit on the inner surface of the shell	Brown ring disease (BRD)	14 108 146 610 611
Clams (*Mercenaria mercenaria*), Bivalve molluscs	1. *Vibrio tubiashii*		1. Larvae cease swimming, digestion of tissue	1. Bacterial necrosis and systemic disease	321 762
Mussel (*Protothaca jedoensis* Lischke)		*Shewanella japonica*			397
Oyster					
Eastern oyster (*Crassostrea virginica*)	1. *Roseobacter* spp. strain CVSP 2. *Vibrio tubiashii*	2. *Shewanella colwelliana*	1. Lesions, mantle retraction, conchiolin deposits inside shell 2. Promotes settlement of oyster larvae	1. Juvenile oyster disease (JOD) 2. Normal flora	104 105 814 815

Species	Organism	Description	Summary	Ref
Japanese oyster (*Crassostrea gigas*), Pacific oyster (*Crassostrea gigas*)	1. *Alteromonas* species (not speciated) 2. *Listonella anguillarum* 3. *Nocardia crassostreae* 4. *Photobacterium damselae* ssp. *damselae* 5. *Vibrio splendidus* biovar II 6. *Vibrio splendidus* 7. *Vibrio tubiashii* 8. *Vibrio vulnificus* serovar E (some strains are avirulent for Taiwanese eels) 9. *Aeromonas media* 10. *Aeromonas* spp. 11. *Alteromonas* spp. 12. *Pseudomonas* spp. 13. *Vibrio* spp. Total bacterial count in soft tissue = 2.9×10^4 colonies per gram, and in haemolymph = 2.6×10^4 colonies per ml (Ref 596)	1. Larval deaths 2. Failure of veliger larvae to maintain themselves in the water column 3. Focal areas of brown discoloration on the mantle, or green-yellow nodules on abductor muscle, gills, heart and mantle 5. Organism isolated from gonads of broodstock. High bacterial load in larvae, bacteria in shell margin, loss of cilia and velum. 7. Larvae cease swimming, digestion of tissue	1, 2, 4. Mortality 3. Nocardiosis 5. Mass mortality. Disease resembles bacillary necrosis 6. Mortality. Disease is stress-associated 7. Bacterial necrosis and systemic disease 9. Probiotic against *V. tubiashii*	26 222 270 280 294 321 466 467 596 721 762 798
Mediterranean oysters	*Vibrio lentus*			513
Scallop				
Argopecten purpuratus	*Aeromonas hydrophila* *Vibrio alginolyticus*		Larval deaths	650
Pecten maximus	1. *Vibrio pectenicida* 2. *Roseobacter gallaeciensis*	1. Affects larvae	1. Vibriosis	662 470
Pike, Northern pike (*Esox lucius* Linnaeus)	*Yersinia ruckeri*			209
Mud skipper	*Listonella anguillarum*		Fish kill	135
Mullet				
Black mullet, Grey mullet, Flathead mullet (*Mugil cephalus* Linnaeus)	1. *Edwardsiella tarda* 2. *Eubacterium tarantellae* 3. *Lactococcus garvieae* (tentative identification)	1. Septicaemia, focal suppurative or granulomatous lesions, cutaneous ulcerations 2. Organism recovered from brain, liver, kidney, blood 3. Fish lethargic, exophthalmia, congestion and haemorrhage in spleen and kidney, white spots, ascites, granulomas, macrophages and bacteria in organs	1. Edwardsiellosis 2. Mortality 3. Septicaemia, chronic meningitis	157 343 764

continued

Table 1.1. *Continued.*

Host	Pathogen	Normal flora	Tissue site	Disease status	Ref
Borneo mullet, Largescale mullet (*Liza macrolepis* Smith)	*Streptococcus iniae*		Susceptible to infection	Streptococcosis	848
Silver mullet (*Mugil curema* Valenciennes)	*Vibrio harveyi*		Organism isolated in pure culture	Haemorrhagic septicaemia	23
Striped mullet	*Photobacterium damselae* ssp. *piscicida*			Fish pasteurellosis	
Wild mullet (*Liza klunzingeri* Day)	*Streptococcus agalactiae*		Haemorrhages on body especially in eye, mouth, operculum and fins. Organism cultured from brain, eye and blood	Streptococcosis, epizootic	242
ORNAMENTAL FISH					
Black acara (*Cichlasoma bimaculatum* Linnaeus)	*Mycobacterium chelonae abscessus M. fortuitum, M. simiae*			Mycobacteriosis	474
Comets (*Calloplesiops altivelis* Steindachner)	*Flavobacterium columnare*		Ulcers	Mortality	135
Danio, Sind danio (*Danio devario* Hamilton)	*Edwardsiella ictaluri*			Enteric septicaemia	
Blue discus fish (*Symphysodon aequifasciatus* Pellegrin)	1. *Aeromonas hydrophila* 2. *A. janadaei* 3. *Mycobacterium fortuitum*		2. Organism in liver 3. Fin lesions	1, 2. Deaths 3. Mycobacteriosis	116 135
Electric blue hap (*Sciaenochromis ahli* Trewavas)	*Aeromonas janadaei*			Deaths	135
Firemouth cichlid (*Thorichthys meeki, Cichlasoma meeki* Brind)	*Mycobacterium chelonae abscessus*			Mycobacteriosis	474
Flying fox (*Epalzeorhynchos kalopterus* Bleeker)	*Streptococcus iniae*		Isolated in pure growth	Deaths	135
Golden shiner (*Notemigonus crysoleucas* Mitchill)	1. *Streptococcus agalactiae* group B 2. *Edwardsiella ictaluri*			1. Streptococcosis	627

Species	Organism	Clinical signs	Disease/condition	Ref.
Goldfish (*Carassius auratus* Linnaeus)	1. Atypical *Aeromonas salmonicida* 2. *Aeromonas salmonicida* ssp. *nova* 3. *Flavobacterium branchiophilum* 4. *Mycobacterium chelonae abscessus* 5. *Vibrio cholera* (non-01)	1, 2. Ulcerative skin lesions, haemorrhages 3. Bacteria on gill surface, flared opercula, hyperaemic, swollen gill tissue, excess mucus production 5. Septicaemia	1, 2. Goldfish ulcer disease (GUD) 3. Bacterial gill disease 4. Mycobacteriosis 5. Mortalities, opportunistic infection, stress-related	40 474 602 639 695 825
Guppy (*Poecilia reticulata* Peters, *Lebistes reticulatus*)	*Mycobacterium fortuitum*	Fin lesions, emaciated, swam in circles	Mycobacteriosis	116
Molly				
Balloon molly (*Poecilia* spp.)	*Escherichia vulneris*	Exophthalmia in eyes, pale gills, empty digestive tract, opened operculum, yellow liver	Mortality	51
Black molly (*Poecilia sphenops* Valenciennes)	*Flavobacterium columnare*	White spots on the back, head and skin ulcers	Mortality	214
Silver molly (*Poecilia* spp.)	*Escherichia vulneris*	Pale liver, bent or curved body, thinness	Mortality	51
Oscar (*Astronotus ocellatus* Agassiz, *Apistogramma ocellatus*)	*Mycobacterium fortuitum*	Skin lesions on head	Mycobacteriosis	
Rams	*Streptococcus agalactiae* (*S. difficile*) group B	Organism in pure growth in liver	Deaths	135
Rosy barbs (*Puntius conchonius* Hamilton)	*Edwardsiella ictaluri*	Moribund, organism in internal organs	Mortality of 40%	374
Siamese fighting fish (*Betta splendens* Regan)	1. *Edwardsiella tarda* 2. *Mycobacterium fortuitum*	1. Moribund, single or multiple cutaneous ulcers of 1 mm diameter. Organism in internal organs 2. Lesions	1. Mortality of 70% 2. Mycobacteriosis	374 633
Tetra				
Black skirted tetra (*Hyphessobrycon* spp.)	*Edwardsiella tarda*	Septicaemia, focal suppurative or granulomatous lesions, cutaneous ulcerations	Edwardsiellosis	374 548

continued

Table 1.1. *Continued.*

Host	Pathogen	Normal flora	Tissue site	Disease status	Ref
Neon tetra (*Hyphessobrycon innesi* Myers – scientific name, *Paracheirodon innesi* Myers – valid name)	1. *Flavobacterium columnare* 2. *Mycobacterium fortuitum*		1. Skin discoloration, white areas of necrosis. Organism in skin, gills, muscle and inner surface of scales 2. Lesions	1. Muscle infections 2. Mycobacteriosis	116 543 656
Serpae tetra	*Streptococcus agalactiae* (*S. difficile*) group B		Isolated in pure heavy growth	Deaths	135
Three-spot gourami (*Trichogaster trichopterus* Pallas)	*Mycobacteria* species			Mycobacteriosis	672
Otter, European otter (*Lutra lutra*)	1. *Brucella pinnipediae* 2. *Staphylococcus lutrae* 3. *Vagococcus lutrae*		1. Organism isolated from tissues 2. Organism in liver, spleen, lymph node 3. Organism found in blood, liver, lungs, spleen	1. Brucellosis 2. 3. Pathogenicity not known	262 264 267 477
Pacific staghorn sculpin (*Leptocottus armatus* Girard)	*Mycobacterium scrofulaceum*		Livers white and friable	Mycobacteriosis	474
Perch					
European perch (*Perca fluviatilis* Linnaeus)	1. *Aeromonas salmonicida* ssp. *achromogenes* 2. *Yersinia ruckeri*		1. Large open lesions, descalation. No typical 'furuncles' seen 2. Disease is associated with poor environmental conditions	1. Mortality 2. Yersiniosis	534 772
Silver perch (Bidyan perch) (*Bidyanus bidyanus* Mitchell)	Atypical *Aeromonas salmonicida*		Cutaneous ulcers	Skin disease, ulcerative dermatitis	825
White perch (*Morone americana*, Gmelin, *Roccus americanus*)	*Photobacterium damselae* ssp. *piscicida*		Organism isolated from internal organs	Fish pasteurellosis, massive mortality	708
Yellow perch (*Perca flavescens* Mitchill)	*Mycobacterium chelonae*		Granulomatous peritonitis and hepatitis	Mycobacteriosis	204
Pike Northern pike (*Esox lucius* Linnaeus)	Atypical *Aeromonas salmonicida*		Skin ulcers	Infection	323

Species	Organism	Description	Effect	Ref.
Pirarucu (*Arapaima gigas* Cuvier), a gigantic tropical freshwater fish	*Salmonella arizonae*	Organism isolated from liver, spleen, heart, kidney, bloody exudate in body cavity, corneal opacity	Septicaemia	447
Plaice, European plaice (*Pleuronectes platessa* Linnaeus)	*Listonella anguillarum* serotype 07	Lesions	Vibriosis	712
Platies, Southern platyfish (*Xiphophorus maculatus* Gnther)	*Flavobacterium columnare*	White spots on the back and head, and skin ulcers	Mortality	214
Porpoise	See under SEA MAMMALS			
Prawn	See under Shrimp			
Rabbitfish				
Marbled spinefoot (*Siganus rivulatus* Forsskål)	1. *Mycobacterium marinum* 2. *Pseudomonas putrefaciens* 3. *Streptococcus iniae*	1. White-yellow nodules in the spleen 2. Skin discoloration, and focal necrosis becoming haemorrhagic, abdominal ascites, exophthalmia, oedema and necrosis in kidneys 3. Systemic disease with diffuse visceral haemorrhages	1. Mycobacteriosis – infection spread from cage fish to wild rabbitfish 2, 3. Mortality	218 260 666 853
White-spotted spinefoot (*Siganus canaliculatus* Park)	*Streptococcus iniae*	Ascites, hepatomegaly, splenomegaly	Streptococcosis, mass mortality	848
Red Sea fish, Wild fish, Striped piggy (*Pomadasys stridens* Forsskål), Variegated lizardfish (*Synodus variegatus* Lacepède)	*Streptococcus iniae*	Organism cultured from blood. No marked gross signs of disease	Morbidity	183
Redfish, Red drum (*Sciaenops ocellatus* Linnaeus)	1. *Eubacterium* species (tentative identification) 2. *Streptococcus iniae*	1. Organism isolated from brain, liver, kidney and blood 2. Skin lesions, exophthalmia, eye degeneration, necrotic areas of gill rot. Organism in heart and spleen	1, 2. Mortality	343 183
REPTILES				
Snakes	*Edwardsiella tarda*	Isolated from faeces	Carrier status or part of the normal flora	399
Tortoise	*Corynebacterium testudinoris*	Mouth lesions		180

continued

Table 1.1. *Continued.*

Host	Pathogen	Normal flora	Tissue site	Disease status	Ref
Turtles (Caspian terrapin – *Mauremys caspica*, Eastern box turtle – *Terrapene carolina carolina*, Mississippi map turtle – *Malaclemys kohni*, Northern diamondback terrapin – *Malaclemys terrapin terrapin*, Painted turtle – *Chrysemys picta*, Red-eared turtle – *Chrysemys scripta elegans*, Stinkpot turtle – *Sternotherus odoratus*, Travancore crowned turtle – *Melanochelys trijuga coronata*)		*Edwardsiella tarda*, *Salmonella durham*	Cloacae	Carrier status	606
Chelonia mydas, *C. caretta*, *Eretmochelys imbricata*	1. *Aeromonas hydrophila* 2. *Dermatophilus chelonae* 3. *Flavobacterium* spp. 4. *Mycobacterium* spp. 5. *Pseudomonas* spp. 6. *Vibrio alginolyticus*	7. *Acinetobacter calcoaceticus* 8. *Bacillus* spp., *Micrococcus* spp., *Moraxella* spp., *Proteus* spp.	2. Skin 4. Lung lesions 7. Oral cavity 8. Part of normal flora on skin, oral cavity and trachea	1. Associated with bronchopneumonia, ulcerative stomatitis, and traumatic ulcerative dermatitis caused by biting 2. Skin lesions, skin abscess, scabs 3. Associated with traumatic ulcerative dermatitis caused by biting, ulcerative stomatitis, bronchopneumonia and keratoconjunctivitis–ulcerative blepharitis 5. Associated with traumatic ulcerative dermatitis caused by biting, ulcerative stomatitis, adenitis, peritonitis, bronchopneumonia and keratoconjunctivitis–ulcerative blepharitis	300 301 529

Host	Organism	Findings	Significance	No.
			6. Associated with traumatic ulcerative dermatitis caused by biting, bronchopneumonia and osteomyelitis	614
Turtles (*Pseudemis scripta*)	*Aeromonas hydrophila*		Infection	
Hawaiian green turtle	*Citrobacter freundii, Hafnia alvei, Klebsiella oxytoca, Photobacterium damselae, Pseudomonas fluorescens, Pseudomonas putrefaciens, Pseudomonas stutzeri,* non-haemolytic *Streptococcus* spp., *Vibrio alginolyticus, Vibrio fluvialis*	Nasal swabs and cloacal swabs from apparently healthy turtles	Pathogenicity not determined. Most likely normal flora from these sites.	5
Leatherback turtle (*Dermochelys coriacea*)	*Photobacterium damselae* ssp. *damselae*	Fluid in body cavity, calcareous nodules in lung parenchyma, lungs congested	Valvular endocarditis and septicaemia	590
Loggerhead sea turtle (*Caretta caretta*)	*Aerococcus viridans*	Gas in diverticulum, with multifocal granulomatous serositis. Green mucoid fluid and fibrinonecrotic membrane on mucosa.	Oesophageal diverticulum	755
Ridley sea turtle (*Lepidochelys olivacea*)	*Vibrio mimicus*	Reduced viability of eggs	Cause of food poisoning in humans	4
Rhynchopelates oxyrhynchus Temminck and Schlegel, Coastal fish – Japan (*Therapon oxyrhynchus*)	*Vibrio hollisae*	Intestinal contents	Non-pathogenic	580
Roach (*Rutilus rutilus* Linnaeus)	1. *Aeromonas salmonicida* ssp. *achromogenes* 2. Atypical *Aeromonas salmonicida* 3. *Yersinia ruckeri*	1. Large open lesions, areas of descalation. No typical 'furuncles' seen 2. Cutaneous ulcers 3. Disease is associated with poor environmental conditions	1. Mortalities 2. Ulcerative dermatitis 3. Yersiniosis	352 534 772 825 826
Rockfish, Schlegel's black rockfish (*Sebastes schlegeli* Hildendorf)	Atypical *Aeromonas salmonicida*	Ulcer on the trunk. Organism isolated from the kidney and brain	Mortality	385 403

continued

Table 1.1. *Continued.*

Host	Pathogen	Normal flora	Tissue site	Disease status	Ref
Rockling Fourbeard rockling (*Enchelyopus cimbrius* Linnaeus)	Atypical *Aeromonas salmonicida*			Skin ulcers	832
Rotifer (*Brachionus plicatilis*)		1. *Vibrio rotiferianus* 2. *V. neptunius*	1, 2. Isolated from rotifer flow-through culture system. Pathogenicity not known		305 740
Rudd (*Scardinius erythrophthalmus* Linnaeus)	1. *Aeromonas salmonicida* ssp. *achromogenes* 2. *Yersinia ruckeri*				209 318
SALMONIDS					
Arctic Char (*Salvelinus alpinus* Linnaeus)	1. *Aeromonas salmonicida* ssp. *salmonicida* 2. Atypical *Aeromonas salmonicida* 3. *Serratia liquefaciens*	4. *Carnobacterium* spp.	1, 2. Skin ulcers 3. Redness and swelling around anus, ascites, haemorrhagic internal tissues	1. Furunculosis 2. Infection 3. Mortality 4. Non-pathogenic	323 352 648 715
Salmon					
Atlantic salmon (*Salmo salar* Linnaeus)	1. *Aeromonas salmonicida* ssp. *salmonicida* 2. Atypical *Aeromonas salmonicida* 3. *Flavobacterium psychrophilum* 4. *Listonella anguillarum* 5. *Moritella viscosa* 6. *Mycobacterium chelonae* 7. *Nocardia seriolae* 8. *Pasteurella skyensis* 9. *Photobacterium damselae* ssp. *damselae* 10. *Pseudomonas anguilliseptica* 11. *Renibacterium salmoninarum* 12. *Serratia liquefaciens* 13. *Streptobacillus moniliformis*-like organism 14. *Tenacibaculum maritimum* 15. *Vagococcus salmoninarum*	22. *Carnobacterium inhibens* 23. *Carnobacterium* spp. 24. *Lactobacillus* spp. 25. *Photobacterium iliopiscarium* 26. *Vibrio pacinii*	1. Skin ulcers 2. Organism from head kidney, cutaneous lesions 3. Erosion of fins 5. Skin ulcers, internal dark-brown petechiae or ecchymotic haemorrhage 6. Granuloma-like nodules in tissues 7. Granulomatous lesions on body 8. Multifocal areas of coagulative necrosis in kidney, liver, spleen 10. Petechial haemorrhages on skin, mouth and anus, peritoneum and liver 11. White nodules in kidney 12. Swelling of kidneys, bleeding at the anus and intestine, gastroenteritis 13. Organism intracellular in tissues, endothelial cells of kidney glomeruli enlarged	1. Furunculosis 2. Similar to furunculosis 3. Morbidity rather than mortality 4. Vibriosis 5. Winter ulcers 6. Mycobacteriosis 7. Nocardiosis 8. Mortality 9. Pathogenic in challenge experiments 10. Haemorrhagic disease 11. Bacterial kidney disease 12, 13. Mortality 14. Bacterial stomatitis (mouth rot) in smolts 15. Mortality 16. Blindness 17. Possible pathogen	49 100 107 117 132 133 137 144 232 306 330 352 411 412 416 450 506 519

Species	Organisms	Signs	Disease/Notes	Page
(continued)	16. *Varracalbmi* 17. *Vibrio logei* 18. *Vibrio ordalii* 19. *Vibrio salmonicida* 20. *Vibrio wodanis* 21. *Yersinia ruckeri* serotype I	14. Yellow-coloured plaque on teeth and oral cavity 15. Peritonitis, haemorrhagic ascites, retained eggs, engorged testes, languid swimming 16. Deep skin lesions, eye lesions, haemorrhagic and pyogranulomatous lesions in gills, kidney, liver and pseudobranch 17. Skin lesions 21. Haemorrhagic musculature 25. Intestine	18. Mortality 20. Opportunistic infection in 'winter ulcer' disease 21. Enteric redmouth (ERM) 22. Inhibits growth of *L. anguillarum* and *A. salmonicida* 23. Potential probiotic 24. Normal microflora 25. Non-pathogenic	538 564 599 605 611 622 648 682 683 712 767 771 824 828
Chinook salmon (*Oncorhynchus tschawytscha* Walbaum)	1. *Edwardsiella tarda* 2. *Flavobacterium branchiophilum* 3. *F. columnare* 4. *Mycobacterium neoaurum* 5. *Mycobacterium* spp. 6. *Renibacterium salmoninarum* 7. *Tenacibaculum maritimum* 8. *Yersinia ruckeri*	2. Bacteria attached to gill epithelium 3. Gill lesions 4. Organism present in eyes, exophthalmia, and ocular lesions 6. Gill lesions	1. Edwardsiellosis 2. Bacterial gill disease (BGD) 3. Mortality 4. Panophthalmitis 5. Mycobacteriosis 6. Bacterial kidney disease (BKD) 7. Mortality 8. ERM	36 53 88 137 154 209 245 802
Pacific salmon, Coho salmon (*O. kisutch* Walbaum)	1. *Aeromonas salmonicida* (oxidase-negative strain) 2. *Flavobacterium columnare* 3. *F. psychrophilum* 4. *Listonella anguillarum* 01 5. *Renibacterium salmoninarum*	1. Fingerlings affected. Soft kidneys, occasional fish with haemorrhage on fin, otherwise no external signs	1. Furunculosis 3. Cold-water disease 4. Vibriosis 5. BKD	168 712 748 749 757 765
Cherry salmon (*Oncorhynchus masou* Brevoort). Known as Yamame in Japan	*Flavobacterium branchiophilum*		BGD	802
Sockeye salmon (*O. nerka* Walbaum)	1. *Flavobacterium branchiophilum* 2. *Yersinia ruckeri*	1. Bacterium attaches to gill epithelium	1. BGD 2. ERM	137 802

continued

Table 1.1. *Continued.*

Host	Pathogen	Normal flora	Tissue site	Disease status	Ref
Salmonids – Trout					
Brook trout (*Salvelinus fontinalis* Mitchill)	1. Atypical *Aeromonas salmonicida* 2. *Edwardsiella tarda* 3. *Flavobacterium branchiophilum* 4. *Nocardia* spp. 5. *Yersinia ruckeri*		1. Skin ulcers 3. Bacteria attach to gill epithelium 4. Necrosis, thrombosis in kidney, spleen, gills	1. Skin ulcers and septicaemia 2. Acute bacterial septicaemia 3. BGD 4. Nocardiosis 5. ERM	137 604 765 824
Brown trout, Sea trout, Steelhead trout (*Salmo trutta trutta* Linnaeus)	1. *Aeromonas salmonicida* ssp. *salmonicida* 2. Atypical *Aeromonas salmonicida* 3. *Flavobacterium branchiophilum* 4. *Flavobacterium columnare* 5. *Hafnia alvei* 6. *Listonella anguillarum* 02 7. *Pseudomonas anguilliseptica*		1, 2. Skin ulcers 3. Skin around dorsal fin blanched, but no ulceration seen 5. Organism and lesions in the kidney 6. Lesions 7. Petechial haemorrhages on skin, mouth and anus, peritoneum and liver and at base of fins	1. Furunculosis 2. Skin ulcers and septicaemia 3. Columnaris disease, chronic mortality 4. BGD 5. Opportunistic pathogen 6. Vibriosis 7. Haemorrhagic disease	88 352 652 712 802 824 828
Rainbow trout (*Oncorhynchus mykiss* Walbaum) Previous name (*Salmo gairdneri* Richardson)	1. *Aeromonas salmonicida* ssp. *salmonicida* 2. Atypical *Aeromonas salmonicida* 3. Atypical *Aeromonas salmonicida* (growth at 37°C) 4. *Aeromonas sobria* 5. *Carnobacterium piscicola* 6. *Clostridium botulinum* 7. *Escherichia vulneris* 8. *Edwardsiella tarda* 9. *Flavobacterium branchiophilum* 10. *F. psychrophilum* 11. *Janthinobacterium lividum* 12. *Klebsiella pneumoniae* 13. *Lactococcus garvieae* 14. *Lactococcus piscium* 15. *Listonella anguillarum* 16. *Micrococcus luteus* 17. *Moritella viscosa*	31. *Citrobacter freundii*	1, 2, 3. Skin ulcers 4. Pathogenic for fingerlings in virulence assay (750) 5. Bilateral exophthalmia, periocular haemorrhages, ascites fluid and haemorrhages in liver, swimbladder, muscle and intestine 6. Toxin in serum and intestinal contents 7. Haemorrhagic lesions on skin, bloody exudate in digestive tract, liver yellow and hyperaemic, gonads haemorrhagic 8. 9. Bacteria attach to gill epithelium 12. Fin and tail 15. Lesions 16. Organism isolated from kidney, spleen and ascites fluid 17. Skin lesions, petechial haemorrhage in liver and peritoneal membranes	1. Furunculosis 2, 3 Skin ulcers, septicaemia 4. Haemorrhagic septicaemia 5. Chronic disease with low level mortality 6. Botulism 7. Mortality 8. Edwardsiellosis 9. BGD 10. 11. Mortality in fry, possible opportunistic infection 12. Fin and tail disease 13. Mortality 14. Pseudokidney disease, Lactobacillosis 15. Vibriosis, septicaemia 16. Isolated from moribund fish 17. Mortality	296 40 43 48 51 73 76 81 82 107 135 137 141 195 205 233 234 352 268

Species	Organism	Location / tissue	Finding / remarks	Page
	18. *Mycobacterium marinum*	18. Lesions in kidney	18. Mortality, mycobacteriosis	542
	19. *Nocardia asteroides*	19. Lesions in kidney	19. Nocardiosis	564
	20. *Photobacterium damselae* ssp. *damselae*	20. Organism present in head kidney	20. Pathogenic in challenge experiments and natural infection	579
	21. *Plesiomonas shigelloides*	21. Petechial haemorrhages in intestine	21. Possible opportunistic pathogen	604
	22. *Pseudomonas anguilliseptica*	22. Petechial haemorrhages on skin, mouth and anus and base of fins, peritoneum and liver	22. Haemorrhagic disease	618
	23. *Renibacterium salmoninarum* 01, 02, 03, 04	23. White nodules in kidney	23. BKD	640
	24. *Serratia marcescens, S. plymuthica*	25. Exophthalmia, ascites fluid, lesions on fins, organism in kidney and liver	24. Mortality in fingerlings	682
	25. *Staphylococcus warneri*	26, 27. Organisms in brain and eye	25. Diseased and dying trout – opportunistic infection	707
	26. *Streptococcus agalactiae* (*S. difficile*)	28. Loss of equilibrium, haemorrhage around eye and gill, lesions on body, congestion in spleen and liver. Peritonitis, haemorrhagic ascites, retained eggs, engorged testes, languid swimming	26, 27. Meningoencephalitis, septicaemia	712
	27. *Streptococcus iniae*	30. Haemorrhage around mouth and intestines	28. Chronic disease with mortality	716
	28. *Vagococcus salmoninarum*		29. Vibriosis	748
	29. *Vibrio ordalii*		30. ERM	750
	30. *Yersinia ruckeri*		31. Opportunistic pathogen	752
				753
				757
				802
				824
				828
				835
				853
Amago trout	*Pseudomonas chlororaphis*	Haemorrhages, increased ascites	Infection	332
Sardine Pacific sardine, South American pilchard (*Sardinops sagax* Jenyns)	*Tenacibaculum maritimum*	Gliding bacteria seen as a tan-coloured pseudomembrane over the body	Disease	154

SEA BIRDS

Species	Organism	Location / tissue	Finding / remarks	Page
Crane (Sandhill crane – *Grus canadensis*)	*Edwardsiella tarda, Salmonella hartford, S. java*	Large intestine		823
Eagle, Bald eagle (*Haliaeetus leucocephalus*)	*Edwardsiella tarda*	Cloacal swab	Bird was sick but no other evidence of infection	823
Gull, Ring-billed gull (*Larus delewarensis*)	*Edwardsiella tarda*	Large intestine		823
Heron (Great blue heron – *Ardea herodias*)	*Edwardsiella tarda*	Large intestine		823
Loon (Common loon – *Gavia immer*)	*Edwardsiella tarda*	Intestinal content	Haemorrhagic enteritis, intestinal content dark and tarry	823

continued

Table 1.1. *Continued.*

Host	Pathogen	Normal flora	Tissue site	Disease status	Ref
Pelican Brown pelican (*Pelecannus occidentalis carolinensis*)	*Edwardsiella tarda*		Intestinal content, lung, liver	Haemorrhagic enteritis, intestinal content dark and tarry	823
Penguin					
Gentoo penguin from sub-Antarctic region		*Salmonella enteritidis, S. havana, S. typhimurium*		Carrier status	612
Macaroni penguin (*Eudyptes chrysolophus*)	*Burkholderia (Pseudomonas) pseudomallei*		Pin-point, white lesions in liver and lung, fluid in airsacs, organism in spleen, liver, heart blood	Melioidosis	516
Penguins (*Aptenodytes patagonica, Eudyptes crestatus, Pyoscelis papua, Spheniscus demersus, Spheniscus humboldti*)		*Providencia friedericiana*	Isolated from faeces of captive penguins		559
Shad Gizzard shad (*Dorosoma cepedianum* Lesueur)	*Aeromonas sobria*		Isolated in pure culture from kidney, liver, spleen Isolates pathogenic for fingerling trout in virulence assay	Epizootic in spawning females	750
Sea bream	See under Bream				
Sea grass		*Vibrio aerogenes* *Vibrio diazotrophicus*	Present in sediment		319 692
Seahorse (*Hippocampus kuda* and *Hippocampus* species)	*Vibrio harveyi*		External haemorrhages, haemorrhagic liver, ascites fluid	Mortality	11 135
Sea dragon, Leafy sea dragon *Phycodurus equis*	*Vibrio harveyi*		Isolated in pure culture from heart blood, liver	Death. Pathogenicity of the organism not determined	135
Seal. See under SEA MAMMALS					
Sea Lion. See under SEA MAMMALS					
SEA MAMMALS (Order Cetacea)	Sea mammals include the three mammalian orders considered to be true aquatic mammals. These are *Cetacea* (whales, dolphins and porpoises), *Pinnipedia* (seals, sea-lions and walruses), and *Sirenians* (sea cows) (Foster *et al.*, 2002).				
Dolphin					
Dolphin fish – see Mahi-mahi					

Species	Organism	Findings	Disease/condition	Ref.
Atlantic white-sided dolphin (*Lagenorhynchus acutus*)	1. *Brucella cetaceae* 2. *Helicobacter cetorum*	1. Oesophageal ulceration, abortion, hepatic and splenic coagulative necrosis 2. Organism found in glandular mucosa of main stomach	1. Brucellosis 2. Gastric ulceration	262 327 267 327 328
Amazon freshwater dolphin (*Inia geoffrensis*)	*Streptococcus iniae*	Subcutaneous abscesses	Golf ball disease	625
Tursiops aduncas	*Pasteurella multocida*	Intestinal haemorrhage and bacteraemia. Source of infection was contamination from local bird rookery	Enteritis, mortality	726
Tursiops gephyreus	*Helicobacter* spp.	Present in dental plaque	May act as a reservoir for gastric infection and gastric ulcers	303
Atlantic bottlenose dolphin (*Tursiops truncatus*)	1. *Brucella* species 2. *Burkholderia pseudomallei* 3. *Clostridium perfringens* 4. *Edwardsiella tarda* 5. *Erysipelothrix rhusiopathiae* 6. *Helicobacter cetorum* 6. *Mannheimia haemolytica* 8. *Photobacterium damselae* ssp. *damselae* 9. *Photobacterium damselae* ssp. *piscicida*	1. Abortion 2. Oedema, haemorrhage and nodules in lungs 3. Abscess in dorsal muscle 4. Purulent mastitis 5. Ulcers on skin seen as rhomboid plaques, lungs congested and oedematous. Organism in all organs 6. Organism cultured from faeces 7. Haemorrhagic tracheitis	1. Brucellosis 2. Melioidosis 3. Clostridial myositis 4. Mastitis 5. Erysipelas, septicaemia 6. Ulcers in oesophagus and forestomach – gastritis 7. Septicaemia 8. Wound infection 9. Pasteurellosis	349 247 312 292 328 726
Common dolphin (*Delphinus delphis*)	1. *Brucella cetaceae* 2. *Helicobacter* spp. 3. *Pasteurella multocida* 4. *Staphylococcus delphini*	1. Subcutaneous lesion 2. Organisms found in glandular mucosa of main stomach 3. Intestinal haemorrhage 4. Purulent skin lesions	1. Brucellosis 2. Gastric ulceration 3. Enteritis 4. Skin infection	327 267 404 658 726 778
Pacific white-sided dolphin (*Lagenorhynchus obliquidens*)	1. *Burkholderia pseudomallei* 2. *Helicobacter cetorum*	1. Oedema, haemorrhage and nodules in lungs 2. Organism cultured from faeces	1. Melioidosis 2. Ulcers in oesophagus and forestomach – gastritis	349 328
Striped dolphin (*Stenella coeruleoalba*)	1. *Brucella cetaceae* 2. *Cryptococcus neoformans* (yeast) 3. *Actinobacillus delphinicola*	1. Meningitis 2. Pulmonary infection 3. Isolated from various tissues	1. Brucellosis 2. Pulmonary cryptococcosis 3. Pathogenicity not determined	278 262 263 267

continued

Table 1.1. *Continued.*

Host	Pathogen	Normal flora	Tissue site	Disease status	Ref
Porpoise Harbour porpoise (*Phocoena phocoena*)	1. *Brucella cetaceae* 2. *Erysipelothrix rhusiopathiae* 3. *Streptococcus dysgalactiae* ssp. *dysgalactiae* Lancefield Group L 4. *Vagococcus fessus*	5. *Actinobacillus delphinicola* 6. *Actinomyces marimammalium* 7. *Actinobacillus scotiae* 8. *Arcanobacterium pluranimalium* 9. *Phocoenobacter uteri*	1. Lymph nodes 2. Cutaneous lesions 3. Intestine, kidney, lung, spleen 4, 5. Isolated from various tissues 6. Lung 7. Liver, lung, brain and spleen of stranded porpoise 9. Uterus	1. Brucellosis 2. Erysipelas 3. Septicaemia, bronchopneumonia, myocarditis, pyelonephritis 4, 5, 6, 7, 8, 9. Pathogenicity not determined	262 263 265 266 267 369 370 404 480 658 686 727
Sea lion					
Sea lion (*Zalophus californianus*)	1. *Burkholderia pseudomallei* 2. *Clostridium perfringens* 3. *Edwardsiella tarda* 4. *Escherichia coli* 5. *Pasteurella multocida*	6. *Salmonella heidelberg, S. newport, S. oranienburg*	1. Oedema, haemorrhage and nodules in lungs 2. Infection in muscle with gas and pus 3. Abscesses in lungs, sanguino-mucopurulent exudate in trachea and terminal bronchi 4. Verrucous, grey-red lesions in atrioventricular valves 5. Yellow, pus-filled fluid in pleural cavity 6. Recovered from healthy animals	1. Melioidosis 2. Clostridial myositis 3. Bacterial pneumonia, respiratory infection Pathogenicity not conclusive 4. Endocarditis 5. Mortality 6. Non-pathogenic or carrier status	298 312 349 430 435 804
Seals (Pinnipedia)					
Common seal or Harbour seal (*Phoca vitulina*)	1. *Brucella pinnipediae* 2. *Mycoplasma phocicerebrale* 3. *Mycoplasma phocirhinis* 4. *Streptococcus phocae* 5. *Vagococcus fessus*	6. *Arcanobacterium (Corynebacterium) phocae* 7. *Arthrobacter nasiphocae* 8. *Atopobacter phocae* 9. *Mycoplasma phocidae*	1. Gastric lymph node, spleen 2. Isolated from brain, heart, lung, nose, throat 3. Isolated from pus in lung 4. Lesions in lung, exudate in bronchi 5. Organism isolated from liver and kidney in pure growth 6. Organism in nasal passage	1. Brucellosis 2, 3. Associated with respiratory disease 4. Pneumonia – opportunistic infection in seal morbillivirus infection 5, 6, 7, 8. Pathogenicity not known	404 182 295 449 267 369 479 613

Species	Organism	Clinical findings	Disease / outcome	Ref.
Grey seal (*Halichoerus grypus*)	1. *Aeromonas hydrophila* 2. *Brucella pinnipediae* 3. *Burkholderia pseudomallei* 4. *Streptococcus phocae* 5. *Actinomyces marimammalium* 6. *Arcanobacterium phocae*	1. Organism in lung and liver. Isolate positive for aerolysin gene, cytotoxin and haemolysin 2. Isolated from lung 3. Oedema, haemorrhage and nodules in lungs 4. Lesions in lung, exudate in bronchi 5. Isolated from intestine 7. Organism isolated from intestine and lymph nodes. The animal had lymphadenopathy and pulmonary haemorrhage 9. Isolated from respiratory tract	9. Avirulent 1. Septicaemia 2. Brucellosis 3. Melioidosis 4. Pneumonia – opportunistic infection in seal morbillivirus infection 5, 6. Pathogenicity not clear. Isolated from seals with septicaemia and pneumonia	658 660 700 454 349 262 370 267 613 700
Harp seal (*Phoca groenlandica*)	*Brucella* spp.	Organism in lymph nodes	Brucellosis	261
Hooded seal (*Cystophara cristata*)	1. *Brucella pinnipediae* 2. *Actinomyces marimammalium*	2. Isolated from lung	1. Brucellosis, stress-related 2. Pathogenicity not determined	262 370 267
Northern fur seal (*Callorhinus ursinus*)	1. *Salmonella adelaide, S. heidelberg, S. newport, S. oranienburg* 2. *Acinetobacter calcoaceticus, Actinobacillus* spp., *Aeromonas eucrenophila, Alcaligenes faecalis, E. coli, Enterobacter* spp., *Klebsiella* spp., *Moraxella* spp., *Pseudomonas fluorescens, Pseudomonas* spp., *S. epidermidis, Streptococcus* spp. 3. *Bacillus* spp., *Corynebacterium* spp., *E. coli, Listeria* spp., *Moraxella* spp., *Neisseria cuniculi, Proteus mirabilis, Staphylococcus epidermidis, Streptococcus* spp.	1. Recovered from healthy animals 2. Rectum 3. Oropharynx	1. Non-pathogenic or carrier status 2, 3. Normal flora	298 779
Ringed seal (*Phoca hispida*)	*Brucella* species	Organism in lymph nodes	Brucellosis	261

continued

Table 1.1. *Continued.*

Host	Pathogen	Normal flora	Tissue site	Disease status	Ref
South Georgian Antarctic fur seal		*Salmonella enteritidis, S. havana, S. newport, S. typhimurium*		Carrier status	612
Seal	*Mycoplasma*			Secondary infection due to seal morbilivirus	
Southern elephant seal (*Mirounga leonina*)		*Facklamia miroungae*	Nasal swab	Isolated from apparently healthy juvenile animal	368
Whale					
Beluga whale, White whale (*Delphinapterus leucas*)	1. *Helicobacter cetorum* 2. *Mycobacterium marinum*		1. Inappetence and lethargy. Oesophageal and forestomach ulcers 2. Pyogranulomatous dermatitis and panniculitis	1. Gastritis 2. Mycobacteriosis	329 327 328 111
False killer whale (*Pseudorca crassidens*)	*Burkholderia pseudomallei*		Oedema, haemorrhage and nodules in lungs	Melioidosis	349
Killer whale (*Orcinus orca*)	1. *Burkholderia pseudomallei* 2. *Candida* spp. 3. *Clostridium perfringens*		1. Oedema, haemorrhage and nodules in lungs 2. Necrotic skin lesions, blowhole lesions, can become systemic 3. Toxaemia, oedematous lymph nodes, liquefied muscle	1. Melioidosis 2. Infection 3. Clostridial myositis	349 312 726
Minke whale (*Balaenoptera acutorostrata*)	1. *Brucella* species 2. *Granulicatella balaenopterae*		1. Isolated from liver and spleen 2. Isolated from pure growth from liver and kidney of beached whale	1. Disease status not known. Whale caught during commercial fishing 2. Pathogenicity not determined	171 179 478
Pacific pilot whale (*Globicephala scammoni*)	*Candida* spp.		Infection of nares usually as a secondary infection	Infection	726
Sowerby's beaked whale		*Actinobacillus delphinicola*	Isolated from various organs	Pathogenicity not determined	263
Sea Urchin		*Vibrio diazotrophicus*	Part of gut flora		319
Sharks					
Brown shark (*Carcharhinus plumbeus*)	*Photobacterium damselae* ssp. *damselae*				314

		Organism isolated from organs	Mortality	
Nurse shark (*Orectolobus ornatus*). Animal held in captivity.	*Photobacterium damselae* ssp. *damselae*			618
Sandbar shark	*Vibrio (carchariae) harveyi*	Isolated from kidney		316
Smooth dogfish (*Mustelus canis* Mitchill) and Spiny dogfish (*Squalus acanthias* L)	*Alteromonas* spp., *Photobacterium* spp., *Pseudomonas* spp., *Shewanella putrefaciens*, *Vibrio* spp.	All isolated from lesions in the head kidney of healthy sharks	The role of the lesions, from which these organisms were cultured, in morbidity and mortality is unknown	109
Blacktip shark (*Carcharhinus limbatus*), Lemon shark (*Negaprion brevirostris* Poey), Nurse shark (*Ginglymostoma cirratum*), Tiger shark (*Galeocerdo curvier*)	1. *Aeromonas salmonicida*; *Alteromonas* spp.; *Moraxella* spp.; *Neisseria* spp.; *Photobacterium damselae* ssp. *damselae*; *Photobacterium damselae* ssp. *piscicida*; *Plesiomonas shigelloides*; *Vibrio alginolyticus*; *Vibrio harveyi*; 2. *Vibrio harveyi*	3. *Photobacterium damselae* ssp. *damselae*	1. Isolated from various tissue sites; 2. Active splenic and hepatic disease on histological examination	1. Pathogenicity not assessed; 2. Isolated from experimentally infected Lemon sharks but did not show clinical disease. Stress-related; 3. Not recovered from experimentally inoculated Lemon sharks	315 316
Sheatfish, Wels catfish (*Silurus glanis* Linnaeus)	1. *Flavobacterium branchiophilum*; 2. *F. columnare*		1. Bacterium attaches to gill epithelium; 2. Organism isolated from kidney, skin blanched and ulcerated	1. BGD; 2. Columnaris disease	88 251 802
Shrimp				
Black tiger prawn (*Penaeus monodon*)	1. *Vibrio harveyi*; 2. *Vibrio parahaemolyticus*; 3. *Vibrio splendidus* II	1, 2, 3. Infects hepatopancreas where there is an inflammatory response in the intertubular sinuses	1, 2, 3. Mortalities	363 410
Brine shrimp (*Artemia* species)	*Vibrio proteolyticus*	Affects microvilli, disrupts gut epithelial cell junctions, devastates cells and tissues in the body cavity	Deaths	135 788
Chinese shrimp (*Penaeus chinensis*)	1. *Vibrio alginolyticus*; 2. *Vibrio harveyi*; 3. *Vibrio vulnificus* serogroup E; 4. *Vibrio pacinii*	1, 2. Affects larval development from zoea stage onwards. Inactivity, anorexia, opaqueness and larvae settle to the bottom of the tank	1, 2, 3. Vibriosis, mortality; 4. Isolated from healthy shrimp	98 306 777

continued

Table 1.1. *Continued.*

Host	Pathogen	Normal flora	Tissue site	Disease status	Ref
Fairy shrimp (*Branchipus schaefferi* Fisher, *Chirocephalus diaphanus* Prévost, *Streptocephalus torvicornis* Waga)	*Aeromonas hydrophila*		Black nodules on thoracic appendages, the cercopods and antennae	Black disease	220
Giant freshwater prawn (*Macrobranchium rosenbergii*)	1. *Aeromonas caviae* 2. *A. veronii* ssp. *veronii* 3. *Lactococcus garvieae*		3. Yellowish-white spots in muscle, whitish muscle and swollen, yellow hepatopancreas, fluid accumulation between cuticle and muscle tissue	1. 2. Pathogenic in virulence studies 3. Mortality	156 723
Kuruma prawn (*Penaeus* [*Marsupenaeus*] *japonicus, Penaeus stylirostris*)	1. *Vibrio harveyi* 2. *Vibrio penaeicida*	3. *Acinetobacter* spp., *Alteromonas* spp., *Bacillus* spp., *Corynebacterium* spp., *Flavobacterium* spp., *Micrococcus* spp., *Moraxella* spp., *Pseudomonas* spp., *Staphylococcus* spp.	1. Black spots on exoskeleton 2. Septicaemia 3. Isolated from apparently healthy shrimp, but not from diseased shrimp	1. Mortality 2. Syndrome 93 3. Non-pathogenic	187 388 198 23
White shrimp (*Penaeus vannamei* Boone)	1. *Mycobacterium peregrinum* 2. *Vibrio harveyi*	*Vibrio xuii*	1. Multifocal, melanized nodular lesions on carapace 2. Black spots on exoskeleton	1. Opportunistic infection 2. Mortality 3. Pathogenicity not known	23 551 740
Snakehead fish (*Channa striatus* Fowler)	*Mycobacterium poriferae*, later identified by PCR to be *M. fortuitum*		Internal nodular lesions	Mycobacteriosis	633 756
Snook (*Centropomus undecimalis* Bloch)	*Vibrio harveyi*		Opaque cornea	Infection	453
Sole (*Solea senegalensis* Kaup)	*Photobacterium damselae* ssp. *piscicida*		Haemorrhagic septicaemia and white areas of granuloma in the kidney, liver, spleen	Fish pasteurellosis	855
Dover sole (*Solea solea*)	*Tenacibaculum maritimum*		Blistering of the skin between the caudal and marginal fins, loss of dermal tissues, which progresses to necrotic ulcers	Black patch necrosis	90 539
Spadefish Atlantic spadefish (*Chaetodipterus faber* Broussonet)	*Vibrio harveyi*		Bilateral exophthalmia, haemorrhages in and around eyes, corneal opacity	Mortality	23

Sponge. Marine sponge (*Halichondria bowerbanki*). Also known as Crumb-of-bread sponge	*Mycobacterium poriferae*			608
Squid				
Loligo pealei	1. *Shewanella pealeana* 2. *Vibrio logei*	1. Associated with the reproductive organ, the accessory nidamental gland, in females 2. Symbiont in light organ	1. 2 Normal flora	257 492
Hawaiian sepiolid squid (*Euprymna scolopes*)	*Vibrio fischeri*	Light organ	Light organ symbiont	257
Sepiola affinis, S. robusta	*Vibrio fischeri Vibrio logei*	Light organ	Symbiotic bacteria in light organ	257
Stingray (*Dasyatis pastinaca*) Animals held in captivity	*Photobacterium damselae* ssp. *damselae*	Organism isolated from organs	Mortality	618
Striped jack, White Trevally (*Pseudocaranx dentex* Bloch and Schneider)	1. *Photobacterium damselae* ssp. *piscicida* 2. *Pseudomonas anguilliseptica*	1. Haemorrhagic septicaemia and white areas of granuloma in the kidney, liver, spleen 2. Haemorrhages in mouth, nose, operculum and brain. Organism in kidney	1. Fish pasteurellosis 2. Mortalities	465 567
Sturgeon				
Adriatic sturgeon (*Acipenser naccarii* Bonaparte)	*Lactococcus garvieae*	Inappetence, irregular swimming, some bilateral exophthalmia and abdominal ascites	Mortality	669
Siberian sturgeon (*Acipenser baerii baerii* Brandt)	*Yersinia ruckeri*	Can be infected with or without clinical signs	Yersiniosis	797
Tench (*Tinca tinca* L.)	*Mycoplasma mobile*	Isolated from gills	Pathogenicity not stated	439 440
Tilapia				
Nile tilapia (*Oreochromis* sp., *O. niloticus niloticus* Linnaeus) Also known as St Peter's fish	1. *Edwardsiella tarda* 2. *Streptococcus agalactiae* 3. *Streptococcus iniae*	1. Septicaemia, focal suppurative or granulomatous lesions, cutaneous ulcerations 3. Central nervous system involvement, lethargy, erratic swimming	1. Edwardsiellosis 2, 3. Streptococcosis	233 442

continued

Table 1.1. *Continued.*

Host	Pathogen	Normal flora	Tissue site	Disease status	Ref
Tilapia (*Saratherodon* (*Tilapia*) *aureus*)	1. *Edwardsiella ictaluri* 2. *Streptococcus iniae*		2. Loss of orientation, exophthalmia, petechiae around anus, mouth and pectoral fins, fluid in peritoneal cavity, enlarged organs	1. Slightly susceptible 2. Mortality	621 627
Tilapia (*Sarotherodon niloticus*) Also known as Nile tilapia	1. *Lactococcus garvieae* 2. *Streptococcus iniae* 3. *Vibrio cholerae* non-01		1. Dermal haemorrhage and exophthalmia, epicarditis, peritonitis, pale-coloured liver, splenomegaly, nodule formation in gonads 2. Loss of orientation, exophthalmia, petechiae around anus, mouth and pectoral fins, fluid in peritoneal cavity, enlarged organs	1, 2. Systemic infection – streptococcosis 3. Farm mortality. May be opportunist	507 550 621
Turbot (*Scophthalmus maximus* Linnaeus)	1. Atypical *Aeromonas salmonicida*. Also an oxidase-negative strain 2. *Chryseobacterium scophthalmum* 3. *Listonella anguillarum* 01 and 02β, 02α 4. *Mycobacterium chelonae* and *M. marinum* 5. *Photobacterium damselae* ssp. *damselae* 6. *Photobacterium damselae* ssp. *piscicida* 7. *Pseudomonas anguilliseptica* 8. *Serratia liquefaciens* 9. *Streptococcus parauberis* 10. *Vibrio splendidus* biotype I	11. *Enterovibrio norvegicus* 12. *Vibrio cholerae* non-01 and *Vibrio mimicus* did not cause deaths in virulence assay 13. *Vibrio neptunius* 14. *Vibrio scophthalmi*	1. Skin ulcers 2. Swollen intestines, haemorrhages in eye, skin and jaw, gill hyperplasia 4. Granulomas in organs 7. Organism in head–kidney and spleen 8. Swollen kidney and spleen, yellow nodules, foci of liquefaction necrosis 9. Lesions, haemorrhage in the anal and pectoral fins and petechiae on the abdomen, exophthalmia and pus in the eyes 9. Haemorrhages in mouth, abdominal distension, reddish fluid in peritoneal cavity 11, 13, 14. Part of gut flora	1. Mortality 2. Gill disease, haemorrhagic septicaemia 3. Vibriosis 4. Mycobacteriosis 5. Mortality 6. Fish pasteurellosis 7. Haemorrhagic septicaemia 8. Opportunist pathogen, mortality 9. Streptococcosis, hepatomegaly, mucohaemorrhagic enteritis 10. Epizootic in juveniles 11, 14. Normal flora 13. Pathogenicity not known	268 281 224 149 31 96 475 507 557 617 673 712 740 741 748 749 751 754 791 832

Species	Organism	Observation	Disease/Condition	Page
Turbot (*Colistium nudipinnis* Waite) Brill (*C. gunther*). Both species are also known as flatfish	*Vibrio campbellii*-like, *Vibrio splendidus* I	Organism in brain, kidney, liver, which showed lesions and haemorrhages	Acute mortality in juveniles, opportunistic infection	221
Yellowtail (*Seriola quinqueradiata* and *S. purpurascens* Temminck and Schlegel) Also called Rudderfish and Japanese amberjack	1. *Lactococcus garvieae* 2. *Listonella anguillarum* 3. *Mycobacterium* spp. 4. *Nocardia seriolae* 5. *Photobacterium damselae* ssp. *damselae* 6. *Photobacterium damselae* ssp. *piscicida* 7. *Streptococcus iniae*	1. Erosion of tail fin, redness of anal fin, petechiae inside operculum, exophthalmia 4. Abscesses in epidermis, formation of tubercules in gills, kidneys, spleen 5. Organism in spleen and kidney, toxin produced 6. Bacterial colonies of white-grey colour seen on spleen and kidney 7. Organism isolated from brain	1. Streptococcosis 3. Mycobacteriosis 4. Nocardiosis 5. Mortality 6. Fish pasteurellosis, mortality 7. Septicaemia	459 462 424 455 464 236 233 235
Whale. See under SEA MAMMALS				
Whitefish (*Coregonus* sp.), Cisco (*Coregonus artedi* Lesueur), Lake whitefish (*Coregonus clupeaformis* Mitchill), Peled (*Coregonus peled* Gmelin) Family *Salmonidae*	1. *Aeromonas salmonicida* ssp. *salmonicida* 2. Atypical *Aeromonas salmonicida* 3. *Pseudomonas anguilliseptica* 4. *Yersinia ruckeri*	3. Petechial haemorrhages on skin, mouth and anus, peritoneum and liver 4. Disease is associated with poor environmental conditions	1. Skin ulcers, furunculosis 2. Skin ulcers 3. Haemorrhagic disease 4. Yersiniosis	323 352 772 828
Wolf-fish (*Anarhichas lupus* Linnaeus), Spotted wolf-fish (*A. minor* Olafsen)	1. Atypical *Aeromonas salmonicida*	2. *Carnobacterium divergens* 3. *Carnobacterium* spp.	1. Atypical furunculosis 2, 3. Part of intestinal microflora	648
Wrasse (*Labridae*) Cleaner fish	Atypical *Aeromonas salmonicida*	Haemorrhagic internal organs, bloody exudate	Furunculosis	468

Bacterial gill disease (BGD)

Bacterial gill disease is caused by the bacterium *Flavobacterium branchiophilum*, which are large filamentous Gram-negative rods. The bacterium attaches to the epithelial surface of the gill (Snieszko, 1981; Ostland *et al.*, 1994).

Bacterial kidney disease (BKD)

The disease agent is *Renibacterium salmoninarum*, which affects salmonids. The disease begins in a chronic form that develops full expression when the fish are 1 year old. Systemic granulomatous lesions are seen in all organs but particularly in the kidney, where grey, necrotic abscesses occur throughout the tissue, causing enlargement and necrosis of the kidney (OIE, 2000a).

Bacterial stomatitis (mouth rot)

This condition is seen as yellow-coloured plaque on the teeth and the oral cavity in Atlantic salmon smolts in their first year in seawater. The causative organism was identified as *Tenacibaculum maritimum*. Some differences were noted between these strains and the reference strains. The strains from mouth rot had an optimum temperature of 18–25°C, which is slightly lower than the reference strains, and they also had an optimum for media containing 70% seawater. Many of the strains from mouth rot had α- and β-glucosidase activity, an unusual finding for *T. maritimum* isolates (Ostland *et al.*, 1999b).

Black patch necrosis

The disease affects Dover sole and is caused by *T. maritimum*. Initially there is a slight blistering of the skin surface, which develops into loss of epithelium and necrotic ulcers (Bernardet *et al.*, 1990).

Brown ring disease (BRD)

This is a disease caused by *Vibrio tapetis*, and may cause mass mortality in Manila clams (*Ruditapes philippinarum*). It was first reported in France in 1987 where it caused high mortality in cultured stocks. The disease is characterized by a ring of brown deposit of several layers seen between the pallial line and the edge of the shell. A shell repair process occurs and this can be seen as white calcified areas that cover the brown deposit (Paillard and Maes, 1994).

Enteric redmouth disease (ERM)

This disease is caused by the Gram-negative bacterium *Yersinia ruckeri*. It is an economically serious disease in the rainbow trout farming industry of many countries. Clinical signs are haemorrhagic areas around the mouth, intestines and other organs.

There are a number of modes of transmission of the disease, including birds, wild fish and carrier fish (Willumsen, 1989). The organism also forms biofilms on fish tanks, which is a source of reinfection (Coquet *et al.*, 2002).

Enteric septicaemia of catfish (ESC)

The disease is caused by the bacterium *Edwardsiella ictaluri*, and is a major disease problem for the commercial channel catfish industry. Ornamental species and salmonids are susceptible and it has also been reported in sea bass *Dicentrarchus labrax* (Hawke *et al.*, 1981). Disease resistance is variable among channel catfish species, with blue catfish (*Ictalurus furcatus*) and Red River strain showing the most disease resistance (Wolters and Johnson, 1994). Outbreaks of the disease occur in the temperature range of 18–28°C, although low-level mortality and carrier status may be seen at temperatures outside this range.

In the acute form, the disease expresses as acute septicaemia. Petechial haemorrhages occur around the mouth, the throat and the fins, and internally in the liver and other organs. The organism crosses the intestinal mucosa into the internal organs. A chronic form of the disease occurs as a meningoencephalitis with behavioural changes, and ulceration or 'hole-in-the-head' (Hawke *et al.*, 1981; OIE, 2000b).

Furunculosis

The causative organism is *Aeromonas salmonicida* spp. *salmonicida*, which causes high rates of mortality in salmonid fish. The disease is characterized by boil-like inflammatory lesions, known as furuncles, which can penetrate deep into the musculature. These clinical signs are not always present (OIE, 2000a).

Pasteurellosis

The disease known as fish pasteurellosis is caused by *Photobacterium damselae* spp. *piscicida* (previously *Pasteurella piscicida*). It can affect many cultured fish species throughout Japan, USA and Europe. It usually causes high mortality, with very few external or clinical signs of disease, although a dark body colour may be seen. The spleen usually shows white nodules or tubercles from which the organism can be cultured (Kusuda and Yamaoka, 1972; Hawke *et al.*, 1987; Toranzo *et al.*, 1991; Baptista *et al.*, 1996; Candan *et al.*, 1996; Fukuda *et al.*, 1996).

Petechiae

Petechiae (pinpoint-sized areas of haemorrhage) on the underside and on the abdomen of fish may indicate septicaemia and generalized bacteraemia.

Rainbow trout fry syndrome

This condition has been reported from the UK, where diseased fish have anaemia, exophthalmia, pale gills and increased pigmentation of the skin. The abdomen is swollen with ascites fluid in the peritoneal cavity and the kidney is swollen. As yet no definitive organism has been deemed to be the causative organism. A number of bacteria have been suggested to be associated with the condition. These include *Flavobacterium columnare* (previously *Cytophaga columnaris*), *Janthinobacterium* spp., *Micrococcus luteus* and *Planococcus* spp. (Austin and Stobie, 1992b).

Skin and tail rot

Skin and tail rot may be associated with bacteria such as *Pseudomonas* spp., *Aeromonas* spp., *Flavobacterium* or *Flexibacter* spp. However, the condition is usually due to poor water quality and is therefore regarded as primarily a management problem.

Skin ulcers

Skin ulcers are seen as boil-like or pimple-like and convex. They may be caused by a variety of bacteria including typical *Aeromonas salmonicida* and the many species of atypical *A. salmonicida*. When the disease cause is *Listonella anguillarum*, ulcers may be seen as boils with red putrefying and liquefying flesh.

Streptococcosis

This disease is caused by the Gram-positive cocci, *Streptococcus agalactiae*, a group B, β-haemolytic streptococcus. Clinical signs exhibited with this infection include abnormal behaviour such as erratic swimming, whirling on the surface, and C-shaped curvature of the body while swimming at the surface. The eye may be opaque, exophthalmia may be seen and haemorrhages may be present. Haemorrhagic areas are also seen on the head and body, in particular around the mouth, snout, operculum and fins. There may be a haemorrhagic enteritis.

Streptococcus iniae infection

Clinical signs of the disease may vary according to the species of fish infected. Chronic infection seems to occur at a temperature of 25°C, whereas a more acute form of the disease is seen when the temperature ranges from 28 to 32°C (Yuasa *et al.*, 1999).

1.3 Bacteria and Relationship to Host

Table 1.2 lists the bacteria that may be pathogens or saprophytes of fish and other aquatic animals.

Table 1.2. Bacterial pathogens and saprophytes of fish and other aquatic animals.

Bacteria	Disease	Disease signs	Host/Isolation site	Distribution	Ref
Abiotrophia balaenopterae spp. nov.	See *Granulicatella balaenopterae* com. nov.				179
Abiotrophia elegans (nutritionally variant *Streptococci*)	See *Granulicatella elegans*				
Achromobacter xylosoxidans ssp. *denitrificans* (previously *Alcaligenes denitrificans*)					659
Acinetobacter baumannii Genospecies 2	Human infections		Isolated from human patients and environment		110
Acinetobacter calcoaceticus Genospecies 1	1. Environmental organism 2. Oral cavity flora		1. Isolated from soil 2. Turtles	2. Part of normal flora	110 300 301
Acinetobacter haemolyticus Genospecies 4	Pathogenicity not known for fish	Lesions in organs	Isolated from Atlantic salmon, channel catfish, environment and human clinical samples	Norway, USA	110
Actinobacillus delphinicola	Pathogenicity not determined	Isolated from various tissues	Sea mammals – harbour porpoise, striped dolphin, Sowerby's beaked whale	Scotland	263
Actinobacillus scotiae	Pathogenicity not determined	Organisms in liver, lung, brain, spleen	Isolated from stranded porpoise	Scotland	265
Actinomyces marimammalium	Pathogenicity not determined	Isolated from various tissues, (lung, liver, spleen, intestine) with other organisms	Dead hooded seal, dead grey seal, dead harbour porpoise	UK	370
Aequorivita antarctica, A. lipolytica, A. crocea, A. sublithincola	Environmental isolates (member of *Flavobacteriaceae* family)		Isolated from marine environment – seawater, sea ice	Antarctica	113
Aerococcus viridans	Associated with infection in oesophageal diverticulum	Organism in oesophageal lesions.	Loggerhead sea turtle	Spain	755
Aerococcus viridans var. *homari* (heavy growth)	High mortality, gaffkemia disease, fatal septicaemia	May see pink or red coloration in the haemolymph. Organisms multiply in the hepatopancreas, haemolymph and to a lesser extent in the heart and skeletal muscle	Lobster, seawater and sediment in lobster ponds. Exist free-living in marine benthos. Crabs and other crustaceans may act as reservoirs or carriers of the organism	Norway, Canada	299 719 827

Aeromonas allosaccharophila (HG15) (previously HG 14 in 1995)	1. Diseased elvers 2. Diarrhoeic stools	1. Diseased elvers 2. Faeces	1. Spain 2. South Carolina, USA	527
Aeromonas bestiarum (HG2) (formerly genomospecies DNA group 2, *A. hydrophila*)	Pathogenicity in doubt		USA	13
Aeromonas caviae (HG4)	1. Septicaemia, mortality when organism present in high numbers 2. Infection, gastroenteritis	1. Dermal ulceration, present in hepatopancreas 2. Infection	1. Freshwater ornamental fish, Atlantic salmon, octopus, giant freshwater prawn, turbot larvae 2. Human	Ubiquitous in the environment Turkey, Kenya, Taiwan	21 723
Aeromonas culicicola	Presence in aquatic species not known		Isolated from mosquito	India	624
Aeromonas encheleia (HG16) (previously HG11)	Non-pathogenic		Healthy eels, freshwater	Spain	427 379 241
Aeromonas eucrenophila (previously *A. punctata* ssp *punctata*) (HG6)	Non-pathogenic		Ascites of carp, drinking water, well water	Europe, Germany	379
Aeromonas group 501 (formerly enteric group 501) (HG12)	See *A. schubertii*				
Aeromonas hydrophila ssp. *hydrophila* (HG1) (usually isolated in heavy reasonably pure growth for it to be considered the primary pathogen)	1. Haemorrhagic septicaemia, peritonitis, redsore disease, fin rot, red-fin disease, mortality. Associated with the fungal disease epizootic ulcerative syndrome, caused by *Aphanomyces invadans*, in catfish and snakehead fish (Thailand, Philippines). Opportunistic infection in seal morbillivirus infection 2. Opportunistic and primary pathogen 3. Black disease 4. Pathogenicity unclear 5. Gastroenteritis	1. Erosive or ulcerative dermal lesions, haemorrhage on fins and trunk, swelling of anus, erythema 2. Red-leg frog disease 3. Black nodules on thoracic appendages 4. Isolated in cases of abortion	1. Freshwater and ornamental fish, ayu, channel catfish, walking catfish, tilapia, trout, turtles, eels, reptiles, grey seal, occasionally marine fish – cultured Atlantic salmon, sea bass, scallop larvae. Increase in organic matter and temperatures above 18°C aid proliferation of *A. hydrophila*. Found in fresh, brackish and coastal waters 2. Frogs, farm-raised bullfrogs 3. Fairy shrimps 4. Cattle, horses, pigs 5. Humans	1, 2. Ubiquitous in the environment worldwide 3. Algeria, Germany, Spain	227 220 21 454 29 300 301 456 497 530 614 650 783
Aeromonas hydrophila ssp. *dhakensis* group BD-2 (HG1)	Diarrhoea	Cytotoxic and haemolytic properties	Isolated from diarrhoeal children	Bangladesh	383

continued

Table 1.2. *Continued.*

Bacteria	Disease	Disease signs	Host/Isolation site	Distribution	Ref
Aeromonas jandaei (HG9) (previously HG9 *A. sobria*)	1. Pathogen 2. Clinical significance	1. Tissues 2. Isolated from blood, wound, diarrhoeal stools	1. Eel 2. Humans	1. Spain 2. USA	143
Aeromonas media (HG5)	1. Environmental organism 2. Clinical, gastroenteritis		1. River water. Probiotic properties against *V. tubiashii* 2. Humans	1. UK	15 294
Aeromonas popoffii	Environmental organism		Drinking water reservoirs	Finland, Scotland	380
Aeromonas salmonicida ssp. *salmonicida* (HG3) 'typical' *A. salmonicida* (produces brown pigment)	Goldfish ulcer disease (GUD), furunculosis in trout and salmon	Dermal ulceration showing typical umbonate furuncles. Organisms penetrate to underlying tissues, kidney, coelomic fluid, spleen, intestine	Many fish; goldfish, carp, silver perch, trout, Atlantic salmon, marine flounder, greenback flounder, eel, catfish, carp, cleaner fish Frogs, after feeding from contaminated trout	Highly virulent strain from North America, Europe, UK, USA. Not present in Australia	468 531 584
Aeromonas salmonicida ssp. *salmonicida* (non-pigmented strains)	Diseased salmon	Head kidney	Farmed Atlantic salmon Injection experiments reveal the non-pigmented strain produced a higher mortality than pigmented strains	Norway	450
Aeromonas salmonicida ssp. *achromogenes*	'Atypical' *A. salmonicida* Variety of pathologies, carp erythrodermatitis, goldfish ulcer disease, ulcer disease of flounder	Large open skin lesions surrounded by areas of descalation with softened and haemorrhagic dermis	Salmonids and non-salmonids, Atlantic cod, silver bream, perch, roach Found in fresh water, brackish water and marine environments	Worldwide: Australia, England, Central and Northern Europe, Iceland, Japan, North America, South Africa	186 534 830
Aeromonas salmonicida ssp. *masoucida*	1. 'Atypical' *A. salmonicida*	Superficial skin lesions	1. Salmonids – (sakuramasu – *Oncorhynchus masou* and pink salmon – *O. gorbuscha*)	Japan	438 830
Aeromonas salmonicida ssp. *nova*	GUD. Cutaneous ulcerative disease in goldfish	Cutaneous ulcers	Goldfish, salmonids/non-salmonids, eel, carp, marine fish	England, Japan, USA. Worldwide. Australian goldfish strains are thought to belong to this subspecies. Australian salmonids are susceptible	144 535 695 761 824 825
Aeromonas salmonicida ssp. *pectinolytica*	Environmental organism		Polluted river	Argentina	615
Aeromonas salmonicida ssp. *smithia*	'Atypical' *A. salmonicida*	Superficial skin lesions	Non-salmonids	England	47 830

Organism	Disease	Signs	Host	Location	Ref.
Aeromonas salmonicida (atypical strains)	Mortality; Variety of pathologies, carp erythrodermatitis, GUD, ulcer disease of flounder, ulcer head disease of eels, ulcerative disease	Skin lesions, necrosis, haemorrhagic ulcers, umbonate furuncles and swelling on head of eels. Sometimes underlying tissues affected, particularly in farmed fish	Blenny, carp, chub, cleaner fish, dab, eels, sand eels, flounder, goldfish, minnow, rainbow trout (salt water), roach, rockfish	Baltic Sea, Denmark, England, Finland, Japan, Norway, South Africa, USA	17 352 331 107 403 468 584 832
Aeromonas salmonicida atypical strains Oxidase-negative	Deaths, ulcerative disease	Lethargy, skin ulcers	Turbot, and flounder from a salt-water fish farm, coho salmon	Denmark, Baltic Sea, USA	153 617 832
Aeromonas salmonicida atypical strains Growth at 37°C	Death and morbidity	Skin ulcers	Carp, goldfish, roach Garden ponds, aquaria, rivers	England	40
Aeromonas schubertii (HG12) (previously called enteric group 501)	1. Environmental organism 2. Septicaemia, gastroenteritis, diarrhoea	2. Abscesses, wounds, pleural fluid, blood	2. Humans, often immunocompromised	2. USA, Puerto Rico and southern coastal states of USA	348 2
Aeromonas sobria (HG7) (now also called *A. veronii* ssp. *sobria*)	1. Peritonitis, epizootic ulcerative syndrome. Pathogenicity in doubt 2. Diarrhoea, renal failure, cellulitis, necrotizing gastroenteritis in adult	1. Peritonitis 2. Faeces. Production of cytotoxin	1. Freshwater ornamental fish, gizzard shad. May be found in the intestines of healthy fish 2. Infant, adult with alcoholic liver cirrhosis	Ubiquitous in the environment USA	21 259 393 452 750
Aeromonas trota (HG14) (previously called *A. enteropelogenes*)	1. Non-pathogenic for fish 2. Diarrhoea	2. Stool samples	1. Found in marine, estuarine and freshwater environments 2. Humans	South-East Asia (Bangladesh, India, Indonesia, Thailand), worldwide	142 178 382
Aeromonas veronii ssp. *sobria* (HG8)	1. Environmental organism 2. Humans		1. Ubiquitous in the environment 2. Humans – most pathogenic of *Aeromonas* taxa	Worldwide	259 393
Aeromonas veronii ssp. *veronii* formerly known as enteric group 7 (HG10) (previously called *A. ichthiosmia*)	1. Mortality when organism present in high numbers 2. Diarrhoea, wounds, cholecystitis	1. Hepatopancreas 2. Faeces, wound site	1. Giant freshwater prawns 2. Humans Found in freshwater	1. Taiwan	347 2 178 381 723
Alcaligenes faecalis homari	See *Halomonas aquamarina*				

continued

Table 1.2. *Continued.*

Bacteria	Disease	Disease signs	Host/isolation site	Distribution	Ref
Allomonas enterica	Environmental organism		Isolated from contaminated river water and human faeces	Russia	418
Alteromonas species	Bacterial necrosis and septicaemia	Necrosis, septicaemia			
Alteromonas citrea	See *Pseudoalteromonas citrea*				
Alteromonas colwelliana	See *Shewanella colwelliana*				
Alteromonas communis	See *Marinomonas communis*				
Aquaspirillum spp.	Report of an association in epizootic ulcerative syndrome – caused by the fungus. *Aphanomyces invadans*	*Aquaspirillum* induced slight dermomuscular necrotic lesions	Catfish – mild virulence only. Snakehead fish are not infected when challenged with *Aquaspirillum* species	Thailand	497
Arcanobacterium bernardiae	Isolated from clinical sources especially blood and abscesses		Human-derived strains		274 636
Arcanobacterium (*Corynebacterium*) *phocae*	Pathogenicity not determined	Tissues and fluids	Seals. Recovered in mixed growth from seals with septicaemia and pneumonia	Scotland	613 636
Arcanobacterium pluranimalium	Pathogenicity not determined	Isolation site not stated	Dead harbour porpoise, dead sallow deer	UK	480
Arcanobacterium pyogenes	Associated with a variety of pyogenic conditions	Mucus membranes, tissues	Occurs in humans and animals	Worldwide	636 641
Arthrobacter agilis	Environmental organism		Water, soil, human skin		
Arthrobacter nasiphocae	Possibly normal flora	Nasal cavity	Common seal (*Phoca vitulina*)		182
Arthrobacter rhombi	Pathogenicity not determined	Organism isolated from internal organs	Greenland halibut (healthy)	Greenland	600
Atopobacter phocae	Pathogenicity not determined	Intestine, lymph nodes, lung	Isolated from dead seal	Scotland	479
Bacillus cereus	Branchio-necrosis	Found on necrotic gills of carp	Carp, striped bass	Russia, USA	74 634
Bacillus mycoides	Mortality	Ulcers on dorsum, focal necrosis of epaxial muscle	Channel catfish Organism is ubiquitous in soil and has been implicated in disease in humans and parrots	Poland, USA	307
Bacillus subtilis	Part of bacterial flora in branchio-necrosis		Carp	Poland	634

Benechea chitinovora (not on the list of recognized bacterial names) previously called *Bacillus chitinovorus*	Ulcerative shell disease, shell rot, spot disease, rust disease. Mildly contagious, chronic self-limiting disease	Affects the chitinous plates of the carapace (dorsal shell) and plastron (ventral shell). Shell becomes pitted and early lesions have blotchy dark coloration	Free-ranging and captive turtles (spiny soft-shelled turtles, red-eared sliders, striped-necked musk, South American side-neck turtle, Eastern painted turtle)	USA	806
Bordetella bronchiseptica	1. Bronchopneumonia, secondary pathogen in phocine morbillivirus (distemper) infection	Lung, trachea	1. Seals 2. Bronchopneumonia in dogs, laboratory animals, cats, rabbits, horses, turkeys, monkeys, humans, associated with atrophic rhinitis in pigs	Europe, Scotland, Denmark, UK	642
Brevundimonas (Pseudomonas) diminuta	Environmental organism				685
Brevundimonas (Pseudomonas) vesicularis	Environmental organism		Found in streams		685
Brucella abortus	Brucellosis	Positive serology	Weddell seals (*Leptonychotes weddellii*)	Antarctica	592
Brucella abortus, B. melitensis, B. suis and rarely *B. canis*	Brucellosis		Generally host specific – *B. abortus* (cattle), *B. canis* (dogs), *B. melitensis* (goats), *B. neotomae* (desert wood rats), *B. ovis* (sheep), *B. suis* (pigs, reindeer, hares) **Zoonotic.** Use biological safety cabinet	Worldwide	185
Brucella cetaceae (previously part of *Brucella maris* sp. nov. biovar I & II)	1. Brucellosis, abortion, infection 2. Brucellosis	1. Aborted fetus, subcutaneous lesions, spleen, lung 2. Illness and positive blood culture	1. Dolphins (Atlantic white-sided dolphin, common dolphin, striped dolphin, bottlenose dolphin), harbour porpoise, whale 2. Human infection **Zoonotic.** Work with all suspect tissues and *Brucella* cultures in a biological safety cabinet	Canada, Europe, Scotland, USA	261 404 125 267 172 658
Brucella pinnipediae (previously part of *Brucella maris* spp. nov. biovar I & II)	1. Brucellosis, abortion, infection	1. Aborted fetus, subcutaneous lesions, spleen, lung	1. Seals (common seal, grey seal, hooded seal, harp seal, ringed seal), European otter **Possibly zoonotic.** Work with all suspect tissues and *Brucella* cultures in a biological safety cabinet	Canada, Europe, Scotland, USA	261 404 125 267 172 658

continued

Table 1.2. Continued.

Bacteria	Disease	Disease signs	Host/Isolation site	Distribution	Ref
Brucella species	Brucellosis	Organism in liver and spleen of whale, and in lymph nodes of seals	Harp seal, ringed seal, minke whale	Canada, Norway	261 171
Budvicia aquatica	Environmental organism		Isolated from river and drinking water	Czechoslovakia, Sweden	591
Burkholderia (Pseudomonas) cepacia	Environmental organism		Freshwater, soil	Ubiquitous	298
Burkholderia (Pseudomonas) pseudomallei	Melioidosis	Septicaemia, abscesses in lung, spinal column, liver, kidney	Cetaceans in oceanarium – (false killer whale, bottlenose dolphins, white-sided dolphins, sea lion, grey seal), sheep, penguin, goats, dog, galah, cockatoo, humans Found in soil and water **Zoonotic**. Use biological safety cabinet	Hong Kong. Disease of tropical and subtropical regions (Australia, South-East Asia)	349 516
Candida spp. (yeast)	Opportunist infection		Cetaceans – particularly dolphins		726
Carnobacterium alterfunditum	Environmental organism		Lakewater	Antarctica	412
Carnobacterium divergens	Normal intestinal microflora of healthy fish	Intestine and digestive tract	Atlantic salmon juveniles, Atlantic cod, Arctic charr, saithe	France, Norway	176
Carnobacterium divergens Strain 6251	Potential probiotic against *A. salmonicida* ssp. *salmonicida*, *L. anguillarum*, *M. viscosus*	Present in intestine	Arctic charr (*Salvelinus alpinus* L.)		649
Carnobacterium funditum	Environmental organism		Lakewater	Antarctica	412
Carnobacterium gallinarum	Environmental organism		Isolated from ice slush around chicken carcasses		176
Carnobacterium inhibens Strain K1	Normal intestinal microflora of healthy fish	Found in digestive tract	Inhibits growth of *L. anguillarum* and *A. salmonicida* in Atlantic salmon		412 411
Carnobacterium mobile	Environmental organism		Processed chicken meat		176
Carnobacterium (Lactobacillus) piscicola	Pseudokidney disease, Lactobacillosis. Post-stripping peritonitis. Seen in fish, 1 year or older, following stress such as handling and spawning. Most strains are opportunistic and possess low virulence; however, other strains have higher virulence and cause high mortality	Epicarditis, peritonitis, visceral granulomas, distension of abdomen, ascites fluid, blood or blisters under the skin. Collect samples from kidney, spleen, swimbladder. Virulent strains cause bilateral exophthalmia, periocular and liver haemorrhages, and ascites fluid	Salmonid fish, cutthroat trout, rainbow trout, chinook salmon, cultured striped bass, channel catfish, brown bullhead catfish Less virulent in striped bass and channel catfish	Australia, Belgium, Canada, France, UK, USA	353 73 176 752

Organism	Disease/status	Clinical signs/isolation	Host/source	Geographic location	Reference
Catenibacterium spp.	See *Eubacterium* spp.				
Cellulophaga (*Cytophaga*) *lytica*	Environmental isolate		Marine, beach mud	Costa Rica	163
Chromobacterium violaceum Pigmented and non-pigmented strains	1. Environmental isolate 2. Wound infection, septicaemia, abscesses		1. Found in soil and water 2. Humans	1. Tropical and subtropical regions 2. Australia, Malaysia, Senegal, Taiwan, USA, Vietnam	482 599
Chryseobacterium (*Flavobacterium*) *balustinum*	Flavobacteriosis		Marine fish	USA	802
Chryseobacterium (*Flavobacterium*) *gleum*	Non-pathogenic to fish		Found in human clinical specimens		366
Chryseobacterium indologenes (previously *Sphingobacterium* and *Flavobacterium indologenes*)	1. Systemic infection 2. Clinical sources	1. Torticollis, gross lesions, granulomas, enlarged organs	1. Farmed bullfrogs (*Rana castesbeiana*) 2. Human specimens and hospital environment	USA	530 844
Chryseobacterium (*Flavobacterium*) *indoltheticum*					557
Chryseobacterium (*Flavobacterium*) *meningosepticum*	1. Non-pathogen for fish 2. Pericarditis, septicaemia 3. Systemic infection 4. Meningitis in infants	2. Isolated from pericardium, liver, eye lesion 3. Torticollis, gross lesions, granulomas, enlarged organs	1. Reported from fish blood and marine mud 2. Birds (chickens, pigeon, finch) 3. Farmed bullfrogs (*Rana castesbeiana*) 4. Human pathogen	USA, worldwide	530 773
Chryseobacterium (*Flavobacterium*) *scophthalmum*	Gill disease, haemorrhagic septicaemia. 100% mortality in juveniles	Swollen gill lamellae (hyperplasia), haemorrhagic septicaemia, swollen intestines containing yellow fluid, haemorrhages in eye, skin, jaw	Healthy and diseased turbot Coastal waters	Scotland	556 557
Citrobacter freundii	1. Opportunistic infection, low virulence for trout 2. Systemic infection	1. Haemorrhagic spots on skin, eye and fins. Organism isolated from kidney, liver, spleen 2. Torticollis, gross lesions, granulomas, enlarged organs	1. Angel fish in aquaria, rainbow trout, sunfish, carp Commonly isolated from soil, water, sewage, food, and from organs of diseased and healthy animals including mammals, birds, reptiles and amphibians 2. Farmed bullfrogs (*Rana castesbeiana*)	India, Japan, UK, USA, worldwide	425 530 675 753

continued

Table 1.2. *Continued.*

Bacteria	Disease	Disease signs	Host/isolation site	Distribution	Ref
Clostridium botulinum Type E	Morbidity and mortality		Farmed trout, salmon, coho salmon *C. botulinum* can be a commensal in trout intestine and on gills. Toxin usually not produced in live fish, but is produced in dead fish as part of the decomposition process. Found in lake sediment. Bacterium produces toxin in an anaerobic environment	Britain, Canada, Denmark, USA	141
Clostridium perfringens Type A	1. Enterotoxaemia 2. Myositis at injection site	1. Gut content 2. Abscess in muscle	Captive whales, dolphins, seals	UK	312
Colwellia maris previously *Vibrio* strain ABE-1	Environmental organism		Psychrophilic, found in seawater	Japan	849
Corynebacterium aquaticum	1. Exophthalmia 2. Clinical infection	1. Organism seen in brain, haemorrhages in eyes	1. Striped bass, rainbow trout. Organism found in natural fresh and distilled water 2. Reported to cause infection in immunocompromised patients 3. Found in freshwater sources	1. USA 2. Worldwide	73 75
Corynebacterium phocae	See *Arcquobacterium phocae*				
Corynebacterium testudinoris	Associated with necrotic mouth lesions	Mouth lesions	Tortoise	Scotland	180
Cryptococcus lupi (yeast)	Environmental organism		Soil	Antarctica	55
Cryptococcus neoformans var. *gatii* (yeast)	Death, septicaemia	Organism isolated from lung, brain, lymph nodes	Dolphin **Zoonotic** organism Isolated from bat guano and associated with eucalyptus trees	Australia Tropics and southern hemisphere	278 135
Cytophaga aquatilis	See *Flavobacterium hydatis*				
Cytophaga arvensicola	Environmental isolate		Soil	Japan	89
Cytophaga aurantiaca	Environmental isolate		Swampy soil	Germany	92
Cytophaga columnaris	See *Flavobacterium columnare*				
Cytophaga fermentans	Environmental isolate		Marine mud	California	89 162
Cytophaga hutchinsonii	Environmental isolate		Soil		89
Cytophaga johnsonae	See *Flavobacterium johnsoniae*				
Cytophaga latercula	Environmental isolate		Marine	USA	163

Organism					
Cytophaga marinoflava	Environmental isolate		Seawater	Scotland	92
Cytophaga psychrophila	See *Flavobacterium psychrophilum*				
Dermatophilus chelonae	Dermatophilosis	Skin lesions, skin abscess, skin scabs	Testudines (Chelonians) – turtle and tortoise	Australia	529
Dermatophilus congolensis	Dermatophilosis (referred to as 'lumpy wool' and mycotic dermatitis in infected sheep)	Subcutaneous nodules and lesions containing caseous material	1. Aquatic species – crocodiles, bearded dragon, polar bears, seals 2. Humans, horses, sheep, blue-tongue lizard, cat, cattle, goats, deer, monkeys, pigs, rabbits, owls, foxes, giraffe, gazelle	Australia	419 308 699
Dietzia maris (previously *Rhodococcus maris*)	Microflora	Skin and intestinal flora	Carp, soil	USSR	573
Edwardsiella hoshinae	Part of normal flora	Faeces	Birds (puffin, flamingo) and reptiles (monitor, lizard), water	Worldwide	317
Edwardsiella ictaluri Strains with a limited tolerance for oxygen have been isolated (Mitchell and Goodwin, 2000)	1. Enteric septicaemia of catfish 2. Environment	1. Initial infection in brain. Petechial haemorrhage under jaw and belly, head lesion, gills, kidney infected	1. Freshwater ornamental fish, cultured channel catfish (*Ictalurus punctatus*), *Danio*, white catfish, green knifefish, bass, cyprinids and *Anguillidae*. Catfish are the most susceptible species 2. Isolated from organically polluted waters, urine and faeces of man, and intestinal microflora of snakes	1. Thailand, USA, Vietnam	194 334 374 426 500 547 627
Edwardsiella tarda (*E. anguillimortifera* is the senior synonym, however, *E. tarda* is conserved for use. Originally named *Paracolobactrum anguillimortiferum*)	1. Edwardsiellosis, redpest, emphysematous putrefactive disease of catfish, fish gangrene 2. Systemic infection 3. Human clinical samples, diarrhoea	1. Septicaemia, and ulcerative dermatitis, intestinal infection. Lesions and abscesses in muscle. Trout may have mucopurulent discharge from cloaca, congested spleen, enlarged liver, petechial haemorrhages on serosal fat and gills. Eels show abscessed or ulcerative lesions in kidney or liver 2. Torticollis, gross lesions, granulomas, enlarged organs	1. Alligators, angelfish, black mullet, bass, catfish, eels, freshwater and ornamental fish, flounder, goldfish, marine mammals, ostrich, rainbow trout, brook trout, sea lion, snakes, tilapia, turtles. Also part of normal flora in many aquatic animals (fish, frogs, amphibians, reptiles, mammals, captive little penguins, toads, turtles), and inhabitants of aquatic environment and surface water. Found in bile of healthy pigs 3. Has been isolated from abscesses, meningitis, wounds, urine, blood, faeces, spinal fluid	Ubiquitous in environment, Asia, Australia, Canada, Japan, USA	374 530 606 640 711 800 804 823

continued

Table 1.2. *Continued.*

Bacteria	Disease	Disease signs	Host/isolation site	Distribution	Ref
Empedobacter brevis (previously *Flavobacterium breve*)	1. Environmental 2. Clinical	2. Eye, urine, blood culture, bronchial secretion	1. Fish, canal water. Maybe pathogenic for laboratory animals 2. Humans	England, Ireland, Switzerland, Czechoslovakia	363 775
Enterobacter agglomerans	See *Pantoea agglomerans*				
Enterococcus faecalis (reported as *E. faecalis* ssp. *liquefaciens*)	1. Possible pathogen. Identity in doubt 2. Systemic infection	1. Bacteria in liver and kidney, ulcers on fins	1. Rainbow trout, catfish, brown bullhead 2. Crabs Part of normal intestinal flora of man and animals. May cause nosocomial infections	1. Italy, Croatia 2. French Mediterranean coast. Worldwide	
Enterococcus faecium	Normal flora. May cause nosocomial infections		Part of normal intestinal flora of man and animals	Worldwide	
Enterococcus seriolicida	See *Lactococcus garvieae*				731
Enterovibrio norvegicus	Normal flora		Present in gut of turbot larvae	Norway	741
Erysipelothrix rhusiopathiae	1. Non-pathogenic for fish 2. Erysipelas 3. Erysipeloid, skin disease, polyarthralgia, septic arthritis, renal failure, peritonitis	2. Systemic disease and skin disease 3. Skin disease, septicaemia	1. Parasitic on fish, lobster 2. Dolphins, pigs, kangaroos, emus, cattle, sheep, dogs, horses, avian species, crocodile 3. Human pathogen, occupational exposure	Worldwide	292 408 229
Escherichia coli	Endocarditis	Lesions in heart valves	Sea lion	Korea	435
Escherichia vulneris	1. Septicaemia, mortality 2. Wounds	1. Abnormalities and haemorrhages in gills, liver, kidney, spleen	1. Freshwater fish – rainbow trout, balloon molly, silver molly, caucasian carp 2. Humans, also isolated from faeces. Found in water of fish ponds and fish faeces	1. Turkey	51
Eubacterium spp. (identification is tentative (Udey *et al.*, 1977). Initially identified as *Catenabacterium* (Henley and Lewis, 1976)	Mortality	Organism recovered from brain, liver, kidney and blood	Grey mullet, redfish	USA	343
Eubacterium tarantellae (original spelling *E. tarantellus*)	Meningitis. Possible secondary pathogen	Organism isolated from brain tissue	Striped mullet	Florida	764
Facklamia miroungae	Normal flora		Nasal cavity	Southern elephant seal	UK	368
Flavobacterium aquatile	Environmental organism		Deep well water	Kent, UK	92

Organism	Disease	Clinical signs/lesions	Host	Location	Ref.
Flavobacterium branchiophilum (*Flavobacterium branchiophila*)	Bacterial gill disease (BGD). High mortality	Anorexia, suffocation. Lesions and white-grey spots on gills. Organisms seen on gill surface	Freshwater fish (goldfish), salmonids	Canada, Europe, Hungary, Japan, Korea, USA	604 802
Flavobacterium cauliformans	Environmental organism		Lakewater		533
Flavobacterium columnare (previously called *Cytophaga columnaris*, and *Flexibacter columnaris*)	Columnaris disease, saddleback disease, systemic disease in severe cases	Yellow/brown lesions on gills. Ulcers and necrosis on body surface, 40% of infections eventually penetrate to internal tissues and organs	Freshwater fish – (channel catfish, blue catfish, carp, white bass, large-mouth bass, barramundi, fathead minnow), black bullhead, salmonids (brown trout), black comets, mollies, eels, neon tetra, platies. Occurs where water temperature exceeds 14°C, particularly 25–32°C	Australia, France, Hungary, Japan, USA, worldwide	89 90 88 135 211 214 543
Flavobacterium flevense (previously *Cytophaga flevensis*)	Environmental organism		Lakewater	The Netherlands	89 533
Flavobacterium frigidarium	Environmental organism		Marine sediment	Antarctica	376
Flavobacterium gillisiae	Environmental organism		Environment	Antarctica	533
Flavobacterium hibernum	Environmental organism		Freshwater lake	Antarctica	532 533
Flavobacterium hydatis (*Cytophaga aquatilis*)	Gill disease. Pathogenicity not determined	Isolated from gills of diseased salmon	Cultured salmonid	Europe, USA	720
Flavobacterium johnsoniae previously *Cytophaga johnsonae* (includes previously named *Flexibacter aurantiacus*)	False columnaris disease, gill disease, skin disease	Dermal and gill lesions	Barramundi; salmonids, koi and other fish. Isolated from soil and mud	Australia, Europe, France, UK, USA	89 145
Flavobacterium meningosepticum	See *Chryseobacterium meningosepticum*				
Flavobacterium mizutaii, formerly (*Sphingobacterium mizutae*)	Meningitis. Pathogenicity not determined	Spinal fluid	Premature infant. Natural habitat not known	Japan	844
Flavobacterium (*Cytophaga*) *pectinovorum*	Environmental organism		Soil	England	92
Flavobacterium psychrophilum (previously *Flexibacter psychrophilus* and *Cytophaga psychrophila*)	Bacterial cold-water disease, peduncle disease, rainbow trout fry syndrome, fin rot	Erosion of the peduncle area, ulcers on scales, leads to penetration into tissues	Fish, especially fry and young fish, coho salmon, chinook salmon, rainbow trout, carp, eels, cyprinids, Japanese sweetfish. Occurs where water temperature is below 15°C	Australia, Canada, Chile, Denmark, England, France, Germany, Japan, Korea, Spain, northern USA	89 90 168

continued

Table 1.2. *Continued.*

Bacteria	Disease	Disease signs	Host/isolation site	Distribution	Ref
Flavobacterium (Cytophaga) saccharophilum	Environmental organism		River water	England	533
Flavobacterium scophthalmum	See *Chryseobacterium scophthalmum*				
Flavobacterium (Cytophaga) succinicans	Pathogenicity not confirmed	Isolated from superficial lesions on diseased fish	Salmon Found in freshwater	USA	92 162
Flavobacterium tegetincola	Environmental organism		Associated with cyanobacterial mats	Antarctica	533
Flavobacterium (Cytophaga) xanthum	Environmental organism		Mud pool	Antarctica	533
Flexibacter aggregans	Environmental organism		Marine environment, beach sand	Ghana	162
Flexibacter canadensis	Environmental organism		Soil	Canada	162
Flexibacter columnaris	See *Flavobacterium columnare*				
Flexibacter elegans	Environmental organism		Freshwater, hot spring		162
Flexibacter flexilis ssp. – *algavorum – iolanthe – pelliculosus*	Environmental organism		Found in freshwater, lily pond	Costa Rica	162
Flexibacter litoralis	Environmental organism		Marine and freshwater	California	162
Flexibacter maritimus	See *Tenacibaculum maritimum*				551
Flexibacter ovolyticus	See *Tenacibaculum ovolyticum*				551
Flexibacter polymorphus	Environmental isolate		Found in marine environment	Mexico, USA	494
Flexibacter psychrophilus	See *Flavobacterium psychrophilum*				
Flexibacter roseolus	Environmental isolate		Hot spring	Costa Rica	162
Flexibacter ruber	Environmental isolate		Hot spring	Iceland	89
Flexibacter sancti	Environmental isolate			Argentina	89
Flexibacter tractuosus	Environmental organism		Marine and freshwater	Vietnam	162
Granulicatella adiacens and *G. elegans* (previously *Abiotrophia adiacens* and *A. elegans* referred to as nutritionally variant *Streptococci* (NVS)	Clinical isolates. *Streptococci* that show satellite growth around other colonies	Normal flora of mouth, intestinal and urogenital tract. May cause endocarditis, conjunctivitis, otitis media	Humans. Require pyridoxal hydrochloride analog (Vitamin B_6) or L-cysteine HCl (*G. elegans*) for growth		421 179 653

Organism	Pathogenicity	Comments	Host	Location	
Granulicatella balaenopterae com. nov. (previously *Abiotrophia balaenopterae*)	Pathogenicity not determined	Isolated in pure growth from liver and kidney	Beached minke whale	Scotland	179 478
Haemophilus piscium	Re-classified as 'atypical' *A. salmonicida*. Most closely related to *A. salmonicida achromogenes*				50
Hafnia alvei	1. Haemorrhagic septicaemia. Mortalities 2. Intestinal disorders, pneumonia, meningitis, abscesses and septicaemia		1. Cherry salmon, rainbow trout, brown trout 2. Humans Ubiquitous in the environments of soil, sewage and water	1. Bulgaria, England, Japan	313 652
Halomonas aquamarina (synonymous with *Alcaligenes faecalis homari, Deleya aesta, D. aquamarina, A. aquamarinus*. Transferred to genus *Halomonas* as *H. aquamarina*)	Moribund	Softened shells, opaque areas on carapace. Organisms isolated from haemolymph	Lobsters	USA	45 719 8
Halomonas cupida (previously *Alcaligenes cupidus* and *Deleya cupida*)	Mortalities		Black sea bream fry	Japan	463
Halomonas elongata	Environmental organism		Hypersaline environments	The Netherlands	795
Halomonas halodurans	Environmental organism		Hypersaline environments	USA, The Netherlands, Pacific Ocean	336
Halomonas marina (previously *Pseudomonas marina* and *Deleya marina*)	Environmental organism		Marine environment		66
Halomonas venusta (previously *Alcaligenes venustus*)	Clinical infection		Human – caused by fish bite. Present in marine waters	Maldive Islands	66 310
Helicobacter cetorum	1. Organism found in dental plaque 2. Gastric ulceration	1. Potential reservoir for gastric infections 2. Organism in glandular mucosa and the main stomach	1. Captive dolphins (*Tursiops gephyreus*) 2. Dolphins, beluga whale	USA	303 327 329
Hydrogenophaga (*Pseudomonas*) *palleronii*	Environmental organism		Present in water	Germany, Russia	834

continued

Table 1.2. *Continued.*

Bacteria	Disease	Disease signs	Host/Isolation site	Distribution	Ref
Hydrogenophaga (Pseudomonas) pseudoflava	Environmental organism		Present in water, mud, soil	Germany	39 834
Iodobacter fluviatilis (previously *Chromobacterium fluviatile*)	Environmental organism		Found in freshwater	Antarctic lakes, England, Scotland, Ubiquitous	502
Janthinobacterium lividum	Anaemia	Exophthalmia, pale gills, internal symptoms	Rainbow trout Found in soil and spring water	Scotland	48
Klebsiella pneumoniae	1. Fin and tail disease 2. Microflora		1. Rainbow trout 2. Mammalian tissue	UK Worldwide	205
Klebsiella planticola *Klebsiella trevisanii*	See *Raoultella planticola*				256 228
Klebsiella ornithinolytica	See *Raoultella ornithinolytica*				228
Klebsiella terrigena	See *Raoultella terrigena*				228
Klebsiella oxytoca					228
Lactobacillus spp., especially a *Lactobacillus plantarum*-like isolate	Normal intestinal microflora of healthy fish	Intestine and digestive tract	Atlantic cod, Atlantic salmon, rainbow trout, wolf-fish, Arctic charr	France, Norway	
Lactobacillus piscicola	See *Carnobacterium piscicola*				
Lactococcus garvieae Biotypes 1–13 (previously *Enterococcus seriolicida*, *Streptococcus* type I and *Streptococcus garvieae*)	1. Lactococcosis, haemorrhagic septicaemia, haemorrhagic enteritis, meningoencephalitis 2. Subclinical mastitis 3. Infection, osteomyelitis	1. Bilateral exophthalmia, darkening of skin, congestion of intestine, liver, kidney, spleen, brain, distended abdomen, bloody ascites fluid in peritoneal cavity. Organism seen in heart, gills, skin, spleen, eyes, kidney 2. Milk 3. Blood, skin, urine, faeces	1. Farmed rainbow trout, eel, yellowtail, prawns, turbot, Adriatic sturgeon Found in seawater, mud, intestines of wild fish e.g. Spanish mackerel, black scraper (Biotypes 1, 2, 3, 4, 5, 6, 10) 2. Cows, buffalo (Biotypes 4, 7, 8, 9) 3. Humans (Biotypes 1, 2, 10, 11, 12, 13)	Australia, Europe, France, Italy, Israel, Japan, North America, Saudi Arabia, Spain, South Africa, Taiwan, UK, USA	236 238 237 156 174 157 464 669 731 780
Lactococcus piscium	Lactobacillosis, pseudokidney disease		Rainbow trout	North America	835

Organism	Disease	Clinical signs	Host	Location	Ref.
Listonella anguillarum (Serovars 01, 02, 08, 09) previously *Vibrio anguillarum* biotype I. Most of the outbreaks are caused by serotypes 01 and 02	Vibriosis, systemic disease, ulcerative disease, necrosis	Red spots on ventral and lateral areas of fish, ulcerative skin lesions. Organisms in blood and haemopoietic tissues	1. Fish, molluscs – (larval and juvenile), ayu, flatfish (turbot, sole, halibut), plaice fry, winter flounder, sole, halibut), lobster, eel, salmonids, (rainbow trout), sea bream, octopus; 2. Shrimps; 3. Crabs	1. Worldwide; 2. Indo-Pacific & East Asia; 3. UK	341 222 561 563 620
Listonella pelagia (previously *V. pelagia* I & II)	Mortalities	Erosion on fins and tail. Haemorrhages on fins and organs	Juvenile farmed turbot	Spain	
Listonella ordalii	See *Vibrio ordalii*				
Mannheimia haemolytica (previously *Pasteurella haemolytica*)	1. Ulcerative stomatitis; 2. Disease; 3. Haemorrhagic tracheitis		1. Reptiles; 2. Sheep, goats, cattle; 3. Dolphins	1. USA	709 726
Marinilabilia salmonicolor biovar *agarovorans* (previously *Cytophaga salmonicolor* and *C. agarovorans*)	Environmental organism		Marine mud	California	92 89 162
Marinobacter hydrocarbonoclasticus (*Pseudomonas nautica*)			Found in marine environments		69
Mesophilobacter marinus	Environmental organism		Found in seawater	Japan, Indian Ocean	583
Micrococcus luteus	Mortality	Pale gills, ascites fluid, gastroenteritis, internal haemorrhages	Rainbow trout fry	England	43
Moraxella spp.	Associated with mortality		Striped bass	USA	72
Moritella japonica	Environmental organism		Seabed sediment	Japan	585
Moritella marina (*Vibrio marinus*)	Skin lesions	Isolate from surface skin lesions	Atlantic salmon, seawater and sediment of north Pacific Ocean	Iceland, Norway, Pacific Ocean, Scotland	82 99 766
Moritella viscosa (previously *Vibrio viscosus*)	Winter ulcer disease	Skin lesions, haemorrhages on internal organs	Atlantic salmon, lumpsucker, rainbow trout. Found in cold water	Coldwater in Iceland, Norway, Scotland	81 82 132 506

continued

Table 1.2. *Continued.*

Bacteria	Disease	Disease signs	Host/isolation site	Distribution	Ref
Mycobacterium spp.	Mycobacteriosis, systemic disease	Lesions on skin and kidneys. Nodules in organs, softening of spleen, kidney, liver. Granulomas in tissues	Many species of freshwater, salt water and ornamental fish, freshwater snails, frogs, reptiles, turtles, Pacific green sea turtles, New Zealand fur seals, freshwater crocodiles in Australia **Zoonotic**	Worldwide	592 737
Mycobacterium abscessus	Granulomas, systemic disease	No obvious clinical signs of piscine mycobacteriosis. Occasional external granuloma around buccal cavity and vent, internal granulomas	Japanese medaka, freshwater tropical fish, black acaras, goldfish, firemouth cichlid, oscar **Zoonotic**	USA	474 736
Mycobacterium chelonae	Mortality, granulomas, emaciation, exophthalmos, keratitis, skin ulcers, abnormal swimming behaviour	Multiple greyish-white miliary granuloma-like nodules in tissues, kidney, liver, spleen	Atlantic salmon (*Salmo salar*), yellow perch, snake, turbot, turtle **Zoonotic**	Australia, Canada, Portugal, Shetland Islands, Scotland, worldwide	133 375 204 673 737
Mycobacterium fortuitum (previously *M. piscium* and *M. salmoniphilum*)	Septicaemia, fish may be emaciated, exophthalmia, inflammation of skin	Granulomas on skin and in tissues. Seen as whitish spots on liver, kidney, heart, spleen	Ornamental finfish – (black acara, comets, discus fish, gourami, guppy, neon tetra, oscar, Siamese fighting fish), Atlantic salmon **Zoonotic** – skin ulcers and diffuse pulmonary disease in humans	Australia, South Africa, Thailand, worldwide	116 375 474 633
Mycobacterium marinum	Mycobacteriosis. Dermatitis and panniculitis in captive white whale	Lesions in kidney and on skin. Nodular lesions may be systemic in all organs. Granulomas in organs	Freshwater trout, freshwater ornamental fish, marine fish, rabbitfish, sea bass, striped bass, turbot, captive white whale. Skin lesions in humans **Zoonotic**	Australia, Israel, Portugal, USA, worldwide, ubiquitous	35 111 135 218 339 474 673
Mycobacterium neoaurum	Panophthalmitis	Ocular lesions, nodules in muscle and organs. Organism isolated with a *Rhodococcus* species	Atlantic salmon, chinook salmon.	Canada	53
Mycobacterium peregrinum	Mycobacteriosis. Shrimp appeared healthy apart from black lesions on carapace	Multifocal, melanized nodular lesions in carapace	Pacific white shrimp (*Penaeus vannamei*) Causes skin infection in humans. Found in water and soil	USA	551

Organism	Condition	Lesions/signs	Host / source	Location	Page
Mycobacterium poriferae This isolate has since been identified by PCR as *M. fortuitum*	Mycobacteriosis	Internal nodular lesions	Freshwater snakehead fish (*Channa striatus*) Previously reported from a marine sponge	Thailand	608 633 756
Mycobacterium scrofulaceum	Mycobacteriosis	Lesions in kidney and liver. Liver white and friable	Pacific staghorn sculpin	USA	474
Mycobacterium simiae	Mycobacteriosis	Lesions in kidney and liver	Black acara Found in environmental water **Zoonotic**	USA	474
Mycobacterium spp. not identified to known species	Mycobacteriosis	External dermal ulcers and nodules in internal lesions	Wild striped bass	Chesapeake Bay (USA)	337
Mycobacterium species *Mycobacterium triplex*-like	Granulomatous dermatitis	Florid skin nodules – soft, gelatinous, grey- and tan-coloured around head and trunk	Green moray eels, spotted moray eels	USA	345
Mycoplasma alligatoris	Epizootic pneumonia, polyserositis and multifocal arthritis	Organism found in trachea, lung, joint fluid, cerebrospinal fluid (CSF)	American alligator	USA	128
Mycoplasma crocodyli	Exudative polyarthritis	Swollen joints. Also found in lungs	Crocodiles	Zimbabwe	441
Mycoplasma mobile	Red disease	Gills	Tench	USA	439 440
Mycoplasma phocicerebrale (previously *M. phocacerebrale*)	Associated with respiratory disease	Isolated from, brain, nose, throat, lungs, heart	Seals	North Sea	295 449
Mycoplasma phocidae (*Mycoplasma phocae* – name correction is not legitimate)	Avirulent	Respiratory tract	Harbour seals	USA	449 660
Mycoplasma phocirhinis (previously *M. phocarhinis*)	Associated with respiratory disease	Isolated from nose, throat, trachea, lung, heart	Seals	North Sea	295 449
Mycoplasma testudinis	Non-pathogenic		Cloaca of tortoise	UK	350
Myroides (Flavobacterium) odoratimimus			Clinical specimens, hospital environments		774
Myroides odoratus (previously *Flavobacterium odoratum*)			Clinical specimens (urine, wound swabs, leg ulcer), hospital environments	UK, Czechoslovakia	362 774

continued

Table 1.2. *Continued.*

Bacteria	Disease	Disease signs	Host/isolation site	Distribution	Ref
Nocardia asteroides	Nocardiosis		Neon tetra, rainbow trout, large mouth bass, Formosa snakehead. Also causes infections in cats, cattle, dogs, fish, goats, humans, marine mammals	Argentina, Taiwan	155
Nocardia brasiliensis and *N. transvalensis*	Actinomycete mycetoma				661
Nocardia crassostreae sp. nov.	Nocardiosis	Brown discoloration on mantle, green or yellow nodules on abductor muscle, gills, heart and mantle	Pacific oysters	Canada, USA	270
Nocardia flavorosea	Environmental organism		Soil isolate	China	165
Nocardia nova			Human pathogen		805
Nocardia salmonicida	Nocardiosis		Blueback salmon		391
Nocardia seriolae (previously *N. kampachi*)	Nocardiosis	Abscesses and light-yellow-coloured nodules in epidermis and tubercles and granulomas in gills, kidney, liver, heart and spleen	Cultured fish – rudderfishes, yellowtail, Japanese flounder, sea bass	Japan, Taiwan	155 424 455
Nocardia spp.	Septicaemia, mortality	Small white spot lesions on dermis, muscle, gills and organs. Also found in internal organs	Atlantic salmon, farmed chinook salmon, marine fish, freshwater ornamental fish. Isolated from soil and plants	Worldwide, Australia, Canada, Japan, India, Taiwan, USA	117
Oceanomonas baumannii	Environmental organism	Degrades phenol	Estuary of River Wear	UK	130
Oceanomonas (Pseudomonas) doudoroffii	Environmental organism		Marine environment		69 130
Pantoea (Enterobacter) agglomerans	1. Pathogenicity in doubt – possible opportunist 2. Humans	1. Haemorrhages in eyes, dorsal musculature 2. Wounds, blood, urine	1. Dolphin fish (mahi-mahi) 2. Humans. Also found in enteric tract 3. Found on plant surfaces, seeds, water. Also reported in enteric tract of deer without disease signs	USA, ubiquitous	325 291 249
Pantoea dispersa	Environmental organism		Plant surfaces, seeds, soil, environment	Ubiquitous	291
Pasteurella multocida	1. Pneumonia, death, pleurisy (fowl cholera) 2. Enteritis (contamination from nearby bird rookery)	1. Exudate in lungs, fluid in pleural cavity 2. Intestinal haemorrhage	1. Alligator, Californian sea lion, penguins 2. Dolphins Respiratory disease in sheep, goats, cattle, rabbits	USA Worldwide	430 520 709 726

Organism	Disease/condition	Signs	Host/source	Location	Ref.
Pasteurella piscicida	See *Photobacterium damselae* ssp. *piscicida*				745
Pasteurella skyensis	Mortality (low virulence)	Exhibit signs of loss of appetite, morbidity	Sea-farmed Atlantic salmon (*Salmo salar* L.)	Scotland	100 416
Pasteurella testudinis	Multifocal bronchopneumonia and commensal	Abscesses, lung lesions	Californian desert tortoise	USA	709
Pedobacter heparinus (previously *Cytophaga* and *Sphingobacterium heparinum*) (*Cytophaga heparina*)	Environmental isolate		Soil. Degrades heparin		89 163
Pedobacter (*Sphingobacterium*) *piscium*	Environmental organism		Associated with frozen fish	Japan	728
Phocoenobacter uteri gen nov. spp. nov.	Pathogenicity unknown	Uterus	Harbour porpoise	UK	266
Photobacterium angustum	Environmental organism		Marine environment		67
Photobacterium damselae ssp. *damselae* (previously *Vibrio damselae*, EF-5)	1. Vibriosis, systemic disease, granulomatous ulcerative dermatitis, deaths 2. Wound infections	1. Skin ulcers in region of pectoral fin and caudal peduncle 2. Soft tissue infection occurs due to production of cytolysin	1. Bream, barramundi, damselfish, dolphins, eel, octopus, oysters, penaeid prawns, sharks, shellfish, shrimps, stingray, trout, rainbow trout, turbot, turtles, seahorses, yellowtail. May be part of microflora in healthy carcharhinid sharks and marine algae. Australian native and introduced fish 2. Humans	Australia, Denmark, Europe, Japan, Spain, USA	268 429 504 506 555 590 618 705 745
Photobacterium damselae ssp. *piscicida* (previously *Pasteurella piscicida*, *Flavobacterium piscicida*, *Pseudomonas piscicida*)	Pasteurellosis, fish pseudotuberculosis	Bacterial colonies and white nodules in spleen, kidney	Bass, Japanese flounder, gilt-head sea bream, sea bass, striped bass, sole, white perch, yellowtail. Farmed and wild fish	Europe, France, Greece, Italy, Japan, Malta, Portugal, Scotland, Spain, Taiwan, Turkey, USA Not in Australia	289 57 60 140 273 333 518 745 855

continued

Table 1.2. *Continued.*

Bacteria	Disease	Disease signs	Host/isolation site	Distribution	Ref
Photobacterium fischeri	See *Vibrio fischeri* – homotypic synonym				
Photobacterium histaminum	Considered a later subjective synonym of *Photobacterium damselae* ssp. *damselae*				437 595
Photobacterium iliopiscarium (previously *Vibrio iliopiscarius*)	Non-pathogenic	Intestine	Herring, coal fish, salmon and cod living in cold waters		599 767
Photobacterium leiognathi	Non-pathogenic	Light organ	Microflora of the luminous organ of a sea fish, *Leiognathus*		643
Photobacterium logei	See *Vibrio logei*.				
Photobacterium phosphoreum	Environmental organism May cause spoilage of seafood		Marine environment. Symbiotic association with marine animals in light organs of teleost fishes		67
Photobacterium profundum	Environmental isolate		Isolated from deep sea sediment		586
Planococcus citreus	Environmental isolate		Motile Gram-positive coccus isolated from seawater, marine clam and frozen boiled shrimp		326
Planococcus kocurii	Environmental isolate		Skin of North Sea cod, fish curing brine, frozen boiled shrimp, frozen prawn	Japan	326
Planococcus spp. presumptive identification	Mortality	Pale gills, ascites fluid, gastroenteritis, internal haemorrhages	Rainbow trout fry	England	43
Planomicrobium okeanokoites (previously *Planococcus okeanokoites* and *Flavobacterium okeanokoites*)	Environmental organism		Isolated from marine mud	Japan	566
Plesiomonas shigelloides (previously *Aeromonas shigelloides*)	Possible opportunist pathogen	Emaciation, petechial haemorrhages in intestine	African catfish, eel, gourami, rainbow trout, sturgeon. Captive penguin, aquatic reptiles, ubiquitous in environment	Australia, Germany, Portugal	195 443
Providencia rettgeri (also known as *Proteus rettgeri*)	Septicaemia	Organism isolated from internal organs, ulcerative external lesions	Silver carp Associated with poultry faeces	Israel	79

Providencia rustigianii (previously *P. friedericiana*)	Normal flora	Faeces	Penguins (*Aptenodytes patagonica, Eudyptes crestatus, Pyoscelis papua, Spheniscus demersus, Spheniscus humboldti*)	Penguins in zoo in Germany	559
Pseudoalteromonas (*Alteromonas*) *antarctica*	Environmental organism		Muddy soils and sediments in coastal areas	Antarctica	115
Pseudoalteromonas bacteriolytica	Red-spot disease	Organism produces a red pigment on the *Laminaria* and induces damages to the seed supply	Culture beds of *Laminaria japonica*	Japan	677
Pseudoalteromonas (*Alteromonas*) *citrea*	1. Environmental isolate 2. Microflora		1. Marine surface water 2. Far-eastern mussel (*Crenomytilus grayanus* and *Patinopecten yessoensis*), molluscs, ascidians, sponges	1. Mediterranean Sea, France 2. Sea of Japan, Bering Sea	285 396
Pseudoalteromonas (*Alteromonas*) *denitrificans*	Environmental organism			Norway fiord coast	239
Pseudoalteromonas (*Alteromonas*) *distincta*	Microflora		Marine sponge	Komandorskie Islands, Russia	394
Pseudoalteromonas elyakovii (previously *Alteromonas elyakovii*)	1. Spot disease 2. Microflora	1. Fronds	1. Spot-wounded fronds of *Laminaria japonica* 2. Far-eastern mussel (*Crenomytilus grayanus*)	1. Sea of Japan 2. Troitsa Bay (Russia), Sea of Japan	679
Pseudoalteromonas (*Alteromonas*) *espejiana*	Environmental isolate		Marine environment	Californian coast	150
Pseudoalteromonas flavipulchra (previously *Pseudoalteromonas* and *Alteromonas aurantia*)	Environmental organism		Surface seawater	Mediterranean, France	398 286
Pseudoalteromonas (*Alteromonas*) *luteoviolacea*	Environmental organism			Mediterranean, France	287
Pseudoalteromonas maricaloris	Environmental organism		Marine sponge (*Fascaplysinopsis reticulata*)	Coral Sea	398
Pseudoalteromonas (*Alteromonas*) *nigrifaciens*	Non-pathogenic		Isolated from seawater and mussels	Japan	395

continued

Table 1.2. *Continued.*

Bacteria	Disease	Disease signs	Host/isolation site	Distribution	Ref
Pseudoalteromonas piscicida (previously *Alteromonas piscicida*, *Pseudomonas piscicida*, *Flavobacterium piscicida*)	Diseased eggs, mortality	Eggs	Damselfish (*Pomacentridae*)	Greece, Japan, USA	288 77 134 572
Pseudoalteromonas rubra (previously *Alteromonas rubra*)	Environmental organism		Marine water	Mediterranean, France	283
Pseudoalteromonas ulvae	Environmental isolate	Inhibits germination of marine algal spores and settlement of invertebrate larvae	Marine environment. Found on the surface of marine alga *Ulva lactuca*	Australia	231
Pseudoalteromonas (Alteromonas) undina	Environmental isolate		Marine environment	Californian coast	150
Pseudomonas anguilliseptica	Red spot disease of eels. Associated with winter disease in sea bream. Septicaemia	Haemorrhagic and ulcerative lesions, haemorrhages in eye, nose, operculum, brain, liver, kidney	Eels, rainbow trout, marine fish, sea bass, sea bream, ayu, salmon, herring, striped jack, turbot	Denmark, Finland, France, Japan, Scotland, Spain, Taiwan. Not in Australia	225 96 465 719 541 799 828
Pseudomonas chlororaphis	1. Mortality 2. Opportunist pathogen	Haemorrhages, increased ascites fluid	1. Amago trout 2. Crayfish	1. Japan 2. UK	332
Pseudomonas fluorescens	1. Mortalities, septicaemia. Opportunist pathogen 2. Associated with 'atypical BGD', water temperature <10°C	1. White nodules in spleen, abscesses in swim bladder. Fin or tail rot 2. Bacteria on gills, bacterial cells adhere to lamellar epithelium	1. Freshwater ornamental fish – carp, red oscar, tilapia, goldfish, sea bream, rainbow trout 2. Cultured salmonids (rainbow trout, chinook salmon, Atlantic salmon)	1. Worldwide, Japan 2. Canada, Chile, Norway	460 548 604 693
Pseudomonas plecoglossicida	Bacterial haemorrhagic ascites – heavy mortality	Bloody ascites	Cultured ayu	Japan	582
Pseudomonas pseudoalcaligenes	Mortalities	Skin ulceration	Rainbow trout	Scotland	42
Pseudomonas putida	Haemorrhagic ascites		Atlantic salmon, yellowtail	Japan. Ubiquitous in environment	461
Pseudomonas putrefaciens	See *Shewanella algae, S. baltica* and *S. putrefaciens*				

Organism	Status	Signs	Environment / Host	Location	Ref.
Pseudomonas stanieri	Environmental organism		Found in marine environment		69
Pseudomonas stuzeri (*Pseudomonas perfectomarina* is a junior synonym)	1. Lesions	1. Lesions	1. Octopus 2. Human, clinical material Found in wastewater, water and soil and marine environments	1. USA 2. Denmark, Argentina, Spain, USA	69 359
Psychroflexus (*Flavobacterium*) *gondwanensis*	Environmental organism		Antarctic environment	Antarctica	533
Rahnella aquatilis	1. Environmental organism 2. Clinical pathogen – rare. Bacteraemia, wound infection, urinary infection, respiratory infection		1. Isolated from water and snails 2. Humans	1. France, ubiquitous, USA 2. Korea and elsewhere	124 151 506 599
Raoultella ornithinolytica (previously known as ornithine-positive *Klebsiella oxytoca*)	Environmental organism				228
Raoultella planticola (previously *Klebsiella planticola* and *K. trevisanii*)	Environmental organism		Found in unpolluted water (drinking and surface water), soil, plants and occasionally mammalian tissue	Europe	256 402 228
Raoultella terrigena (previously *Klebsiella terrigena*)	Environmental organism		Found in unpolluted water (drinking and surface water), soil, plants and occasionally mammalian tissue. May be involved in nosocomial infections and neonatal infections	Europe	402 228
Renibacterium salmoninarum	Bacterial kidney disease	Exophthalmia, blisters on flank, ulcers, abscesses, lesions in organs. Grey-white enlarged necrotic abscesses seen in kidney	Salmonids – brown trout, rainbow trout, brook trout, chinook salmon, coho salmon, Atlantic salmon, ayu	Canada, Chile, Denmark, France, Germany, Iceland, Japan, Spain, USA, Yugoslavia, not in Australia	564 671
Rhodococcus (*luteus*) *fascians*	Microflora	Skin and intestinal flora	Carp, soil	USSR	573
Rhodococcus spp.	Panophthalmitis, ocular oedema. Pathogenicity not determined	Ocular lesions, nodules in muscle and organs. Found in association with *Mycobacterium neoaurum*	Chinook salmon	Canada	53
Roseobacter gallaeciensis	Normal flora	Bacterial flora on larvae of scallop	Scallop *Pecten maximus*	Europe	662

continued

Table 1.2. *Continued.*

Bacteria	Disease	Disease signs	Host/Isolation site	Distribution	Ref
Roseobacter species strain CVSP	Juvenile oyster disease (JOD)	Reduced growth rate, fragile and uneven shell margins, cupping of left valve Internally – mantle retraction, lesions, proteinaceous deposits (conchiolin) inside shell	Hatchery reared oysters (*Crassostrea virginica*)	USA	104 105
Salegentibacter (Flavobacterium) salegens	Environmental organism		Hypersaline lake	Antarctica	533
Salinivibrio (Vibrio) costicola ssp. *costicola*	Environmental organism		Hypersaline environments	Canary Islands, Spain	279 371
Salmonella arizonae	Septicaemia	Organism isolated from organs	Pirarucu (tropical freshwater fish) in zoo aquarium	Japan	447
Salmonella durham	Carrier state	Cloacal swabs	African mud turtle, yellow-spotted Amazon turtle, helmeted terrapin, Eastern box turtle, Northern diamondback terrapin, Mississippi map turtle, Travancore crowned turtle, Caspian terrapin, painted turtle, red-eared turtle, stinkpot turtle	USA	606
Salmonella enteritidis, S. havana, S. newport, S. typhimurium	Carrier state	Faeces	Gentoo penguins, macaroni penguins, grey-headed albatross, Antarctic fur seals	Bird Island South Georgian archipelago – Antarctica	612
Salmonella heidelberg, S. newport, S. oranienburg	Carrier state in healthy animals, or mild gastroenteritis	Rectal swabs	Californian sea lion pups, Northern fur seal pups	USA	298
Salmonella group O type B	Death and illness		Sea turtles (*Chelonia mydas*)	Western Australia	592
Serratia fonticola	Environmental organism		Water, spring-water, soil, wild birds, gut content of slugs and snails	Ubiquitous	290 558
Serratia liquefaciens	Mortalities, septicaemia. Opportunistic pathogen	Kidney, spleen, liver affected. Bloody ascites, haemorrhages in tissues. Redness and swelling around anus in Arctic charr	Arctic charr, Atlantic salmon, turbot.	France, Scotland, USA (mid-Atlantic region)	500 538 715 791
Serratia marcescens	Pathogenicity in doubt – opportunist	Kidney	Bass, trout, white perch	USA polluted river	76
Serratia plymuthica	Pathogenicity in doubt – opportunist. Isolated from moribund trout	Organisms in internal organs	Rainbow trout. Association with pollution by domestic sewage	Scotland, Spain	42 579

Organism					Reference
Shewanella algae previously identified as *Shewanella (Pseudomonas) putrefaciens* group IV, Gilardi biovar 2, CDC biotype 2	1. Environmental isolate 2. Associated with human faeces, skin ulcers, otitis media, and found in bacteraemia		1. Red algae, marine environment 2. Pathogenic to humans	1. Japan 2. Worldwide, Canada, Denmark, Sweden, USA	433 360 588 792
Shewanella baltica previously *Shewanella (Pseudomonas) putrefaciens* Owen's group II	Environmental isolate		Oil brine	Baltic Sea	851
Shewanella benthica			Intestine, holothuria (*Psychrobotes longicauda*)		112
Shewanella (Alteromonas) colwelliana (organism designated LST in original reference)	Autochthonous	Contributes to oyster larvae sediment. Adheres to surfaces such as oyster shell, glass, plastic	Estuarine oyster water. Eastern oyster, (*Crassostreae virginica*)	UK	814 815
Shewanella frigidimarina	Environmental isolate		Ice, ice algal biomass, cyanobacterial mat	Antarctica	112
Shewanella gelidimarina	Environmental isolate		Ice	Antarctica	112
Shewanella hanedai (*Alteromonas hanedai*)	Environmental isolate		Sediment, marine environment	Arctic	409
Shewanella japonica	Environmental isolate		Found in seawater and the mussel (*Protothaca jedoensis*). Digests agar	Coastal area of Sea of Japan	397
Shewanella oneidensis (formerly *Shewanella* sp. MR-1)	Environmental isolate		Lakewater	Oneida Lake, Lake Michegan, USA and Black Sea	782
Shewanella pealeana	Normal flora, symbiotic	Accessory nidamental gland	Squid (*Loligo pealei*)		492
Shewanella putrefaciens previously *Pseudomonas putrefaciens* Owens Group I, Gilardi biovar 1 & 3, CDC biotype 1	Septicaemia	Haemorrhagic necrosis on body. Frayed fins and exophthalmia	Rabbit fish Found in aquatic environments including marine, sediments, oil fields, spoiling fish	Saudi Arabia Antarctica, UK, worldwide	433 665 666
Shewanella woodyi	Non-pathogenic		Squid ink and seawater	Alboran Sea	521
Sphingobacterium multivorum (previously *Flavobacterium multivorum* and group lik, biotype 2)	Clinical sources	Spleen	Human pathogen	UK, USA	364 365 844

continued

Table 1.2. Continued.

Bacteria	Disease	Disease signs	Host/isolation site	Distribution	Ref
Sphingobacterium (Flavobacterium) spiritivorum F. yabuuchiae is a junior synonym	Clinical isolates		Humans		365 844
Sphingomonas (Pseudomonas) paucimobilis	Environmental organism		Clinical, hospital environment, water		361
Staphylococcus aureus	Eye disease	Cornea red to opaque	Silver carp. Also found in marine environments	India	320 688
Staphylococcus delphini	Skin lesions	Purulent skin lesions	Dolphins in captivity	Europe	778
Staphylococcus epidermidis	1. Staphylococcosis 2. Skin microflora	1. Ulceration and haemorrhages on fins	1. Red sea bream, yellowtail. Also found in marine and estuarine waters 2. Humans	1. Japan, Taiwan	320
Staphylococcus hominis	Not a known aquaculture pathogen		Reported from marine and estuarine environments and human sources		320 444
Staphylococcus lutrae	Pathogenicity not determined. May be pathogenic	Liver, spleen, lymph node	Isolated from European otters	UK	264
Staphylococcus spp. – S. capitis, S. cohnii, S. epidermidis, S. haemolyticus, S. saprophyticus, S. simulans, S. xylosus	Microflora	Skin	Humans		320 444 681
Staphylococcus warneri	1. Diseased and dying trout. Opportunistic infection 2. Associated with septicaemia, endocarditis, conjunctivitis, urinary tract and wound infections	1. Ulcerated lesions on fins and exophthalmia, ascitic fluid. Isolated from liver and kidney	1. Rainbow trout 2. Humans	1. Spain	296 320 444
Stappia stellulata-like Strain M1 (previously Agrobacterium stellulatum)	Potential probiotic in JOD		May prevent colonization of juvenile oysters by Roseobacter strain CVSP	USA	105
Streptobacillus moniliformis or possibly from the Fusobacterium group. Identification for this disease is not conclusive (Maher et al., 1995)	Bacterial disease	Intracellular organisms seen in tissues. Enlargement of endothelial cells of kidney glomeruli	Atlantic salmon. Seen in pneumonia in koalas, cervical abscess in guinea pigs, polyarthritis in mice, tendon-sheath arthritis in turkey. Causative agent of rat-bite fever in humans	Ireland, worldwide	169 519 611

Organism					Ref.
Streptococcus type I	Now recognized as *Lactococcus garvieae*				
Streptococcus agalactiae group B	Mortality, exophthalmia, haemorrhage	Haemorrhagic areas on body, mouth, fins. Organism in internal organs	Bluefish, cultured seabream, wild mullet, striped bass, sea trout, bull minnows. Aquarium fish – rams. Causes mastitis in cattle, infections in many other animal species, and neonatal meningitis in humans	Australia, Kuwait, Chesapeake Bay, Alabama USA	71 135 242 637
Streptococcus agalactiae group B, Type 1b capsular antigen (*Streptococcus difficile*)	Septicaemia, meningoencephalitis	Organism isolated from brain	Carp, rainbow trout, tilapia, freshwater ornamental fish, frogs, mice, bovines, humans	Australia, Israel	233 776
Streptococcus difficile	Confirmed as *Streptococcus agalactiae* group B, type Ib capsular antigen				776
Streptococcus dysgalactiae ssp. *dysgalactiae*, group C	Animal pathogen		Animals – bovine mastitis		790
Streptococcus dysgalactiae ssp. *dysgalactiae* serovar L	Infections and septicaemia with resulting bronchopneumonia, myocarditis, osteomyelitis, pyelonephritis, abscesses	Isolated in almost pure culture from lung, kidney, intestine, spleen	Isolated from harbour porpoises stranded or caught in fishing nets. Infections in cattle, dogs, pigs and other animals	Baltic Sea, North Sea	727 790
Streptococcus dysgalactiae subsp. *equisimilis*, group C, G or L	Animal and human pathogen		Animals and humans		790
Streptococcus iniae (*Strep shiloi* is a junior synonym)	1. Deaths, meningoencephalitis 2. Golf ball disease in dolphins 3. Systemic infection 4. Wounds when cleaning fish. Deaths	1. Culture organism from the brain. Also kidney and liver 2. Subcutaneous abscess 3. Torticollis, gross lesions, granulomas, enlarged organs 4. Localized cellulitis, ulcers	1. Barramundi; European sea bass, gilthead sea bream, puffer fish, rabbit fish, reef fish, rainbow trout, stingray, tilapia, cultured ayu, flying fox aquarium fish, freshwater and salt-water fish 2. Freshwater Amazonian dolphin 3. Farmed bullfrogs (*Rana castesbeiana*) 4. Humans – especially elderly people and fish handlers Marine and brackish water	Australia, Bahrain, Barbados, Canada, China, Israel, Japan, South Africa, Thailand, USA	223 233 235 135 127 621 625 626 530 848
Streptococcus milleri (identification not certain. *S. milleri* not a recognized name)	Disease	Ulcers on flank and tail	Koi carp	UK	41 87

continued

Table 1.2. *Continued.*

Bacteria	Disease	Disease signs	Host/isolation site	Distribution	Ref
Streptococcus parauberis (previously *Streptococcus uberis* genotype II)	1. Streptococcosis, mucohaemorrhagic enteritis 2. Mastitis	1. Haemorrhages in anal and pectoral fins, and eyes. Organisms isolated from liver, kidney, spleen 2. Milk	1. Cultured turbot 2. Bovine	Spain	224 754
Streptococcus phocae spp. nov.	Opportunistic in viral infections		Seals	Norway	700
Streptococcus porcinus (previously *Streptococcus infrequens*, Lancefield groups E, P, U, V)	Infections	Cervical lymph nodes and milk	Pigs	Worldwide	175
Streptomyces salmonis (*Streptoverticillium salmonis* basonym)	Streptomycosis		Salmonids	USA	41
Tenacibaculum maritimum (previously *Cytophaga marina* and *Flexibacter maritimus*)	Marine columnaris, erosive skin disease, gill lesions and ulcerative dermatitis, black patch necrosis, bacterial stomatitis (mouth rot)	Eroded mouth and fins, ulcerated skin lesions	Marine fish, especially Atlantic salmon, chinook salmon, sea bass, red sea bream, black bream, rock bream, northern anchovy, striped trumpeter, Dover sole, flounder, yellowtail	Atlantic, Canada, Europe, France, Japan, Scotland	89 90 91 154 551 605 725 801
Tenacibaculum (*Flexibacter*) *ovolyticum*	Opportunist pathogen Larvae and egg mortality	Dissolves chorion and zona radiata of the egg shells	Atlantic halibut eggs and larvae	Norway	324 725
Vagococcus fessus	Pathogenicity not determined	Organism isolated from liver, kidney	Isolated from a dead harbour seal and porpoise	Europe	369
Vagococcus fluvialis	1. Environmental organism 2. Clinical samples	2. Blood, peritoneal fluid, wounds	1. River water, chicken faeces, domestic animals (pig, cattle, cats, horse) 2. Humans	UK	177 629 732
Vagococcus lutrae	Pathogenicity not determined	Organism found in blood, liver, lungs, spleen	Otter	UK	477

Organism	Disease/relationship	Details	Host/association	Location	Ref.
Vagococcus salmoninarum	Lactobacillosis, pseudokidney disease, peritonitis, septicaemia	Haemorrhages on gills and in ocular region, peritonitis, heart lesions, enlarged spleen, listless behaviour	Atlantic salmon, brown trout, rainbow trout	Australia, France, North America, Norway	542 682 807
Varracalbmi spp. nov.	Eye lesions (bloody eye) and blindness	Lesions in kid, liver, gills, pseudobranch. Organism isolated from eyes. Skin ulcers	Atlantic salmon	Norway	771
Vibrio aerogenes	Environmental organism		Associated with sea grass sediment, found in shallow coastal and oceanic seawaters	Taiwan	692
Vibrio aestuarianus	Environmental organism		Found associated with shellfish (oyster, clam, crab) and estuarine waters	USA	747
Vibrio agarivorans	Environmental organism		Seawater. Abalone – pathogenicity not known	Australia, Mediterranean Sea (Spain)	135 514
Vibrio alginolyticus	1. Systemic disease, ulcerative disease, necrosis, eye disease, vibriosis, scallop larval mortality 2. Wound infections, external otitis, cellulitis	1. Organism isolated from organs and eye, scallop larvae 2. Wounds on exposure to seawater	1. Australian native and introduced fish. Molluscs (larval and juvenile), red abalone, South African abalone, scallop larvae, eel, rainbow trout, sea bream, turbot, turtles, seahorse. Shrimps 2. Humans	Australia, Chile, Mexico, worldwide Indo-Pacific & East Asia	30 300 301 650
Vibrio anguillarum	See *Listonella anguillarum*				
Vibrio brasiliensis	Pathogenicity not known. Likely to be normal flora		Isolated from bivalve larvae (*Nodipecten nodosus*)	Brazil	740
Vibrio calviensis	Environmental organism		Seawater	Western Mediterranean Sea, France	216
Vibrio campbellii	Environmental organism		Seawater	Florida, Puerto Rico	819 820
Vibrio campbelli-like	Mortalities in juveniles	Organism in brain, kidney, liver. Organs had lesions and haemorrhage	Hatchery-reared turbot and brill	New Zealand	221
Vibrio carchariae	Is a junior synonym of *V. harveyi*. Slight differences of phenotype between the strains are reported in the literature				619

continued

Table 1.2. *Continued.*

Bacteria	Disease	Disease signs	Host/isolation site	Distribution	Ref
Vibrio cholerae (non-01) Heiberg groups I, II	1. Septicaemia 2. Gastroenteritis	1. Skin haemorrhages 2. Diarrhoea, vomiting	1. Ayu, goldfish, shellfish. Isolated from fresh and estuarine waters 2. Humans – consumption of raw shellfish	1. Australia, Japan 2. Africa, Asian continent, Europe, USA, UK	196 255 275 434 507
Vibrio cholerae El Tor (serogroups 01 & 0139)	Cholera		Humans Found in aquatic environment – surface waters	Serogroup 0139 in Bangladesh, Indian subcontinent only. Pandemic strain	275 392 9
Vibrio cholerae-like	Death		Lobsters	USA. (Organism doesn't grow in temperatures greater than 25°C)	
Vibrio cincinnatiensis	1. Environmental organism 2. Bacteraemia, meningitis		1. Environment 2. Human pathogen	2. USA	120
Vibrio coralliilyticus spp. nov. (previously *V. coralyticus* YB)	Pathogen of coral	Tissue lysis and death. Small white spots seen on coral at 3–5 days and complete destruction of tissue after 2 weeks	Coral (*Pocillopora damicornis*). Also isolated from oyster larvae from the Atlantic Ocean	Atlantic Ocean, Indian Ocean and Red Sea	83 84
Vibrio cyclitrophicus (originally spelt *V. cyclotrophicus*)	Environmental organism			Eagle Harbor, Washington, USA	338
Vibrio diabolicus	Environmental organism		Isolated from polychaete annelid (*Alvinella pompejana*) in deep-sea vent	Pacific Ocean	635
Vibrio diazotrophicus	Environmental organism		Gastrointestinal tract of sea urchins and marine animals, surface of reeds, occurring in estuarine environments, seawater and sediments	Canada, Scotland, England	319
Vibrio fischeri (synonymous with *Photobacterium fischeri*)	1. Pathogenicity in doubt 2. Symbiont	1. White nodules on skin, haemorrhagic ulceration, tumours in pancreas and bile duct 2. Present in light organ of sepiolid squid	1. Sea bream, turbot 2. Sepiolid squid (*Euprymna scolopes*)	1. Spain 2. Hawaii	106 267 342 486

Organism			
Vibrio fluvialis (previously group F (Furniss et al., 1977) and group EF6 (Huq et al., 1980))	1. Mortality 2. Environment 3. Acute watery, cholerae-like diarrhoea. Toxin produced	1. Plaice fry (*Pleuronectes platessa*) 2. Found in aquatic, particularly estuarine environment. Isolated from marine molluscs and crustacea 3. Humans	1. Denmark 2. Worldwide 3. Bangladesh, India, Kenya, Middle East, Philippines, Spain, Tanzania, Tunisia	276 378 485 620 687
Vibrio furnissii (formerly *V. fluvialis* biovar 2 (aerogenic biogroup), group F, group EF-6)	1. Pathogenicity in doubt 2. Environmental organism 3. Food poisoning, gastroenteritis	1. Haemorrhages in intestine 3. Diarrhoea	1. Eel 2. River water, animal faeces, marine molluscs and crustacea 3. Humans Found in aquatic, particularly estuarine environment	1. Spain 2. UK, worldwide 3. Bahrain, Bangladesh, Indonesia, Peru, USA	123 276 240 485 687
Vibrio gazogenes	Environmental organism			67
Vibrio halioticoli spp. nov.	Non-pathogenic	Main gut microflora	Japan	678
Vibrio harveyi (organism previously identified as *V. carchariae* is said to be a junior synonym of *V. harveyi*)	1. Mortality in sharks and abalone, necrotizing enteritis in summer flounder, gastroenteritis in grouper, systemic disease, ulcerative disease, necrosis, vibriosis 2. Wound (shark bite)	1. White spots on foot of abalone. Lose ability to adhere. Necrotic degeneration and vacuolation in the lesion. Flounder have extended abdomen and red anus. Fish – organism from kidney, eye. Shrimps – black spots on exoskeleton. Organism in lesions, eye, hepatopancreas	1. Abalone, shark, flounder, grouper, milkfish, molluscs, oysters (larval and juvenile), black tiger prawns, penaeid prawns and shrimps, salmonids, sea bass, sea bream, seahorse, sharks, octopus. Australian native and introduced fish 2. Humans	1. Australia, Europe, Indonesia, Indo-Pacific & East Asia, Japan, Philippines, South America, Taiwan, Thailand, Venezuela, USA, worldwide	23 11 410 576 581 619 710 734 735 847
Vibrio hollisae (previously Group EF-13)	1. Non-pathogenic for fish 2. Food poisoning, gastroenteritis, wound infection, bacteraemia	2. Diarrhoea Has thermostable haemolysin similar to *V. parahaemolyticus*	1. Present in fish intestine 2. Humans	1. Japan 2. USA	346 555 580
Vibrio ichthyoenteri	Mortalities, intestinal necrosis	Opaque intestines	Japanese flounder larvae	Japan	389
Vibrio lentus	Pathogenicity not stated		Associated with Mediterranean oysters	Spanish Mediterranean coast	513
Vibrio logei	1. Possible pathogen 2. Non-pathogenic	1. Skin lesions 2. Symbiont in light organ	1. Atlantic salmon 2. Squid (*Sepiola*)	1. Iceland 2. Western Pacific	257

continued

Table 1.2. *Continued.*

Bacteria	Disease	Disease signs	Host/isolation site	Distribution	Ref
Vibrio mediterranei	1. Non-pathogenic 2. Probiotic properties 3. Coral bleaching – see under *V. shilonii* Appears to be a heterogeneous species		1. Plankton, sea sediments, seawater 2. Some strains are a probiotic for turbot larvae 3. Coral bleaching	Spain, Mediterranean coast	631
Vibrio metschnikovii	1. Non-pathogenic for fish 2. Peritonitis and bacteraemia	1. Isolated from fowl with choleraic disease	1. Found in marine and freshwater environment, especially rivers, estuaries and sewage. Isolated from clams, cockles, oysters, lobster 2. Humans	Worldwide	483
Vibrio mimicus (previously *Vibrio cholerae* Heiberg group V)	1. Systemic disease and opportunist 2. Contaminant on sea turtle eggs 3. Food poisoning, gastroenteritis, otitis media	1. Haemolymph, inflammation of pericardium 3. Diarrhoea, ear infection	1. Crustaceans – penaeid prawns, yabbies, marron, freshwater crayfish. Barramundi 2. Reduced viability of turtle eggs and human pathogen following consumption of eggs 3. Humans – consumption of raw shellfish. Found in fresh and brackish water	1, 3. Asia, Australia, Bangladesh, Mexico, New Zealand, Guam, Canada, worldwide 2. Costa Rica	4 161 135 210 230 507
Vibrio mytili	1, 2. Non-pathogenic	1. Normal flora	1. Mussels 2. Humans	Spain	632
Vibrio natriegens	Environmental isolate		Salt marsh mud, water, oysters	UK, USA	820
Vibrio navarrensis	Environmental isolate		Environmental – sewage and surface water	Spain	768
Vibrio neptunius	Microflora		Dominant microflora in a recirculating system for rotifers. Isolated from healthy and diseased bivalve larvae (*Nodipecten nodosus*), gut of turbot larvae, rotifers		305
Vibrio nereis	Environmental isolate		Seawater		820

Species	Disease/pathogenicity	Symptoms/organs	Host/source	Location	Ref.
Vibrio nigripulchritudo	Environmental isolate		Seawater		819 820
Vibrio ordalii (previously *Vibrio anguillarum* biotype 2 and *Listonella ordalii*)	Vibriosis, bacterial necrosis and systemic disease	Organisms on muscle, skin, gills, digestive tract, heart, liver, kidneys, spleen. Necrosis and haemorrhage	Molluscs – (larval and juvenile), salmon	Australia, Japan, North-West Pacific, USA, worldwide	680
Vibrio orientalis	Environmental isolate		Luminous marine bacterium	China	846
Vibrio pacinii spp. nov.	Pathogenicity not known		Isolated from shrimp larvae (*Penaeus chinensis*)	China	306
Vibrio parahaemolyticus (previously *Beneckea parahaemolytica*)	1. Septicaemia, mortality 2. Withering syndrome 3. Food poisoning, gastroenteritis, wound infection, bacteraemia	1. External haemorrhage, tail rot, organism in internal organs 2. Organisms found in haemolymph 3. Diarrhoea, vomiting	1. Crustaceans (shrimps, marron), mullet, octopus, Iberian toothcarp 2. Cultured small abalone 3. Humans	1. Indo-Pacific and East Asia, Spain 2. Taiwan	10 135 272 417 499
Vibrio pectenicida	Pathogenic	Moribund scallop	Scallop larvae, shrimps, prawn	East Asia, Indo-Pacific, France	470
Vibrio pelagius	See *Listonella pelagia*				
Vibrio penaeicida (previously *Vibrio* PJ) Proposed causative agent of Syndrome 93	1. Vibriosis of prawns 2. Septicaemia, associated with Syndrome 93	1. Brown spots (nodules) in lymphoid organs and gills 2. Haemolymph	1. Kuruma prawns. Also isolated from healthy prawns 2. Shrimp (*Penaeus stylirostris*)	1. Japan 2. New Caledonia	388 187 676
Vibrio proteolyticus, strain CW8T2	Pathogen	Affects microvilli, gut cells, devastates cells in body cavity	Juvenile *Artemia* spp. (brine shrimp), used as live diets for aquaculture species	Europe	67 788
Vibrio rotiferianus	Pathogenicity not known		Isolated from rotifer flow-through culture system	Belgium	305
Vibrio rumoiensis	Environmental isolate		Isolated from drain pool of fish product processing plant	Japan	850
Vibrio salmonicida	Coldwater vibriosis, Hitra disease	Septicaemia, haemorrhages in the integument surrounding the organs. Organism found in blood and kidney	Atlantic salmon, diseased cod	Canada, Faroe Islands (Denmark), Iceland, Norway, Scotland	198 232
Vibrio scophthalmi	Non-pathogenic	Part of gut microflora	Juvenile turbot	Spain	149 254

continued

Table 1.2. *Continued.*

Bacteria	Disease	Disease signs	Host/isolation site	Distribution	Ref
Vibrio shilonii (previously *V. shiloi*, *Vibrio* species AK-1). Latest information suggests this is a later synonym of *V. mediterranei* (Thompson *et al.*, 2001a,b)	Coral bleaching	Adheres and penetrates into the epidermis of the coral, *Oculina patagonica*. Culture from the coral mucus. The organism is intracellular and non-culturable	Bacterium binds to the coral at elevated temperatures and disrupts the symbiotic process with the photosynthetic micro-algae endosymbionts (zooxanthellae). Heat-stable toxin produced	Mediterranean Sea	59 458 742
Vibrio splendidus	Mortality in stressed oysters		Oyster larvae (*Crassostrea gigas*)	France	466 467
1. *Vibrio splendidus* I. 2. *Vibrio splendidus*-like	1. Septicaemia. Mortalities in juveniles 2. Mortalities	1. Haemorrhages at mouth, anus and fins. Organisms in brain, kidney, liver 2. Skin haemorrhages, fin and tail rot erosion, decay of soft tissue between fins. Sample from kidney and spleen	1. Gilt-head sea bream, rainbow trout, turbot larvae, shrimps (prawns). Juvenile hatchery-reared turbot and brill 2. Plaice fry	1. France, New Zealand, Scotland, Spain 2. Denmark	281 254 31 221 620
Vibrio splendidus II	Bacillary necrosis, mortality	Larvae become inactive and settle to the bottom of the tank	Japanese oyster larvae, cupped oyster (*Crassostrea gigas*)	Japan, France	721 798
Vibrio tapetis (previously *Vibrio* P1 or VP1 or VTP)	Brown ring disease, high mortality	Brown conchiolin deposit on inner surface of shell between pallial line and shell margins	Cultured Manila clams	England, France, Portugal, Spain	108 146 14 609 610
Vibrio trachuri. Latest information suggests this is a junior synonym of *V. harveyi* (Thompson *et al.*, 2002b)	Disease	Haemorrhages in organs, exophthalmia	Horse mackerel	Japan	400 743
Vibrio tubiashii	Bacterial necrosis and systemic disease	Larvae cease swimming, tissue digestion	Molluscs (oyster and clam) – larval and juvenile, Pacific oyster	Australia, UK, USA	321 294
Vibrio viscosus	See *Moritella viscosa*				82 506

Vibrio vulnificus biotype I	1. Environmental 2. Infection, systemic disease, ulcerative disease, necrosis 3. Wound infection, septicaemia following consumption of raw seafood especially oysters. Septicaemia has a 50% fatality rate	1. Microflora of intestines or gills of fish, mussels, crabs and seabirds 3. Ingestion of contaminated seafood, damage to the intestinal wall. Wound infection	1. Atlantic coastal regions, Denmark, Europe, Indo-Pacific and East Asia, Australia, Mexico 2. Australia, Indo-Pacific and East Asia, Europe, Japan, USA 3. Australia, Belgium, Denmark, Germany, Japan, Holland, Sweden, USA	358 597 746
Vibrio vulnificus biotype 2 serovar E lipopolysaccharide-based O serogroup, O4	1. Vibriosis, haemorrhagic septicaemia 2. May cause illness and wound infection in humans following handling of eels	1. Organism isolated from gills, mucus, intestinal contents, spleen, and kidney	1. Eels – virulent and avirulent strains are found. Cultured shrimps. Found in brackish water, oysters and sediment 2. Humans	201 356 34 25 597 746
Vibrio vulnificus biotype 3	Severe wound infection and bacteraemia	Wounds and blood	Denmark, Europe, Japan, Norway, Spain, Sweden, Taiwan Israel	101 565
Vibrio wodanis	Non-pathogenic, may be an opportunist in winter ulcer disease	Atlantic salmon, rainbow trout, haddock	Cold water of Iceland, Norway	82 506
Vibrio xuii	Pathogenicity not known	Isolated from shrimp culture water and bivalve larvae (*Nodipecten nodosus*)	China	740
Weissella strain DS-12	Potential probiotic	Part of intestinal microflora of flounder. Species of *Weissella* are found in fermented foods such as fish and sausages	Korea	139
Yersinia aldovae previously *Y. enterocolitica* group X2	Environmental organism	Isolated from drinking water, river water, soil and fish		85
Yersinia bercovieri previously *Y. enterocolitica* biogroup 3B	Environmental organism	Isolated from freshwater and soil		812
Yersinia enterocolitica ssp. *enterocolitica*	Generally non-pathogenic for mammals	Sucrose negative strains from small rodents. Found in terrestrial and aquatic habitats	Japan, Europe, Canada, USA	422 121

continued

Table 1.2. *Continued.*

Bacteria	Disease	Disease signs	Host/isolation site	Distribution	Ref
Yersinia frederiksenii	Environmental organism		Found in water. Fish, humans, mammals and birds may be healthy carriers		423
Yersinia intermedia	Environmental organism		Atlantic salmon. Found in water. Fish and mammals may be healthy carriers	Australia	423
Yersinia kristensenii	Environmental organism		Found in water, soil, animals. Fish, humans and mammals may be healthy carriers Sucrose-negative strains	Australia, Denmark, France, Germany, Greece, Japan, Norway, UK, USA	86 423
Yersinia mollaretii (previously *Y. enterocolitica* biogroup 3A)	Environmental organism		Isolated from freshwater and soil		812
Yersinia rohdei	Environmental organism		Surface water, human faeces, dog faeces	California, Germany	12
Yersinia ruckeri	1. Enteric redmouth disease (ERM) – severe form of the disease. Yersiniosis – milder form of the disease. Systemic disease 2. Salmonid blood spot 3. Non-pathogenic	1. Reddening of throat and mouth, haemorrhages on gills, fin. Isolated from kidney 2. Blood spots in eyes and haemorrhages in musculature of the caudal peduncle. Organisms in liver and kidney 3. Microflora of intestinal contents	1. Juvenile Atlantic salmon, wild Atlantic salmon, channel catfish, goldfish, brown trout, hatchery-reared rainbow trout, salmonids, eel, otter, freshwater ornamental, and marine fish. Disease is usually associated with poor environmental conditions 2. Hatchery-reared trout and Atlantic salmon. Partially cross reacts with serovar I and was termed serovar I'; however, proved to be serovar III 3. Seabirds	1. Australia, Bulgaria, Canada, Finland, France, Germany, Greece, Italy, New Zealand, Norway, South Africa, Scotland, Spain, Switzerland, Turkey, UK, USA 2. Australia	250 311 167 203 137 207 500 657 718
Zobellia galactanovorans	Environmental organism	Associated with red alga (*Delesseria sanguinea*)	Marine environment	Brittany, France	61
Zobellia uliginosa (previously *Cytophaga* and *Flavobacterium uliginosum*)	Environmental organism		Marine environment		61

'Pathogenicity not determined' means that the organism was isolated from a dead or diseased animal, but that Koch's postulates for disease was not determined.

The information is presented in tabular form so that it is readily accessible in a summarized form.

The bacterial names are listed in alphabetical order. The disease column indicates whether the organism is a saprophyte, an environmental organism or is a pathogen of fish and other aquatic animals. The next column indicates the disease signs that may be seen with an infection caused by this organism, and then the aquatic species is listed, as is the geographical distribution of the organism. The final column lists the reference number.

1.4 Taxonomy and Disease Status of Bacteria

New bacteria from the aquatic world are being discovered, identified and named at an increasing rate. Some are then reassigned to a different or newly described genus, while others change their species name as more research and study on the taxonomy follows. The following section gives brief details of the current taxonomic status of a genus or species. Because some of this information is likely to become out of date, microbiologists are encouraged to keep up with such information, which is found at taxonomic websites. Some addresses and suggested reading can be found under Further Reading and Other Information Sources, p. 278. Refer to these and other texts for a comprehensive coverage of diseases.

Abiotrophia spp.

These bacteria, which are known as nutritionally variant streptococci (NVS), satellite around other bacterial growth, or require L-cysteine or pyridoxal hydrochloride (vitamin B₆) for growth. *A. elegans* requires L-cysteine for growth. *A. adiacens*, *A. defectiva* and *A. elegans* are isolated as part of the normal flora from the oral cavity, intestinal tract and genital tract of humans. They have also been isolated in cases of endocarditis, otitis media and post-surgery wound infections. They have been included in this book because an organism found in a minke whale was originally placed into this genus. It has since been placed into the genus *Granulicatella* along with some other previously named *Abiotrophia*

species. *G. balaenopterae*, found in the minke whale, does not require additional growth factors and grows in air or 5% CO_2.

Aequorivita spp.

Strains from this genus belong to the family *Flavobacteriaceae* and the order *Cytophagales*, and have been isolated from the marine environment in Antarctica. They can grow on MA 2216.

Aeromonas spp.

The taxonomy of *Aeromonas* spp. (family *Aeromonadaceae*) is in a continual state of flux as new species are described and the descriptions of the existing taxa are refined. There is a phenotypic and a genotypic classification. Genotypic classification is based on DNA–DNA hybridization with 16 DNA HG recognized at present. The designation of the genospecies is dependent on the type strain within the HG. Problems exist for the diagnostic microbiologist in the phenotypic identification of strains. Phenotypic diversity of strains occurs within a DNA HG and strains from different HG may be biochemically similar. In this book the phenotypic tests are recorded for the type strains. A table (Table 4.2) is also presented of the phenotypic tests according to HG (Abbott *et al.*, 1992; Kaznowski, 1998). An average of the results from these two studies is reported. Results from different references can be difficult to assess. Abbott *et al.* (1992) stated that when using 22 strains of HG1, including the type strain *A. hydrophila* ATCC 7966, 100% were positive for LDC. This was also supported by research that recognized a new subspecies within the *A. hydrophila* group, with the recognition of *A. hydrophila* ssp. *hydrophila* and *A. hydrophila* ssp. *dhakensis* (Huys *et al.*, 2002b). This study reported results for *A. hydrophila* ssp. *hydrophila* of the type strain LMG 2844, which is equivalent to ATCC 7966. However, Kaznowski (1998), using one strain only (the type strain ATCC 7966), stated that 0% of HG1 are positive for LDC. Likewise, other researchers have found that the type strain ATCC 7966 is negative for LDC (Nielsen *et al.*, 2001). One of the reasons that there are variable reports for LDC may be due to misinterpretation of the test, or the use of different methods to detect LDC. In the conventional tube test, LDC is a pale purple colour, particularly in comparison with the ADH tube. Also, a stronger result may be obtained after 48 h incubation rather than at 24 h incubation.

HG8 and HG10 can be separated by biochemical tests; they are genotypically identical and are named *A. veronii* biogroup *sobria* and *A. veronii* biogroup *veronii*, respectively. The type strain of the non-motile *A. salmonicida* belongs in HG3, even though HG3 corresponds mainly with the phenospecies of *A. hydrophila*.

Aeromonas allosaccharophila and *Aeromonas encheleia*: The status of these species is controversial (Working group on *Aeromonas* taxonomy. *International Journal of Systematic Bacteriology*, 1999, 49: 1946).

Aeromonas bestiarum was shown to be virulent to the common carp in pathogenicity studies (Kozińska *et al.*, 2002).

A. caviae and *A. veronii* may cause 100% mortality in giant freshwater prawns when either organism is present in high numbers; in the order of 3.8×10^6 and 3.7×10^5 cells/g of body weight, respectively (Sung *et al.*, 2000).

A. hydrophila is an opportunistic organism following a primary stressor; however, the pathogenesis and virulence of this organism remains unclear. A number of toxins are produced, including haemolysins, cytotoxins, enterotoxin, a cytolytic haemolysin and aerolysin. Aerolysin is a channel-forming toxin that kills cells, is lethal to mice and possesses haemolytic and cytotoxic activities. It is the main contributor to pathogenicity (Chopra *et al.*, 1993). The aerolysin gene is not found in non-haemolytic strains of *A. hydrophila*, non-haemolytic strains of *A. caviae*, or strains of *A. sobria* (Pollard *et al.*, 1990).

A dominant *A. hydrophila*, which was negative for the phenotypic tests of LDC and cellobiose fermentation, was found to be associated with disease outbreaks in aquaculture in China (Nielsen *et al.*, 2001). However, this is in contrast to other findings that have suggested that LDC-negative isolates are less virulent than LDC-positive isolates, where 78% ($n = 23$) of isolates from diseased mammals and 100% ($n = 4$) of isolates from moribund fish were LDC-positive (Lallier and Higgins, 1988). Although the number of isolates tested from moribund fish was small ($n = 4$), strains of *A. hydrophila* in these cases were also indole-negative, did not produce enterotoxin,

and were negative for agglutination with 0.2% acriflavine.

Virulence studies in strains isolated from walking catfish suggested that there was a scale of virulence ranging from avirulent, weakly virulent, to strongly virulent. All virulent strains possessed haemolysin genes, and were able to lyse red blood cells of a wide variety of animal species. The majority of avirulent or weakly virulent strains either did not possess any of the three haemolysin genes tested (AHH_1, AHH_5, ASA) or had combinations of only one or two of the genes. These strains were generally unable to lyse the red blood cells of as many animal species. The virulence strains were able to cause infection at a dose rate of 10^4 CFU/ml, whereas avirulent strains did not cause disease at this dose (Angka *et al.*, 1995).

A recent article reports a subgroup within *A. hydrophila* HG1, and these are referred to as *A. hydrophila* ssp. *dhakensis* ssp. nov. These strains were isolated from cases of diarrhoea in children from Bangladesh. They were differentiated from *A. hydrophila* ssp. *hydrophila* by genetic tests and the following phenotypic tests. *A. hydrophila* ssp. *dhakensis* is negative for acid production from L-arabinose, and negative for utilization of methyl α-D-mannoside, L-fucose, and L-arabinose, whereas *A. hydrophila* ssp. *hydrophila* is positive for acid production from L-arabinose and positive for utilization of methyl α-D-mannoside, L-fucose, and L-arabinose (Huys *et al.*, 2002b).

A. ichthiosmia is a later synonym of *Aeromonas veronii* (Huys *et al.*, 2001).

Aeromonas sobria is now called *Aeromonas veronii* ssp. *sobria*. Some commercial systems still refer to it as *A. sobria*. Both terminologies are stated in the tables here.

Aeromonas trota is a junior synonym of *A. enteropelogenes*. *A. enteropelogenes* was published first; however, because the name *A. trota* is more widely used, this name has been retained (Huys *et al.*, 2002a).

Aeromonas veronii biovar *sobria* is being retained so that there is continuity in the literature and with biochemical identification schemes. This organism was previously called *A. sobria* but is actually genetically *A. veronii* (Working group on *Aeromonas* taxonomy, 1999, IJSB, 49: 1946). This species has been shown to be non-pathogenic to the common carp when tested in virulence studies (Kozińska *et al.*, 2002).

Aeromonas salmonicida ssp. *salmonicida*

These strains are characterized by pigment production and by being non-motile. They belong to *Aeromonas* DNA HG3. These virulent strains are termed 'typical' *A. salmonicida*, possess an A-protein layer and autoagglutinate in distilled water. However, these phenotypic tests are not reliable indicators of the *in vivo* virulence of *A. salmonicida* (Bernoth, 1990; Olivier, 1990). *A. salmonicida* ssp. *salmonicida* causes the disease known as furunculosis, so called because lesions in the dermis are seen as characteristic umbonate ulcers. The organism penetrates into the underlying tissues and organs. Furunculosis affects mainly salmonid fish.

The atypical *Aeromonas salmonicida*

The term 'atypical' is used to describe strains of *A. salmonicida* that show variations in biochemical reactions, may have slow growth, be slow to produce pigment, be negative for oxidase, and that may produce a variety of extracellular proteases. The pathogenicity of atypical strains shows great variability, and virulence mechanisms differ for those described for 'typical' *Aeromonas salmonicida*. An extracellular A-layer has been detected, an extracellular metallo-protease and a different iron utilization mechanism has been described for atypical strains compared to typical strains (Wiklund and Dalsgaard, 1998). Current thought suggests that the skin ulceration seen in the disease may be due to only one well-defined biotype of atypical *A. salmonicida*, yet these organisms cannot be isolated consistently. The organisms are most commonly isolated in the healing stage of the ulcer, suggesting that the infection is not lethal (Wiklund *et al.*, 1999).

A number of subspecies have been named, including *A. salmonicida* ssp. *masoucida*, ssp. *smithia*, ssp. *achromogenes*, ssp. *nova*, the latter isolated from non-salmonid fish. The subspecies of *masoucida* and *smithia* have not been reported since their initial isolation by Kimura (1969) and Austin *et al.* (1989). There are many strains of *A. salmonicida* reported in the literature that are atypical and do not fall into any of the above-mentioned subspecies (Wiklund and Dalsgaard, 1998). There is some contention about the validity of retaining *A salmonicida* ssp. and research

continues in order to understand and enable definitive identification of this diverse species. The atypical strains infect a wider variety of fish species than does *A. salmonicida* ssp. *salmonicida*.

Alteromonas genus

The genus *Alteromonas* was divided into two genera: *Pseudoalteromonas*, containing 11 species, and *Alteromonas*, which contains *A. macleodii* as the type strain. They grow on marine agar 2216 (Difco).

Brucella spp.

Brucella species are small cocco-bacilli that are intracellular in the cells of mammals. Traditionally they are classified according to host, as they are generally host-specific in their disease. Studies suggest that the *Brucella* genus is monospecific and that only one species, *Brucella melitensis*, should be recognized in the genus. It is proposed that the other six species be recognized as biovars thus: *B. melitensis* biovar *abortus*, *B. melitensis* biovar *canis*, *B. melitensis* biovar *melitensis*, *B. melitensis* biovar *neotomae*, *B. melitensis* biovar *ovis*, and *B. melitensis* biovar *suis* (Verger *et al.*, 1985). However, it is practical to retain the current species delineation to avoid confusion (Wayne *et al.*, 1987).

Recent research into the *Brucella* species isolated from marine mammals indicates that the monospecific species theory does not apply when these isolates are considered (Moreno *et al.*, 2002). The *Brucella* strains isolated from marine mammals appear to be a distinct species based on IS711-based DNA fingerprinting (Bricker *et al.*, 2000). The name *Brucella maris*, comprising three biovars, was suggested for the isolates from seals, porpoises, dolphins, an otter and a whale (Jahans *et al.*, 1997). PCR-restriction fragment length polymorphism of two outer membrane protein genes, *omp2* and *omp2b*, indicates that the isolates from sea mammals are a separate group from those isolated from terrestrial mammals. The isolates from the sea mammals appear to be a more heterogeneous group than those from terrestrial mammals and, therefore, instead of naming all aquatic isolates as *B. maris*, it is proposed that the isolates are named according to their host species in a similar manner to terrestrial isolates. As yet this has not been approved by the International Committee on Systematic Bacteriology. However, the proposal is that the aquatic strains are now named *B. pinnipediae* for strains isolated from seals, and *B. cetaceae* for strains isolated from cetaceans (whales, dolphins and porpoises) (Cloeckaert *et al.*, 2001; Foster *et al.*, 2002).

Carnobacterium spp.

Carnobacterium species can be differentiated from *Lactobacillus*, as *Carnobacterium* species fail to grow on acetate agar (Rogosa medium, Oxoid; Rogosa *et al.*, 1951). Discrepancies have been reported between conventional tube tests and the API 50CH system with results for inulin, lactose, sorbitol and galactose (Baya *et al.*, 1991). The *Carnobacterium* species are phylogenetically closer to the genera *Enterococcus* and *Vagococcus* than to members of the genus *Lactobacillus*.

   Carnobacterium (*Lactobacillus*) *piscicola* may have been reported as *Corynebacterium* spp., *Lactobacillus* spp., *Listeria* spp. or *Vagococcus salmoninarum* in earlier reports in the literature.

Cytophaga spp.

This genus now contains only two cellulolytic species: *C. hutchinsonii*, the type species, and *C. aurantiaca*. Their closest relative is the genus *Sporocytophaga*, containing *S. myxococcoides*, which is cellulolytic and produces microcysts (Bernardet *et al.*, 1996).

   Cytophaga aurantiacus, strain NCIMB 1382 described by Lewin and Lounsbery (1969), is thought to be a *Flavobacterium johnsoniae* (Bernardet *et al.*, 1996).

Edwardsiella tarda

This name is conserved for use, although *Edwardsiella anguillimortifera* is the more senior synonym. The organism can be found in the environment in pond water, lakes, rivers, soil and from the cloacae of crocodiles, lizards, seagulls, snakes, tortoises and waterfowl, and may be part of the normal flora or exist in a carrier status. It is a pathogen of many aquatic species including

catfish, trout, eels, and also causes enteritis in humans. Because it is widespread in the environment, the isolation of the organism from some aquatic animals may make it difficult to assess its pathogenic status. However, site of isolation, clinical signs, disease status and histopathological examination should enable a correct diagnosis (Wallace *et al.*, 1966; Otis and Behler, 1973; White *et al.*, 1973; Miyashita, 1984; Humphrey *et al.*, 1986; Reddacliff *et al.*, 1996; Uhland *et al.*, 2000).

Enterococcus spp.

Enterococcus faecalis was previously known as *Streptococcus liquefaciens* and *Streptococcus faecalis*.

   Enterococcus seriolicida was previously described as *Streptococcus* species biotype 1 from Australian trout in Victoria and Tasmania, and is now recognized as *Lactococcus garvieae*.

Enterovibrio norvegicus

This organism has recently been isolated and identified to be a member of the *Vibrionaceae*. It resembles a *Vibrio*, however, is resistant to the vibriostatic agent 0/129 but grows slowly after 3 days on TCBS. It is part of the intestinal microflora of turbot larvae (Thompson *et al.*, 2002a).

Flavobacterium/Flexibacter/Cytophaga

The *Flavobacterium* genus represents predominantly gliding, yellow-pigmented bacteria that belong to the *Flexibacter–Bacteroides–Flavobacterium* phylum. *Flavobacterium* is the type genus of the Family *Flavobacteriaceae*, which includes the genera *Bergeyella*, *Capnocytophaga*, *Chryseobacterium*, *Empedobacter*, *Gelidibacter*, *Myroides*, *Ornithobacterium*, *Polaribacter*, *Psychroflexus*, *Psychroserpens*, *Riemerella*, *Weeksella*, and misclassified *Cytophaga* and *Flexibacter* species. Members of *Flavobacterium* genus are aerobic, Gram-negative rods, produce yellow-pigmented colonies, are motile by gliding, do not hydrolyse cellulose and are widely distributed in soil and freshwater habitats. Some are pathogenic for fish (Bernardet *et al.*, 1996).

Flavobacterium columnare: Genetic diversity has been reported amongst strains of *Flavobacterium columnare*, which enables division into three groups or genomovars. However, the phenotypic characteristics are identical for the three genomovars and there is no evidence to classify the groups as new species or subspecies (Triyanto and Wakabayashi, 1999). A genomovar is defined as phenotypically similar but genetically distinct (Ursing *et al.*, 1995).

Flexibacter aurantiacus (Lewin and Lounsbery, 1969) strains NCIMB 1382 (previously *Cytophaga aurantiaca*) and NCIMB 1455 (*Flexibacter psychrophilus*) are now thought to be *Flavobacterium johnsoniae* strains, and have been shown to be different from the *bona fide* strains of *Cytophaga aurantiaca* and *Flavobacterium psychrophilum* (Bernardet *et al.*, 1996).

The *Flexibacter* genus contains one species, *F. flexis*. The only genuine members of the *Cytophaga* genus are *C. hutchinsonii* (the type species), and *C. aurantiaca* (excluding strains NCIMB 1382 and NCIMB 1455).

Hafnia alvei

Hafnia alvei is known to cross-react with *Yersinia ruckeri* antisera (Stevenson and Airdrie 1984; and personal observation).

Haemophilus piscium

Strains originally identified as *H. piscium* have been included with strains of *Aeromonas salmonicida*. Phenotypic evidence suggests that the strain is synonymous with *A. salmonicida* ssp. *achromogenes*, yet molecular studies involving ribotyping, RAPD and PCR grouping did not support this (Austin *et al.*, 1998).

Helicobacter spp.

Since *Helicobacter pylori* was first isolated as a cause of gastric ulcer in humans (Marshall and Warren, 1984), the genus has expanded to include some 30 species isolated from the gastrointestinal tract of many different animals. The organisms are microaerophilic and the cells are fusiform, curved or spiral, and possess flagella

in different numbers and location according to the species.

Listonella spp.

There is contention in the scientific community regarding the validity of placing *V. anguillarum* into the genus *Listonella* as proposed by MacDonell and Colwell (1985). There is, however, agreement of the separation of *V. anguillarum* biotypes 1 and 2 into two species, as *V. anguillarum* and *V. ordalii*, respectively (Austin *et al.*, 1997). In this book, *V. anguillarum* is referred to as *Listonella anguillarum*.

Listonella anguillarum

This pathogen causes disease in a wide range of fish. It is of particular economic importance to the salmonid industry, as salmonids are particularly susceptible to disease when being transferred from freshwater to salt water.

Listonella anguillarum isolates can be divided into ten serotypes (European serotype designation) based on the detection of the O antigen. Of these, only serotypes 01 and 02 have been isolated from diseased fish, following outbreaks at a distribution rate of 70% and 15%, respectively. In 75% of vibriosis infections in feral fish, serotype 02 is isolated. The other serotypes have only been isolated from the environment (Sørensen and Larsen, 1986). Serotype 01 strains tend to be arabinose-positive. Isolates from serotype 01 are a homogeneous group, whereas isolates from serotype 02 can be further divided into groups 02α and 02β based on differences detected by double immunodiffusion, dot blot assay and enzyme-linked immunoabsorbant assay (Toranzo and Barja, 1990). *Vibrio ordalii*, previously classified as *L. anguillarum* biotype II, cross-reacts with serotype 02 antiserum (Toranzo *et al.*, 1987).

Listonella pelagia strain NCIMB and NCMB 1900: Recent phylogenetic analysis of NCMB 1900 has indicated that this strain is more likely to be *Vibrio natriegens*, and that the original strain deposited at NCIMB may have been lost. Phenotypic tests for strain NCIMB 1900 corresponded to those expected for *V. natriegens*, which is positive for acid production from L-arabinose, melibiose, and L-rhamnose, with a negative result for lactose fermentation and alginate degradation (Macián *et al.*, 2000). The phenotypic results recorded as *L. pelagia* NCIMB 1900 by Lunder *et al.* (2000) have been recorded as *V. natriegens* in this book.

Listonella ordalii = Vibrio ordalii

Both names are used in the literature, although *Vibrio ordalii* is officially used in taxonomy databases.

Moritella marina and Moritella viscosa

Moritella marina and *Moritella viscosa* are closely related and have a sequence similarity of 99.1% based on 16S rDNA sequencing (Benediktsdóttir *et al.*, 2000).

Mycoplasma spp.

Mycoplasma phocicerebrale is a legitimate correction to the previously reported name of *M. phocacerebrale*. The correction of *Mycoplasma phocae* to *M. phocidae* is not legitimate (de Vos and Truper, 2000). *Mycoplasma phocirhinis* is a legitimate correction from *M. phocarhinis*.

Nocardia spp.

Members of the *Nocardia* produce branched substrate hyphae that fragment *in situ* into rod-shaped to coccoid non-motile elements. Aerial hyphae are numerous and are always present but some may only be seen microscopically. Hyphae on a 3-day-old colony may only be 1 mm in length.

Photobacterium spp.

The species that should be contained in this genus is still in dispute. There is agreement of some species rightfully belonging to the genus (*P. angustum*, *P. phosphoreum*, *P. leiognathi*). However, other species are in dispute, such as *V. fischeri*, *V. logei* and *V. damselae* (Lunder *et al.*, 2000) although placing *V. damselae* into the genus *Photobacterium* is generally agreed upon.

It is agreed that *Photobacterium damselae* ssp. *piscicida* belongs to the genus

Photobacterium rather than *Pasteurella* where it was first placed. However, there is still controversy as to whether *P. damselae* ssp. *piscicida* is a subspecies of *P. damselae* or a species in its own right (Gauthier *et al.*, 1995b; Thyssen *et al.*, 1998).

Photobacterium histaminum is a later subjective synonym of *P. damselae* ssp. *damselae* (Kimura *et al.*, 2000).

Photobacterium fischeri and *Vibrio fischeri* are homotypic synonyms.

Providencia rettgeri and Proteus rettgeri

Providencia rettgeri and *Proteus rettgeri* are homotypic synonyms, which means they share the same type strain (Brenner *et al.*, 1978).

Pseudoalteromonas species

See also under *Alteromonas*. The genus *Pseudoalteromonas* was formed to include most of the previously named *Alteromonas* species with the exception of *A. macleodii*. The majority of *Pseudoalteromonas* are associated with eukaryotic hosts and are frequently isolated from marine waters. The majority grow on MA 2216 and show optimal growth at 23°C.

Shewanella putrefaciens

Strain NCIMB 400 has been reclassified as *Shewanella frigidimarina*.

Sphingobacterium spp.

Members of the *Sphingobacterium* genus have a low G+C content (39–42 mol %) and contain sphingolipids.

Streptococcus spp.

Streptococcus agalactiae is a Lancefield group B streptococcus and is non-haemolytic. There are nine serogroups of group B streptococcus based on their capsular polysaccharide.

Streptococcus dysgalactiae consists of two species. *Streptococcus dysgalactiae* ssp. *dysgalactiae* strains are α- or non-haemolytic and are Lancefield group C. Strains belonging to *Streptococcus dysgalactiae* ssp. *equisimilis* are all β-haemolytic and may have the Lancefield group C, G or L antigen (Vieira *et al.*, 1998).

Streptococcus difficile has now been identified as *Streptococcus agalactiae* type Ib (Vandamme *et al.*, 1997).

Vagococcus genus

The genus *Vagococcus* was proposed to accommodate Gram-positive cocci that were motile and reacted with Lancefield group N antisera (Collins *et al.*, 1989).

Vibrio spp.

Members of the Family *Vibrionaceae* are *Colwellia*, *Listonella*, *Moritella*, *Photobacterium* and *Vibrio*. Those isolated from fish grow well on ordinary plate media.

Vibrio carchariae and *V. harveyi*: *V. carchariae* is a junior synonym of *V. harveyi* and the two organisms were indistinguishable by amplified fragment length polymorphisms, DNA : DNA hybridization, ribotyping (Pedersen *et al.*, 1998) and 16S ribosomal DNA sequencing (Gauger and Gómez-Chiarri, 2002).

Vibrio cholerae: A biotyping scheme was based on the fermentation of sucrose, D-mannose and L-arabinose. These biotypes are known as Heiberg types I–VIII. A classification scheme was suggested where *V. cholerae* isolated from cholera patients was classified as 01, and all other strains were classified as non-01. Non-01 strains cause gastroenteritis and septicaemia and other infections in humans, and have also been reported to cause infections in terrestrial and aquatic animals. There are two biotypes, El Tor and Classical; these cannot be differentiated serologically and belong to serovar 01. Serovar 01 causes the classical cholera epidemics seen in humans in some countries. Serotype 0139 Bengal is also an epidemic strain of cholera (Furniss *et al.*, 1978; Albert *et al.*, 1993).

Vibrio anguillarum: See notes on *Listonella anguillarum*.

Vibrio natriegens: There are some reports in the literature identifying strain NCIMB and NCMB 1900 as *Listonella pelagia*. Sequencing has shown that strain 1900 corresponds to *Vibrio natriegens* (Macián *et al.*, 2000). See also notes under *L. pelagia*.

Vibrio trachuri: Isolated from Japanese horse mackerel (*Trachurus japonicus*) and identified as a new species of *Vibrio* (Iwamoto *et al.*, 1995a). However, further testing has suggested that this species is a junior synonym of *Vibrio harveyi* (Thompson *et al.*, 2002b).

Vibrio vulnificus: Biotyping and serotyping of *Vibrio vulnificus* is a little confusing at present. Biotype 1 isolates are indole-positive and associated with human infection, whereas biotype 2 isolates are indole-negative and associated with infection in eels (Tison *et al.*, 1982). However, in Denmark and Sweden 85% of biotype 2 isolates showing pathogenicity for eels are indole-positive (Høi *et al.*, 1998b). Biotype 2 strains are serovar E and these can be further divided into 03 and 04 based on LPS-O antigen. Some biotype 2, serovar E, strains from Taiwan have been found to be avirulent for eels. These strains are positive for mannitol fermentation (Amaro *et al.*, 1999). *Vibrio vulnificus* biotype 3 has been put forward by the Centre for Disease Control, USA, as a possible identification for *Vibrio* species isolated from wound infections in humans in Israel (Bisharat *et al.*, 1999). It is different from currently recognized *V. vulnificus* species because of negative reactions in cellobiose, citrate, lactose, salicin and ONPG. It is positive by PCR for the *V. vulnificus* haemolysin gene (Bisharat *et al.*, 1999; Nair and Holmes, 1999). In the conventional biochemical identification tables and the API 20E tables, the various *V. vulnificus* strains and their different reactions are listed so as to assist the microbiologist when making identification.

Vibrio wodanis has a sequence similarity of 98.8% with *Vibrio logei* strain CIP 103204 based on 16S rRNA sequencing (Benediktsdóttir *et al.*, 2000).

Yersinia spp.

Yersinia enterocolitica: *Y. enterocolitica* was divided into the subspecies *Y. enterocolitica* ssp. *enterocolitica* and *Y. enterocolitica* ssp. *palearctica*. *Yersinia mollaretii* and *Y. bercovieri* were formerly *Y. enterocolitica* biogroups 3A and 3B, respectively (Wauters *et al.*, 1988).

Yersinia ruckeri: An original serotyping scheme identified six serovars (Hagerman serovars). Most naturally occurring outbreaks in diseased fish and rainbow trout are caused by serovar I called the Hagerman strain (McCarthy and Johnson, 1982). Some Australian isolates of *Y. ruckeri* that cross-reacted with serovar I were termed serovar I'. Serovar II may be significant when it is associated with ERM in chinook salmon (Cipriano *et al.*, 1986). Serovar II and serovar V isolates are sorbitol-positive. Serovar III is found in Australia. Initially it was suggested that sorbitol fermentation was associated with pathogenicity; however, this is not a reliable indicator of virulence (Stevenson and Airdrie 1984; Cipriano *et al.*, 1986).

More recently it has been shown that *Y. ruckeri* can be divided into clonal groups based on biotype, serotype and outer membrane proteins, and that clonal groups may relate to virulence of the organism. There are two biotypes based on hydrolysis of Tween 20 and Tween 80, and motility. Biotype 2 strains are non-motile, and do not hydrolyse Tween 80 (Davies and Frerichs, 1989). A serotyping scheme based on heat-stable O-antigens identified serotypes 01, 02, 05, 06 and 07 in Europe, North America, Australia and South Africa (Davies, 1990). All serotypes occur in Europe and North America, whereas in Australia and South Africa only serotype 01 has been identified. This study suggested that the Australian isolate described as serotype III was serotype 01 by their scheme. Five outer membrane protein profiles were identified, which enabled differentiation of isolates within a serotype. Serotype 01 consisted of six clonal groups. Clone 2 was found in disease outbreaks in the UK and clone 5 was found in disease outbreaks in mainland Europe, North America and South Africa. Clones 3, 4 and 6 do not appear to be associated with disease outbreaks. Clones 1 and 3 were found from Australian isolates, with clone 3 also being found in Finland, France, West Germany and the USA. Clone 4 was from Norwegian isolates, and clone 6 from isolates in Finland, France, Norway and Canada (Davies, 1991).

Hafnia alvei is known to cross-react with *Yersinia ruckeri* antisera (Stevenson and Daly, 1982; Stevenson and Airdrie 1984; and personal observation).

ERM outbreaks caused by *Yersinia ruckeri* are generally a result of poor environmental conditions, resulting in low oxygen level, high water temperature and poor water quality, all of which stress the fish, making them more susceptible to disease.

2

Bacteriological Culture Techniques: Microscopy, Culture and Identification

Successful isolation and accurate identification of a suspected pathogen depends upon the use of standardized bacteriological culture methods. It is important to work through these methods in a precise and logical step-by-step manner. Basically the bacteria are grown on culture media, then inoculated into biochemical identification tests and those results are recorded and compared to standard results. Table 2.1 sets out the steps for the isolation and identification of an organism and also directs the reader to the location of the technique or method in this manual.

Table 2.1.   Outline of steps for culture and identification.

Day	Activity	Method or technique
Day 1	1. Sample collection and preparation	2.1
	2. Inoculation of sample to primary isolation plate media (or broths where appropriate)	2.2, Table 2.2, Table 2.3
	3. Incubate at appropriate temperature and atmosphere	2.2, Table 2.2, Table 2.3
Day 2 (24 h)	1. Examine culture plates	2.3
	2. Select suspect colonies and subculture to BA or MSA-B to obtain pure growth (secondary plates)	Table 2.4 for cultural and microscopic appearance
	3. Re-incubate primary plates	2.2, Table 2.2 and 2.3
	4. Incubate all plates at appropriate temperature and atmosphere as before	
Day 3 (48 h)	1. Re-examine primary plates for slow-growing pathogens	2.3
	2. Check that subcultures on secondary plates are pure	
	3. Perform primary identification tests	2.4, Chapter 3, and media (Chapter 7)
	4. Inoculate appropriate biochemical identification set	2.5, 2.6
	5. Re-incubate primary plates as before	2.2, Table 2.2 and 2.3
Day 4 (72 h)	1. Re-examine primary plates for growth of slow-growing pathogens. Re-incubate if disease suggests a pathogen that requires more than 3 days for growth	2.3
	2. Examine biochemical identification set/s and record results at 24 h incubation	Table 3.1
Day 5 (96 h)	1. Examine biochemical identification set/s and record results for 48 h incubation. Add reagents for tests for indole, methyl red, nitrate, Voges-Proskaüer	Table 3.1 Chapter 3 and Tables 4.1 to 4.22 for biochemical results (Biochem set), and
	2. Interpret results from appropriate identification table	Tables 4.23 to 4.31 for results for API kits

BA, blood agar; MSA-B, marine salt agar.

2.1 Specimen Collection and Submission

Samples should be collected from live sick animals or from recently dead animals. Animals that have been dead for more than 6 h usually have grossly contaminated organs due to the overgrowth of post-mortem invading bacteria. It is very difficult for the laboratory to work with these samples. Unfortunately, it is not always possible to get the samples to the laboratory within the ideal time, due to large distances between some laboratories and the aquaculture farm.

Ideally all samples for bacteriological examination should be collected before the start of antimicrobial treatment. Samples collected from aquatic animal species after commencement of treatment are unlikely to grow any pathogenic bacteria.

Samples may be submitted as swabs, tissues or whole fish. The most appropriate samples for the identification of a systemic disease may be kidney, haemolymph or blood. For skin disease, swabs or samples of skin are appropriate. Refer to Tables 1.1 and 1.2 for tissue sites where the bacteria are likely to be located.

Transport medium and transport to the laboratory

If swabs have been collected, then they must be placed into a transport medium to prevent the swab drying out and for the organisms to remain viable. Amies transport medium is the medium of choice for transport of swabs to the laboratory.

Fish, tissues and swabs in Amies transport medium should be transported to the laboratory as soon as possible after collection, and preferably on ice or at 4°C.

Stuarts medium may be unsuitable for transport and survival of *Vibrio* species. Numbers of viable cells of *Vibrio alginolyticus*, *Vibrio harveyi* and *Photobacterium damselae* ssp. *damselae* were greatly reduced after 24 h and were almost non-viable at 48 and 72 h when tested from Stuarts media kept at 4°C. However, there was no reduction of viable cells when all three organisms were tested from Amies transport media held at 4°C for 24, 48 and 72 h (H. McLetchie and N. Buller, unpublished).

Sample preparation

Swabs

Swabs require no sample preparation, and are inoculated directly to the appropriate media.

Tissues

Aseptically remove a piece of infected tissue from the fish and place into a McCartney bottle. Macerate the tissue with flame-sterilized scissors, and use a sterile, cotton-tipped swab to inoculate to appropriate media. Tissues that have not been aseptically removed from the fish need to be surface-sterilized. This can be done either by rinsing the piece of tissue in 3–4 washes of sterile water or saline, or for larger pieces of tissue (walnut-sized), dip in 70% alcohol and flame briefly for 3–5 s. The tissue is then placed into a sterile McCartney bottle or appropriate container and macerated with flame-sterilized scissors. A sterile, cotton-tipped swab is used to sample the tissue and inoculate to appropriate agar plates, and to make smears for Gram stain.

Lesions

Always collect material from the edge of the lesion, including some of the immediate tissue, to ensure sampling for true invading bacteria. Sampling from the middle of the lesion may culture secondary invading bacteria that will confuse the results. Also, the invading bacterium may no longer be viable at the centre of the lesion.

Whole fish

For small fish, the whole fish is surface-sterilized by rinsing the fish in 3–4 changes of sterile distilled water or normal saline. For larger fish, surface-sterilize by either flaming with 70% alcohol or washing the area with 3–4 washes of sterile distilled water or normal saline. Using flamed-sterilized forceps and scissors, make an incision from behind the gills and to the mid-line. Next cut along the mid-line towards the anus. Lift back the flap of flesh to expose the internal organs. Select the required organs for analysis and remove with flame-sterilized forceps and scissors. Chop tissues as before and inoculate to plates.

2.2 Culture and Incubation

Culture media

All specimens should be cultured to a general-purpose medium. For freshwater specimens use BA and MCA. TSA can also be used; however, the addition of blood can improve the isolation of some organisms such as *Aeromonas salmonicida*, *Carnobacterium* and *Streptococci*. For marine specimens, use a medium containing Na ions, BA, and MCA. Marine salt agar (MSA-B), which is prepared from TSA with 5% horse blood and 2% NaCl (see Media section), is a good general-purpose medium for marine organisms. Marine 2216, commercially available from Difco, and based on ZoBell's (1941) original formula, is also suitable. MSA-B may give improved growth, particularly on primary culture with commonly isolated *Vibrio* species. MA 2216 is more suitable when a complex mixture of electrolytes is required by some marine organisms.

Isolation and growth conditions: media, temperature and time of incubation

All specimens should be inoculated to the general-purpose medium. This medium is then incubated at 25°C for 2–5 days as indicated in Table 2.2 for general culture conditions. However, specimens from a particular environment such as a cold water environment or a marine environment may need a particular incubation temperature or NaCl may be required in the medium. These variations are listed in Table 2.3.

Some organisms have special growth requirements or may be better detected with the use of a selective and/or enrichment medium if they are available for that organism. Such requirements for specific organisms are detailed in Table 2.3.

The preparation of all media mentioned in Tables 2.2 and 2.3 is listed in Chapter 7.

2.3 Examination of Culture Plates

Many bacteria from fish and aquatic animals need 48 h before colonies appear, or before they are of a suitable size for examination and subculture. At 24 h incubation, the colony appearance of different *Vibrio* species may not be distinguishable; however, at 48 h incubation, the differences in physical appearance between different species may be more apparent. The addition of blood into a medium such as MSA-B, for *Vibrio* species, improves the ability to differentiate between the colonies of different species, and to determine haemolysis. Generally, a bacterium that is present in a moderate to heavy growth or as the predominant organism will be significant. Bear in mind the age of the samples at the time they were received at the laboratory. Specimens older than 24 h from the collection date may be overgrown with commensal bacteria.

Select appropriate colonies for subculture. These may be given an individual number (e.g. #1, #2, etc). For example, selected colonies are circled and numbered on the underside of the Petri dish using a felt pen, and then picked off and subcultured to fresh plates. Freshwater samples are subcultured to BA, MCA and TCBS (the latter for suspect *Vibrio* species). Samples from marine sources are subcultured to BA, MCA, MSA-B and TCBS. Organisms that require special media for growth are subcultured to the appropriate media (see Table 2.3). Some organisms from marine environments may need to be cultured to MA 2216 media. Incubate a further 24 h or as required for sufficient growth. Pure subculture growths are used for the inoculation of biochemical identification tests.

Microscopic and cultural characteristics

Table 2.4 details microscopic and colony appearance, plus the results of the primary tests of Gram

Table 2.2.   General culture.

Fish, aquatic animals	Media – freshwater animals	Media – salt water animals	Temperature optimum (range)
All specimens, tissues	BA, MCA	MSA-B, BA, MCA	20–25°C, 2–5 days
Cold water	BA, MCA	MSA-B, BA, MCA	15–22°C, 3–5 days
Tropical	BA, MCA	MSA-B, BA, MCA	25°C, 2–5 days

BA, blood agar; MCA, MacConkey agar; MSA-B, marine salt agar containing blood.

Table 2.3. Specific culture requirements of organisms.

Bacterium	Media for freshwater animals	Media for salt water animals	Temperature, atmosphere, time
Abiotrophia balaenopterae	BA		37°C, 5% CO_2 or air, 24–48 h
Actinomyces marimammalium	BA		Air or 5% CO_2, 24–72 h
Aequorivita spp.		MA 2216 NA, TSA, AO with either 2.5% NaCl or 35 g/l seawater salts added	20°C, 3–7 days
Aerococcus viridans var. *homari*		NA+5% blood, TSA, BA, BHIB SIEM selective media	30°C, 24–48 h
Aeromonas eucrenophila, A. encheleia, A. popoffii	TSA, BA		28°C, 24–48 h
Aeromonas salmonicida	BA, TSA-B Examine for brown pigment at 5–7 days from a subculture on FA. Isolation of 'atypical' strains is improved using BA or TSA-B as the primary isolation medium	TSA with 0.5% NaCl	15–22°C, 3–5 days Incubation temperature is critical, as differences in biochemical reactions have been noted at different temperatures (Hahnel and Gould, 1982)
Alcaligenes faecalis var. *homari*		Marine 2216	18°C for 7 days
Alteromonas		Marine 2216 Growth is poor on TSA and NA even with added salt	Strict aerobes, 25°C, 24–48 h
Aquaspirillum spp.	TSA, BA		25°C, 2–4 days
Arcanobacterium spp.	BA		37°C, 5% CO_2 or air, 24–48 h
Arthrobacter spp.	BA		37°C, 5% CO_2 or air, 24–48 h
Beneckea chitinovora	BA		25°C, 24–48 h
Bordetella bronchiseptica	BA, CFPA, MCA		37°C, 48–72 h
Brucella spp. **Use biological safety cabinet**	TSA, BA, Farrell's medium (FM) Isolates from seals either don't grow on FM or appear after 14 days		37°C, 5–10% CO_2, 3–5 days. Isolates from seals and otters, and some isolates from cetacean species require 10% CO_2
Burkholderia pseudomallei **Use biological safety cabinet**	BA, glycerol plates, selective broth, Ashdown's medium		37°C, 1–4 days
Carnobacterium inhibens, Carnobacterium piscicola	BA, TSA, BHIA		15–25°C, 48–72 h
Chryseobacterium (Flavobacterium) meningosepticum	TSA		30–37°C, 48 h

Chryseobacterium scophthalmum	TSA, AO	Medium K, MA 2216, AO-M	25°C, 48 h
Clostridium botulinum	Anaerobe. ANA plates. Identified by testing for toxin production after inoculation into Robertson's cooked meat broth		Incubate broths for 6 days at 30°C prior to testing for toxicity
Corynebacterium aquaticum	BA, TSA, BHIA		25°C, 48–72 h
Cytophaga hutchinsonii	Dubos medium supplemented with 30% (wt/vol) D-cellobiose		22°C, 48–72 h
Cytophaga latercula		Marine Flexibacter medium (Lewin and Lounsbery, 1969)	
Dermatophilus chelonae	BA, poly plate		25°C, 2–5 days.
Dermatophilus congolensis	BA, poly plate		37°C, 2–3 days
Edwardsiella ictaluri	BA, TSA, EIM		25–37°C, aerobic and anaerobic. Strains with limited tolerance to oxygen have been isolated (Mitchell and Goodwin, 2000)
Edwardsiella tarda	BA, MCA, DCA, SS agar, strontium chloride B enrichment broth		37°C, 1–2 days
Edwardsiella spp.	BA, BHIA, NA		25–28°C, 2–5 days
Empedobacter (Flavobacterium) brevis	TSA		30°C, 2–3 days
Enterococcus spp.	BA, TSA, BHIA		25°C, 48–72 h
Enterovibrio norvegicus		MA 2216 or NA and TSA supplemented with 1.5% NaCl	27–28°C, 2 days
Erysipelothrix rhusiopathiae	BA, Woods selective broth, Packers plate		37°C, 48 h. Subculture Woods broth to BA and Packers plates at 24 and 48 h
Eubacterium tarantellae	BHIB or BA (anaerobe), ANA		20–25°C, 7 days, anaerobic conditions
Facklamia miroungae	BA	MSA-B	37°C, 24–48 h
Flavobacterium branchiophilum	AO (no growth on media containing >0.2% NaCl)		18°C, 5–7 days
Flavobacterium (Flexibacter) columnare	AO, Shieh medium with added tobramycin, TSA, NA, BA, TYG agar, Hsu-Shotts agar		22–30°C, 5–7 days
Flavobacterium (Cytophaga) psychrophilum	TYG agar, Shieh medium, AO agar (no growth on TSA). Sample from mucus and kidney. Culture dilutions of sample tissue to improve isolation rate. Use of FPM medium may improve isolation rate		14–18°C, 5–7 days

continued

Table 2.3. *Continued.*

Bacterium	Media for freshwater animals	Media for salt water animals	Temperature, atmosphere, time
Flavobacterium gillisiae, F. tegetincola, F. xanthum	NA, TSA, R2A	MA 2216	20°C, 2–7 days
Flexibacter spp. – marine origin		AO-M (AO medium prepared with artificial seawater (Sigma) at 38 g/l)	25°C, 2–7 days
Flexibacter polymorphus	Flexibacter medium of Lewin (1974)		30°C, 2–7 days
Flexibacter roseolus	NA		25°C, 2–7 days
Flexibacter ruber	NA		25°C, 2–7 days
Halomonas elongata		TSA+8% salt	30°C, 24–48 h
Helicobacter spp.	TSA + 5% blood, BA, Brucella agar, Skirrow's medium (VPT) Homogenize stomach tissue in 1 ml Brucella broth (Difco) containing 5% faecal calf serum. Place 100 µl to plates and/or filter through a 0.45 or 0.8 µm filter before inoculating to plates (Butzler *et al.*, 1973; Harper *et al.*, 2000)		37°C microaerophilic atmosphere of N$_2$, H$_2$, CO$_2$ (80:10:10) for 2–4 weeks Commercial systems are available for generation of correct atmosphere (e.g. MGC Anaero Pak™, Campylo from Mitsubishi Gas Chemical Company)
Lactobacillus spp., *Lactococcus* spp.	BA, TSA, BHIA		25–30°C, 48–72 h
Listonella anguillarum		TSA + 1% NaCl, MSA-B, MA 2216, VAM	22°C, 24–48 h
Moritella viscosa		TSA + 2% NaCl	15°C, 4–9 days
Mycobacterium abscessus	Middlebrook 7H10-ADC medium		25°C, 7–28 days
Mycobacterium chelonae	BHIA, TSA, AO	MSA-B	15–22°C, 5 weeks. (growth is usually seen within 7 days)
Mycobacterium marinum, M. scrofulaceum, M. intracellulare, M. fortuitum	BA, Lowenstein-Jensen (BBL)		20–23°C, 7–14 days Will grow in 5–7 days at room temperature on bovine BA
Mycobacterium neoaurum	BA		25°C, 3–4 days
Mycobacterium peregrinum	BA, Middlebrook 7H11 agar (Difco)		37°C in CO$_2$, 4–7 days
Mycoplasma species	Mycoplasma agar and Mycoplasma broth		37°C in CO$_2$. Subculture broth to agar medium every 3–4 days for 2 weeks. Examine plates at 3–4 day intervals
Myroides (Flavobacterium) odoratus	TSA		30°C, 48–72 h

Organism	Media	Conditions
Nocardia spp. - general (*N. asteroides, N. brasiliensis, N. nova, N. pseudobrasiliensis, N. otitidiscaviarum, N. seriolae, N. vaccinii*)	BA	25–30°C, 14 days. Colonies begin to appear after 5 days
Nocardia crassostreae	BHIA	BHIA + 1% NaCl	28°C, 14 days
Nocardia seriolae	BA, BHIA, Lowenstein–Jensen medium		25–37°C for 7–30 days
Pasteurella skyensis		TSA + blood + 1.5% NaCl. MA 2216 + blood (no growth without blood or 1.5% NaCl)	22–30°C, in air, 48 h
Pedobacter heparinus, P. piscium	NA, PY		28°C, 48 h
Photobacterium damselae ssp. *damselae*	BA	MSA-B, TSA + 2% NaCl, MA 2216	22–25°C, 24–48 h
Photobacterium damselae ssp. *piscicida*		NB with 2–3% NaCl, BHIA with 2% NaCl, MA 2216, or MSA-B	22°C, 24–48 h
Pseudoalteromonas spp.		MA 2216	15–30°C, 1–5 days
Pseudoalteromonas antarctica	TSA, TSB	MA 2216, MSA-B	15°C, 5 days
Pseudoalteromonas maricaloris	TSA	MA 2216	25–35°C, 5 days
Pseudomonas anguilliseptica	BA, NA, TSA, BHIA. No growth on Pseudomonas isolation agar (Difco)		20–25°C, 7 days
Pseudomonas fluorescens	BA, Pseudomonas selective agar (Becton, Dickinson Co)		10–25°C, 2–5 days
Renibacterium salmoninarum	KDM2, KDMC, SKDM		15–18°C, 20–30 days
Roseobacter strain CVSP		SWT, MA 2216	23°C, 5–7 days
Salegentibacter salegens		MA 2216, NA, TSA, R2A	20°C, 2–7 days
Shewanella benthica, S. colwelliana, S. hanedai, S. gelidimarina, S. pealeana, S. woodyi		MA 2216, MSA-B	25°C, 48–72 h
Shewanella (*Alteromonas*) *hanedai*		TSA + 0.05 g/l yeast extract, MA 2216	15°C, 48–72 h
Sphingobacterium (*Flavobacterium*) *multivorum, S. spiritivorum*	NA, TSA, PY		28–30°C, 48 h
Staphylococcus lutrae	BA		37°C, 24 h
Staphylococcus warneri	BA, TSA		22–25°C, 48–72 h

continued

Table 2.3. *Continued.*

Bacterium	Media for freshwater animals	Media for salt water animals	Temperature, atmosphere, time
Stappia stellulata-like strain M1		SWT, MA 2216	23–25°C, 2–7 days
Streptobacillus moniliformis	BA. Requires addition of 20% serum to broth for growth		25–37°C, 5–7 days
Streptococcus iniae	BA		25°C, 24–48 h
Streptococcus parauberis	BA, TSA		22–37°C, 24 h
Tenacibaculum (Flexibacter) maritimum		AO media containing 30% seawater, AO containing ASW at 38 g/l. Will not grow on AO with NaCl alone. NaCl and KCl ions required for growth. Ca^{2+}, Mg^{2+} enhance growth. Isolation of organism may be improved by diluting sample material in artificial seawater and culturing dilutions to AO + ASW medium (Ostland *et al.*, 1999). TYG-M, HSM	25°C for 2–5 days (range 15–34°C)
Tenacibaculum (Flexibacter) ovolyticum	BA, TSA	As for *T. maritimum*, MA 2216	19°C, 2–5 days
Vagococcus salmoninarum	BA, TSA		22–25°C, 48 h
Varracalbmi		MSA-B	4–22°C (optimum 15°C), 48 h
Vibrio agarivorans		MSA-B, MA 2216	25°C, 48 h
Vibrio coralliilyticus		MA 2216, TSA + 2% NaCl	30°C, 24–48 h
Vibrio halioticoli		MA 2216 with or without 0.5% sodium alginate	25°C, 24–48 h
Vibrio hollisae		TSA + 1% NaCl, MSA-B, MA 2216	25°C, 37°C, 24–48 h
Vibrio mimicus	BA, TSA	MSA-B	25°C, 37°C, 24–48 h
Vibrio ordalii, Vibrio parahaemolyticus		TSA + 1% NaCl, MSA-B, MA 2216	22°C, 24–48 h
Vibrio salmonicida		NA + 5% blood, TSA + 1.5% NaCl, MSA-B	15°C, 3–5 days
Vibrio vulnificus	BA	Cellobiose-polymyxin B colistin agar (CPC) selective agar, CCA, VVM, MA 2216, MSA-B, TSA or NA with 0.5% (w/v) NaCl. A *V. vulnificus*-specific probe is available	25°C, 24 h
Yersinia ruckeri	TSA, BA, MCA, XLD (*Yersinia* selective agar, (YSA, Oxoid) is not suitable for *Y. ruckeri*)		25°C, 24–48 h Note that motility and citrate will be negative at 37°C but positive at 25°C

Table 2.4. Microscopic and cultural characteristics.

Bacterium	Gram	Morphology	βH	TCBS	Colony characteristics	Cat	Ox	Identification sets
Acinetobacter spp.	Neg	Predominantly diplococcal forms 1.0 × 0.7 μm from plate growth	–		Colonies circular, smooth, translucent to slightly opaque, butyrous to mucoid non-pigmented. 0.5–2 mm at 24 h at 30°C	+		Biochem set
Acinetobacter haemolyticus	Neg	Predominantly diplococcal forms 1.0 × 0.7 μm from plate growth	βH		Colonies on TSA are circular, convex, smooth, slightly opaque and may have a sticky consistency	+	–	Biochem set
Actinobacillus delphinicola	Neg	Pleomorphic rods	– w	NG	Colonies on BA in 10% CO_2 are 0.75–1 mm at 24 h, round, smooth grey. Blood or serum enhances growth. Growth at 42 but not at 22°C	–	+	Biochem set, API 20E, API 20NE, API-ZYM. Use a heavy inoculum
Actinobacillus scotiae	Neg	Pleomorphic rods	–	NG	Colonies 0.5 mm round, grey, on BA at 24 h. Requires 10% CO_2. May be weakly haemolytic on sheep blood	–	+	Biochem set, API 20E
Actinomyces marimammalium	Pos	Straight or slightly curved rods, some branching	–		On BA, colonies are 0.5 mm grey, entire, convex at 48 h. Growth at 37°C in air with 5% CO_2	–		API Coryne, API rapid ID 32 Strep
Aequorivita spp.	Neg	Rods 0.5–20 × 0.2–0.3 μm			On MA 2216, yellow or orange colonies, compact, circular, convex, smooth with an entire edge, non-spreading, butyrous consistency. Growth on NA + 2.5% NaCl			Biochem set, API 20E, API-ZYM (some reactions)
Aerococcus viridans var. homari	Pos	Tetrad-forming coccus	α	NG	αH on BA. Examine with India ink for encapsulated cocci	+	+	Biochem set, API rapid ID 32 Strep
Aeromonas culicicola	Neg	Rod			On BA colonies small, grey, metallic sheen and β haemolytic at 24 h	+	+	Biochem set, API 20E, API 50CH
Aeromonas encheleia	Neg	Straight rods	βH	NG	Non-pigmented colonies on TSA at 24 h	+	+	Biochem set
Aeromonas eucrenophila	Neg	Straight rods		NG		+	+	Biochem set
Aeromonas hydrophila	Neg	Small rods	βH	Weak	3–5 mm shiny, cream, becoming pale brown/green with age	+	+	Biochem set, API 50CH, API 20E, API 20NE, API-ZYM

continued

Table 2.4. *Continued.*

Bacterium	Gram	Morphology	βH	TCBS	Colony characteristics	Cat	Ox	Identification sets
Aeromonas media	Neg	Rods 1 × 2 µm		NG	On TSA, cream, shiny, smooth, round, raised, 2 mm after 2 days at 22°C. A diffusible brown pigment is produced	+	+	Biochem set
Aeromonas spp.	Neg	Rods 1–3 µm	βH	Weak growth, yellow	2–3 mm, grey, cream, shiny, round on BA and TSA	+	+	Biochem set or API 20E
Aeromonas salmonicida ssp. salmonicida	Neg	Small rods to cocco-bacilli, 1–2 µm	v	NG	Brown diffusible pigment on FA or TSA. Rough and smooth phase, colonies, 0.5–3 mm at 72 h. No growth at 37°C. FA is preferred for pigment production	+	+	Biochem set. API 20E gives variable results at 25°C, improved results at 30°C. FAT and agglutination test available
Aeromonas salmonicida ssp. nova	Neg	Cocco-bacilli to small rods	βH	NG	On BA, slow growing, friable colony, 0.5 mm reaching 4–8 mm on day 8. Colony slides across plate when pushed with loop. Pronounced zone of haemolysis, similar to a group C Strep. On TSA a brown-water-soluble pigment is not seen until day 8 at 25°C. No growth at 37°C	+	+, v	Biochem set, API 20E, API-ZYM. FAT available, may cross-react with *Aeromonas sobria*
Aeromonas salmonicida ssp. pectinolytica	Neg	Straight rods singly, in pairs or short chains	βH	ND	Colonies develop in 24 h at 35°C on BA and TSA. Production of brown diffusible pigment	+	+	Biochem set
Aeromonas veronii	Neg		βH	NG	Colonies 2 mm, βH on BA at 24 h	+	+	Biochem set, API 20E
Allomonas enterica	Neg	Straight or slightly curved rods			Growth on media containing 3–5% NaCl. Light brown non-diffusing pigment after 2–5 days at 25 or 37°C	+	+	Biochem set
Alteromonas spp.	Neg	Rods		NG	Growth on MA 2216 as small pale yellow colonies	+	+, v	Biochem set
Arcanobacterium phocae	Pos	Cocco-bacilli and short rods, singly, pairs and clusters. Non-acid fast	βH		Colonies on BA at 24 h, white, tiny, round, with large zone of βH	v		Biochem set, API 50CH, API-ZYM
Arcanobacterium pluranimalium	Pos	Straight to slightly curved, non-branching, slender rods	αH		α haemolysis on BA, small colonies	+		API Coryne, API rapid ID32 Strep
Arcanobacterium pyogenes	Pos	Coccobacilli and short rods, singly, pairs and palisade formation with short diphtheroid forms with clubs, 0.2–0.9 × 0.3–2.5 µm	βH		Pinpoint, haemolytic colonies at 24 h on BA. At 48 h colonies 0.5 mm, circular, opaque white with zone of haemolysis 2–3 times the diameter of the colony	−		Biochem set. Litmus milk reaction = acid, clot and reduction at 48 h using heavy inoculum

Arthrobacter agilis	Pos	Cocci in pairs and tetrads, 0.8–1.2 μm diameter	NH	Colonies on agar are smooth, matt, entire, with a rose-red-pigment that is water-insoluble	+	+	Biochem set
Arthrobacter nasiphocae	Pos	Irregular-shaped rods, non-spore forming. Some coccoid forms	NH	Strict aerobe. Growth on BA as circular, entire, convex, greyish-white, 1 mm at 24 h and 37°C	+		Biochem set, API CORYNE, API-ZYM
Arthrobacter rhombi	Pos	Single coryneform rods, short or ovoid in shape		Yellow-whitish colonies, 1 mm 48 h, BHIA with 1% NaCl	+	+	Biochem set, API Coryne
Atopobacter phocae	Pos	Short irregular rods	NH	Colonies pin-point, grey, smooth on BA at 24 h at 37°C in CO_2	–	–	Biochem set, API rapid ID 32 Strep, API-ZYM
Bacillus cereus	Pos	Rods	βH	On BA colonies are white-grey, ground-glass appearance, 2 mm at 24 h. Become slightly yellow with age	+		Biochem set
Bacillus mycoides	Pos	Rods in chains		Rhizoid colonies with counterclockwise filamentous swirling pattern on BA	+		Biochem set
Bordetella bronchiseptica	Neg	Thin rods with tapered ends, singly and pairs. May have long filamentous forms	NH	On BA and CFPA colonies are 1 mm at 48 h and may be haemolytic or non-haemolytic. Depending on phase variation, rough and smooth colonies occur and may be rough, translucent with a raised centre and undulating outer margin, or smooth, opaque and pearl-like	+	+	Biochem set
Brevundimonas diminuta	Neg	Short rods 1–4 × 0.5 μm	αH	Growth BA, NA, TSA, 30–37°C, 2 days. Pantothenate, biotin, cyanocobalamin are required as growth factors	+	+	Biochem set, API 50CH, API 20NE, API-ZYM
Brevundimonas vesicularis	Neg		αH	Yellow colonies on BA, TSA, NA	+	+	Biochem set
Brucella abortus	Neg	Small coccoid rod		Growth at 24 h is scant, occurring where inoculum is heaviest	+	+	See below
Brucella spp.	Neg	Coccobacilli. May stain faintly		Growth on primary culture 7–14 days. 10% CO_2 required for some strains. Growth on TSA or FM. Colonies are raised, convex, entire edge and shiny surface, honey-coloured and transparent in transmitted light	+		Positive by MAF stain. Phenotypic tests, serotyping – best performed by a specialist laboratory **Work with suspect tissues** and *Brucella* cultures **in a biological safety cabinet**

continued

Table 2.4. *Continued.*

Bacterium	Gram	Morphology	βH	TCBS	Colony characteristics	Cat	Ox	Identification sets
Budvicia aquatica	Neg	Straight rods			Small colonies on NA, 0.5 mm 24 h at 30°C. Translucent with smooth entire edges	+	–	Biochem set
Burkholderia pseudomallei	Neg	Oval to round cells with bipolar staining at 3–4 days. Cells may be mistaken for spores as only the periphery stains			Colonies 0.5–1 mm at 24 h, and 3–4 mm at 3–4 days. White, convex and smooth, with white sheen, becoming uneven and umbonate and wrinkled. Broth culture is turbid with wrinkled pellicle	+	+	Biochem set, API 20E **Work in a biological safety cabinet**
Carnobacterium inhibens	Pos	Rods. 0.2 × 0.5–1.2 μm	αH		1–2 mm off-white colonies on BA	–	–	Biochem set
Carnobacterium (Lactobacillus) piscicola	Pos	Very small rod, 1.1–1.4 × 0.5–0.6 μm. Diplococcoid in tissues, chains seen in broth cultures		NG	1–1.5 mm, grey/white, strep-like at 25°C for 48 h. Very little greening of the agar. Growth on BA, BHIA, TSA	–	–	Biochem set, API 20 Strep, API 50CH
Chromobacterium violaceum	Neg		NH or βH		Pigmented strains are deep purple, round, slightly raised on BA. Non-pigmented strains show β-haemolysis on BA	+	+	Biochem set, API 20E, API 20NE
Citrobacter freundii	Neg	Rods		NG	2 mm grey colonies at 24 h on BA	+	–	Biochem set, API 20E
Chryseobacterium balustinum	Neg	Rods, 0.5 × 1.0–3.0 μm	–	NG	Mucoid, yellow colonies on AO	+	+	Biochem set, API-ZYM, API 20NE
Chryseobacterium (Flavobacterium) gleum	Neg	Non-sporing rods with rounded ends	NH	NG	On NA circular, entire, viscid become mucoid and translucent after 5 days. Bright yellow pigment	+	+	Biochem set, API-ZYM, API 20NE
Chryseobacterium indologenes	Neg	Straight rods 0.5 × 1.3–2.5 μm			Colonies on heart infusion agar are 1 mm at 24 h at 30°C. On 0.3% agar growth spreads out in a flower-like growth pattern	+	+	Biochem set, API 20NE, API-ZYM
Chryseobacterium meningosepticum	Neg		NH		Colonies on BA are about 1 mm in diameter and surrounded by a zone of greenish discoloration at 24 h	+	+	Biochem set, API-ZYM, API 20NE
Chryseobacterium scophthalmum	Neg	Short rods, 2.0 × 0.8 μm	NH	NG	Smooth, round, shiny, orange pigmented colonies, 2–3 mm after 2 days at 25°C on MSA-B, MA 2216, TSA. Fresh isolates show gliding motility on Anacker-Ordal medium. Gliding ability lost after storage	+	+	Biochem set, API-ZYM, API 20NE

Organism		Morphology			Colony/growth description			Identification
Clostridium botulinum	Pos	3.4–7.5 × 0.7 μm oval, subterminal spores	βH		Semi-opaque to translucent, matt colony, 1–3 mm, irregular lobate margin and raised centre	–		API rapid 32 A
Colwellia maris	Neg	Curved rods; 0.6–1 × 2–4 μm			Marine 2216 agar, optimal growth at 15°C. Growth range 0–22°C, but no growth at 25°C	+	+	Biochem set
Corynebacterium aquaticum	Pos	Rods 0.5–0.8 × 1–3 μm. Slight pleomorphism with some club-shaped forms and angular arrangements of cells	βH		On BA and TSA colonies round, raised, entire, opaque, slightly viscid with yellow non-diffusible pigment after 48 h at 25°C	+	–	Biochem set, API Coryne
Corynebacterium phocoae	Pos	Coryneform rods	NH		Shiny, round colonies 1 mm on BA 24 hr, 37°C	+		Biochem set, API Coryne, API 50CH, API-ZYM
Corynebacterium testudinoris	Pos	Diphtheroid rods			Colonies on BA are yellow-pigmented	+		API Coryne, API-ZYM
Cryptococcus neoformans var. *gattii*	Pos	Budding yeasts	NH		Slow growing. Colonies 0.5–1 mm cream dull, 3 days on BA. Colonies change from cream to dirty yellow, light tan colour after 7 days on BA	+		India ink for capsule stain, urease positive. Growth on Strep selective media (Oxoid), and SAB plates **Use biological safety cabinet**
Dermatophilus	Pos	Branching filaments containing rows of cocci (zoospores). Zoospores are motile	βH		Colonies on BA in CO_2 are grey-white to grey-yellow, adherent and pit the agar. At 48 h, colonies are tiny, rough, granular, raised. Further incubation produces colonies up to 2–3 mm, umbonate, molar tooth crater forms and become mucoid	+		Biochem set. Make impression smears from underside of freshly removed scabs. Stain with Giemsa stain. Filaments are best stained with Giemsa rather than Gram stain
Edwardsiella hoshinae		Straight rods			Growth on NA, BA, MCA. Colonies flat or slightly convex, 1–2 mm at 24–48 h, 30 and 37°C	+	–	Biochem set, API 20E
Edwardsiella ictaluri	Neg	Rod to coccobacilli, 0.75 × 1.5–2.5 μm	w	NG	Slow growing, 1–2 mm, round, non-pigmented, pale grey, 48 h at 28–30°C. May be a greenish tinge and slight haemolysis under the colony (pale lemon colour on a loop). Musty smell	+	–	Biochem set, API 20E

continued

Table 2.4. *Continued.*

Bacterium	Gram	Morphology	βH	TCBS	Colony characteristics	Cat	Ox	Identification sets
Edwardsiella tarda	Neg	Rod. 1 × 1–2 μm		NG	0.5 mm round, grey colonies at 24–48 h (pale lemon colour on a loop). Growth on BA, MCA (NLF), SS agar. Colonies will be smaller in size than *Salmonella* on the selective media	+	–	Biochem set, API 20E
Empedobacter brevis	Neg	Rods			Yellow colonies on NA and BA, 0.2–2.5 mm at 24 h and 30°C, low convex, entire edge. At day 7 on BA may see slight α-haemolysis	+	+	Biochem set
Enterovibrio norvegicus	Neg	Cells 0.8 × 1.0–1.2 μm		G	On MA 2216 colonies are beige, smooth, round, raised, 1 mm at 48 h at 28°C	+	+	Biochem set, API 20E, API-ZYM
Erysipelothrix rhusiopathiae	Pos	Gram-variable rods, pleomorphic, tangled forms, 2–5 μm	NH αH	ND	Growth on BA at 24–48 h, 0.5 mm grey-green colonies small zone of α-haemolysis	–		Biochem set, API rapid ID 32 Strep, API Coryne. *Note:* add a few drops of sterile serum to inoculating medium to improve reactions in Biochem set
Eubacterium spp.	Pos	Long chains of pleomorphic rods. No spores. In older cultures cells may have 'ovoid' bodies	βH		On BA under anaerobic conditions, colonies are haemolytic, translucent, slowly spreading, flat, and contoured with filamentous edges. Anaerobic atmosphere required	–	–	Biochem set
Eubacterium limosum	Pos	0.6–0.9 × 1.6–4.8 μm. Rods may have swollen ends	NH		Punctate, circular, convex, entire colonies that are translucent to slightly opaque. Less than 1 mm at 48 h. Anaerobic	–		Biochem set, API rapid 32 A
Eubacterium tarantellae	Pos	Very long filamentous, unbranched rods (10 μm)	βH	NG	2–5 mm, translucent colonies, slightly rhizoid and mucoid. Anaerobic	–		Biochem set
Facklamia miroungae	Pos	Ovoid cells (0.8–0.9 μm) in pairs and short chains	NH	ND	Colonies on BA 0.5 mm at 24 h, 37°C. Circular, entire, shiny, convex and grey-coloured	–		API rapid ID 32, API-ZYM
Flavobacterium aquatile	Neg				Colonies on AO media low, convex, round	+	+	Biochem set, API-ZYM
Flavobacterium branchiophilum	Neg	Long thin rods 0.5–8 × 10 μm. Smears from colonies may show circular, slightly refractile, 'cysts'	NH	NG	On AO media, light yellow, smooth, round, raised, colonies, 0.5–1 mm, 5 days at 18°C. No gliding motility. Growth on 20-fold dilute TSA but not on 'full-strength' TSA	V	+	Biochem set, API-ZYM, API 20E

Organism	Gram	Cell morphology			Colony characteristics			Identification methods
Flavobacterium (Flexibacter) columnare	Neg	Long thin rods (4–8 µm). Filamentous	NH	NG	Bright yellow, flat, dry, rhizoid, slow spreading growth 5 days 20–25°C. Adheres strongly to agar	V	+	Biochem set, API-ZYM, API 20E, API 50CH. Yellow pigment changes to pink in 3% NaOH
Flavobacterium flevense	Neg				On AO media, low convex, round colonies sunken into agar	+	+	Biochem set, API-ZYM
Flavobacterium frigidarium	Neg	Rods 0.8–2 × 0.5–0.7 µm, singly and in pairs			Growth on AO, flat round yellow colonies with entire margins. Grows on NA, TSA and MA 2216	+	+	Biochem set, API-ZYM
Flavobacterium gillisiae	Neg	Rods 2–5 × 0.4–0.5 µm			Orange-pigmented colonies, butyrous, circular and convex with entire edge. Growth on MA 2216, NA, TSA, R2A	+	–	Biochem set, API 20E
Flavobacterium hibernum	Neg	Rods 0.7 × 1.8–13 µm			Growth on TSA. Yellow, mucoid colonies at 25°C, gelatinous at 4°C	+	–	Biochem set, API 20E, API 20NE
Flavobacterium hydatis (Cytophaga aquatilis)	Neg	Rods, 8.0 × 0.5 µm	NH	NG	Mucoid, yellow-orange colonies with flat spreading filamentous margins. Incubate 14°C for 14 days	+	–	Biochem set, API-ZYM, API 20E
Flavobacterium (Cytophaga) johnsoniae	Neg	Long thin rods	NH	NG	On AO media, pale yellow, 5–10 mm, flat, smooth, rhizoid with spreading, filamentous margins, 5 days. On BA, yellow, round, smooth, 1–2 mm, 2 days	V	+	Biochem set, API-ZYM, API 20E
Flavobacterium pectinovorum	Neg				On AO media, colonies low round with entire margins	+	+	Biochem set, API-ZYM
Flavobacterium (Cytophaga) psychrophilum	Neg	Slender, flexible rods 1–7 × 0.5 µm		NG	Smooth, glossy, bright yellow colonies with thin spreading edges, 5 days at 5–25°C. Does not adhere to agar	V	+	Biochem set, API-ZYM, API 50CH
Flavobacterium saccharophilum	Neg				On AO media, colonies flat, spreading, sunken into agar	+	–	Biochem set, API-ZYM
Flavobacterium succinicans	Neg				On AO media, colonies flat, spreading, with filamentous margins			Biochem set, API-ZYM
Flavobacterium tegeticola	Neg	Rods 2–5 × 0.4–0.5 µm			Yellow-pigmented colonies, butyrous, circular, convex, entire. Growth on MA 2216, NA, TSA, R2A	+	–	Biochem set, API 20E

continued

Table 2.4.   *Continued.*

Bacterium	Gram	Morphology	βH	TCBS	Colony characteristics	Cat	Ox	Identification sets
Flavobacterium xanthum	Neg				Growth on MA 2216, NA, TSA, R2A	+	+	Biochem set, API 20E
Flexibacter polymorphus	Neg	Filaments multicellular. At end of each cell is a refractile granule of lipid material			Marine agar. Cobalamin required for growth. Peach-coloured pigment	–	+	Biochem set, API-ZYM, API 50CH
Granulicatella adiacens *Granulicatella elegans*	Pos	Cocci, including pleomorphic ovoid cells, cocco-bacilli, rod-shaped cells	α		Strains are nutritionally deficient, and satellite around other bacterial growth. Grow on BA with Staph streak or pyridoxal disc, or BA supplemented with 20 μg pyridoxal HCl per ml. For *G. elegans* use L-cysteine HCl (0.01%)	–	–	Biochem set, API 20 Strep, API rapid ID 32 Strep
Granulicatella balaenopterae	Pos	Coccus, single cells and short chains	α	NG	Growth on BA, 0.2 mm colony in air or CO_2. Is not nutritionally deficient like other strains in the genus	–		Biochem set, API-ZYM, API 20 Strep, API rapid ID 32 Strep
Hafnia alvei	Neg	Short rods			White to yellow non-mucoid colonies. Growth on BA, NA, MCA, DCA	+	–	Biochem set, API-ZYM, API 20E
Halomonas aquamarina (previously *Alcaligenes faecalis homari*)	Neg	Straight or curved rods 1.5 μm, bipolar staining			Colonies on MA 2216, off-white translucent, raised smooth 2–3 mm at 24 h at 18°C and 37°C. Slight tendency to spread	+	+	Biochem set
Halomonas elongata	Neg	Rods singly or paired			Colonies 2 mm at 24 h, smooth, glistening, opaque white. After 24 h colonies spread due to flexuous filaments, but are no larger than 4 mm. Requires 8% NaCl for growth			Biochem set. Add 8% salt to medium
Halomonas halodurans	Neg	Rods single, pairs or chains			On MA 2216, colonies are smooth, glistening, translucent white, convex, 1–2 mm in 24 h at 20 or 30°C			Biochem set
Halomonas venusta	Neg	Rod			At 48 h colourless, mucoid colonies on BA, MCA	+	+	Biochem set, API 20NE

Organism	Gram	Morphology	Haemolysis	Growth	Colony description			Tests
Helicobacter cetorum	Neg	Fusiform to slightly spiral			On BA, Skirrow's media (VPT), or TSA + blood, colonies are pin-point and may grow as a thin spreading film across the plate, 5–14 days 37°C	+	+	Biochemicals, API Campy
Hydrogenophaga (Pseudomonas) pseudoflava	Neg	Ovoid cells. Rods of 2.5 µm in older cultures		NG	On NB agar yellow-pigmented colonies 2–4 mm at 3 days, slightly irregular shape with an undulating margin	W	+	Biochem set
Iodobacter fluviatile	Neg	Small rod, 0.7 × 3.0–3.5 µm		NG	Violet-coloured colonies, thin spreading margin on low nutrient agar such as ¼ strength NA. Non-spreading on NA	+	+	Biochem set
Lactococcus garvieae	Pos	Coccobacilli 0.7–1.4 µm, paired cocci, short chains	αH	NG	1 mm grey/white, round, with greening under colony. Strep-like on BA. Growth on NA, TSA, BA	–	–	Biochem set, API 20Strep, API rapid ID 32 Strep, API 50CH. Strep group = D neg
Lactococcus piscium	Pos	Coccobacilli				–	–	Biochem set
Listonella (Vibrio) anguillarum	Neg	Short rods, curved or straight, rounded ends, occurring singly and in pairs, pleomorphic. 0.5–0.7 × 1–2 µm. Rapid motility	βH	Y	At 2 days colonies are 2 mm, glistening cream-colour in young colonies and greenish-pale brown in older colonies on MSA-B. Haemolysis under colony. On MSA-B, *V. cholerae* and *V. mimicus* have similar appearance but grow more quickly. On NA colonies are off-white to buff-coloured, translucent or opaque, circular, shiny, 1–2 mm	+	+	Biochem set, API-ZYM, API 20E, API 20NE
Listonella pelagia	Neg		NH		Grey-translucent colonies on MSA-B	+	+	Biochem set
Mannheimia haemolytica	Neg	Pleomorphic rods, long and short forms	βH		Grey colonies 1–2 mm on BA at 24 h. β-haemolysis seen	+	+	Biochem set, API-ZYM
Marinilabilia salmonicolor	Neg				Gliding, yellow to pale pink-pigmented colonies	+	+	Biochem set, API-ZYM
Mesophilobacter marinus	Neg	Coccobacilli, pleomorphic	NH	NG	Growth on MA 2216 and NA. Colonies circular sometimes irregular, convex, glistening, opaque, pale yellow-brown. Describes marine coccobacilli that morphologically resemble the *Acinetobacter-Moraxella* group	+	+	Biochem set
Moritella marina	Neg	Curved or straight rods	βH	NG	Greyish-cream, translucent, circular, convex colonies on MSA-B	+	+	Biochem set, API-ZYM, API 20E

continued

Table 2.4. *Continued.*

Bacterium	Gram	Morphology	βH	TCBS	Colony characteristics	Cat	Ox	Identification sets
Moritella viscosa	Neg	Long cells when grown in liquid media. Short or elongated curved rods on solid medium	βH	NG	Creamy-yellow colonies on MSA-B and TSA+NaCl, 0.5 mm 24 h at 15–22°C. Colonies are viscous and adhere to the medium. Form long threads when lifted from the plate. Light haemolysis seen underneath colony	+	+	Biochem set, API-ZYM, API 20E, API 50CH
Mycobacterium abscessus	AFB				Growth on Middlebrook 7H10-ADC medium in 7 days. Growth on MCA	+		Biochem set, or send to specialist laboratory
Mycobacterium chelonae	Pos and AFB	Pleomorphic rods, 2–7 × 0.2–0.5 μm	NH	NG	Colonies on MSA-B are circular, smooth, pale-cream at 7 days and 15°C. Growth on MCA, TSA, BHIA	+	–	Biochem set, or send to specialist laboratory. PCR primers available.
Mycobacterium marinum	Pos and AFB	Pleomorphic rods, 2–7 × 0.2–0.5 μm	NH	NG	Colonies on MSA-B are circular opaque, 0.2 mm at 7 days, 22°C. When grown in the light, colonies are yellow			Biochem set, or send to specialist laboratory. PCR primers available
Mycobacterium neoaurum	Gram stain = neg, AFB	AFB 3–4 × 0.6 μm	NH		Yellow colonies on BA at 25°C in 5–7 days. No growth at 37°C. At 8 days, colonies 0.4 mm, smooth, round clear, with slight yellow colour. No growth on MSA	+		Send to specialist laboratory
Mycobacterium peregrinum	AFB	Acid fast rods			Buff-coloured colonies on Middlebrook 7H11 medium in 4 days under CO_2			Send to specialist laboratory
Mycobacterium poriferae	AFB	Acid fast rods			Rapid growth with smooth colonies on Middlebrook 7H11 medium			Send to specialist laboratory
Mycobacterium spp.	Weak pos non-staining, acid-fast	Pleomorphic medium to long rods, non-branching, 1.5–3.0 μm	NH	NG	0.5 mm off-white, 5–7 days granular	+		Biochem set, or send to specialist laboratory **Use biological safety cabinet**
Mycobacterium triplex-like	AFB	Coccoid and rod forms. Beaded AFB rods	αH		Growth on BA after 12 weeks at 25°C produces rough, dry colonies with a raised centre and flat border			Nested PCR
Mycoplasma mobile					Cells glide in wet preparation as seen by dark ground illumination. Cells are elongated. Optimum temperature is 25°C			Mycoplasma set

Myroides (*Flavobacterium*) *odoratimimus*	Neg	Rods 0.5 × 1–4 µm	NH		Yellow-pigmented colonies. Growth on MCA, NA and TSA. No gliding or swarming. Fruity odour	+	+	Biochem set, API-ZYM, API 20NE
Myroides (*Flavobacterium*) *odoratus*	Neg	Rods 0.5 × 1–4 µm	NH		Yellow-pigmented colonies, 3–4 mm on MCA, NA and TSA at 24 h. No gliding or swarming. Fruity odour	+	+	Biochem set, API-ZYM, API 20NE
Nocardia crassostreae	Pos	Branched hyphae, fragment into rods and cocci			Dry wrinkled, waxy colonies, pale yellow. No aerial hyphae	+		Biochem set, acid fast
Nocardia salmonicida	Pos				Branched orange, substrate mycelium, with white to pink aerial growth. Filamentous colony margins			Biochem set
Nocardia seriolae	Pos	Branching vegetative mycelia that break up into non-motile rods. Coccoid and long slender multi-septate rod forms			Macroscopic aerial mycelia are not found. Dry, waxy, wrinkled, colonies that are yellowish-orange on yeast extract–malt extract agar (ISP no2 Difco). Appear as white colonies on BHIA and yellow colonies on LJM			Biochem set, weakly acid fast
Nocardia spp.	Pos weak acid-fast	Coccoid to oval cells, long slender multi-septate rods, branching. 5–50 µm	N	NG	<1 mm, white, cream, rough or with very short dense mycelium. Growth 3–7 days. BA and NA	V, +		Biochem set
Nocardia spp. (Australian strain)	GPR	Branching rods			Colonies appear at 5 days. Dry, rough, tan/yellow, adherent, 1–2 mm			Biochem set. Modified acid fast stain = positive
Pantoea (*Enterobacter*) *agglomerans*	Neg	Non-encapsulated, non-spore forming, broad, straight rod, 0.5–1.0 × 1.0–3.0 µm	NH	NG	Colonies on NA are smooth, convex, entire, and translucent and may have a yellow pigment. Growth on BA, MSA-B, and MA 2216. Growth at 37°C but not at 44°C	+	–	Biochem set, API 20E, API 50CH
Pasteurella multocida	Neg	Cocco-bacilli	NH		Growth on BA, 1–2 mm at 24–48 h. Colony size and appearance vary from different animal species. Colonies may be smooth or mucoid, dark grey with greenish appearance. Distinctive musty sweetish smell	+	+	Biochem set, API-ZYM, API 20E
Pasteurella skyensis	Neg		NH or weak		Growth TSA-B+1.5% NaCl, at 48 h at 22°C. Colonies circular, entire, low, convex, grey, 0.5 mm. No growth without 1.5% NaCl or blood. No growth at 37°C	–	+ weak	Biochem set, API-ZYM. Add 1.5–2% NaCl to sets and use heavy inoculum in Biochem set. Read after 4 days

continued

Table 2.4. *Continued.*

Bacterium	Gram	Morphology	βH	TCBS	Colony characteristics	Cat	Ox	Identification sets
Pasteurella testudinis	N	Pleomorphic rod 0.2 × 1.5–2 μm	βH		BA white, mucoid, 0.5–1 mm 24–48 h, 20–37°C	+	+	Biochem set
Pedobacter heparinus	Neg	Non-sporing rods, 0.4–0.5 × 0.5–1.0 μm	βH slow		Growth on NA, 1–3 mm at 48 h, circular, low convex, smooth, opaque. A yellow or creamy white non-fluorescent pigment is produced. Creamy white colonies on PY	+	+	Biochem set, API-ZYM, API 50CH
Pedobacter piscium	Neg	Non-sporing rods 0.4–0.5 × 0.5–1.0 μm			Colonies on NA are circular, entire, low convex at 2 days. A yellow or creamy white, non-fluorescent pigment is produced	+	+	Biochem set, API-ZYM, API 50CH
Phocoenobacter uteri	Neg	Pleomorphic rods	NH		Growth on BA with and without CO₂: Colonies entire, circular, low convex, smooth, grey 0.5 mm at 48 h at 37°C	–	+	Biochem set, API 20E, API 20NE, API-ZYM
Photobacterium angustum	Neg	Short rods	NH		White colonies on MA 2216, MSA-B	+	+	Biochem set, API 20E, API 50CH
Photobacterium damselae ssp. *damselae*	Neg	Rods, relatively pleomorphic	βH	G	Entire, smooth, greyish-white, translucent colonies, 2–3 mm on MSA-B	+	+	Biochem set, API-ZYM, API 20E, API 20NE, API 50CH
Photobacterium damselae ssp. *piscicida*	Neg	Small rods to coccobacilli, 1–1.5 μm, bipolar staining	βH	NG	1–2 mm grey/yellow, shiny, 72 h. Growth on BA and NA with 0.5% NaCl, MSA-B, MA 2216	+	+	Biochem set, API-ZYM, API 20E, API 20NE, API 50CH
Photobacterium iliopiscarium	Neg	Pleomorphic rods, straight and curved	NH	G	Colonies small colourless to greyish opaque with entire edge. May take up to 14 days for colonies to appear	+	+	Biochem set, API 20E, API 50CH
Photobacterium leiognathi	Neg	Short rods or coccobacilli, 1–2.5 × 0.4–1 μm	NH	NG	Colonies on MSA-B, off-white, translucent, circular, smooth, shiny. Luminous at 3 days	+	–	Biochem set, API 20E, API 50CH
Photobacterium phosphoreum	Neg	Short rods	NH		White colonies on MSA-B and MA 2216	+	–	Biochem set, API 20E
Planococcus kocurii	Pos	Spherical cells 1.0–1.2 μm, singly, pairs and tetrads. Motile			Colonies on peptone yeast extract agar are circular, smooth, convex, and yellow-orange pigmented	+		Biochem set
Planomicrobium okeanokoites	Neg, v	Rods 0.4–0.8 × 1–20 μm			Cells bright orange to yellow. Requires 3% NaCl	+	W	Biochem set. Add NaCl (3%)

Plesiomonas shigelloides	Neg	Straight, variable length rod	NH	W, G	Growth on BA and MCA 1–2 mm 24 h at 25–37°C. Pungent smell	+	+	Biochem set, API-ZYM, API 20E
Providencia rettgeri	Neg	Rod 0.6 × 1.0–1.5 μm	NH		Colonies on BA, TSA circular, discrete, convex, glistening, creamy-white 24 h at 37°C	+	–	Biochem set, API 20E
Providencia rustigianii	Neg	Non-sporing rods 0.5 × 1–3 μm	NH		On BA, colonies 1–2 mm at 24 h, glossy, semitranslucent, smooth. Orange-red on MCA	+	–	Biochem set, API 20E
Pseudoalteromonas genus	Neg	Straight or curved, non-spore forming rods 0.2–1.5 × 1.8–3 μm. Single flagella			Strict aerobes, growth on MA 2216 at 20°C. Negative for ADH. All positive for gelatin and DNase	+, w, v	+	Biochem set
Pseudoalteromonas antarctica	Neg	Rods, 0.9 × 1–3 μm, become larger and filamentous with age (10 μm)			Growth at 4–30°C on TSA, MSA-B, MA 2216. Colonies round, smooth, beige, convex, mucoid, 1–2 mm in 5 days at 15°C. Growth factors not required. Aerobic	+	+	Biochem set, API 20NE, API-ZYM
Pseudoalteromonas flavipulchra (Pseudoalteromonas aurantia)	Neg	Straight rods 0.7–1.5 × 1.5–4 μm			On MA 2216, colonies 1 mm, bright pale yellow 24 h, 23°C. At 5 days, orange-brown to green-brown 4 mm. On MSA-B, bright, mucoid, dark-brown colonies 2 days, then medium turns black, and haemolysis is seen. Poor growth on TSA and NA even with 2% NaCl	+	+	Biochem set
Pseudoalteromonas bacteriolytica	Neg	Rods with rounded ends, 0.6–0.9 × 1.9–2.5 μm			Growth on MA 2216, red pigment may or may not be present	+w	+	Biochem set, API 20NE
Pseudoalteromonas citrea	Neg	Straight rods 0.7–1.5 × 1.5–4 μm	βH slow		On MA 2216 colonies 0.5 mm 24 h at 23°C, bright, whitish, becoming 4 mm and lemon-yellow up to 4 days. On MSA-B, 6–7 mm bright, mucoid, whitish colonies turning black around the colony after 2–3 days. Ring of haemolysis seen after 5 days. Also growth on NA with added NaCl	+	+	Biochem set, some API-ZYM reactions
Pseudoalteromonas denitrificans	Neg	Rods 2–4 × 0.5–0.7 μm			On MA 2216, colonies have pink to red pigment	–	+	Biochem set
Pseudoalteromonas elyakovii	Neg	Rods with rounded ends, 0.5–0.8 × 1.8–4.0 μm			Growth on MA 2216 as beige-coloured colonies, round, circular, smooth, convex	+	+	Biochem set

continued

Table 2.4. *Continued.*

Bacterium	Gram	Morphology	βH	TCBS	Colony characteristics	Cat	Ox	Identification sets
Pseudoalteromonas espejiana	Neg	Straight rods, 0.2–1.0 × 2.0–3.5 μm			Growth on MA 2216	+	+	Biochem set
Pseudoalteromonas flavipulchra	Neg	Rods, single cells, 0.5–1.5 μm			Growth MA 2216 with orange colonies	+	+	Biochem set
Pseudoalteromonas luteoviolacea	Neg				Growth on MA 2216. Colonies 3–5 mm at 4 days at 25°C, regular, convex, opaque, violet-coloured	–	+	Biochem set
Pseudoalteromonas maricaloris	Neg	Rods, single cells, 0.7–0.9 × 1.0–1.2 μm			Growth on MA 2216, as round colonies 2–3 mm, circular, regular, convex, translucent, smooth, lemon-yellow pigment at 48 h. Optimum 25–35°C, 0.5–10% NaCl	+	+	Biochem set
Pseudoalteromonas piscicida	Neg	Ovoid, Gram-negative rods			On MA 2216 colonies are 3–6 mm, after 2 days at 28°C. Raised centre with light orange to white pigment becoming darker orange at the edges. Water-soluble pigment diffuses into the agar	+	+	Biochem set, carbohydrates using MOF, API 50CH
Pseudoalteromonas rubra	Neg	Straight or slightly curved rods, 2–4 × 0.8–1.5 μm	βH		Growth on MA 2216 bright pinkish-white at 24 h at 23°C. Centre turns red and sometimes blue at 4 days, size 6–7 mm with 2 or 3 concentric rings. With blood added to MA 2216, colonies are bright, mucoid, red almost black with β-haemolysis. Colonies produce a diffusible brown pigment with an odour of hydrogen cyanide	+	+	Biochem set, API-ZYM
Pseudoalteromonas ulvae	Neg	Rods 1.75–2.5 × 1–1.5 μm			Growth on TSA with 2% NaCl, MSA-B, MA 2216. Colonies on TSA are white, whereas on MA 2216 medium colonies are purple at 48 h, 23°C	+	+	Biochem set, API 20E
Pseudoalteromonas undina	Neg	Curved rods 0.7–0.9 × 1.8–3.0 μm			Growth on MA 2216	+	+	Biochem set
Pseudomonas anguilliseptica	Neg	Long rods, slightly curved, 5–10 μm	NH	NG	On BA, TSA, BHIA, NA colonies <1 mm, round, shiny, pale-grey at 4–7 days. No growth on Pseudomonas isolation agar (Difco)	+	+	Biochem set

Pseudomonas fluorescens	Neg			Rods	Growth on BA, light-grey colonies 1.5 mm 24 h at 25°C. 3–5 mm at 48–72 h	+	+	Biochem set, API 20E
Pseudomonas plecoglossicida	Neg	βH		Rods 0.5–1 × 2.5–4.5 μm	Growth on TSA, BA, MSA-B, 25°C	+	+	Biochem set, API-ZYM, API 20NE
Pseudomonas pseudoalcaligenes	Neg			Short rods	Cream-coloured colonies, 'gummy' consistency	+	+	Biochem set
Pseudomonas stutzeri	Neg	NH	NG	Straight and slightly curved rods, some bipolar staining	There may be rough and smooth colonies. Colonies 0.5 mm, grey, wrinkled, dry and buff-coloured. May be slightly yellow in colour	+	+	Biochem set
Renibacterium salmoninarum	Pos	βH	NG	Small bacilli, singly and in pairs (0.3–1.5 μm)	Colonies 2 mm at 20 days at 15–18°C, smooth, cream to granular on KDM2 medium. On cysteine serum agar colonies are circular, convex, white to creamy yellow and of varying sizes. On Loeffler coagulated serum a creamy matt growth is seen. On Dorset egg medium, growth appears as a raised shiny yellow layer	+	–	Biochem set, API-ZYM Haemolysis reported against salmonid RBC. Does not cross-react with Lancefield group G antisera. Cysteine required for growth
Rhodococcus (luteus) fascians	Pos			Straight or slightly curved rods 0.6–1 × 3–6 μm in angular or parallel arrangement	Weak growth on NA colonies yellow, raised, glistening, smooth. Growth on Lowenstein-Jensen medium is abundant and yellow-orange	+		Biochem set
(Rhodococcus maris) Dietzia maris	Pos			Short ovoid rods 0.6–1 × 1–2 μm	Weak growth on NA, colonies yellow, raised, glistening, smooth	+		Biochem set
Rhodococcus spp.	Pos			Rods, slightly club-shaped, 2–3 × 0.6 μm	Growth on BA and MSA-B in 3–4 days at 25°C. No growth at 37°C (except for *R. equi*). At 8 days colonies slightly domed, round, smooth, dry, deep creamy yellow	–	–	Biochem set
Roseobacter gallaeciensis	Neg			Ovoid rod, 0.7–1 × 1.7–2.5 μm	On MA 2216 at 25°C, colonies are circular 0.5 mm, smooth, convex, brownish colour, with regular edges. At 7 days, colonies are 2 mm with diffusible brown pigment produced	+	+	Biochem set
Roseobacter spp. CVSP	Neg			Rods, 0.25 × 1.0 μm	Growth on SWT, MA 2216, 1 mm round, non-mucoid at 5 days. Appearance of pink pigment at 7 days	w+	+	Biochem set

continued

Table 2.4. *Continued.*

Bacterium	Gram	Morphology	βH	TCBS	Colony characteristics	Cat	Ox	Identification sets
Salegentibacter salegens	Neg	Rod-shaped cells, singly, pairs, occasional chains			Growth on MA 2216, NA, TSA	+	+	Biochem set
Salinivibrio costicola	Neg	Curved rods 0.5 × 1.5–3.0 µm	βH		Colonies circular, convex, cream-coloured at 2 days at 37°C. Media with 0.5–20% NaCl, MA 2216, MSA-B	+	+	Biochem set, add NaCl
Serratia fonticola	Neg	Rods 0.5 × 3.0 µm			Growth on NA at 30 and 37°C	+	–	Biochem set, API 20E
Serratia liquefaciens	Neg	Rods	NH	NG	Growth on BA, TSA, MCA, in 48 h at 21°C. Non-pigmented	+	–	Biochem set, API 20E
Serratia plymuthica	Neg	Rods	NH		Red colonies on TSA 24–48 h at 22°C	+	–	Biochem set, API 20E
Shewanella algae	Neg	Short straight rods	βH	NG	Colonies yellow-orange or brown after 2 days, 37°C. Growth on SS agar, MA 2216	+	+	Biochem set, API 20E
Shewanella colwelliana	Neg	Rod 1–3 µm long. On solid media and in late phase cells become helical, filamentous – 20 µm	NH		Colonies on MA 2216 are circular, 1 mm, convex with undulate margins. At 7 days, 5 mm with irregular appearance. In broth culture (2216) produces red-brown pigment	+	+	Biochem set
Shewanella frigidimarina	Neg	Rods curved or straight, 1–2.5 × 0.5–0.8 µm			Growth on MA 2216 as tan-pigmented colonies, 3–5 days at 10°C. Mucoid colonies after 10 days incubation. No NaCl requirement. Optimal temp = 20–22°C	+	+	Biochem set, API 20E
Shewanella gelidimarina	Neg	Rods curved or straight, 1.5–2.5 µm, singly and pairs			Growth on MA 2216 as tan-pigmented colonies, 3–5 days at 10°C. Mucoid colonies after 10 days incubation. NaCl requirement. Optimal temp = 15–17°C	+	+	Biochem set, API 20E
Shewanella japonica	Neg	Rods 1–2 × 0.6–0.8 µm	βH		Growth on MA 2216 as circular, smooth and slightly pink colonies, and MSA-B at 25°C. Degrades agar	+	+	Biochem set
Shewanella oneidensis	Neg	Rod 2–3 µm × 0.4–0.7 µm	NH		Growth on MA 2216, MSA-B. Pale pink to beige-coloured colonies, 1–4 mm, circular, smooth, convex	+	+	Biochem set
Shewanella pealeana	Neg	2.0–3.0 × 0.4–0.6 µm	NH		Growth on MA 2216 in 2 days at 25°C. Colonies opaque salmon-coloured with mucoid surface	+	+	Biochem set

Organism	Gram	Morphology			Colony description			Tests
Shewanella putrefaciens	Neg	Rods	NH		A reddish-brown or pink pigment is produced	+	+	Biochem set, API-ZYM, API 20E, API 20NE
Shewanella woodyi	Neg			NG	Growth on MA 2216, as pink-orange pigment colonies at 20°C	+	+	Biochem set, API 20E
Sphingobacterium multivorum	Neg	Non-sporulating rod 0.5–1.0 × 1.4–2 μm			Yellow colonies on NA, circular, low, convex	+	+	Biochem set, API-ZYM
Sphingobacterium spiritivorum	Neg	Non-sporing rods 1 μm, singly or pairs	NH		Colonies on NA, yellow, low convex, smooth 30–37°C	+	+	Biochem set, API-ZYM
Sphingomonas (Pseudomonas) paucimobilis	Neg	Non-sporing rods 1 μm	NH		Growth on NA and BA. Yellow-pigmented colonies circular, low, convex at 2 days at 22°C	+	+	Biochem set
Staphylococcus delphini	Pos	Cocci 0.8–1 μm, mostly in clusters but also singly and paired	βH	NG	On NA, colonies 5–7 mm circular, smooth, opaque to translucent with incubation	+	–	Biochem set. Coagulase test. API-ZYM
Staphylococcus hominis	Pos	Cocci in tetrads, occasionally pairs	NH		Colonies 3–5 mm, smooth, dull, opaque, slightly umbonate with bevelled edge. Pigment pale yellow to grey-white	+	+	Biochem set, Coagulase, DNase
Staphylococcus lutrae	Pos	Cocci singly, pairs and clusters	βH		Colonies 1.5–2 mm, smooth, round 24 h, 37°C	+	–	Coagulase test, DNase test, Biochem set
Staphylococcus warneri	Pos	Cocci 0.5–1.2 μm diameter in pairs and singly, occasionally tetrads	NH		Colony growth in 24 h at 37°C, 3–5 mm, smooth, round, sticky. Most strains are bright yellow-orange or with a yellow ring around the edge of the colony. 20% of strains are non-pigmented	+	+	Biochem set, Coagulase, DNase
Stappia stellulata-like strain M1	Neg	Motile rods			Growth on MA 2216, SWT at 23°C. Mucoid colonies with light-brown pigment at 3 days. Poor growth under anaerobic conditions. Forms star-shaped aggregates when grown in liquid media	+	+	Biochem set, API 20NE
Streptobacillus moniliformis	Neg	Pleomorphic rods frequently in chains, tangled filaments with bulbous swellings	NH		Clear, non-haemolytic colonies, 0.5 mm at 48 h on BA in CO_2 and 37°C. In broth culture produces discrete fluff-ball-like colonies	–	–	Biochem set

continued

Table 2.4. *Continued.*

Bacterium	Gram	Morphology	βH	TCBS	Colony characteristics	Cat	Ox	Identification sets
Streptobacillus moniliformis-like organism. ID not conclusive (Maher *et al.*, 1995)	Neg	Coccobacilli 0.4–0.6 μm in tissues. Older cultures show filamentous cells with possible branching and some terminal swellings	βH		Growth on BHIA containing fetal calf serum and NaCl. Colonies 0.1 mm at 4–8 days, aerobically at 15–22°C. Colonies off-white, friable, convex, granular, 'bread-crumb' appearance. Older colonies were concave 'molar tooth' appearance. βH after 7–14 days	–	–	Biochem set. Weakly acid-fast
Streptococcus agalactiae group B	Pos	Cocci in small chains	βH		Growth on BA. Colonies 1 mm, pale grey at 24 h, zone of β-haemolysis			Biochem set, API 20 Strep, API rapid ID 32 Strep, Strep grouping antisera (group B)
Streptococcus agalactiae group B, Type 1b (*S. difficile*)	Pos	Cocci in small chains with variable diameter	NH		Cells adhere strongly to agar. On BHIA colonies 1 mm at 48 h. Optimal growth at 30°C. Tests must be performed at this temperature. No growth at 37°C unless grown in microaerophilic atmosphere (5% O_2, 10% CO_2, 85% N_2)	–		Strep grouping (group B), API 20 Strep, API rapid ID 32 Strep, API 50CH, Biochem set
Streptococcus dysgalactiae ssp. *dysgalactiae* serovar L	Pos	Cocci in chains	βH		Colonies 1 mm light grey on BA with zone of haemolysis at 24 h	–		API rapid ID 32 Strep. Streptex A-G = negative. Lancefield group L = positive, Bacitracin = sensitive
Streptococcus species (Group B)	Pos	Cocci in chains	βH	NG	0.5–1 mm grey-white, zone of haemolysis on BA	+		Biochem set, API rapid ID 32 Strep, Streptex A-G
Streptococcus iniae	Pos	Coccus 0.3–0.5 μm. long chains	βH		Growth on BA, NA, BHIA, TSA. Colonies on BA 1 mm white, umbonate, opaque centre spot at 24 h. β-haemolysis surrounded by diffuse ring of α haemolysis. Haemolysis may be variable	–		API rapid ID 32 Strep, API 20 Strep, API 50CH, Biochem set. Reactions at 25°C are slower than at 37°C
Streptococcus parauberis	Pos	Short rods to coccobacilli in pairs or short chains	αH		Colonies 1.5–2 mm at 24 h, round, whitish and slightly α-haemolytic	–		Biochem set, API 50CH, API rapid ID 32 Strep. Incubate at 37°C for 24 h
Streptococcus phocae	Pos	Cocci 1 μm in diameter, singly, pairs or chains	βH		On BA colonies circular, entire, smooth, glistening non-pigmented. 0.8 mm at 24 h, 37°C	–		Biochem set, API 20 Strep, API 50CH
Streptococcus porcinus	Pos	Spherical to ovoid cells in small chains	βH		On BA colonies are small, elevated, entire and haemolytic	–		Biochem set, API 20 Strep, API 50CH

Species	Gram	Morphology	Haem.	Growth	Characteristics			Biochem
Streptomyces salmonis	Pos	Mycelia		ND	Brick-red to orange substrate mycelia and white to pink and yellow shades of aerial mycelia			Biochem set
Tenacibaculum (Flexibacter) maritimum	Neg	Slender, flexible rods, 0.5 × 2–30 μm, occasionally up to 100 μm long		NG	On AO-M medium, pale yellow or orange, flat, thin irregular colonies with uneven edges. Slow spreading growth, 5 days. Colonies never larger than 5 mm. Colonies adhere strongly to the agar. Growth variable on MA 2216. Brown pigment on AO-M with 0.5% tyrosine	V	+	Biochem set, API-ZYM, API 20E, API 50CH
Tenacibaculum (Flexibacter) ovolyticum	Neg	Long slender rods 0.4 × 2–20 μm	NH	NG	Pale yellow colonies, gliding motility. On MA 2216 colonies light brownish yellow, flattened and elevated with regular edges. Rapid loss of viability, therefore see an area of lysis in the centre of a 5-day-old colony with viable cells at the edge	+	+	Biochem set, API-ZYM
Vagococcus fessus	Pos	Cells coccoid, elongated in direction of the chain, singly, pairs, short chains	αH	ND	Growth on BA, 37°C, 5% CO_2, small αH colonies. Cells motile	–		Biochem set, API rapid ID 32 Strep, API-ZYM
Vagococcus fluvialis	Pos	Cells ovoid, elongated in direction of the chain, singly, pairs, short chains	αH	ND	On BA, colonies slightly larger than *Enterococcus* species and αH. Improved growth in CO_2	–		Biochem set, API rapid ID 32 Strep, API-ZYM, Lancefield group N = positive
Vagococcus lutrae	Pos	Coccus, single and in chains. Cell slightly elongated	αH		Small 0.1–0.2 smooth colonies on BA after 24 h at 37°C and 5% CO_2. Motile	–		Biochem set, API rapid ID 32 Strep, API-ZYM
Vagococcus salmoninarum	Pos	Coccobacilli. Forms chains in liquid media	αH	NG	Colonies 0.5–1 mm, white-grey, glistening. Strep-like. Slight greening of agar after 2–3 days	–	ND	Biochem set, API 20 Strep, API rapid ID 32 Strep, API 50CH, Strep grouping
Varracalbmi	Neg	Slender, straight rods 1.7–3.5 μm	Neg or αH	NG	1 mm opaque, grey, convex, adherent colonies which leave an imprint in the agar. αH seen after 1 week	–		Biochem set
Vibrio aerogenes	Neg	Straight to slightly curved rods 0.6–0.8 × 2–3 μm			Colonies flat, circular, off-white after 2 days on PY agar	+	+	Biochem set
Vibrio aestuarianus	Neg	Straight or curved rod, 0.5 × 1.5–2 μm	βH	Y	Growth on MSA, TSA containing 0.5% NaCl		–	Biochem set. Addition of NaCl to set, optional

continued

Table 2.4. *Continued.*

Bacterium	Gram	Morphology	βH	TCBS	Colony characteristics	Cat	Ox	Identification sets
Vibrio agarivorans	Neg	Rod 2–4 × 0.4–0.6 μm		G	Non-pigmented colonies on MSA-B and MA 2216 that produce a shallow pit in the medium	+	+	Biochem set, API 20E, API-ZYM. Add NaCl to set
Vibrio alginolyticus	Neg		NH	Y	Grey colonies. Swarm across MSA-B and MA 2216 in 24 h at 25°C	+	+	Biochem set, API 20E, API-ZYM, API 20NE. Add NaCl to sets
Vibrio brasiliensis	Neg	Rod, 2.5–3 × 1 μm		Y	Growth on TSA+2% NaCl. Colonies beige, translucent, convex, round, smooth, 2–3 mm after 48 h at 28°C	+	+	Biochem set, API 20E, API ZYM
Vibrio calviensis	Neg	Slightly curved or straight rods. 0.25–1.0 × 0.75–2.5 μm		G	Growth on MA 2216. Colonies brownish, translucent, circular, smooth, convex, entire at 25–30°C. No growth at 37°C	+	+	Biochem set, API-ZYM. Add NaCl to set
Vibrio cholerae	Neg	Slightly curved rods	W, βH	Y	Colonies round smooth 2–3 mm greenish-grey on MSA-B	+	+	Biochem set, API 20E, API-ZYM, API 20NE
Vibrio cholerae 0139	Neg	Curved rods	βH	Y	Colonies greyish-opaque with darker centres	+	+	Biochem set. Resistant to vibriostatic compound 0/129
Vibrio cincinnatiensis	Neg	Rod 0.7–2 μm		Y	Colonies 1–2 mm cream, round, smooth, glossy, 24 h at 25 and 35°C	+	+	Biochem set. Add NaCl to set
Vibrio corallilyticus	Neg	Rods 1.2–1.5 × 0.8 μm		Y	On MA 2216, colonies are 3 mm at 3 days, cream-beige, round, entire, smooth	+	+	Biochem set, API 20NE. Add NaCl to sets (3%)
Vibrio cyclitrophicus	Neg	Rods 0.6 × 1.5–2.5 μm			Colonies 4 mm, cream-coloured, circular, flat on MA 2216	+	+	Biochem set. Add NaCl to set
Vibrio diabolicus	Neg	Straight rod 0.8 × 2 μm		Y	Non-pigmented, 2 mm at 3 days on MA 2216 and swarming (9 mm with added glucose)	+	+	Biochem set, API 20E, API 20NE, API 50CH, API-ZYM. Add NaCl to sets
Vibrio diazotrophicus	Neg	Short rods, 0.5 × 1.5–2 μm	NH	Y	Colonies on MA 2216 are flat, circular off-white	+	+	Biochem sets. Addition of NaCl to set, optional
Vibrio (Photobacterium) fischeri	Neg	Rods 0.5 × 1–1.5 μm. Singly or pairs, rounded ends, straight or curved	βH, V	NG or G	Grey or off-white, translucent colonies on MSA-B, 1–2 mm. Become pale yellow, and luminous at 3 days	+	+	Biochem set, API-ZYM, API 20E, API 50CH. Addition of NaCl to set, optional
Vibrio fluvialis	Neg	Short rods, straight or curved, singly or pairs, may be pleomorphic	V	Y	Colonies on BHIA are opaque, shiny smooth, round, doomed, may be mucoid and are 2–3 mm at 24 h, 30°C. Growth on BHIA, MSA-B	+	+	Biochem set, API 20E. Add NaCl to sets

Species		Morphology			Colony characteristics			Notes
Vibrio furnissii	Neg	Straight to lightly curved rod	V	Y	Colonies on BHIA are opaque, shiny smooth, round, doomed, may be mucoid and are 2–3 mm at 24 h, 30°C. Growth on BHIA, BA, MSA-B	+	+	Biochem set, API 20E, API-ZYM. Add NaCl to sets
Vibrio halioticoli	Neg	Rods, 0.6–0.8 × 1.7–2.0 μm	ND	G	On MA 2216, beige, circular, smooth, convex colonies	+	+	Biochem set. Add NaCl to set
Vibrio harveyi	Neg	Short rod, straight or slightly curved, rounded ends occurring singly or in pairs	NH	Y	Grey-coloured, off-white, raised shiny, slow spreading growth on MSA-B, may be mucoid. Luminous after 3 days	+	+	Biochem set, API-ZYM, API 20E, API 20NE, API 50CH. Add NaCl to set
Vibrio hollisae	Neg	Rod, some slightly curved	W βH	NG or weak	On BA and MSA-B, colonies 1–2 mm opaque. Haemolysis after 7 days	+	+	Biochem set. Add NaCl to set
Vibrio ichthyoenteri	Neg	Short rods 1.6–2.5 × 0.6–0.8 μm		Y, w	Non-pigmented colonies	+	+	Biochem set. Add NaCl to set
Vibrio lentus	Neg	1.5–3 × 0.8–1 μm		G	Colonies on MA 2216 at 24 h at 22°C are 0.3–0.5 mm, round, opaque, non-pigmented	+	+	Biochem set. Add NaCl to set
Vibrio (Photobacterium) logei	Neg		NH		Yellow-opaque colonies on MSA-B	+	+	Biochem set, API-ZYM. Addition of NaCl to sets, optional
Vibrio mediterranei	Neg	Rods, 1–2 × 0.5 μm	NH	Y	Colonies on marine agar circular, translucent, non-pigmented. On MSA-B, colonies are 2–3 mm at 48 h, creamy and mucoid	+	+	Biochem set. Add NaCl to set
Vibrio metschnikovii	Neg	Short rods, curved or straight, singly, pairs or short chains, 0.5 × 1.5–2.5 μm	βH	Y or NG		+	–	Biochem set. Addition of NaCl to sets, optional
Vibrio mimicus	Neg	Curved rods	βH	G	Colonies round, smooth, 2–3 mm greenish-grey on MSA-B and BA	+	+	Biochem set, API 20E, API 20NE, API-ZYM. Addition of NaCl, optional
Vibrio mytili	Neg	Coccobacilli.		Y	Growth MA 2216, TSA in 24 h. Colonies round and non-pigmented	+	+	Biochem set, API-ZYM, API 20E, API 20NE, API 50CH. Add NaCl to sets
Vibrio navarrensis	Neg	Rods 1–2 × 0.8–1 μm		Y	Colonies 2–3 mm round opaque non-pigmented 24 h on NA with 2% NaCl	+	+	Biochem set, API 20NE, API 20E. Add NaCl to sets
Vibrio neptunius	Neg	Slightly curved rod, 2.3–3 × 1 μm		Y	Growth on TSA + 2% NaCl. Smooth, rounded, beige colonies 3 mm at 48 h and 28°C	+	+	Biochem set, API 20E, API ZYM. Add NaCl to sets

continued

Table 2.4. *Continued.*

Bacterium	Gram	Morphology	βH	TCBS	Colony characteristics	Cat	Ox	Identification sets
Vibrio ordalii	Neg	2.5–3 × 1 µm, curved	βH	NG	Circular convex off-white to grey colonies, 1–2 mm. 4–6 days, 22°C	+	+	Biochem set, API-ZYM, API 20E. Add NaCl to sets
Vibrio orientalis	Neg		NH	Y	White-opaque colonies on MSA-B	+	+	Biochem set, API-ZYM. Add NaCl to sets
Vibrio pacinii	Neg	Rods		Y	Non-pigmented, translucent colonies on MA 2216			Biochem set, API 20E
Vibrio parahaemolyticus	Neg	Pleomorphic with straight or curved rods	βH	G	Grey colonies on BA, MSA-B, MA 2216. 1–2 mm at 24 h, 3–4 mm 48 h. Swarm across plate	+	+	Biochem set, API 20E. Add NaCl to sets
Vibrio pectenicida	Neg	Curved rod		NG	Smooth, non-pigmented, swarm after 48 h, MSA-B, MA 2216	+	+	Biochem set, API 20E. Add NaCl to sets
Vibrio penaeicida	Neg	Straight or slightly curved short rods, 1.5–2 × 0.5–0.8 µm		G	Round circular, low, convex, cream-coloured on MA 2216, MSA-B	+	+	Biochem set, API 50CH, API 20NE, API 20E. Add NaCl to sets
Vibrio proteolyticus	Neg			Y/G	Flat spreading colonies that will completely cover an MSA-B plate in 24 h. Colony colour darkens with age	+	+	Biochem set
Vibrio rotiferianus	Neg	Curved rods, 0.8–1.2 × 2.0–3.5 µm		Y	Growth on MA 2216. Translucent non-pigmented colonies	+	+	Biochem set, API 20E, API ZYM
Vibrio rumoiensis	Neg	Rods 0.5–0.9 × 0.7–2.1 µm. Under EM, blebs seen on cell surface			Growth on MA 2216 and PYS-2 medium Colonies circular and colourless, 48 h at 30°C	+	+	Biochem set. Add 3% NaCl to set
Vibrio salmonicida	Neg	Curved rods, 0.5 × 2–3 µm	NH	G	Colonies on MSA-B are small, grey, non-pigmented	+	+	Biochem set, API 50CH, API 20E, API-ZYM. Add NaCl to sets
Vibrio scophthalmi	Neg	Short rods	NH	Y	Round non-pigmented colonies on TSA with 1.5% NaCl, MSA-B, MA 2216	+	+	Biochem set, API 20E. Addition of NaCl optional
Vibrio shilonii	Neg	Rods 2.4 × 1.6 µm		Y	On MA 2216 colonies have a slightly serrated edge	+	+	Biochem set, API 20NE. Use inoculum of 3% NaCl
Vibrio splendidus I	Neg	Rod	βH	Y	White-opaque colonies on MSA-B	+	+	Biochem set, API 20E, API 50CH, API-ZYM
Vibrio splendidus II	Neg	Rod	βH	G	White-grey opaque colonies on MSA-B	+	+	Biochem set, API-ZYM, API 50CH, API 20E

Species	Gram	Morphology			Colony/growth description			Identification
Vibrio tapetis	Neg	Coccobacilli 1–1.5 × 0.5 µm		G	Growth on MSA-B, MA 2216, TSA with 2% NaCl. At 48 h, colonies circular, translucent, non-pigmented	+	+	Biochem set, API 20E, API 20NE. Add NaCl to set
Vibrio tubiashii	Neg	Short rods straight or curved, 0.5 × 1.5 µm	βH	Y	White-grey, opaque colonies on MSA-B, may be mucoid	+	+	Fish set, API 50CH, API 20NE, API 20E. Add NaCl to sets
Vibrio vulnificus	Neg	Curved rods 2–3 µm in length	βH	G	Growth on MSA-B, MA 2216. Colonies 2–4 mm at 48 h and 25°C, light grey, slight green colour in heavy growth	+	+	Biochem set, API 20E. Add 3% NaCl to sets
Vibrio wodanis	Neg	Short or elongated rods	βH		Colonies on MSA-B, 2–3 mm at 2 days, round, opaque, yellow with butyrous consistency. Haemolysis after 2 days	+	+	Biochem set, API-ZYM. Add NaCl to sets
Vibrio xuii	Neg	2–3 × 1 µm		Y	Growth on TSA+2% NaCl. Beige colonies 3–4 mm, round, smooth, convex at 48 h and 28°C	+	+	Biochem set, API 20E, API ZYM
Yersinia kristensenii	Neg	Rods			Growth on NA, BA in 24 h at 28°C, 1–2 mm. Cultures may have strong, musty or cabbage-like odour	+	−	Biochem set, API 20E
Yersinia ruckeri	Neg	Rods, 1–3 µm	NH	NG	Colonies 2–3 mm, off-white, grey, shiny, raised, entire. Typical *Enterobacteriaciae*-like colony. Musty *Pasteurella*-like smell. Dark centre to colony at 48 h. Growth on MCA, XLD. Australian strains tend to be 3–4 mm with irregular edges after 48 h incubation at 25°C	+	−	Biochem set, API-ZYM. API 20E may not differentiate from *Hafnia alvei*. Results of motility and urease dependent on temperature. Incubate at 25°C for best results
Zobellia galactanovorans	Neg	Rods 0.3–0.4 × 3–8 µm with rounded ends			Growth on MA 2216 as yellow spreading colonies at 30°C. Hydrolyse agar	+	+	Biochem set, API 20E, API 20NE, API-ZYM. Addition of NaCl, optional
Zobellia uliginosa	Neg				On MA 2216 agar colonies are orange with tenacious slime. Spreading activity is weak or negative. Hydrolyse agar	+	+	Biochem set, API 20E, API 20NE, API-ZYM. Addition of NaCl, optional

Where NaCl is required for optimal growth, add NaCl to the set at a final concentration of 2% NaCl unless stated otherwise.

reaction, oxidase and catalase. The 'Identification sets' column lists the identification sets that can be performed for identification of that organism.

2.4 Biochemical Identification Tests

Biochemical identification of a bacterium proceeds through a number of steps. A pure subculture of the organism is used to perform primary identification tests and to inoculate biochemical identification sets (secondary identification tests), composed of either in-house prepared media, or commercial identification sets such as API, available from bioMérieux. The identification sets are incubated at the appropriate temperature and for the appropriate time for reactions to occur. The results are recorded on a laboratory work sheet or, in the case of the commercial identification kits, on the supplied recording sheets. See Chapter 3 for the interpretation of biochemical identification sets. The media and reagents used in the biochemical identification are detailed in Chapter 7, along with information on growth characteristics and reagent reactions.

The tests described in this chapter and Table 3.1 form the 'Biochem set'.

Primary tests

The primary tests include microscopic examination of smears, in particular the Gram stain, catalase, oxidase, presence of haemolysis, motility, and growth on MCA. More information on the performance of these tests and their interpretation is found in Chapter 3, and methods of media and reagent preparation are described in Chapter 7.

Secondary tests: biochemical identification sets

Biochemical identification is achieved by secondary identification tests, which are the main tests used to identify an organism to species level.

Tubed media

CARBOHYDRATE FERMENTATION.   L-arabinose, glucose, inositol, lactose, maltose, mannitol, mannose, salicin, sorbitol, sucrose, trehalose, xylose. The carbohydrates are commonly called 'sugars'.

DECARBOXYLASES.   ADH, LDC, ODC, decarboxylase control tube.

OTHER BIOCHEMICAL TESTS.   Aesculin, motility, MRVP, nitrate, oxidative fermentative tubes, ONPG, TSI for H_2S, indole, urea. See next section for inoculation methods.

GROWTH AT DIFFERENT TEMPERATURES.   Use either a TSB or glucose tube and incubate at 37°C, 40°C or desired test temperature.

GROWTH IN 10% NaCL.   To test an organism's ability to grow in the presence of 10% NaCl, dispense equal volumes of TSB and 20% NaCl into a sterile McCartney bottle or bijou bottle to give a final concentration of 10% NaCl. Inoculate with the organism to a concentration of McFarland opacity tube 1. Incubate at 25°C or optimum temperature and after 24 h incubation observe for evidence of growth as seen by an increase in the turbidity of the broth.

Plate media

DNase plate, Gelatin/salt plate, MCA, TCBS. Use a MSA-B or BA plate for purity check of the inoculum.

Discs

Ampicillin 10 µg, 'Vibrio discs' 0129 10 µg and 150 µg. The discs are placed on a lawn inoculum of the bacterium on either a BA or MSA-B plate according to the organism's growth requirements. The inoculum should be prepared to a density of McFarland tube 1. Use for Listonella, Moritella, Photobacterium, Vibrio and Aeromonas species. The Vibrio discs (vibrio static agent pteridine 0/129) differentiates between Vibrio species and Aeromonas species, the latter being resistant to both concentrations. A zone of 9 mm is classified as susceptible for the 0/129 150 µg disc (Bernardet and Grimont, 1989). Also see Chapter 7, 'Preparation of Media for Culture and Identification'.

2.5 Inoculation of Biochemical Identification Sets

Examine the subculture plate and, if the growth is pure, use to inoculate biochemical identification sets. For marine samples, NaCl needs to be added to the set at a final concentration of 2%. Many enzymes from a variety of *Vibrio* species will not be active at lower NaCl concentrations and false negative results will be obtained. This may apply to indole, VP, ADH, ODC and LDC reactions. Some tests need a heavy inoculum. These are the decarboxylases and ADH, urea and citrate. Incubate the set at the appropriate temperature for a minimum of 48 h. ODC, LDC, ADH, MR, VP, and indole usually require 48 h incubation as a minimum for a positive reaction even if reasonable growth is seen in the tubes.

To inoculate the tubes, plunge a sterile Pasteur pipette three-quarters of the way into the tubed media, releasing 3–4 drops of inoculum as the pipette is withdrawn.

After inoculating all the tubes, place one drop of the inoculum on to a purity plate and streak out for isolated colonies. After suitable incubation, check that the growth is pure.

Freshwater isolates

For bacteria isolated from freshwater sources, inoculate bacterium into sterile normal saline or sterile distilled water (usually 10 ml) to opacity McFarland 3. Add 3–5 drops of inoculum to each tube. Some media require a heavy inoculum (opacity approx McFarland 6). These are aesculin, ADH, LDC, ODC, the decarboxylase control, MRVP, citrate and urea.

Marine isolates and addition of NaCl

Fish organs are at physiological saline level. Therefore, be aware that some bacteria may adapt to this NaCl concentration. By subculturing an organism to BA and MSA-B, or culturing to a 0% and 3% salt plate, the salt requirement or preference can be determined. However, many of these organisms produce more accurate results of biochemical tests when the NaCl is added to the media, usually at a final concentration of 2%. Therefore check the optimal NaCl concentration range of the suspect organism from the NaCl column in the tables for biochemical results (Tables 4.1 to 4.22).

For bacteria isolated from the marine environment, the optimal final salt concentration in the medium is all-important. Insufficient salt concentration will lead to false negative results even though good growth may be seen in the tube. An example of this is *V. parahaemolyticus*, which will give a negative indole result when an inoculum of physiological saline is used, but a positive result when a final concentration of 2% NaCl is achieved in the tryptone water (indole test) medium (*See photographic section*).

Tubed media can be prepared with or without added NaCl. However, a laboratory may want to keep media preparation to a minimum and prepare all media without additional NaCl. At the time of inoculation, 500 µl of a 20% sterile NaCl stock solution is added to all liquid media (5 ml) (aesculin, indole test, ADH, ODC, LDC, nitrate, MRVP). A sterile 20% salt stock solution can be prepared in a Schott bottle to which an autoclavable 2 ml volumetric dispenser is added, and set to 500 µl. An example of a volumetric dispenser is a Socorex Calibrex 520, 2 ml with 0.05 ml divisions.

Paraffin oil overlay

ADH, LDC, ODC and control and one tube of the OF media is overlaid with sterile paraffin oil, to approximately 5 mm depth. For ease of use, paraffin oil can be sterilized in a Schott bottle to which a 2 ml volumetric dispenser is added (Socorex, Calibrex 520, 2 ml with 0.05 ml divisions).

Incubation

Plate and tubed media are incubated at the appropriate temperature and duration for the appropriate bacterium; 25°C and 2–5 days is used as a general rule. Refer to incubation guidelines in Table 2.3.

2.6 API Identification Systems

API 20E	bioMérieux, Marcy l'Etoile, France
API 20NE	bioMérieux, Marcy l'Etoile, France
API 50CH	bioMérieux, Marcy l'Etoile, France
API rapid A	bioMérieux, Marcy l'Etoile, France
API rapid ID 32 STREP	bioMérieux, Marcy l'Etoile, France
API 20 Strep	bioMérieux, Marcy l'Etoile, France
API Coryne	bioMérieux, Marcy l'Etoile, France
API ZYM	bioMérieux, Marcy l'Etoile, France

Identification using API 20E

There have been a number of reports in the literature about the failure of the API 20E system to identify many *Aeromonas* and *Vibrio* species (Santos *et al.*, 1993). Some of this is due in part to the lack of such information in the API database. However, it is well known that differences in reactions occur between the API system and conventional biochemical tests. This is especially true of the decarboxylases, citrate, urea, indole and VP. Where such differences have been reported in the literature, these have been indicated in the conventional tables (Biochem set). When using the biochemical identification tables in this book, make sure the correct database is referred to, i.e. the conventional database (Tables 4.1 to 4.22), or the API 20E database (Tables 4.23 to 4.25).

MacDonell *et al.* (1982) recommends using a diluent containing 20‰ salt. They used marine salts mix (Instant Ocean) purchased from Aquarium Systems, Mentor, Ohio, which is composed of marine salts with salinities adjusted to 20‰. Kent (1982) recommended suspending bacterial cells in 50% sterile artificial seawater for inoculating the API 20E strips. Artificial seawater salts can also be purchased from Sigma (see Chapter 7, 'Preparation of Media for Cultural Identification'). Alternatively, an inoculum with 2% sterile NaCl works well for the commonly isolated *Vibrio*, *Photobacterium* and *Listonella* species. The optimal NaCl concentration is extremely important for tests such as citrate, urea, MR, VP and indole. For example, *V. parahaemolyticus* gives a negative result for indole when an inoculum of 0.85% NaCl is used, but a positive indole result is achieved when 2% NaCl is used.

Adjust the suspension of cells to an opacity of McFarland number 1 and inoculate the set. It is recommended that the API 20E system is incubated at 25°C for 48 h. The sugars are read at 24 h and the remaining tests are read at 48 h. The decarboxylases (LDC, ODC) and ADH reactions may not be detected until 48 h. As recommended by the manufacturer, a negative nitrate reaction should always have zinc dust added to the reaction cupule to determine a true negative result (see Table 3.1). See Chapter 4 for interpretation and results from the API kits.

3

Interpretation of Biochemical Identification Tests and Sets

3.1 Conventional Media: 'Biochem Set'

Positive sugar fermentation results may be recorded at 24 h. All liquid media must be incubated for at least 48 h, and sometimes longer, for positive results to be achieved. This applies in particular to ADH, LDC, ODC, aesculin, citrate and MRVP. For interpretation see Table 3.1.

3.2 Identification Tests and their Interpretation

This section contains notes on the interpretation of identification tests and any problems that may occur. A description of each test is also found in Chapter 7 'Preparation of Media for Culture and Identification'.

Aeromonas salmonicida fluorescent antibody test (FAT)

A FAT for the Australian strain of *Aeromonas salmonicida* has been noted to cross-react with *Aeromonas sobria*.

Aesculin

A positive result is recorded when the colour is pitch-black and when half or more of the tube is blackened (MacFaddin, 1980). Shades of grey should not be recorded as positive. Results should be read up to 3 days. Some organisms will not grow in the aesculin medium despite the optimal salt concentration. These are *Listonella anguillarum*, *L. pelagia* and *Vibrio ordalii*. These organisms are all negative for aesculin production.

Some bacteria have been reported to produce melanin (Coyne and Al-Harthi, 1992) and in the aesculin medium will cause blackening of the medium (Choopun et al., 2002). Therefore, it is recommended that an aesculin test showing blackening of the medium be tested for true hydrolysis of aesculin by determining the loss of fluorescence in long-wave UV light at 354 nm. Aesculin will fluoresce and therefore the presence of fluorescence will indicate that the aesculin has not been hydrolysed (MacFaddin, 1980).

Some plastic tubes are UV-opaque; therefore, test for fluorescence by holding a Wood's UV lamp over the top of an opened tube, or pour the contents into a Petri dish to observe presence or absence of fluorescence. When using UV light, protect the eyes with UV-opaque glasses.

ADH

Some strains of *Vibrio* may show contradictory results for ADH when using Møller's or Thornley's medium. These are *V. mediterranei*, *V. mytili*, *V. orientalis*, most strains of *V. splendidus* I, and some strains of *V. tubiashii*. They are all positive according to Thornley's method, but negative in Møller's ADH. The glucose in Møller's medium appears to inhibit the reaction due to catabolite repression of the inducible ADH system (Macián et al., 1996). *Photobacterium* species were found

Table 3.1. Interpretation of tests for 'biochem set'.

Test	Reagents added	Result
Carbohydrate fermentation	Nil	Colour change from red to yellow (if using phenol red pH indicator) *See photographic section*
Aesculin	Nil	Dark black colour. All positive reactions should be checked for true hydrolysis, as colour change may be due to melanin production from some organisms such as *V. cholerae*. Test for loss of fluorescence (indicating a positive result) using a UV light such as a Wood's lamp
ADH (Arginine dihydrolase)	Nil	Change of colour from green-brown to bright purple (Møller's method). Thornley's arginine is recommended for some *Vibrio* species *See photographic section*
Citrate	Nil	Change of colour from green to blue *See photographic section*
DNase plate	Flood plate with 1M HCl. Leave 1–2 min and pour off	Examine plate over a dark-coloured tile for zones of clearing around bacterial growth *See photographic section*
ODC, LDC	Nil	Colour change from green-brown to purple/grey colour *See photographic section*
Gelatin/salt plate	Nil	Record growth or lack of growth on 0% or 3% NaCl. Hold plate against a dark background and with transmitted light note whether a zone of gelatin hydrolysis can be seen around the bacterial colony. This can be seen as a zone of clearing on the 0% NaCl side or as a zone of opacity on the 3% NaCl side. Chilling the plate at 4°C, or flooding with ammonium sulphate may improve readability of zones *See photographic section*
H₂S	Nil	Blackening of medium especially along the inoculum line
Indole	Add 3–7 drops of Kovács reagent	Formation of upper pink layer in tube. *See photographic section* The test solution must have NaCl (2% final concentration) added to it if the organism requires NaCl for growth
Motility – tube method	Nil	Motile bacteria grow and spread through the semi-solid gel. Spreading is seen as diffuse growth emanating from the line of inoculation. Non-motile bacteria do not migrate from the line of inoculation.

Test	Reagent/Method	Interpretation
Motility – hanging drop method	Nil	Hanging drop. Place a drop of saline on to a coverslip. Mix in some organisms from a plate or broth culture in logarithmic phase of growth. Invert the coverslip on to a glass slide that has been prepared with 3–4 mm dobs of plasticine or equivalent, so that the drop containing the organisms is suspended from the coverslip. The coverslip may be placed directly on to the microscope slide, but must be examined before the slide has dried. Care must also be taken that the liquid is not pulled from the slide if it comes into contact with the microscope stage apparatus, as the appearance of bacterial cells being drawn to the side of the coverslip may give the appearance of motility. Examine under ×40 or ×100 objective and inspect bacterial cells for definite individual motion that causes the cell to traverse the glass slide. Do not mistake motility for Brownian motion, which is seen as movement of the cell within a small confined area, usually the diameter of the cell
MR (Methyl Red)	Add 3–5 drops of Methyl Red reagent after 48 h incubation (test for Voges-Proskauer (VP) reaction first)	Persistence of red colour The reaction should not be tested less than 48 h after incubation and preferably 3 days because a false-negative reaction may be obtained. Aliquots can be taken and tested daily. A negative reaction should be incubated for 3 days *See photographic section*
VP	Remove 200 µl of medium from MRVP tube and place into a 0.6 ml microfuge tube. Add one drop each of VP I and VP II reagents (can use reagents from API 20E kit)	After 10–20 min, development of pink-red colour indicates a positive reaction. If an aliquot is negative at 24 h, incubate for a further 24 h and repeat test. The temperature of incubation is important for VP as some organisms may be negative at 37°C, but positive at 25°C
Nitrate	Add 5 drops of each reagent, Nitrate A and Nitrate B. Can use in-house prepared reagents, or reagents from API 20E kit	Formation of a red colour is positive For negative reactions add a match-head size amount of zinc dust. Formation of a pink colour confirms a negative nitrate result, whereas no further colour with Zn indicates a positive reaction
OF medium (oxidative fermentative)	Nil	Record formation of yellow colour (glucose fermentation) in tubes. Record growth seen in one or both tubes. Growth in the tube without paraffin oil (open tube) indicates an oxidative organism. Growth in both the open and covered tube, indicates the organism is a facultative anaerobe and grows with and without oxygen
ONPG	Nil	Yellow colour is positive. Clear or no colour is recorded as negative
TCBS (thiosulphate-citrate-bile salts-sucrose agar)	Nil	Record growth of yellow (Y) or green (G) colonies, or record 'no-growth' *See photographic section*
'Vibrio discs' 0/129 10 µg and 150 µg	Nil	Record a zone around the disc as 'sensitive', and no zone as 'resistant'. A zone of 9 mm is classified as susceptible for the 0/129 150 µg disc

Notes: Some organisms will not grow in the aesculin medium despite the optimal salt concentration. These are *L. anguillarum*, *L. pelagia* and *V. ordalii*. These organisms are all negative for aesculin production.

Other tests: For a description of other tests, such as hippurate hydrolysis etc, see under Media (Chapter 7).

to produce alkaline products in Thornley's medium, yet none possessed a constitutive ADH system when tested with more sensitive analytical methods (Baumann *et al.*, 1971; West and Colwell, 1984). Other *Vibrio* species such as *Listonella* (*Vibrio*) *anguillarum*, *Photobacterium* (*Vibrio*) *damselae* ssp. *damselae*, *V. fluvialis* and *V. furnissii* are positive for ADH by both methods. Thornley's method is the recommended method for *Vibrio* species (Macián *et al.*, 1996).

Carbohydrate fermentation

In some reports in the literature, the terms utilization and fermentation have been used interchangeably and it has been difficult to ascertain which result was reported from which test method. Fermentation should be used to describe a result from the breakdown of a carbohydrate or 'sugar' reaction such as used in the conventional biochemical media and the API 20E. A pH indicator in the medium is able to detect the acid change caused by the breakdown products. Utilization should only refer to tests that determine an organism's ability to use a substance as a sole carbon source. There is usually no pH indicator in the medium and growth is observed as an increase in opacity or turbidity of the media. The exception is Simmons citrate medium, which determines an organism's ability to use citrate as a sole carbon source. There are no other nutrients in this medium. On the other hand, Christensen's citrate method contains other nutrients and is not a utilization test for citrate as a sole carbon source (Cowan and Steel, 1970; MacFaddin, 1980).

Carotenoid pigment detection

A few drops of 0.01% aqueous CR (Congo Red) is placed on a few isolated colonies growing on a medium such as AO agar. After 2 min the colonies are rinsed with water. The colonies develop a red colour if the test is positive and the colour can last for a number of hours. CR detects the presence of extracellular galactosamine glycan (Johnson and Chilton, 1966, citing, E.J. Ordal, personal communication).

Catalase

A loopful of bacterial growth is removed from the culture plate and smeared to a glass slide. A drop of 30% hydrogen peroxide is placed on to the bacterial cells and the appearance of bubbles indicates a positive test. When removing cells from an agar medium that contains blood, care must be taken not to transfer any of the blood-containing medium with the bacterial growth, as the presence of blood may lead to a false-positive result.

CR

Used for the detection of carotenoid pigment. See under carotenoid pigment detection.

Decarboxylases

ADH, ODC and LDC should be inoculated with a heavy inoculum of the organism. Incubate for a minimum of 2 days and maximum of 14 days. Add salt to give a final concentration of 2% for marine organisms. Without NaCl many tests may produce a false-negative reaction.

Flexirubin pigment

Flexirubin pigment may be produced by some of the members of the *Flavobacteriaceae* family. For its detection, growth is taken from a culture plate (usually AO medium) and smeared to a glass slide, which is placed on a white background such as a piece of paper. The colony mass is flooded with 20% KOH and examined for an immediate colour change to reddish purple or brown. It may be helpful to place two bacterial masses on the glass slide, one that is flooded with KOH and the other to act as a control for colour differentiation. The KOH can also be used directly on the colony growth on a culture plate; however, the colour change may not be noticeable if there is only a thin layer of growth (Bernardet *et al.*, 2002). A reversion to the original colour occurs when the bacterial cells are flooded with an acidic solution (Reichenbach *et al.*, 1989).

Gliding motility

The demonstration of gliding motility of the *Cytophaga–Flavobacterium–Bacteroides* group may be difficult as much by choice of culture technique and as by the definition of gliding motility. The organism needs to be grown on a low-nutrient-concentration agar such as that of AO, which contains 1.5% w/v agar. Gliding is affected by the amount of surface moisture and best results are obtained with freshly poured plates and incubation in a humid atmosphere. Gliding is observed by direct microscopic observation of the swarming edge of bacterial growth on a thinly poured plate after overnight incubation. A high-powered dry lens is used. Gliding is defined as 'movement that is continuous and regularly follows the long axis of the cells that are predominantly organized in bundles during movement'. It should be noted that spreading colonies are not necessarily an indication of gliding motility, as spreading may result from other forms of surface translocation mechanisms (Henrichsen, 1972). Gliding can usually be suspected if a colony has a rhizoid shape on AO agar; however, this will not be seen if the agar surface is too dry. Nor will gliding motility be seen if bacterial growth is suspended in saline. Gliding motility can also be observed in a hanging drop prepared from a liquid culture (Bernardet *et al.*, 2002).

Haemolysis

Haemolysis is observed on BA. A clear zone of lysis of the red cells around a bacterial colony is referred to a β-haemolysis (βH). A greenish tinge seen around some colonies such as some of the *Streptococci* species, is referred to as α-haemolysis (αH). *Streptococcus iniae* shows complete β-haemolysis on sheep blood agar, but only partial haemolysis when the medium is supplemented with human or bovine blood.

Some *Vibrio* species are haemolytic on BA without salt, but non-haemolytic on salt-containing media such as MSA-B, even though their preferred growth medium is with Na ions. It is suggested that haemolysins may be produced when the organism is under stress.

Indole

A false-negative result may be recorded for salt-requiring organisms if there is an insufficient final concentration of NaCl in the test medium. Growth may be seen in the test medium; however, the organism may not express the enzyme unless the NaCl concentration is optimal. To achieve a 2% NaCl final concentration that is optimal for the majority of marine organisms, add 500 μl of 20% NaCl stock solution to 5 ml of test medium. Refer also to section 2.5 (Inoculation of Biochemical Identification Sets).

KOH

Used for the detection of flexirubin pigment. See under flexirubin pigment.

Luminescence

Luminescence may be detected by growing the organism on an appropriate growth medium. Nutrient broth No. 2 (Oxoid) (25 g), NaCl (17.5 g), KCl (1.0 g), MgCl$_2$.6H$_2$O (4.0 g), agar (12.0 g), distilled water (1000 ml) (Furniss *et al.*, 1978). However, reliance on luminescence needs to be treated with caution, as expression of luminescence appears to be dependent on a number of factors, including media, and ideally should be measured with a luminometer rather than by eye (J. Carson, Department of Primary Industries, Water and Environment, Tasmania, 2003, personal communication). Luminescence is optimal after 18–24 h of incubation at 25°C. It is also dependent upon aerobic conditions, so that broth cultures need to be shaken to aerate the medium before luminescence can be detected. Luminescence is detected by observing the plates or broth culture in a dark room and allowing the eyes to adjust to the dark for 5 min. However, if possible, luminescence should be measured from a broth culture using an instrument such as a Wallac Microbeta Plus liquid scintillation counter, which measures relative light units (Manefield *et al.*, 2000).

Organisms that are positive for luminescence include *Photobacterium leiognathi*, *P. phosphoreum*, *Vibrio fischeri*, *V. logei*, *V. orientalis*,

V. splendidus biovar I (Furniss *et al.*, 1978; Lunder *et al.*, 2000).

Organisms that are negative for luminescence include *Aeromonas salmonicida*, *Enterovibrio norvegicus*, *Listonella anguillarum*, *L. pelagia*, *Moritella marina*, *M. viscosa*, *Photobacterium angustum*, *Photobacterium damselae* ssp. *damselae*, *Plesiomonas shigelloides*, *Vibrio agarivorans*, *V. alginolyticus*, *V. brasiliensis*, *V. calviensis*, *V. campbellii*, *V. coralliilyticus*, *V. diazotrophicus*, *V. fluvialis*, *V. furnissii*, *V. gazogenes*, *V. halioticoli*, *V. ichthyoenteri*, *V. lentus*, *V. metschnikovii*, *V. natriegens*, *V. navarrensis*, *V. neptunius*, *V. nereis*, *V. nigripulchritudo*, *V. ordalii*, *V. pacinii*, *V. parahaemolyticus*, *V. penaeicida*, *V. proteolyticus*, *V. rotiferianus*, *V. salmonicida*, *V. splendidus* biovar II, *V. tapetis*, *V. tubiashii*, *V. vulnificus*, *V. wodanis* and *V. xuii* (Furniss *et al.*, 1978; Lunder *et al.*, 2000; Gomez-Gil *et al.*, 2003a,b). *Vibrio fischeri* NCMB 1281^T was reported to be negative for luminescence (Lunder *et al.*, 2000).

Strains of *V. cholerae*, *V. logei*, *V. salmonicida* and *V. harveyi* are variable for luminescence (Furniss *et al.*, 1978; Lunder *et al.*, 2000).

MR

The MRVP medium (Difco) should be inoculated with a heavy inoculum of the organism. Salt needs to be added if required as a growth factor by the organism under identification. Incubate at either 25 or 37°C according to the organism's requirements (see Tables 2.2, 2.3, 2.4). Incubate for 48 h, and then detect the presence of acetylmethylcarbinol by adding 3–5 drops of MR. Continuation of a red colour indicates a positive result. For a negative reaction the red colour disappears as the MR reagent is added to the reaction tube. The success of this test for a positive result is dependent on incubation time, not the amount of growth, although obviously growth must be seen in the tube to indicate that the organism has grown in the medium. The test can be carried out at 24 h if desired, by removing an aliquot (200 µl) to a microfuge tube and testing for MR by the addition of 1 drop of MR reagent. A negative reaction must be incubated for a further 24 h and re-tested after this time before being classified as negative. MR may take 3 days for a positive reaction.

Oxidase test

Oxidase test strips are available commercially, and these are recommended because of standardization of the reagent. For the manual oxidase test use the oxidase reagent tetramethyl-*p*-phenylenediamine. Prepare a 1% solution in distilled water and in a light-protecting bottle and store at 4°C. For use, place a few drops of the reagent on to some filter paper. Using a wooden matchstick such as an orange stick or a platinum loop, smear bacterial growth on to the moistened filter paper. Development of a purple colour within 10–30 s denotes a positive.

A nichrome loop should not be used, as it causes false-positive reactions. This test should not be performed from organisms grown on media that contain sucrose or nitrates. Therefore TCBS is an unsuitable medium from which to determine an oxidase reaction (Furniss and Donovan, 1974; Jones, 1981).

TCBS

This medium is used to identify an organism as a *Vibrio* species. However, there are some *Vibrio* species that do not grow on TCBS. These are *Moritella* species, *Photobacterium leiognathi*, *Vibrio hollisae* and *Vibrio ordalii*. *Listonella* (*Vibrio*) *anguillarum* may show slow growth on TCBS. Strains of *Vibrio fischeri* may show weak or negative growth.

TCBS is not strictly selective and some other organisms will grow, although at a reduced colony size. Species of *Aeromonas* and *Enterococcus* will grow as small (1 mm) yellow colonies on TCBS plates. *Proteus* species may grow as 1 mm yellow/green colonies on TCBS. *Plesiomonas* does not grow well on TCBS. Some yellow colonies of *Vibrio* species may revert to green colonies after a few days growth as they use up the sucrose in the medium (Oxoid manual).

VP

The MRVP medium (Difco) should be inoculated with a heavy inoculum of bacterium. Salt needs to be added if required by the organism. Incubate at either 25 or 37°C according to the organism's requirements (see incubation Table 2.3). As a general rule use an incubation temperature of 25°C for all bacteria isolated from aquatic environments and aquatic animals. Incubate for 48 h, and then detect acetoin production using the reagents. VP can be tested at 24 h if there is sufficient growth in the test medium. A negative result, however, should be re-incubated for 48–72 h. Test an aliquot by placing 200 µl of the incubated medium into a 0.6 ml microfuge tube. Add one drop each of VP reagent A and VP reagent B. Commercial reagents from the API 20E kit work well with the Difco MRVP medium. Examine for red colorization up to 20 min. Although the two tests can be performed in the same tube, it is preferable to test the reactions at 24, 48 and 72 h. The incubation time influences the production of acetoin.

Vibriostatic agent

Resistance to the vibriostatic agent pteridine (0/129) has been noted in some developing countries. Microbiologists are advised to continue using 0/129 as a means of differentiating *Vibrio* species from *Aeromonas* species, but caution may be required in some countries (Huq *et al.*, 1992; Nair and Holmes, 1999). A zone of 9 mm is classified as susceptible for the 0/129 500 µg disc for *Vibrio* species (Bernardet and Grimont, 1989). A zone size of 22 mm is considered sensitive for *Photobacterium damselae* ssp. *damselae* (Love *et al.*, 1981). *Vibrio cholerae* strain 0139 is resistant to the 0/129 500 µg disc (Albert *et al.*, 1993; Islam *et al.*, 1994).

3.3 Using the Biochemical Identification Tables

The following sections are devoted to the identification of an unknown organism using phenotypic or biochemical tests. The variations in the literature for different strains are also recorded, so that the microbiologist will be aware of the difficulties encountered when identifying particular species. Hopefully this system should not prove too tedious for the user of the manual; it is intended to give the users more confidence in naming an organism and to alert them to potential difficulties.

In some cases, species are relatively newly described and different researchers have obtained different biochemical reactions. Rather than presenting a consensus result, both results are listed. It must be stressed that difficulties arise when assessing results of some phenotypic tests when two different versions of a biochemical test have been used. Where differences were noted from different journal articles, these are either stated in the tables, or are detailed under the notes on interpretation and the problems encountered, section 3.4. Other notes are from the author's experience with particular species. Phenotypic tests are still the primary identification system, and therefore microbiologists must be aware of the difficulties encountered with some tests for particular organisms. Some species have been newly described on one strain only and as more information is gathered about this species it may become evident that the type strain is not in actual fact representative of the species (Janda and Abbott, 2002). Therefore subsequent strains isolated and identified as belonging to this species may show slightly different results in the biochemical tests. Therefore, reporting these differences may assist the user of this manual with identification. For example the type strain of *Vibrio lentus* is negative for mannitol fermentation, yet all other species reported thus far are positive. Likewise with *Vibrio penaeicida*, where the type strain is negative for indole and 50% of other strains are positive. The atypical *Aeromonas salmonicida* group is another example. In this case, different biochemical results are reported from strains isolated from different fish species. For the diagnostic laboratory trying to identify an unknown isolate, it may be difficult to say for certain that the isolate is an atypical *A. salmonicida*. However, with the phenotypic details listed of all isolates from different fish species, it may enable a more definitive identification, or at least an identification can be made with slightly more confidence.

There are limitations to all phenotypic (and genotypic) identification systems (Janda and Abbott, 2002). By detailing the biochemical variations it is hoped that the user of the manual will be assisted in making a more informed identification.

3.4 Interpretation and Identification of Genera and Species

In the biochemical identification tables (Tables 4.1–4.31), where authors have recorded a different result for the same type strain, then these results are listed separately. Rather than present a consensus from the literature, listing the different results provides an indication of those tests that are likely to produce disparate results for the individual organisms. However, for some results a 'v' or variable reaction is recorded. Variable results may have been reported in the literature or the organism itself may be variable in its fermentation of the biochemical.

This next section provides information about discrepancies found in the literature or information that is specific to the identification of a certain organism.

Aeromonas spp.

Aeromonas spp. – motile strains

All strains of motile Aeromonas species are positive for the fermentation of glucose and maltose, and ONPG. All strains are negative for urease and fermentation of inositol and xylose and are resistant to the vibriostatic reagent 0/129.

Aeromonas hydrophila ssp. hydrophila and A. hydrophila ssp. dhakensis

A subspecies of A. hydrophila, A. hydrophila ssp. dhakensis, has been described (Huys et al., 2002b). The subspecies is differentiated from A. hydrophila ssp. hydrophila by the following tests. A. hydrophila ssp. dhakensis is negative for acid production from L-arabinose, and negative for utilization of methyl α-D-mannoside, L-fucose and L-arabinose, whereas A. hydrophila ssp. hydrophila is positive for acid production from L-arabinose and positive for utilization of methyl α-D-mannoside, L-fucose and L-arabinose (Huys et al., 2002b).

There are many reports in the literature of variable results for LDC. Some reports state that A. hydrophila (the type strain ATCC 7966) is negative for LDC (Kaznowski, 1998; Nielsen et al., 2001). However, there are other reports of the type strain ATCC 7966 being positive for LDC (Abbott et al., 1992; Huys et al., 2002b). One of the reasons that there are variable reports for LDC may be due to misinterpretation of the test, or the use of different methods to detect LDC. In the conventional tube test, LDC is a pale purple colour, particularly in comparison with the ADH tube. Also, a stronger result may be obtained after 48 h incubation than after 24 h incubation.

There are also varying reports for utilization of citrate and reaction for MR. Citrate is said to be positive for 60% of strains of A. hydrophila, and MR is positive for 53% of strains (Abbott et al., 1992).

Aeromonas popoffii

Type strain LMG 17541 is indole-negative, but other strains are indole-positive (Huys et al., 1997b).

Aeromonas salmonicida: non-motile Aeromonas – general

Identification of an isolate as A. salmonicida is based on Gram-negative rod, negative motility, catalase positive, oxidase positive, production of acid from glucose, resistance to vibriostatic agent 0/129, and lack of growth at 37°C (Shotts et al., 1980). The initial division for A. salmonicida was based on production of a brown water-soluble pigment on media containing tryptone. However, these criteria are not necessarily reliable as there are many reports of atypical strains producing pigment, and strains that grow at 37°C (Austin, 1993).

Pigment production occurs under aerobic conditions, whereas no pigment production is seen under anaerobic conditions (Donlon et al., 1983). It has also been shown that pigment production can be reduced in the presence of D-glucose. In the case of atypical A. salmonicida strains from goldfish in Australia, which are thought to belong to the subspecies nova, pigment production was intense after 3 days on Columbia agar, and light-brown at 6 days on TSA. Addition of 0.1% w/v of glucose delayed pigment production, and 0.15% (w/v) of glucose completely inhibited pigment production (Altmann et al., 1992).

Differences in biochemical reactions have been noted when Aeromonas salmonicida strains are incubated at different temperatures such as 11, 18 and 28°C (Hahnel and Gould, 1982). For consistency, an incubation temperature of 22°C is recommended and was used in the study by Koppang et al. (2000).

The A-protein layer, a major virulence factor for A. salmonicida ssp. salmonicida, can be detected using TSA supplemented with CBBA where bacterial colonies containing the A-protein layer are seen as blue-coloured colonies, and A-protein-negative colonies appear white (Evenberg et al., 1985). Autoagglutination in distilled water also indicates the presence of the A-protein layer.

However, these tests are not reliable indicators of the *in vivo* virulence of *A. salmonicida* (Bernoth, 1990; Olivier, 1990).

Aeromonas salmonicida – *atypical strains*

The term 'atypical' is used to describe isolates that are slow-growing, have slow or no pigment production, and biochemical characteristics different from *A. salmonicida* ssp. *salmonicida* and the subspecies *achromogenes*, *masoucida*, *salmonicida* and *smithia*. The atypical strains represent a diverse group of organisms, and are difficult to identify to species level because of the range of phenotypes. A recent study highlighted the need for standardized phenotypic tests to reduce inter-laboratory variation, and also suggests that 'atypical' isolates should be defined as any isolate that does not fit the existing classification of *A. salmonicida* ssp. *salmonicida* (Dalsgaard *et al.*, 1998). Therefore, to assist the microbiologist in the identification of these organisms, biochemical tests are reported in this manual from a number of different isolates from different fish species (although the list is not extensive). Results that have been achieved using the standardized recommended methods are indicated in the Table 4.1. See Chapter 7 for media composition.

Pigment production can vary and may depend on the media used. In a study of atypical strains by Hänninen and Hirvelä-Koski (1997), FA detected the highest number of strains with pigment (100%), followed by BHIA (86%) and TSA (74%). *A. salmonicida* ssp. *achromogenes* produced pigment on FA, TSA and BHIA, but not on nutrient agar. The addition of L-tyrosine to TSA or BHIA leads to an increase in pigment detection; however, pigment is still detected earlier on FA, within 3–7 days at 20°C. TSA, BHIA and FA are not recommended for primary isolation, as the number of organisms detected is not as high as when BA is used. They are, however, suitable for subculture (Bernoth and Artz, 1989; Austin, 1993).

Brucella spp.

The *Brucella* species isolated from marine mammals are zoonotic (Brew *et al.*, 1999) and as such should be dealt with in a class III biological safety cabinet.

Isolates from seal and an otter, and some of the strains from cetaceans, have an absolute requirement for 10% CO_2. Most of the strains from cetaceans do not require CO_2 atmosphere for growth (Foster *et al.*, 2002).

Brucella species from aquatic mammals give a numerical profile of 1200004 in the API 20NE. The manufacturer's database will identify this as 'good identification for *Moraxella phenylpyruvica*' (Foster *et al.*, 1996a).

Carnobacterium spp.

Carnobacterium is distinguished from *Lactobacillus* by its ability to grow at pH 9.0 but not on acetate agar (pH 5.4) or at pH 4.5.

When testing *Carnobacterium piscicola* for carbohydrate fermentation, discrepancies between results for some carbohydrates have been reported in the literature. In particular, a weak reaction for a positive fermentation of a carbohydrate tested using Phenol Red Broth base has been reported as negative when Purple Broth basal medium is used in the test. Sorbitol and lactose may be positive when Phenol Red Broth base is used, but negative when Purple Broth basal medium is used (Toranzo *et al.*, 1993).

Carnobacterium, Lactobacillus, S. iniae, Vagococcus *and* Renibacterium *differentiation*

The following characteristics in Tables 3.2 and 3.3 give a guide to the differentiation of these genera and species. *Renibacterium salmoninarum* is differentiated mainly by its slow growth and specific growth requirements. Use KDM2 and KDMC media for culture and isolation. Plates are incubated at 15°C for up to 2 months with initial growth of pin-point colonies visible between 2 and 8 weeks. The other genera all grow within 1–3 days of incubation on general-purpose agar such as BA.

Table 3.3 was kindly supplied by Dr Jeremy Carson (Department of Primary Industries, Water and Environment, Tasmania).

Citrobacter freundii

Strains of *Citrobacter freundii* show variation in reactions for ADH, ODC, fermentation of

Table 3.2. *Carnobacterium, Lactobacillus, Vagococcus* and *Renibacterium* differentiation.

Test	Lactobacillus spp.	Carnobacterium piscicola	Carnobacterium divergens	Lactococcus garvieae	Lactococcus piscium	Vagococcus fluvialis	Vagococcus lutrae	Vagococcus salmoninarum	Renibacterium salmoninarum
Growth on acetate agar (RAA)	+	−	−	−	−	−	−	−	
ADH	v	+	+	+	−	−	−	−	
Acid from mannitol	v	+	−	+	+	+	−	−	−
H₂S	−	−	−	−	−	−	ND	+	
Lancefield group D reaction		Positive		Negative		Weak, delayed reaction	ND	Weak, delayed reaction	
Lancefield group N reaction				Positive[a]	Positive[b]	Positive	ND	Negative	
Streptococcal grouping kit groups A-G (Oxoid)		D		Negative	Negative	D weak reaction	ND	D weak reaction	Negative

[a]Eldar *et al.* (1999) reported a positive result. Teixeira *et al.* (1996) reports as negative.
[b]Schmidtke and Carson (1994) reported a negative result.

Table 3.3. Differential tests for some non-fastidious fish pathogenic Gram-positive cocci and rods.

Test	L. garvieae*	C. piscicola	V. salmoninarum[a]	L. piscium	S. iniae
Gram	+	+	+	+	+
Cell shape	ec	sr	sr/cb	c (sr/cb)[b]	c
Haemolysis	α	–	α	–	α[c], β[c,d]
Catalase	–	–	–	–	–
VP (plate)	89%+	+	–[#]	+	–
H₂S	–	–	+	–	–
Bile-aesculin	+	+	62%+	–	–
PYR	+	+	+	–	95%+[e]
ADH	+	+	–	–	71%+[e]
Aesculin	+	+	+	+	+
Glucose	+	+	+	+	+
Galactose	+	65%+	–	+	+
Lactose	–	35–60%+	–	+	–
Maltose	90%+	94%+	17%+	+	+
Mannitol	90%+	88%+	4%+	+	+
Raffinose	–	29%+	–	+	–
Salicin	98%+	+	87%+	+	+
Sucrose	7%+	+	52%+	+	+
Sorbitol	–	–	4%+	–	–
Trehalose	+	+	87%+	+	+
Glycerol	–	40–100%+	–	–	–
Inulin	–	0–100%+	–	–	–
L-arabinose	–	–	–	– (+)[b]	–
Dulcitol	–	–	–	–	–
Fructose	+	+	+	+	+
Starch	–	0–100%+	13%+	–	+
Xylose	–	–	–	–	–
Adonitol	–	–	–	–	–
Melibiose	–	30%+	–	+ (–)[b]	–

All reactions determined by miniaturized microtitre tray tests (Schmidtke and Carson, 1994) or conventional tests. [#]Type strain is positive; [b]original description reaction in brackets; [c]α on bovine blood, β on sheep blood. ec, Elongated coccus; sr, short rod; cb, cocco-bacilli; c, coccus; PYR, L-Pyrrolidonyl-β-naphthylamide; ADH, arginine dihydrolase. *Lactococcus garvieae (syn. Enterococcus seriolicida).
[a]Schmidtke and Carson, 1994; [b]Eldar et al., 1994; [c]Weinstein et al., 1997; [d]Vuillaume et al., 1987; [e]Dodson et al., 1999.

sucrose, melibiose, amygdalin and salicin (API 20E), regardless of source or geographical location (Toranzo et al., 1994).

Cryptococcus spp. (yeasts)

All Cryptococcus species are positive for urease, whereas Candida species are negative.

Edwardsiella spp.

Edwardsiella hoshinae may smell like Plesiomonas shigelloides, which has a strong, pungent, sweetish smell. Vibrio (carchariae) harveyi ATCC 35084 has a similar pungent smell, but not as strong as P. shigelloides. E. hoshinae was reported as indole-positive by Grimont et al. (1980), but negative or weak results were reported by Farmer and McWhorter (1984). Likewise for TSI, Farmer and McWhorter (1984) reported a negative result, whereas a positive result was recorded by Grimont *et al.* (1980).

Edwardsiella ictaluri will grow on Brilliant Green agar, and Salmonella–Shigella agar. MR and VP are positive and negative, respectively, at both 37 and 20°C. There is no growth at 42°C. The organism grows in 0–1.5% NaCl, but not in 2% NaCl.

Enterococcus spp.

Enterococcus faecalis and *E. faecium* are positive for Lancefield group D antigen, grow at 45°C, and grow in the presence of 6.5% NaCl. Reliable tests for differentiation between the two species are acid production from L-arabinose (*E. faecalis* is negative and *E. faecium* is positive) and pyruvate utilization (*E. faecalis* is positive and *E. faecium* is negative).

Enterococcus seriolicida is now known as *Lactococcus garvieae*.

Family *Flavobacteriaceae*

Many samples from the aquatic environment, be they freshwater or marine, will produce yellow-pigmented colonies on culture. It is important to identify whether these colonies are pathogens or saprophytes. Clinical information will be extremely useful in knowing how much time and effort to put into their identification. Thus, if bacterial cells, usually long and thin, are seen adhering to the surface or epithelium of aquatic animals, then genera of the *Flavobacteriaceae* Family may be suspected. To aid in their identification, it is important for the microbiologist to understand the complexities of this family. The following information will assist in this regard. Much of this information has been taken from a paper by Bernardet *et al.* (2002), setting out their proposed guidelines for describing members of the family *Flavobacteriaceae*.

The *Cytophaga–Flavobacterium–Bacteroides* phylum contains the family *Flavobacteriaceae*, which consists of 18 genera and two unaffiliated organisms – *Cytophaga latercula* and *C. marinoflava* – that remain genetically misclassified at this stage. At various times, this group has been known as the yellow-pigmented rods, *Flavobacterium*-like, or *Cytophaga* group or *Flexibacter* group,

and the genera and species within this group have undergone numerous name changes and reclassifications in recent years. The type genus is *Flavobacterium*.

Genera of the Family *Flavobacteriaceae* consist of a group of halophilic organisms (tolerating saline conditions), some of which are psychrophilic (tolerating cold temperatures). They are described as having cells that are Gram-negative, 1–10 μm long and 0.3–0.6 μm wide, with some species forming filamentous flexible cells and others coiled or helical cells. Some members exhibit a gliding motility and the rest are non-motile. The optimum temperature for growth for all genera is 25–35°C, with some species being psychrophilic or psychrotolerant. Growth is aerobic for most of the genera; however, microaerophilic or anaerobic conditions are required for some genera.

Sphingolipids are absent and this feature differentiates the Family *Flavobacteriaceae* from the Family *Sphingobacteriaceae* (Tables 3.4 and 3.5). None of the genera in the Family *Flavobacteriaceae* digests crystalline cellulose when tested with filter paper, and this characteristic distinguishes them from the genus *Cytophaga*, which only contains species that digest crystalline cellulose. It is important to differentiate the presence of the enzyme cellulase from other enzymes that can degrade cellulose derivatives, such as carboxymethylcellulose or hydroxyethylcellulose, which may be present in some species of the *Flavobacteriaceae* Family. Only cellulase is capable of degrading crystalline cellulose, which is tested using filter paper.

A number of other families are contained under the *Cytophaga–Flavobacterium–Bacteroides* phylum. Like the Family *Flavobacteriaceae* these include genera that produce yellow-pigmented colonies that may be cultured from samples received from the aquatic environment. Some of these species include *Cyclobacterium marinum*, *Cytophaga hutchinsonii*, *Flexibacter flexilis*, *Marinilabilia salmonicolor*, *Pedobacter heparinus*, *Sphingobacterium spiritivorum* and others.

Most of the members of the *Flavobacteriaceae* Family are aerobic except for *Capnocytophaga*, *Coenonia*, *Ornithobacterium* and *Riemerella*. An appropriate isolation technique is culture on to BA and incubation in a carbon-dioxide enriched atmosphere with 5–10% CO_2 or a commercial gas generated atmosphere that achieves 5% O_2, 10% CO_2 and 85% N_2. Growth is

Table 3.4.   Differential characteristics of the genera within the family *Flavobacteriaceae*.

Genus	Host	Pigment type	Seawater requirement	Gliding motility	Atmosphere	Growth on MCA
Aequorivita	Sea water, sea ice	Carotenoid	Variable	Negative	Aerobic	Not tested
Bergeyella	Human	None	Negative	Negative	Aerobic	No growth
Capnocytophaga	Human and dog	Flexirubin	Negative	Positive	Microaerophilic	No growth
Cellulophaga	Marine alga and beach mud	Yes	Variable	Positive	Aerobic	Not tested
Chryseobacterium	Fish, marine mud, human, cow's milk, soil	Flexirubin	Negative	Negative	Aerobic	Variable
Coenonia	Peking duck	Not tested	Negative	Negative	Microaerophilic	No growth
Empedobacter	Human	Flexirubin	Negative	Negative	Aerobic	Positive
Flavobacterium	Fish, water, sea ice, soil, mud, marine lake, Antarctica	Yes	Negative	11 species, positive	Aerobic	Not tested
Gelidibacter	Sea ice	Carotenoid	Positive	Positive	Aerobic	Negative
Myroides	Human	Flexirubin	Negative	Negative	Aerobic	Positive
Ornithobacterium	Turkey	None	Negative	Negative	Microaerophilic	Negative
Polaribacter	Seawater, sea ice, marine lake	Carotenoid	Positive	Negative	Aerobic	Not tested
Psychroflexus	Antarctica	Carotenoid	Variable	Variable	Aerobic	Negative
Psychroserpens	Antarctica	Carotenoid	Positive	Negative	Aerobic	Negative
Riemerella	Duck, pigeon	Variable	Negative	Negative	Microaerophilic	Negative
Salegentibacter	Antarctica	Carotenoid	Negative	Negative	Aerobic	Not tested
Tenacibaculum	Marine algae, marine sponge, fish	Carotenoid, or weak reaction	Variable	Positive	Aerobic	Not tested
Weeksella	Human	None	Negative	Negative	Aerobic	Positive
Zobellia	Red marine alga, marine sediment	Flexirubin	Positive	Positive	Aerobic	Not tested

Some genera only produce one type of pigment, others none, whereas other genera contain species that produce either a flexirubin-type pigment, a carotenoid-type pigment or both. A 'yes' in this column indicates that both or one type of pigment may be produced. MCA, MacConkey agar.

either poor or absent under anaerobic conditions (Bernardet *et al.*, 2002).

Agar digestion is positive for *Cellulophaga*, *Zobellia* and variable for *Flavobacterium*. None of the other genera in the Family *Flavobacteriaceae* digests agarose.

Researchers describing new species in the Family *Flavobacteriaceae* are recommended to follow the guidelines for description and identification of new taxa and species in Bernardet *et al.* (2002).

Flavobacterium columnare

Genetic diversity is reported amongst strains; however, phenotypically they are similar. Genomic group I, which contains the type strain IAM 14301[T], are 50% positive for the nitrate test. Nitrate reductase is also variable for genomic group II, whereas strains in genomic group III are negative for nitrate. Strains for genomic group II

and III are able to grow at 37°C but not at 15°C, whereas strains of genomic group I are variable, with 85% showing growth at 15°C and only 75% at 37°C. The genomovars are distinguishable by PCR (Triyanto and Wakabayashi, 1999). See section 6.1, Molecular Identification by PCR using Specific Primers.

Tenacibaculum maritimum

Reactions for *T. maritimum* in the API ZYM are reasonably consistent in the literature. The API ZYM should be incubated at 22–25°C with overnight incubation.

Hafnia alvei

H. alvei may be differentiated from *Yersinia ruckeri*, as *H. alvei* is positive for xylose fermentation whereas *Y. ruckeri* is negative.

Table 3.5. Further differential characteristics for the genera in the family *Flavobacteriaceae*.

Genus	Aesculin	Catalase	DNase	Gelatin	Glucose acid	Indole	ONPG	Nitrate	Sucrose	Urease
Aequorivita	+	+	75% Neg	+	Neg	NT	NT	Neg	Neg	V
Bergeyella	Neg	+	Neg	+	Neg	+	Neg	Neg	Neg	+
Capnocytophaga	V	V	NT	V	+	Neg	V	V	+	V
Cellulophaga	NT	+	V	V	V	NT	NT	V	V	V
Chryseobacterium	+	+	+	+	+	+	V	V	V	V
Coenonia	+	+	NT	Neg	+	Neg	+	Neg	Neg	Neg
Empedobacter	Neg	+	+	+	+	+	Neg	Neg	Neg	V
Flavobacterium	10/14 =+	+ or weak	V	11/14 =+	V	Neg	V	V	V	V
Gelidibacter	+	+	+	V	+	Neg	Neg	Neg	Neg	Neg
Myroides	Neg	+	+	+	Neg	Neg	Neg	Neg	Neg	+
Ornithobacterium	Neg	Neg	Neg	Neg	V	Neg	+	Neg	Neg	+
Polaribacter	V	+ or weak	NT	V	+	Neg	V	Neg	V	Neg
Psychroflexus	V	+	+	V	V	Neg	Neg	Neg	Neg	V
Psychroserpens	Neg	+	Neg	V	Neg	Neg	V	Neg	Neg	Neg
Riemerella	V	+	NT	+	+	V	Neg	Neg	Neg	V
Salegentibacter	+	+	+	+	V	NT	+	+	V	V
Tenacibaculum	Neg	+	+	+	Neg	NT	NT	V	NT	NT
Weeksella	Neg	+	Neg	+	Neg	+	Neg	Neg	Neg	Neg
Zobellia	+	+	+	+	+	+	+	+	+	Neg

V, variable; NT, not tested; Neg, negative; +, positive; ONPG, *o*-nitrophenyl β-D-galactopyranoside.
10/14 =+ means that 11 of the 14 species in the genus are positive.

Lactococcus spp.

The *Lactococcus* genus was created to accommodate 'lactic' or group N streptococci. There are currently four species, *L. garvieae*, *L. lactis*, *L. piscium*, *L. plantarum* and *L. raffinolactis*. There are varying reports in the literature for group N results for *L. garvieae* and *L. piscium*.

Lactococcus garvieae

Different results have been noted for ribose, depending on the identification system used. Ribose is positive when using the API 50CH system, but negative with the API Rapid ID 32 Strep system (Vela *et al.*, 2000). *L. garvieae* may be biochemically indistinguishable from *Streptococcus parauberis* (Doménech *et al.*, 1996). *Lactococcus garvieae* (*Enterococcus seriolicida*) is phenotypically similar to *L. lactis*; however, they can be differentiated by sensitivity testing to clindamycin. *L. garvieae* is resistant to clindamycin, whereas *L. lactis* is sensitive (Elliott and Facklam, 1996). *L. lactis* is now known as *L. delbrueckii* ssp. *lactis*.

L. garvieae was divided into biotypes (Vela *et al.*, 2000); however, these divisions have been disputed. Further research indicates that *L. garvieae* strains are relatively homogeneous regardless of geographical location or aquatic host. What is important in biochemical identification is that the density of the inoculum be standardized when using the API Rapid ID 32 Strep system and that the bacterial cells to be used for the inoculum are taken from BA plates. Reliable, repetitive results were only obtained when the inoculum density was adjusted using a spectrophotometer, to an optical density of 0.8 at wavelength of 580. Also, cells grown on BA produced reliable results for biochemical tests, whereas growth taken from TSA (Oxoid) produced variability in biochemical tests. However, slight variability was still observed for biochemical tests such as β-galactosidase, hippurate, β-mannosidase, acid from melezitose, N-acetyl-β-glucosaminidase and acid from pullulan (Ravelo *et al.*, 2001).

Listonella anguillarum

Some strains are negative for citrate in both the conventional tube test and in the API 20E. The MR tube may show weak growth and a weak positive reaction at 48 h after incubation and some strains may be negative. Growth on TCBS tends to be slow and there is substantially less growth when compared with the amount of growth on MSA-B or BA plates. Strains isolated from an estuarine environment may grow better in 0.85% NaCl, whereas those strains from a marine environment show optimal growth with 2–3% NaCl, and this can be seen clearly on a gelatin-salt plate. Use growth on this plate or growth on a BA plate, as opposed to growth on MSA-B, as a guide to the salt requirement of the organism when preparing the NaCl for the inoculating fluid for either the conventional biochemical set or the API 20E. Strain NCIMB 2129 is reported to be negative for fermentation of sorbitol and trehalose (Benediktsdóttir *et al.*, 1998). Differences in citrate, indole and MR are reported for NCIMB 2129, NCIMB 6 and ATCC 14181 (Myhr *et al.*, 1991; Benediktsdóttir *et al.*, 1998; Lunder *et al.*, 2000). Serotype 01 strains tend to be positive for L-arabinose (Toranzo and Barja, 1990).

Mesophilobacter marinus

Mesophilobacter marinus describes marine coccobacilli that morphologically resemble the *Acinetobacter* – *Moraxella* group.

Moritella spp.

Moritella (*Vibrio*) *viscosus* often needs prolonged incubation times for growth in test media. Most strains require 1% peptone for optimal growth. Results for salt requirements and temperature growth limits are inconsistent in the literature. One study suggests 1–4% NaCl and a temperature of 25°C (Lunder *et al.*, 2000). A second study suggests 2–3% NaCl with no growth at 4% NaCl, and a temperature range of 4–21°C with no growth at 25°C (Benediktsdóttir *et al.*, 2000).

Photobacterium spp.

Photobacterium damselae *ssp.* damselae

There are some biochemical differences according to the biochemical method used and the amount of salt in the inoculum. The tube urease is positive; however, in the API 20E the urease result seems to depend on the salt concentration in the inoculum. When an inoculum of 0.85% NaCl is used, then the urea gives a positive result, but at an inoculum concentration of 2% NaCl, a negative urea may be obtained. Time of incubation is important also, as urea may be weak or negative at 24 h incubation, but a strong positive at 48 h incubation. *P. damselae* ssp. *damselae* has been grouped into biotypes (Pedersen *et al.*, 1997) and many of these are listed in the conventional biochemical table for *Photobacterium*. Table 3.6 lists other biochemical tests not listed in the conventional biochemical table 4.20.

Photobacterium damselae *ssp.* piscicida

The fermentation reactions may be improved under anaerobic conditions. This is easily achieved by overlaying the medium with sterile paraffin oil. The glucose reaction in the API 20E kit may be very weak or negative. Likewise, the VP reaction may be variable, and research into the biochemical properties of isolates from France, Greece, Italy and Japan found that all isolates tested were negative for VP in the API 20E system (Bakopoulos *et al.*, 1995).

Pseudoalteromonas spp.

Pseudoalteromonas citrea

First described by Gauthier (1977), the type strain is ATCC 29719. Similar strains have been described by Ivanova *et al.* (1998). Some phenotypic tests are slightly different, which is thought

Table 3.6. Additional tests for differentiation of biotypes of *P. damselae*

Biotypes	1	2	3	4	5	6	7	8	9
Test									
Lipase	+	−	+	+	+	+	+	+	+
Cellobiose	+	+	+	+	−	−	+	+	+

Taken from Pedersen *et al.* (1997).

to be due to the different ecological niches. The phenotypic reactions described in this manual are those of the type strain.

Pseudomonas anguilliseptica

Finnish isolates and the type strain are positive for gelatin, whereas Japanese isolates are negative. Motility is only seen when the organism is grown below 20°C (Michel *et al.*, 1992). There are three serotypes of *P. anguilliseptica* based on a heat-labile surface K antigen. Serotypes from eels in Japan are K– and K+; ayu have K+ serotype as do the Finnish isolates (Wiklund and Bylund, 1990). Antisera against *P. anguilliseptica* may show cross-reaction against *Pseudomonas putida* and *Listonella anguillarum*; however, if it is used diluted at 1:5 to 1:10, this may overcome the cross-reaction (Toranzo *et al.*, 1987).

Renibacterium salmoninarum

This organism is the causative bacterium of BKD. It is a fastidious and slow-growing organism and the culture medium used for isolation and growth, KDM2, may give inconsistent performance due to lot-to-lot variations in peptone (Evelyn and Prosperi-Porta, 1989). Consistency in growth can be improved either by using a nurse culture technique, or by supplementation of the growth medium with spent culture medium. The nurse culture technique involves placing a 25 μl drop of nurse culture (prepared from cells suspended in saline or peptone) into the centre of a plate inoculated with the test sample. Once the drop is dry the plates are inverted, sealed to prevent drying, and incubated at 15°C for 21 days. The supplementation with spent culture broth method can be achieved by preparing new KDM2 medium with 1.5% (v/v) of spent culture medium (Evelyn *et al.*, 1990).

Streptococcus **spp.**

Streptococcus agalactiae

Lancefield group B. Different reactions are reported for *S. agalactiae* isolated from human, fish and animal sources, and this may in part be due to the optimum incubation temperature for enzyme reactions. For example, hippurate may be positive at an incubation temperature of 25°C but negative at 37°C for isolates from fish. The conventional VP method shows a negative reaction for *S. agalactiae*, but it is positive in the API rapid ID 32 (Vandamme *et al.*, 1997). Strains of *S. agalactiae* from mammalian sources are β-haemolytic; however, most of the strains isolated from fish were reported to be non-haemolytic and were initially identified as *S. difficile*. They are now identified to be *S. agalactiae* group B, capsular type Ib (Vandamme *et al.*, 1997). These isolates are slow-growing and adhere strongly to the agar plate, which makes it difficult to resuspend the cells in an inoculating fluid. Conventional biochemical tests may need to be incubated for more than 48 h for sufficient growth to occur in the tubes. MR and VP reactions may be weak, and poor growth is seen in the tryptone water (indole test). The ADH test using conventional media may take more than 3 days' incubation before a slight colour change is seen.

For the *S. agalactiae* strains that were isolated from a disease outbreak in seabream and wild mullet in Kuwait, there are some differences in the biochemical reactions compared with the ATCC type strain (Evans *et al.*, 2002). Therefore, the results of the *S. agalactiae* isolates from the seabream and the wild mullet are listed separately in the API rapid ID 32 Table 4.30.

Streptococcus iniae

The organism shows complete β-haemolysis on sheep BA, but only partial haemolysis when the medium is supplemented with human or bovine blood. Variations in haemolysis may be seen between strains.

Strains isolated from either fish or humans have a slightly different biochemical profile.

S. iniae strains isolated from cultured and wild fish in the Red Sea were similar in biochemical profile to other *S. iniae* strains, except for a negative ADH in the majority of strains, and late fermentation of galactose and amygdalin as detected in the API 50CH, which was incubated for 72 h before results were recorded (Colorni *et al.*, 2002). The biochemical profile achieved with the API 20 Strep agrees with other results recorded for fish isolates where 30% of strains may be negative for ADH, and that variation may be seen in the fermentation of mannitol (Dodson *et al.*, 1999).

Streptococcus parauberis

This may be biochemically indistinguishable from *Lactococcus garvieae* (Doménech *et al.*, 1996).

Streptococcus uberis

There are different reports for the reaction to pyrrolidonylarylamidase. *Bergey's Manual of Determinative Bacteriology* (1994) reports a negative reaction, whereas Collins *et al.* (1984) and Doménech *et al.* (1996) report a positive reaction.

Tenacibaculum maritimum

See under *Flavobacteriaceae*.

Vagococcus salmoninarum

The genus *Vagococcus* was proposed to describe motile, Lancefield group N, Gram-positive cocci; however, Schmidtke and Carson (1994) described *V. salmoninarum* as negative for the N antigen.

Vibrio spp.

Vibrio agarivorans

These strains degrade agar and the effect can be seen as small depressions on an agar plate after 3 days' incubation at 25°C. This effect increases over time, and by about 7 days the effect is very marked. This was particularly the case with isolates from abalone in Western Australia. The carbohydrate fermentation reactions may be difficult to read and this effect has been noticed with all strains tested. All carbohydrate fermentation tubes should be compared alongside the glucose reaction, as this will be a bright yellow colour.

The positive reactions are easy to determine, but those that appear as a weak reaction are, in fact, negative reactions. These may appear as a pale yellow colour in the middle of the tube. However, after 48 and 72 h incubation these reactions are quite clearly negative. *V. agarivorans* strains first isolated and reported from the Mediterranean Sea show a positive reaction for aesculin and a negative reaction for gelatin (Macián *et al.*, 2001b). Strains from Western Australia all show a negative reaction for aesculin using the conventional tube test, and a positive reaction for gelatin when the plate method is used. However, the gelatin result in the API 20E is negative after 48 h incubation at 25°C.

Vibrio alginolyticus *and* Vibrio harveyi

These may be difficult to differentiate (Table 3.7). The VP test differentiates them, as does the additional test for fermentation of D-glucuronate (Baumann *et al.*, 1984). *V. alginolyticus* swarms and completely covers an MSA-B plate in 24 h at 25°C, whereas *V. harveyi* has a slow or spreading type of growth. *V. alginolyticus* is urease-negative, whereas *V. harveyi* is usually urease-positive (50% reported as positive).

Vibrio cholerae

Strains of *Vibrio cholerae* are negative for aesculin; however, some strains cause blackening of the medium due to the production of melanin (Coyne and Al-Harthi, 1992). To determine true hydrolysis as opposed to melanin production, the aesculin tube must be tested for loss of fluorescence (Choopun *et al.*, 2002). The aesculin solution will fluoresce under long-wave UV light (354 nm), therefore a loss of fluorescence indicates hydrolysis of aesculin and a positive test result (MacFaddin, 1980). The plastic tubes that the aesculin medium is in may be UV opaque. Therefore the test medium can be poured into a Petri dish and a Wood's UV light held over the dish in a darkened room. An uninoculated

Table 3.7. Differentiation of *V. alginolyticus* and *V. harveyi*.

	VP	10% NaCl	Growth at 42°C	Urease	D-Glucuronate
Vibrio alginolyticus	+	+	+	–	–
Vibrio harveyi	–	–	–	v	+

control and a negative result will show a ring of fluorescence around the edge of the liquid, whereas a true positive aesculin will not show the fluorescence. Protect the eyes with UV-opaque glasses.

VIBRIO CHOLERAE NON-01.   Strains isolated from ayu in Japan were found to be negative for ODC (Kiiyukia *et al.*, 1992). Normally, *V. cholerae* isolates are positive for ODC.

Vibrio coralliilyticus

Six strains of this newly described *Vibrio* species have been characterized. There is some biochemical variation between strains, with two of the six strains positive for ADH as detected in the API 20NE, and four strains negative (Ben-Haim *et al.*, 2003).

Vibrio fischeri

A positive result is obtained for VP in the API 20E system, but VP is negative in MRVP medium (Difco).

Vibrio fluvialis *and* Vibrio furnissii

Both species are indole-negative when tested for indole production in peptone water containing 1% NaCl. When indole is tested using heart infusion broth, 14% of *V. furnissii* are positive, and 4% of *V. fluvialis* are positive. Both species are positive for indole in the API 20E system. The API 20E database may identify these species as *Aeromonas* species (Brenner *et al.*, 1983). *V. furnissii* may show a weak pink indole reaction when approximately 12 drops of Kovács reagent is added to a 48 h tryptone water (5 ml). The colour disappears after a minute and appears as a dirty orange-brown colour. *V. furnissii* will show a negative ADH reaction if no NaCl is added to the reagent tube. ADH will be positive when the final salt concentration is 2% (Buller, 2003). *V. furnissii* strains produce gas in the glucose tube, whereas *V. fluvialis* strains do not produce gas. *V. furnissii* strains (57%) may be positive for L-rhamnose, whereas *V. fluvialis* strains are negative, and 63% of *V. fluvialis* are positive for cellobiose, whereas *V. furnissii* is negative (Lee *et al.*, 1981; Brenner *et al.*, 1983).

Vibrio (carchariae) harveyi ATCC 35084

Citrate and gelatin (plate method) are negative at 24 h and need to be read at 48 h or longer. The culture has a pungent smell similar to *Plesiomonas shigelloides*.

Vibrio harveyi

The strain ATCC 14126 is positive for luminescence and negative for urease, whereas ATCC 35084 is negative for luminescence and positive for urease (Alcaide *et al.*, 2001). Strains NCIMB 1280, ATCC 14126 and ATCC 14129 are reported to be gelatin-positive (Baumann *et al.*, 1984; Benediktsdóttir *et al.*, 1998). Strains ATCC 14126 and ATCC 14129 are reported to be negative for fermentation of sorbitol, whereas strain ATCC 35084 is positive after 2 days (Alcaide *et al.*, 2001; Buller, 2003). Haemolysis is variable, and is reported to be positive against sheep red blood cells (Alcaide *et al.*, 2001). See Table 3.7 for further tests to differentiate *V. harveyi* and *V. alginolyticus*.

Vibrio neptunius

The type strain is positive for ADH; however, other strains are reported to be negative for ADH (Thompson *et al.*, 2003).

Vibrio parahaemolyticus

In the API 20E, strain ATCC 43996 shows a more clearly positive reaction in the arabinose test when incubated for 48 h and prepared with a 2% NaCl inoculum. However, ADH shows a weak positive reaction with a 2% NaCl inoculum and incubation at 48 h. With the conventional tests, indole will record a negative result if normal saline is used as the inoculum. A positive result is recorded with 1.5% and 2% NaCl as a final concentration.

Vibrio mediterranei

At 48 h these are large, creamy, mucoid colonies on MSA-B. After 24 h incubation and an inoculum of 2% NaCl, the ADH in the API 20E is positive and at 48 h incubation LDC may also be positive. However, in the conventional tube tests only LDC is positive after 48 h incubation.

Vibrio proteolyticus

The growth swarms quickly across an agar plate, within 24 h, and could be mistaken for *V. alginolyticus*. In the conventional citrate test a positive result may not be seen until 48 h. A positive urea is seen in the conventional tube test, but is negative in the API 20E at 48 h and with an inoculum of 2% NaCl.

Vibrio scophthalmi

The type strain in positive for chitin hydrolysis, whereas other strains are negative (Farto *et al.*, 1999).

Vibrio scophthalmi and *V. splendidus* biovar I may be difficult to differentiate based on phenotypic tests and this is compounded due to the variation of phenotypic test results between strains in each species (Farto *et al.*, 1999). Indole, ONPG and fermentation of mannitol appear to be the best differentiating phenotypic tests. *V. scophthalmi* is negative for all three, whereas *V. splendidus* biovar I is positive for all three.

Vibrio splendidus

Variations in phenotypic tests are reported between different research groups. One of the tests that seems to be a problem for consistency is ADH, although the majority of reports suggest that ADH is positive. Most *V. splendidus* I strains were found to be ADH-positive when tested using Thornley's method for ADH, yet negative when tested by the more commonly used Møller's arginine method (Macián *et al.*, 1996). *V. splendidus* I type strain (NCMB 1, ATCC 33125) is reported to be positive for ADH by both Lunder *et al.* (2000), and Benediktsdóttir *et al.* (1998). However, the results vary for *V. splendidus* biovar 2 type strain NCMB 2251, which is reported to be negative by Lunder, yet positive by Benediktsdóttir. Both papers report a negative sucrose, whereas the same type strain is reported positive by Farto *et al.* (1999). The strains are also variable for ONPG. Variations such as these make it difficult for the diagnostic laboratory to identify an unknown.

There are two biovars and both have been implicated in disease. Biovar 1 isolates are positive for luminescence and fermentation of mannose, ribose and melibiose, whereas biovar 2 isolates are negative for these tests. In addition, biovar 1

is negative for fermentation of glycerol, and degradation of chitin, with biovar 2 being positive (Benediktsdóttir *et al.*, 1998).

There are reported differences in virulence between strains of *V. splendidus* and it is suggested that valine aminopeptidase as tested by API ZYM is a virulence factor because pathogenic strains produced the enzyme whereas non-pathogenic isolates did not (Gatesoupe *et al.*, 1999).

Vibrio vulnificus

The division of *Vibrio vulnificus* into two main biovars is under contention. The original classification by Tison (1982) was that biovar 1 isolates were positive for indole and ODC and were mainly isolated from human clinical sources. Biovar 2 strains were isolated mainly from diseased eels and were indole- and ornithine-negative. However, there is some overlap between these divisions, and therefore the reactions for the different strains and the geographical site and source of isolation have been included in the conventional identification tables, and in the API 20E tables for profile number. It has been reported that approximately 20% of *V. vulnificus* strains are sucrose-positive (Arias *et al.*, 1998). Note that sucrose medium that is autoclaved for sterility may give false-positive reactions. Filter-sterilized sucrose medium is preferred for more accurate fermentation results.

Vibrio wodanis

A range of salt requirements, 0.5–5% NaCl, exists with these strains (Lunder *et al.*, 2000). One study found that 1 of 16 isolates showed growth at 0.5% NaCl, and 16 of 23 isolates showed growth at 4% (Benediktsdóttir *et al.*, 2000).

Yersinia spp.

Yersinia frederiksenii

These are rhamnose-positive strains, formerly called atypical *Y. enterocolitica*. *Y. intermedia* comprises rhamnose-positive, melibiose-positive, raffinose-positive strains, formerly called *Y. enterocolitica* or *Y. enterocolitica*-like. *Y. kristensenii* refers to sucrose-negative strains.

Yersinia ruckeri

Strains of *Y. ruckeri* may be divided into motile and non-motile types, a finding that is geographically biased. Strains from the UK tend to be non-motile, with occasional non-motile strains reported from Canada and Norway. Non-motile strains lack lipolytic activity when tested using the Tween 20 and Tween 80 tests. Therefore, Shotts-Waltman medium is inappropriate for the differentiation isolation of non-motile strains of *Y. ruckeri* as it contains Tween 80 (Davies and Frerichs, 1989). These authors suggested that non-motile, Tween 80 hydrolysis-negative strains should be referred to as *Y. ruckeri* biotype 2. Results between conventional tube tests and the API 20E system may give variable results. These include citrate utilization, gelatin hydrolysis, VP and nitrate. The nitrate test in the API 20E system may give unreliable results for *Y. ruckeri*, thus a conventional tube nitrate test is recommended. After 24 h incubation at 25°C, citrate and gelatin may be falsely negative, therefore 48 h incubation is recommended. Motility and citrate are negative at 37°C but positive at 25°C. The API 20E may show more VP-positive results than the conventional tube test (Davies and Frerichs, 1989).

Y. ruckeri may be differentiated from *Hafnia alvei* by xylose fermentation. *Y. ruckeri* is negative, whereas *H. alvei* is positive for xylose fermentation.

3.5 Antisera Available

The following are some of the antisera that are available commercially.

Aeromonas salmonicida ssp. *salmonicida*, specific monoclonal antibody (BIONOR Mono AS, BIONOR Aqua, Skien, Norway).

Brucella abortus antiserum (Difco).

Listonella anguillarum, specific monoclonal antibody for serotypes 01, 02, 03 and 04, 05 and 07 for environmental serotypes (BIONOR Mono-Va, BIONOR Aqua, Skien, Norway).

Photobacterium damselae ssp. *piscicida*, specific monoclonal antibody (BIONOR Mono-Pp, BIONOR Aqua, Skien, Norway).

Renibacterium salmoninarum, specific monoclonal antibody (BIONOR Mono-Rs, BIONOR Aqua, Skien, Norway).

Salmonella O and Salmonella H antiserum and specific antiserum for species (Difco).

Staphylase test for identification of *Staphylococcus aureus* (Oxoid).

Streptococcus group A, B, C antiserum (Difco). Antiserum for groups A, B, C, D, F, G (Oxoid).

Vibrio cholerae. Anti-*V. cholerae* 01 serum (Denka Seiken). Bacto-Vibrio-Cholerae antiserum for detecting three serotypes in group 01. These are serotype Ogawa (AB O antigen factors), Inaba (AC O antigen factors) and Hikojima (ABC O antigen factors) (Difco).

Yersinia ruckeri specific monoclonal antibody (BIONOR Mono-Yr, BIONOR Aqua, Skien, Norway).

The BIONOR Mono-Va kit detected serotypes of 01, 02 and 03 of *Listonella anguillarum* and also detected the environmental serotypes of 04, 05 and 07. However, non-specific agglutination against strains of *V. splendidus* and motile *Aeromonas* species occurred. A cellular concentration of 10^8/ml is recommended for positive agglutination (Romalde *et al.*, 1995).

Strains of *Photobacterium damselae* ssp. *piscicida* were detected by the BIONOR Mono-Pp kit, with no cross-reactions detected for *Aeromonas salmonicida*. However, non-specific agglutination occurred from strains of *Actinobacillus pleuropneumoniae*, *Haemophilus parasuis*, *Mannheimia haemolytica* and *Pasteurella multocida* (Romalde *et al.*, 1995).

Renibacterium salmoninarum was successfully detected using the BIONOR Mono-Rs kit; however, strains that lacked the p57 surface protein were not detected (Romalde *et al.*, 1995). No cross-reaction with other Gram-positive organisms such as *Corynebacterium aquaticum*, *Carnobacterium piscicola*, *Enterococcus faecalis* and *Lactococcus garvieae* were detected (Romalde *et al.*, 1995).

The BIONOR Mono-Yr kit detected classical serotypes 01 and 03 of *Yersinia ruckeri*. For the detection of 02 serotypes the test needed to be performed at either 48 h culture growth or from subcultures. The kit is unable to detect the *Y. ruckeri* serotypes 05 and 06 because of the composition of the serotypes included in the kit. A cell concentration of 10^8/ml is needed for positive agglutination. No cross-reaction with other bacterial pathogens was found (Romalde *et al.*, 1995).

4

Biochemical Identification Tables

4.1 Results for Conventional Biochemical Tests – 'Biochem Set'

The following tables (Tables 4.1 to 4.22) are to be used for interpreting the results obtained from in-house or conventional biochemical media. The tables are listed in alphabetical order of the table heading. The organisms in the tables are grouped into those that are considered to be either pathogens in aquatic animals or environmental and saprophytic organisms. The saprophytic, environmental and other species of bacteria are included under 'environmental' so as to assist in differentiating a pathogen from a closely related species and thus obtaining a definitive identification of the unknown organism.

In a number of cases, different or variable biochemical tests were recorded in the literature by different laboratories for the same type strain. Also, differences in a biochemical were recorded for different strains. These results are included here so that the reader will be aware that some test results may not be reliable and that different results may be recorded by different laboratories. In the cases reported in this Manual, the tests that gave variable results appeared to be performed using the same method. It is known that different methods can give different results for the same organism, and therefore it is important to be consistent with the methods used. Where a difference between the conventional biochemical test and the API 20E was reported, this is recorded in the tables, so that the reader is aware of variations for

different test methods for that particular organism. The numbers in the tables report the percentages of positives as reported in the literature when a number of strains were tested. The range of numbers in the 'temp' and 'NaCl' columns refer to the temperature range (°C) and range of salt concentrations (%) at which the organism can grow. A single number indicates the optimum growth temperature or NaCl concentration, respectively.

With the exception of the *Vibrio* tables (Tables 4.21 and 4.22), all other tables list the organisms alphabetically within the groupings of 'pathogenic' or 'environmental' organisms.

The species in the *Vibrio* tables are grouped according to their ODC–LDC–ADH reaction. Therefore all species with a ODC+, LDC+, ADH- result are grouped together. Likewise other ODC–LDC–ADH reaction combinations are grouped together. Use these groupings as a starting point to identify an unknown.

Use the tables in conjunction with the section on interpreting test results and specific reactions noted for some genera and species (see Chapter 3).

An example of a Laboratory Worksheet for recording results is included on page 177.

Use the following schematic (Fig. 4.1) as a guide to which table or tables (Conventional media, 'Biochem set') to use when identifying an unknown. Start with the result for the Gram stain, then cell morphology and, for the Gram-negatives, a further division based on oxidase result.

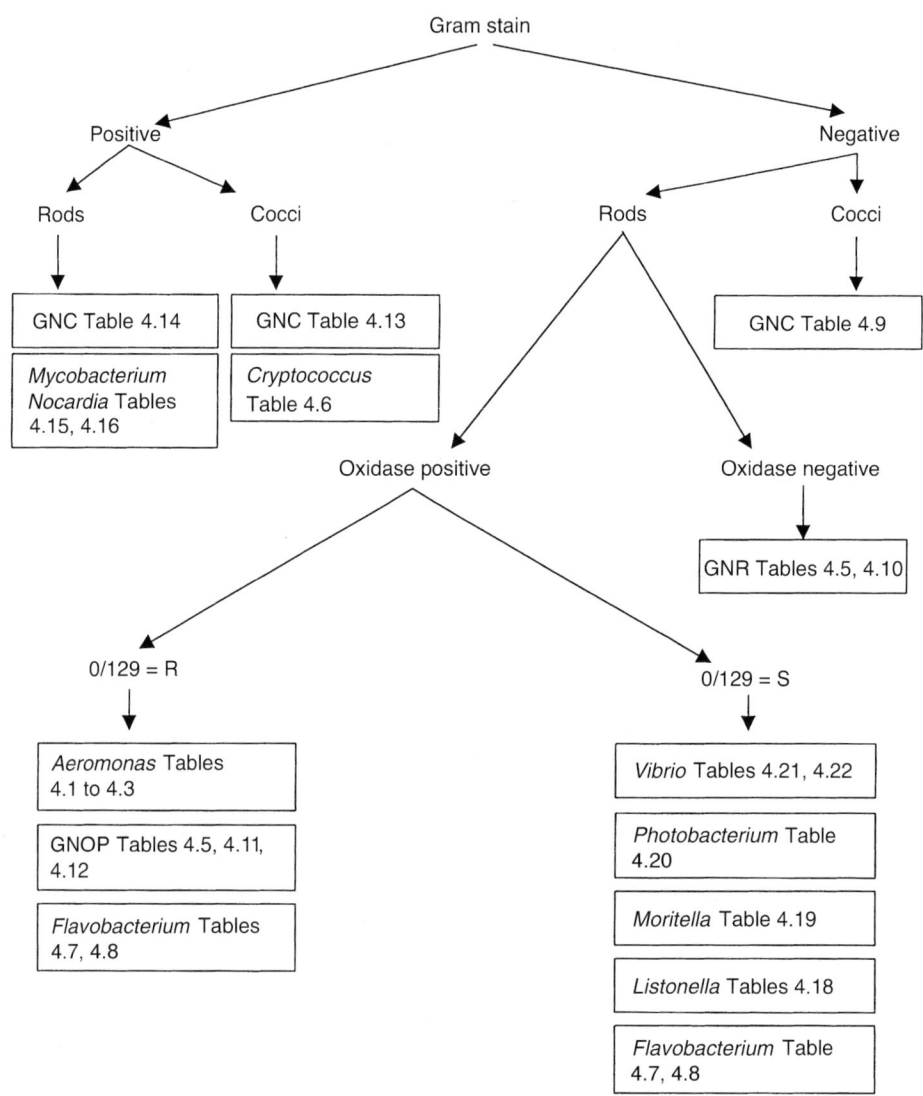

Fig. 4.1. Schematic for using the biochemical identification tables.

4.2 Results for API Kits

Interpretation of API 20E profile

Use the profile index provided by the manufacturer as a guide only. Kent (1982) does not recommend using the profile index provided with the kit because of misidentifications, especially when attempting to identify an organism not found in the commercial database. The API 20E system will even misidentify organisms when the identification is given at the 99% confidence level. For example, Dalsgaard *et al.* (1996) found that the API 20E system misidentified *V. vulnificus*. The system gave a greater than 98% confidence level as *Aeromonas hydrophila*, *Flavobacterium meningosepticum*, *Burkholderia cepacia*, and between 90 and 95% confidence levels as *V. alginolyticus* and *V. parahaemolyticus*. All these 'isolates' were in fact *V. vulnificus*, which they proved using a species-specific DNA probe.

It is recommended that the API 20E system be used with caution. Some additional tests may be required to give a definitive answer. This is particularly so for the identification of species from the genus *Aeromonas*. Adding conventional media of aesculin, MR, and salicin in addition to the API 20E system will improve the likelihood of the correct identification.

The API 20E database listed in this book is a collation from the literature. For some organisms such as *Vibrio vulnificus*, many profile numbers were obtained, indicating the phenotypic strain variation amongst species.

The reactions for the API 20E are listed in three different formats (see Tables 4.23, 4.24 and 4.25). The first format, in Table 4.23, indicates the reactions obtained for the different tests in the same order as the API 20E test strip. The organisms are listed in alphabetical order. The second format, in Table 4.24, lists all organisms in alphabetical order and then the corresponding API 20E profile number. The third format, in Table 4.25, lists the API 20E profile numbers in numerical order. Use this table first when identifying an unknown. Then use Table 4.23 to check the positive and negative reactions obtained. Tables 4.24 and 4.25 should be checked to assess the similarity with close profile numbers and to determine whether additional tests should be done to confirm this initial identification. In some cases the complete 9-digit API 20E profile number was not reported in the journal article, as the tests for growth on MCA and OF were not reported. In these cases the API 20E profile number is recorded in the manual as a 7- or 8-digit profile number.

The numbers in the cells of the tables are the percentage of positive strains recorded with that test result.

Identification using API ZYM

The API ZYM tables are included because many newly described species are tested for the production of enzymes using this kit. It is not generally considered a routine diagnostic kit; however, it is useful for some bacteria from veterinary isolations such as the species from *Pasteurella*, *Actinobacillus* and *Histophilus ovis* and is able to differentiate between species when phenotypic tests may be in doubt (Cousins and Lloyd, 1988). For these organisms the intensity of the colour reactions is consistent for a species and allows species identification. Therefore, the correct inoculum concentration is important. It can also be used to confirm or back-up phenotypic tests when identifying an unknown bacterium isolated from an aquatic animal species.

Species within a genus give very similar results for the API ZYM; however, there are often differences with one or two enzymes between species that can give an indication of the identification of the species. The intensity of the colour reaction is important. Thus it is important to be consistent with the inoculum density. The manufacturer recommends a density of McFarland tube 6.

For *Flavobacterium*, *Flexibacter* and *Cytophaga* species, an incubation time of 12 h is recommended when incubating at a temperature of 18–30°C (Bernardet and Grimont, 1989). For an incubation temperature of 37°C use an incubation time of 4 h as recommended by the manufacturer. In the author's experience isolates from an aquatic environment give better reactions when incubated for 24 h at a temperature of 25°C. However, it must be stressed that if the incubation time and temperature is stated in the tables then use these parameters because the results are reported on these. Some references did not state the time and temperature of incubation that was used, and therefore it is assumed that the manufacturer's recommendations were used.

Improved results may also be obtained if ammonium salt sugar broth is used as the inoculating fluid for the *Flavobacterium/ Cytophaga* group from marine and freshwater sources (Bernardet and Grimont, 1989).

Table 4.1. *Aeromonas salmonicida* (non-motile *Aeromonas* spp.).

Test	Gm	Ox	cat	βH	mot	Pig	ODC	LDC	ADH	Nit	Ind	Cit	urea	mr	vp	aes	G	onpg	OF	arab	glu	inos	lac	malt	man	mano	sal	sor	suc	tre	Xyl	H2S	MCA	TCBS	DNase	temp	NaCl	0129 10	Kf	Amp	Ref
Organism																																									
Pathogenic																																									
Typical																																									
A. salmonicida ssp. salmonicida NCMB 1102 HG3	-	+	+	+	-	+	-	+	+	+	-	-	-	+	-	+	+	+	F	+	+g-	-	-	+	+	+	+	-	-	+	-	-	+	-	+	4–30	0–3	R	S	S	322,450
			No growth 37°C, API 20E LDC may be neg																																						200,475
A. salmonicida ssp. salmonicida ATCC 14174	-	+			-	+		+	+	+	-				-		+		F	-	+		-	+	+	-	-					-									186
A. salmonicida ssp. salmonicida	-r	+	+	+	-	-*	-	-	+	+	-	-	-	-	+	+	+	-	F	+	+g-	-	-	+	+	+	+	-	+	+	-	-	+	-	+	4–30	0.5–3	R			450
'Atypical'			Non-pigmented strain																																						
A. salmonicida ssp. achromogenes NCMB 1110T	-r	+	+	+	-	+ft	-	-	+	+	-	-	-	-	+	-	+	+	F	-	+g-	-	-	+	+	+	-	-	-	-	v	+	+	-	+	4–30	0–4	R	R	R	450,534
			No growth 37°C. API Ind may be neg																																						200,322
A. salmonicida ssp. masoucida	-	+	+	-	-	-	-	+	+	+	-	-	-	+	+	+	+	+	F	+	+g-	-	-	+	+	+	-	-	+	+	+	+	+	-	+	30	0–2	R		R	322,450
			No growth 37°C. In API 20E LDC, VP & H2S may be neg																																						
A. salmonicida ssp. nova	-	+	+	+	-	-	-	-	-	+	+	v	-	-	-	-	-	-	F	-	+g-	-	-	-	+	v	-	-	-	-	-	-	-	-	-	18–25		R	R	R	695
			Haemin requirement																																						
A. salmonicida ssp. smithia	-	+	+	25	-	-	-	-	-	-	-	v	-	-	-	-	+	+	F	-	+	+	-	-	-v	v	-	-	v	+	+	+	-	-	+	5–20	0–2	R	R	R	47,450
			No growth 37°C																																						
American eel	-	+	-		-b	-	-	-	v	+	-	-	-	-	-	-	+	v	F	-	+	-	+	+	-	-	-	-	v	-	-	-	-	-							584
Australian goldfish	-r	+	+		+b	+b	-	-	v	+	-	-	-	-	-	-	-	v	F	-	+g-	-	+	+	+	+	-	-	+	-	-	-	-	-			0.5–3	R			144,825
Australian goldfish	-	+	+		+bt	+bt		-	+	+	-						+								+																135
Baltic sea turbot	-r	-	+		-b	-b		-	v	-	-				>	>	+	-	F	-	+g-		-		-				-					+				R	R	S	200,832
Baltic sea flounder	-r	-	+		-b	-		-	v	>	-				+	+	+	v	F	-	+g-		-		+		+v		+					+				R	S	S	200,832
Baltic sea (Finnish) flounder	-	-	+		-h	-		-	+	>	+				-	+	+	+	F	+	+g-		+		+		-		-			-	+	+				R	S	S	831
Baltic sea dab	-r	+	+		-b	-b		-	-	-	-				+	+	+	-	F	-	+g-		-	+	-		+	+	+	-		-		+w				R	S	S	832
Baltic sea blenny	-r	+	+		-b	+b		-	+	+	-				+	-	+	-	F	-	+g-		+	+	+		-	+	+			+		+				R	R	R	832
Canadian Atlantic cod	-	+			+s				+	+	-				-				F	-	+		-	+	+		+	+	+			+									
Canadian Atlantic salmon Keij strain	-	+			-b	-b			-	+	+				-	-			F	-	+		-	+	+		+		+												186*

Canadian sablefish AS₂	-	+				-		+	+	-		-			F	-	+	-			186*
Danish cod	-	+		+		v	+		-		+		+gv		F		+gv	+		20	0–2	197
Danish sand eel	-	+		+ᵇ	-		-		+		+	+tᵍ⁺		F	+tᵍ⁺		20	0–2	197
Danish turbot	-	-	+	-	-	-		-		+	+	-	+tᵍ⁺		F	-	+tᵍ⁺	-	-		617
English non-salmonid	-	+	+	-	+ᵗ	-	+	+	v	-	+		+	+	v	+	F	+	+gᵖ	+	+	4–37	0–3	40
Norway minnow	-	+		-ᵇ		+w	-		-	+w	-		F	-	+w	+	-			331
South African rainbow trout	-r	+	+	-	+ᵇ		+	+	+	+	+		-	+	+	+	F	+	+	+	+	22		107
pigment positive	+								+			+		+	-			6	3	323
pigment negative									-							56	72	323
Environmental																		
A. salmonicida	-	+	+		-	-v	+	+	+	-		+	-	+	+	+	F	+tᵍ⁺	-	+	35	0	R	R	615
pectinolytica																		

Furunculosis agar (FA) is more sensitive to pigment production than TSA, TSA + t.

P = pigment produced, b = Pigment tested on BA, f = pigment on FA, h = pigment on BHIA, s = pigment production slow, t = Pigment tested on TSA, * = Pigment tested on TSA containing L-tyrosine (TSA + t).

The following methods are recommended so as to standardize testing and reduce discrepancies between different laboratories. Carbohydrate fermentation is performed in phenol red broth (Difco) with 1% carbohydrate. Haemolysis is tested using BA containing horse blood, other methods are from Cowan and Steel (1970) (and in this Manual). For marine isolates the addition of NaCl to a final concentration of 1.5% in 'liquid' tubes such as MRVP, aesculin, nitrate, ADH, LDC, ODC may improve consistency of results. Zones of 20 mm and greater are considered sensitive for Amp and Kf. Incubate at 20°C and read at 7–14 days, Dalsgaard et al. (1998) (ref 200). Refs 197, 450, 831 and 832 also use these methods.

VP at 25°C. Brown pigment

Table 4.2. *Aeromonas* spp. Phenotypic tests according to DNA hybridization groups.

Test	Gm	Ox	cat	βH	mot	pig	ODC	LDC	ADH	Nit	Ind	Cit	urea	vp	aes	G	OF	arab	glu	inos	lac	malt	man	mano	sal	sor	suc	tre	Xyl	DNase	40°C	0129 10	0129 150	Amp
DNA hybridization group																																		
HG 1	-	+	+	+	+	-	-	96	+	+	+	-	-	91	96	+	F	87	+g⁺	-	-	+	96	+	+	-	+	+	-	+	+	R	R	R
HG 2	-	+	+	+	+	-	-	59	94	+	+	+	-	83	+	+	F	+	+g⁺	-	67	+	+	+	+	+	+	+	-	+	-	R	R	R
HG 3	-	+	+	92	+	-	-	55	90	+	+	46	-	90	+	+	F	+	+g⁺	-	54	67	+	+	+	+	+	+	-	+	-	R	R	R
HG 4	-	+	+	10	+	-	-	-	+	+	97	70	-	-	97	+	F	+	+	-	87	+	+	31	+	7	+	+	-	+	87	R	R	R
HG 5	-	+	+	-	-	50	-	-	92	+	+	+	-	-	92	+	F	+	+	-	+	+	+	+	+	+	+	+	-	+	-	R	R	R
HG 6	-	+	+	-	+	-	-	-	87	+	+	+	-	-	+	+	F	87	+	-	+	+	+	+	+	-	+	+	-	+	-	R	R	R
HG 8/10	-	+	+	90	+	-	-	57	+	+	+	91	-	90	10	+	F	14	+g⁺	-	10	+	+	+	+	-	+	+	-	+	+	R	R	R

Gm = Gram reaction, Ox = oxidase, cat = catalase, βH = β haemolysis, mot = motility, pig = brown pigment, ODC = ornithine decarboxylase, LDC = lysine decarboxylase, ADH = arginine dihydrolase, Nit = nitrate reduction, Ind = indole, Cit = citrate, urea = urea hydrolysis, vp = Voges-Proskauer, aes = aesculin hydrolysis, G = gelatin hydrolysis, OF = Oxidative Fermentative, arab = L-arabinose fermentation, glu = glucose fermentation, inos = inositol fermentation, lac = lactose fermentation, malt = maltose fermentation, man = mannitol fermentation, mano = mannose fermentation, sal = salicin fermentation, sor = sorbitol fermentation, suc = sucrose fermentation, tre = trehalose fermentation, Xyl = xylose fermentation, Dnase = hydrolysis of DNA, 40°C = growth at 40°C, 0129/10 = sensitivity to vibriostatic agent 0129 at 10 μg concentration, 0129/150 = sensitivity to vibriostatic agent 0129 at 150 μg concentration, Amp = sensitivity to ampicillin. Numbers refer to percentage of positive strains, - = negative, + = positive reaction. HG = hybridization group.
Data from Abbott et al. (1992), Kaznowski (1998).

Table 4.3. *Aeromonas* spp. – motile.

Test	Gm	Ox	cat	βH	mot	SW	ODC	LDC	ADH	Nit	Ind	Cit	urea	mr	vp	aes	onpg	G	OF	arab	glu	inos	lac	malt	man	mano	sal	sor	suc	tre	Xyl	H_2S	MCA	TCBS	DNase	temp	NaCl	0129 10	0129 150	Amp	Ref
Pathogenic																																									
A. allosaccharophila HG15	–	+	+	+	+	–	–	+	+	+	+	+	–		–	+	+	+	F	+	+g+	–	–	+	+	+	–	–	+	+	–	–	+		+	4-4	0-3	R	R	R	527
A. bestiarum HG2	–	+	+	+	+	–	–	+	+	+	+	+	–	+	+	+	+	+	F	+	+gv	–	+	+	+	+	+	–	+	+	–	+	+		+	4-42		R	R	R	427
A. caviae HG4	–	+	+	10	+	–	30	–	+	+	+	+	–	+	–	+	+	+	F	+	+g–	–	87	+	+	31	+	+	+	+	–	–	+		+	4-37		R	R	R	21, 142
A. hydrophila dhakensis	–	+	+	+	+	–	–	+	+	+	+	60	–	53	+	+	+	+	F	+	+g+	–	–	+	+	+	+	+	+	+	–	–	+	–w		25-42		R	R	R	383
A. hydrophila LMG 2844ᵀ HG1	–	+	+	+	+	–	–	+	+	+	+	+	–	+	+	–	+	+	F	+	+g+	–	–	+	+	+	+	–	+	+	–	–	+	–w	+	4-42	0-2	R	R	R	1, 383, 427
A. jandaei HG9	–	+	+	+	+	–	–	+	+	+	+	75	–	+	+	–	+	+	F	–	+g+	–	–	+	+	+	–	–	–	+	–	–	+	–w	+	4-42	0-3	R	R	R	135, 142, 143
A. veronii ssp. veronii HG10	–	+	+	+	+	–	+	+	–	+	+	+	–	+	+	+	+	+	F	–	+g–	–	–	+	+	+	+	–	+	+	–	–	+	–	+	4-42	0-5	R	R	R	142, 347
Environmental																																									
A. culicicola	–	+	+	+	+	–	–	+	+	+	+	+	–		+	–	+	+	F	–	+g+	–	–	+	+	+	–	–	+	+	–	–			+	4-37		R	R	R	624
A. encheleia HG16	–	+	+	21	+	–	–	–	+	+	+	–	–	v	–	+	+	+	F	–	+g+	–	–	+	+	+	+	–	75	+	–	–	+	–	+	4-37	0-3	R	R	R	241, 379
A. eucrenophila HG6	–	+	+	50	+	–	–	–	+	+	+	v	–	+	–	+	+	+	F	+	+g–	+	+	+	+	+	+	–	+	+	–	–			+	25-37	0-2	R	R	R	1, 379, 420, 427
A. media HG5A/5B	–	+	+	25	–	–	–	–	+	+	+	55	–	+	–	–	+	+	F	+	+g–	–	+	+	+	+	94	–	+	+	–	–	+	–	+	4-37	0-3	R	R	R	15, 294
A. popoffii	–	+	+	–w	+	–	–	–	+	+	–	70	–		+	–	+	+	F	50	+g+	–	–	+	+	+	–	–	–	+	–	50	+	–	+	28-37	0-2	R	R	R	380
A. sobria HG7	–	+	+	+	+	–	–	+	+	+	+	+	–	45	+	+	+	+	F	–	+gv	–	+	+	+	+	v	–	+	+	–	+	+	–	+	30-42	0-5	R	R	R	21, 420

The type strain is indole-negative, other strains are indole-positive

88% VP pos in API 20E

Organism																															temp	NaCl			Ref
A. veronii spp. sobria HG8	−	+	+	+	−	−	+	+	+	50	−	+	+	+	F	−	+	+	+	+g+	−	5	+	+	35	+	+	−	+	+	4-42		R	R	R	21, 142, 427, 456
A. schubertii HG12	−	+	+	+	−	−	+	+	+	−	+	+	+	+	F	−	+	+	+	+g−	−	−	+	+	−	+	−	−	+	−	4-42	0-3	R	R		142, 348
	gel at 22°C, Non-haemolytic sheep blood																																			
A. trota HG14	−	+	+	+	−	−	+	+	+	+	+	+	+	+	F	−	−	+	+	+g+	−	−	+	+	−	+	+	−	+	−	4-42		R	R	R	142
Aeromonas group 501 HG13	−	+	+	+	−	−	+	+	+	−	−	+	+	−	F	−	−	+	+	+g+	−	+	+	+	23	+	+	−	+	−		0-3	R	R		

Table 4.4. Anaerobes.

Test	Gm	Ox	cat	βH	mot	SW	ODC	LDC	ADH	Nit	Ind	Cit	urea	mr	vp	aes	G	onpg	OF	arab	glu	inos	lac	malt	man	mano	sal	sor	suc	tre	Xyl	H₂S	MCA	TCBS	DNase	temp	NaCl	0129 10	0129 150	Amp	Ref
Organism																																									
Pathogenic																																									
Clostridium botulinum E	+r	−	+	+	−	−				−	−		−		−	−		−			+				−		−		−		−	−									141
Rods with oval subterminal endospores with appendages and exosporia																																									
Clostridium perfringens	+r			−						v	−					>		−		−		v	+	+		+	−	v	>	v	−										
Edwardsiella ictaluri oxygen intolerant strain	−	−	+	−	−	−		+	+	+	−	−	−	−	−	−		−	F		+g−	−	+	−		+	+	−	−	−	−	+				25-37					547
Non-motile at 37°C																																									
Eubacterium spp. 841	+r	−	−	−	−	−				−						+	+				+g+	v	+	+			−	−	−	−	−	+				20	0-1				343
Eubacterium spp. 1065	+r	−	−	−	−	−				+						+	+				+g+		+	+	−		−	−	−	−	−	−				20	0-1				343
Eubacterium tarantellae	+r	+	+	−	−	−				−	−		−			−	−				+	+	+	−	−	−	−	−	−	−	−	−		+	25-37		0-1				764
Long unbranched filamentous rods																																									

143

Table 4.5. *Brucella* spp.

Test	Gm	Ox	cat	βH	mot	CO₂	Nit	cit	ind	MR	VP	H₂S	urea	BF	SO	TH	Ala	Asp	Glut	Arg	Orn	Lys	Gal	Rib	Xyl	Ery	Uro	A	M	R	Ref	Host
Organism																												Surface Antigens				
Pathogenic																																
Brucella abortus	–	+	+	v	–		+	–	–	–	–	v	+	+			+	+	+	–	–	–	++	+++	++	+++	+	+			185,404	Cattle
Brucella canis	–	+	+		–	–	+	–	–	–	–	–	+	–			–	–	+	+	++	–	–	+++	–	–	–				185,404	Dogs
Brucella cetaceae	–	+	+		–	–							+	+	+	+	–v	–	+	+	++	++	+	+	+	+	–	+	+	–	267,404	UK dolphin, porpoise
Brucella melitensis	–	+	+		–		+	–	–	–	–	–	+	+			++	++	+++	–	–	–	+	+	+	+++	+	+	+		185,404	Sheep, goat
Brucella neotomae	–	–	+		–		+	–	–	–	–	+	+	–			+	+	++	–	–	+	++	–	+	+++	+				185,404	Desert wood rat
Brucella ovis	–	–	+		–		–	–	–	–	–	+	+	+			+	+	+	–	–	–	–	+	+	–	+				185,404	Sheep
Brucella pinnipediae	–	+	+		–	+						–	+	+	+	+	–v	–	+	–	–	–	–	+	+	+	+–	+	–	–	267,404	Seal, otter
Brucella spp.	–	+	+		–	+						–	+	+	+	+			+					+	+	+		+	–	–	261	Ringed seal, harp seal
Brucella spp.	–	+	+		–	–							+	+	–	+												+			171	Minke whale
Brucella suis 1	–	+	+		–		+	–	–	–	–	+	+	–			–	–	–	++	++	++	++	+++	+++	+++	++	+	+		185,404	Pigs
Brucella suis 2	–	+	+		–		+	–	–	–	–	–	–	–			–	–	++	+	+	+	++	+++	+++	++	+++	+	+		185,404	Pigs, hare
Brucella suis 3	–	+	+		–		+	–	–	–	–	–	+	+			–	–	++	++	++	++	–	+++	+	+++	++	+	+		185,404	Pigs
Brucella suis 4	–	–	+		–		+	–	–	–	–	–	+	+			+	+	++	++	++	++	–	+++	–	+++	++	+	+		185,404	Reindeer
Brucella suis 5	–	+	+		–		–	–	–	–	–	+	+				+	+	++	–	–	+	–	+++	–	+++	+	+	+		404	

Ala = L-alanine, Arg = L-arginine, Asp = L-asparagine, BF = growth on media containing basic fuchsin 20 µg/ml (1/50 000), βH = β haemolysis, CO₂ = carbon dioxide requirement, cat = catalase, cit = citrate, Ery = mesoerythritol, Glut = L-glutamic acid, Gm = Gram reaction, H₂S = hydrogen sulphide, ind = indole, Lys = L-lysine, MR = Methyl Red, Nit = nitrate reduction, Om = DL-ornithine, Ox = oxidase, Rib = D-ribose, SO = growth on medium containing safranin O 100 µg/ml (1/10 000), TH = growth on medium containing thionin 20 µg/ml (1/50 000), Uro = urocanic acid, VP = Voges Proskauer, Xyl = D-xylose, Surface antigens, A, M & R = agglutination with monospecific antisera.

Table 4.6. *Cryptococcus.*

Test	Gm	Ox	cat	βH	mot	Cap	ODC	LDC	ADH	Nit	vp	mr	urea	Cit	Ind	aes	G	onpg	OF	arab	glu	inos	lac	malt	man	mano	sal	sor	suc	tre	Xyl	GT	temp	NaCl
Organism																																		
Pathogenic																																		
Cryptococcus neoformans var. *gattii*	+			–	–	+	+	–	–	+			+					+			+	–	–	–					+	–	–	–	20–42	
Environmental																																		
C. lupi	+													+				–			+	+		+		+	+	+	+	+	+	+	4–25	0–2
Candida spp.	+												–		–																			

No fermentation, only assimilation. Antarctic habitat. Growth on glucose peptone yeast extract

Cap = capsule; GT = germ tube.

Table 4.7. Cytophaga–Flavobacteriaceae–Bacteroides group – pathogenic.

Test / Organism	Gm	Ox	cat	βH	mot	Glid	ODC	LDC	ADH	Nit	Ind	Cit	urea	mr	vp	aes	G	onpg	OF	arab	glu	inos	lac	malt	man	mano	sal	sor	suc	tre	Xyl	H₂S	MCA	CR	KOH	DNase	temp	NaCl	0129 150	NA	TSA	Ref
Pathogenic																																										
Chryseobacterium balustinum	–	+	+	–	–	–v	–	–	–	+v	+	–	–v	–	–	+	+	–v	O	–	+	–	–	–	–	–	–	–	–v	–v	–	–	–	–	+	+v	15–35	0–2	+	NA	+	89,802
Previously Flavobacterium balustinum. Yellow pigment																																										
C. scophthalmum	–	+	+	+	v	v	–	–	–	–	–	–	+	–	–	+	+	–	F	–	+	–	–	+	–	–	–	–	–	–	–	–	–	–	+	+	15–25	0–4			+	556,557
Previously Flavobacterium scophthalmum. Orange pigment																																										
Flavobacterium branchiophilum ATCC 35035	–	+	+	–	–	–	–	–	–	–	–	–	–	–	–	–	+	+	O	–w	–	–	–	+	–	–v	+v	–	+	+	–w	–	–	–	–	–	10–25	0–0.2	S	–	–	92,802
Yellow pigment, no agar hydrolysis. Growth on 1/20 TSA [603]																																										
F. columnare IAM 14301ᵀ	–	+	+	–	–	+	–	–	–	80	–	–	–	–	–	–	+	–	O	–	–	–	–	–	–	–	–	–	–	–	v	+	–	+	+	+	10–37	0–0.5	S	–	–	145,603
Adheres strongly to agar. H₂S from lead acetate medium [135,759] — (other references 111, 139, 298, 299)																																										
F. hydatis	–v	+	+	–	–	+	–	–	–	+	–	–	–	–	–	+	+	+	F	+	+	–	+	+	+	+	+	–	+	–	+	–	–	–	+	+	20–30	0–1	R	+	+	92,720
Previously Cytophaga aquatilis. Mucoid, orange-yellow colonies [89]																																										
F. johnsoniae AHLDA 1714	–	+	+	–	–	+	–	–	–	+	–	+	+s	–	–	+	+	–	F	–	w	–	–	–	–	–	–	–	–	–	–	–	–	+	+	–			S	+	+	135
F. johnsoniae ATCC 17061	–	+	+	–	–	+	–	–	–	+	–	–	+	–	–	+	+	+	O	+	+	–	+	+	+	+	+	–	+	+	+w	–	–	+	+	+	5–30	0–1	S	+	+	145,603
F. johnsoniae DSM 2064	–	+	+	–	–	+	–	–	–	–	–	–	+	–	–	+	+	+	O	+	+	–	+	+								–	–	–	–	+	10–30	0–2	S	+	+	89
F. johnsoniae DSM 29585*	–	+	+	–	–	+	–	–	–	+	–	–	+	–	–	+	+	+	O	+	+	–	+	+								–	–	–	+	+	10–30	0–1	S	+	+	89
F. psychrophilum	–v	+v	+	70	–	w	–	–	2	–	–	–	–	–	–	–	+	–	O	–	–	–	–	–	–	–	–	–	–v	–	–	–	–	–	+	+w	4–20	0–0.8	S	–	–	89,92
Yellow colonies. Growth enhanced in AO media with 0.5% tryptone [90,168]																																										
Tenacibaculum (Flexibacter) maritimus	–	+	+	–	–	+	–	–	–	+	–	–	–	–	–	–	–	+	I	–	–	–	–	–	–	–	–	–	–	–	–	–	–	+	–	+w	14–34	2+	S	–	–	89,801
Pale yellow colonies. No agar hydrolysis. Prepare media with ASW base [90]																																										
T. (Flexibacter) ovolyticus	–	+	+	–	–	+	–	–	–	+	–	–	–	–	–	+	+	–	O	–	–	–	–	–	–	–	–	–	–	–	–	–	–	–	–	+	4–25	3+	S	–	–	324
Pale yellow colonies. NG on TCBS																																										

*This organism is no longer in the DSMZ database.

Numbers refer to percentage of positive strains.

1/20 TSA = TSA prepared at a 1:20 dilution, +s = slow reaction, —w = negative or weak reaction, NG = no growth.

—s = slow reaction, +s = slow reaction, —w = negative or weak reaction, Ov = Oxidative variable results, v = variable reaction, NG = no growth.

Table 4.8. *Cytophaga–Flavobacteriaceae–Bacteroides* group – environmental.

Test / Organism	Gm	Ox	cat	βH	mot	Glid	ODC	LDC	ADH	Nit	Ind	Cit	urea	mr	vp	aes	G	onpg	OF	arab	glu	inos	lac	malt	man	mano	sal	sor	suc	tre	Xyl	H_2S	MCA	CR	KOH	Dase	temp	NaCl	0129 150	NA	TSA	Ref
Aequorivita antarctica	–	+	+	–	–	–	–	–	–	–	–		–v			+	+		O	–	–	–	–	–	–	–	–	–	–	–	–	–	–	–	–	–	0–25	0.5–6				113
Orange-pigmented colonies																																										
Aequorivita crocea	–	+	+	–	–	–	–	–	–	–	–		–			+	+		O	–	–	–	–	–	–	–	–	–	–	–	–	–	–	–	–	+v	0–25	0.5–6				113
Yellow-pigmented colonies																																										
Aequorivita lipolytica	–	+	+	–	–	–	–	–	–	–	–		+v			+	+		O	–	–	–	–	–	–	–	–	–	–	–	–	–	–	–	–	–	0–25	0.5–6				113
Yellow-pigmented colonies																																										
Aequorivita sublithincola	–	+	+	–	–	–	–	–	–	–	–		+			–	+		O	–	–	–	–	–	–	–	–	–	–	–	–	–	–	–	–	–	0–25	1–6				113
Orange-pigmented colonies																																										
Cellulophaga lytica	–	+	+	–	–	+	–	–	–	–	–		–			–	–	+	O		–				–				+				–	–	–	–	15–37	1–2				89
Chryseobacterium (Flavobacterium) gleum	–	+	+	–	–	–	–	–	–	v	+	–	v			+	+	+	–		+		–						–		+	+	+	+	+	+	25–37				+	366
Indole pos with Ehrlichs but negative by Kovács																																										
Chryseobacterium indologenes	–	+	+	–	–	+	–	–	–	v	+		–			+	+	v	O		+		–	+	+				–	+	–	+	+	+	+	+	36					557
Chryseobacterium indoltheticum	–	+	+	–	–	+	–	–	–	–	+		–			+	+	–			+		–	+	+				–			+	+	+			36					557
Chryseobacterium meningosepticum	–	+	+	–	–	–	–	–	–	–	+	v	v		–	+	+	+	O	–	+	–	+	+	v	+	–	–	v	v	–v	w	–	–	–	+	37–42	0–2		+	+	89,802
previously *Flavobacterium meningosepticum*																																										
Cytophaga allerginae	–	+	+	–	–	–	–	–	–	–	–					+	+	–	O		–												–	+	+	–	15–25	0–1				89
Cytophaga arvensicola	–	+	+	–	–	–	–	–	–	–	+					+	+	–	O		–												–	+	+	–	15–37	0–2				89
Cytophaga fermentans	–	+	v	–	–	–	–	–	–	–	–					+	+	–	F		+								+				–	–	–	+	15–25	1–2				89,162
Cytophaga hutchinsonii	–	+	+	–	–	+	–	–	–	–	–		+			v	v	–	O		+												–	–	+	–	15–25	0–1				89
pigmented colonies. Digests cellulose																																										
Cytophaga latercula	–	+	–	–	–	–	–	–	–	+	–		+			+	+	–	O		+								+				+	+	+	+	15–30	1–2				89,494
Bright red pigment																																										
Cytophaga-like	–	+	+	+	–	+	–	–	–	–	–	26	+	–		+	+	+	O		+								–		–	–	–	–		+	15–25	0–4				556
Orange colonies. No agar digestion																																										
Cytophaga marinoflava	–		+			+					+					+	+	+	F		+								+													162

Organism	Notes	References
Empedobacter (Flavobacterium) brevis	Pigmented colonies, fruity odour	89,363
		802
Flavobacterium aquatile		92,802
		89,603
Flavobacterium flevense	Hydrolyses agar	89,533
		92
Flavobacterium frigidarium		376
Flavobacterium gillisiae		533
Flavobacterium hibernum		532,533
Flavobacterium mizutaii		844
Flavobacterium pectinovorum		92,533
Flavobacterium saccharophilum	Hydrolyses agar	92
Flavobacterium succinicans		92,162, 533
Flavobacterium tegetincola		533
Flavobacterium xanthum		533
Flexibacter aggregans	Hydrolyses agar	89
Flexibacter canadensis		89
Flexibacter elegans		162
Flexibacter flexilis	Cells 15–20 μm, Orange-pigmented colonies	89,603
Flexibacter litoralis	Pink pigment	89,162

continued

Table 4.8. *Continued.*

Test	Gm	Ox	cat	βH	mot	Glid	ODC	LDC	ADH	Nit	Ind	Cit	urea	mr	vp	aes	G	onpg	OF	arab	glu	inos	lac	malt	man	mano	sal	sor	suc	tre	Xyl	H_2S	MCA	CR	KOH	Dase	temp	NaCl	0129 150	NA	TSA	Ref
Flexibacter polymorphus	-	+	-	-	-	+											v		O															-	-		22–32	2				89
					Cells >100 µm. Hydrolyses agar. Peach-coloured pigment																																					
Flexibacter roseolus	-	+	-	-	-	-	-	-	-	-						+	-		O		-											-		-	-		22–37	0–2	+	+	+	89,162
					Cells 50–100 µm. Bright-red orange pigment																																					
Flexibacter ruber	-	+	-	-	-	-	-	-	-	-						+	-		O		+								+			-		-	-		22–37	0–2	+	+	+	89,162
					Bright red-orange pigment. Cells 50–100 µm																																					
Flexibacter sancti	-	+	+	-	-	-	-	-	-	-						+	+		O															+	+		15–37	0–1		+	+	89
Flexibacter tractuosus	-	+	+	-	-	-	-	-	-	-						+	+	-	O													–/+		-	-		22–37	0–2		+	+	89,162
					Orange pigment																																					
Marinilabilia salmonicolor	-	+	+	-	-	-	-	-	-	-						+	+		F		+								+			+		-	-		15–37	1–2				89,162
					Salmon-pink colour																																					
Marinobacter hydrocarbonoclasticus	-	+	+	+	+	-			+	+		+				-	-		O		-													-			40–45	2				69
					Previously Pseudomonas nautica																																						
Myroides odoratimimus	-	+	+	-	-	-	-	-	-	-	-	-	+			-	+	-	O		-												+	-	-		18–37					89,362
Myroides odoratus	-	+	+	-	-	-	-	-	-	-	-	-	+			-	+	-	alk	-	-		-	-	-		-	-	-	-	-	-	+	-	+	+	18–37	0–2	+	+	+	89,362
					Previously Flavobacterium odoratum, fruity odour																																					
Pedobacter heparinus	-	+	+	+s	-	-	-	-	-	v	-		v		-	+	-	+	O		+s		+	+	-		-	+	+	+	+	-	-	-	-	+	5–37	0–3	+	+	+	89,163, 728
Pedobacter piscium	-	+	+	-	-	-	-	-	-	-	-					+	-	-	O		-	-	-	-	-		+	-	-	+	-			-	-	+	5–28		+	+	+	728
Salegentibacter salegens	-	+	+	-	-	-										-	-		O		-	-				-	+	-	-	-				-	-		0–30	0–20	R			533
Sphingobacterium multivorum	-	+	+	-	-	-	-	-	-	-	-		+			+ 14	+	+	O	+	+	-	+	+	-	+	+	+	+	+	+	-	+	-	-	60	15–37	0–2	+	+	+	89,802
Sphingobacterium spiritivorum	-	+	+	-	-	-	-	-	-	+	-		+			+	+	+	O	27	+v	+v	+v	+	+	+	+	+	+	+	+	-	+	-	+	+	15–37	0–2	+	+	+	89,365, 802
Zobellia galactanivorans	-	+	+	-	-	+				-	-		-			+	+	+	O	+	+			+	+				+		-		-	+	+	+	13–45	0.5–6				61
Zobellia uliginosa	-	+	+	-	-	+				+	-		-			+	+	+	O	+	+			+	-				+		+		-	+	+	+	13–30	0.5–2				61

alk = alkaline reaction, CR = Congo Red, Glid = gliding motility, KOH = potassium hydroxide, NA = growth on nutrient agar, TSA = growth on tryptone soy agar.

Table 4.9. Gram-negative coccobacilli and cocci.

Test / Organism	Gm	Ox	cat	βH	mot	SW	ODC	LDC	ADH	Nit	Ind	Cit	urea	mr	vp	aes	G	onpg	OF	arab	glu	inos	lac	malt	man	mano	sal	sor	suc	tre	Xyl	H₂S	MCA	TCBS	DNase	temp	NaCl	0129 10	0129 150	Amp	Ref
Pathogenic																																									
Acinetobacter spp.	–cr	+	+	+	–	–	–	–	–	–	–	–	–	–	–	–	–	–	F	–	–	–	–	+	–	+	–	–	–	–	–	–									41
Note: Acinetobacter do not usually produce oxidase																																									
Moraxella spp.	cb	+	+	+	–	–	–			–	–	–	–	–	–	–	–	–		–	–	–	–	–	–	+w	–	–	–			–									72
Haemolysis tested on sheep blood																																									
Taxonomic position of both these organisms is in doubt																																									
Environmental																																									
Acinetobacter baumannii	–	–	+	–	–					–	–	v				–	–	–	O	–	+											–			–	15–37					110
Acinetobacter calcoaceticus	–	–	+	–	–					–	–	+				–	–	–	O	–	+											–			–	15–37					110
Acinetobacter haemolyticus	–	–	+	+	–					–	–	+				–	+	–	O	–	52											–			–	15–37					110
Mesophilobacter marinus	–cb	+	+	–	–					+	v	+	v	+	–	v	v		O	–	+				+							–			–	5–37	1–6				583
Growth on NA, MA 2216																																									

Table 4.10. Gram-negative, oxidase-negative rods.

Test / Organism	Gm	Ox	cat	βH	mot	SW	ODC	LDC	ADH	Nit	Ind	Cit	urea	mr	vp	aes	G	onpg	OF	arab	glu	inos	lac	malt	man	mano	sal	sor	suc	tre	Xyl	H₂S	MCA	TCBS	DNase	temp	NaCl	0129 10	0129 150	Amp	Ref
Pathogenic																																									
Citrobacter freundii	–	–	+	–	+	–	12	–	v	+	–	+	+	+	–	–	–	+	F	+	$+_{g+}$	–	+	+	+	+	–	+	10	+	+	+	+		–	20–37	0–2				414, 425
Edwardsiella hoshinae	–	–	+	–	+	–	+	+	+	+	w+	–	–	+	–	–	–	–	F	30	$+_{gv}$	–	–	+	+	+	75	–	+	+	–	+	+		–	25–40	0–1.5	S			317
Motile at 25°C and 35°C																																									
Edwardsiella ictaluri	–	–	+	–	+	–	+	+	+	+	–	–	–	+	–	–	–	–	F	–	$+_{g+}$	–	–	+	–	+	–	–	–	–	–	–	+		–	20–37	0–1.5				334, 374
Weakly motile at 25°C but not at 35°C. Read MR at 48 h																																									808
Edwardsiella ictaluri oxygen intolerant strain	–	–	+	–	–			+		+		–					–		F	–	$+_{g-}$			+		+			–			+	+			25–37					547
Culture anaerobically. Non-motile at 37°C																																									

149

continued

Table 4.10. Continued.

Test	Gm	Ox	cat	βH	mot	SW	ODC	LDC	ADH	Nit	Ind	Cit	urea	mr	vp	aes	G	onpg	OF	arab	glu	inos	lac	malt	man	mano	sal	sor	suc	tre	Xyl	H₂S	MCA	TCBS	DNase	temp	NaCl	0129 10	0129 150	Amp	Ref
Edwardsiella tarda	−	−	+	+	+	−	+	+	−	+	+	−	−	+	−	−	−	−	F	−	+g+	−	−	+	−	+	−	−	v	−	−	+	+		−	42	0–2				374, 640
					Motile at 25°C and 35°C																																				
Escherichia vulneris	−	−	+	+	+	−	+	+	+	+	−	−	−	+	−	+	−		F	+	+g+	−	+	+	+	+	+	−	−	+	+	−	+	−	−	20–37		R	R		51
Hafnia alvei	−sr	−	+	−	+	−	+	+	−	+	−	−	−	21	87	−	−	+	F	+	+g+	−	−	+	+	+	−	−	−	+	+	−	+	−	−	4–40	0–4	R	PS		313, 652
Halomonas cupida	−	+	+	−	+	−	+	+	+	+	−	+	−	−	−	+	−	−	−	+		+	+	+	+		+	+	+w	+	+	−	+			10–25		S			414
Klebsiella pneumoniae	−	−	+	−	−	−	−	+v	−	+	−	+	+	60	40	+	−	+	F	+	+g+	+	+	+	+	+	+	+	+	+	+	−	+		−	35	0–5	R	R		
					Mucoid																																				
Pantoea agglomerans	−	−	+	−	+	−	−	−	−	+	−	+	−	−	+	+	+	+	F	+	+g−	−	−	+	+	+	+	−	+	+	+	−	+	−	−	4–37	0–6				249, 325
					Opportunist, yellow pigment, MR +ve with 2% NaCl																																				
Providencia (Proteus) rettgeri	−	−	+	+	+	−	−	−	−	+	+	+	+	+	−	+	−	−	F	−	+	+	−	−	+	+	v	−	−	−	−	−	+	−	+	18–37	0–3				79
Salmonella arizonae	−	−	+	+	+	−	+	+	−	+	−	+	−	+	−	−	−	+	F	−	+g+	+	+	+	+	+	−	+	+	+	+	+	+	−	−	15–41	0–6	R	R	R	447
Serratia liquefaciens	−	−	+	+	+	−	+	+	−	+	+	+	−	−	+	+	+	+	F	+	+g+	+	−	+	+	+	+	+	+	+	+	−	+	−	+	4–37	0–5	R	R		250, 290
					Ref 290 is MR +86% at 37°C, 17% at 30°C																																				
Serratia liquefaciens	−	−	+	+	+	−	+	+	−	−	−	+	−	−	+	+	+	+	F	+	+g+	−	−	+	+	+	+	+	+	+	+	−	+	−	+	4–37	0–5	R	R		715
					Arctic char isolates are indole negative, inos negative																																				
Serratia marcescens	−	−	+	v	+	−	+	+	−	+	−	+	−	−	+	+	v	+	F	−	+g+	+	−	+	+	+	−	v	+	+	−	−	+	−	+	4–45	0–8	R	R	R	76
					Red pigment																																				
Serratia plymuthica	−	+	+	−	−	−	−	31	−	+	−	+	−	−	+	+	+	+	F	+	+g−	+	−	+	+	+	+	v	+	+	+	−	−	−	+	4–37	0–8	R	R	R	76, 579
					Red pigment. MR +ve at 37°C only																																				
Streptobacillus moniliformis	−	−	−	−	−	−	−	−	+	−	v	−	−	−	−	−	−	−	F	−	+	−	−	+	−	+	−	−	−	−	−	v				10–25	1–4				519, 611
					ID of organism was not conclusive																																				
Streptobacillus moniliformis	−	−	−	−	−	−	−	−	−	−	−	−	−	−	−	−	−	−	F	−	w	−	−	w	−	−	−	−	−	−	−	−	−			25–35					169
					Reactions of type strain. Requires addition of 20% serum for growth																																				
Yersinia intermedia	−	−	+	−	+	−	+	+	−	+	+	+	+	+	+w	+	+	+	F	+	+	+	−	+	+	+	+	+	+	+	50	−			−			402
					Motile at 25°C but not at 35°C. MR, VP, cit +ve at 35°C, −ve at 25°C																																				

																																	Temp (°C)	NaCl (%)	R	Ref.
Yersinia ruckeri	–	–	+	+	–	+	+	–	85	–	+	–	52	F	–	+	+g–	–	+	+	97	–	+	–	+	+	–	97	–	–	–	–	+	37	0–3		137, 250
Y. ruckeri — Motile at 25°C but not at 35°C. Citrate +ve at 25°C but –ve at 37°C	–	+	+	+	–	+																															500, 657
Y. ruckeri	–	+	+	+	–	+	+	–	+	–	+	–	+	F	+	+	+	+	+	+	+	+	+	+	+	+	+	+	+	+	+	+	+	37	0–5		500
	Australian strain reported by Llewellyn (1980)																																				
Y. ruckeri serovar I																								–													
Y. ruckeri serovar II																	+							+													
Y. ruckeri serovar III																	–							–													
Y. ruckeri serovar IV	These are not strains of *Y. ruckeri*																+							+													
Y. ruckeri serovar V																	+							+													
Y. ruckeri serovar VI	Fish isolates that ferment arab & rhamnose are not *Y. ruckeri*, but may be *Hafnia alvei*																																				
Environmental																																					
Budvicia aquatica	–	+	+	–	–	–	+	+	–	+	+	–	–w	F	+	+	+	–	–	–	–	–	+	+	–	+	+s	–	–	4–37	0–4	–	
Citrobacter diversus	–	+	+	+	+	60	+	+	33	–	+	+	–	F	+	+	+g+	–	+	+	17	+	+	10	+	+	+g+	–	–			R	248
Enterobacter aerogenes	–	+	+	–	–	–	+	–	+	+	+	+	+		+	+	+g+	+	+	+	+	+	+	40	+	+	+	–	–		0–5		414
Enterobacter cloacae	–	+	+	+	–	–	+	+	+	+	+	+	+	F	+	+	+g+	+	+	+	+	+	+	+	+	+	+g+	–	–		0–5	R	414
Enterobacter liquefaciens	–	+	+	–	+	65	+	25	+	+	+	60	+		+	+	+g+	+	+	+	+	+	+	+	+	+	+	+	+		0–5		414
Escherichia coli	–	+	+	+	–	–	+	+	–	+	–	50	40		+	+	+gv	–	+	+	55	+	+	30	+	+	+g	–	–	25–44	0–5		414
Halomonas elongata	+v	+	+	–	66	–	22	11	56	33	+	+	+	F	+	+	+	+	+	+	+	+	+	+	+	+	+	–	–	30	0.1–20	R	795
Halomonas halodurans	–	+	+	–	+	–	–	–	–	–	+	+	–	F	+	+	+	–	–	–	+	–	–	–	–	+	+	+	+	22–37	0.1–20		336
Halomonas marina		–																															
Klebsiella ornithinolytica	–	+	+	+	+	+	+	+	+	+	v	+	v	F	+	+	+	+	+	+	+	+	+	+	+	+	+	+	–	10–37			228
Klebsiella oxytoca	–	+	+	–	+	–	+	+	+	+	–	–	+	F	+	+	+	+	+	+	+	+	+	+	+	+	+	+	–	10–41			228
Klebsiella pneumoniae ozaenae	–	+	+	–	–	–	+	+	–	+	+	–	+		+	+	+g–	–	+	+	+	+	+	+	+	+	+	+	+	25–37	0–5		414
Klebsiella rhinoscleromatis	–	+	+	–	–	–	+	+	–	+	+	–	–	F	+	+	+g+	+	+	+	+	+	+	+	+	+	+	+	+	25–37	0–5		414
Kluyvera ascorbata	–	+	+	+	–	–	+	+	+	+	+	+	+	F	+	+	+g+	–	+	+	41	41	+v	+	+	+	+	–	–	5–			
Kluyvera cryocrescens	–	+	+	+	–	–	+	+	+	+	+	+	+	F	+	+	+g+	–	+	+	41	41	+v	+	+	+	+	–	–	5+			

continued

Table 4.10. *Continued.*

Test	Gm	Ox	cat	βH	mot	SW	ODC	LDC	ADH	Nit	Ind	Cit	urea	mr	vp	aes	G	oONG	OF	arab	glu	inos	lac	malt	man	mano	sal	sor	suc	tre	Xyl	H2S	MCA	TCBS	DNase	temp	NaCl	0129/10	0129/150	Amp	Ref
Pantoea dispersa	–	–	+	+	+	–	–	–	–	+	–	+	–		+	–	$+_s$	+	F	+	$+_{g-}$	+	–	+	+	+	–	–	+	+	+	–			–	30–41	0				291
Yellow pigment. Type strain is ODC +ve																																									
Providencia friedericiana	–	–	+	–	+	10	–	–	–	+	+	w	–	+	–	–	–	–	F	–	$+_{g+}$	–	–	–	–	+	–	–	$+_s$	+	–	–	+		–	10–40	0				559
Motile at 25°C but neg at 37°C																																									
Rahnella aquatilis	–	–	+	+	+rt	–	–	–	–	+	–	+	–	89	+	+	+	+	F	+	$+_{g+}$	–	+	+	+	+	+	+	+	+	+	–	+	–	–	4–37	0	R	R		124
Motile at 25°C but not at 35°C. Man = –ve by API 20E																																									
R. aquatilis Genospecies 2	–	–		31		–	–	–	–	+	–	79	–	+	+	+	34	+	F	+	$+_{g+}$	–	+	+	+	+	+	+	+	+	+	–	+	–	–						124
R. aquatilis Genospecies 3	–	–	+	–	–	–	–	–	–	+	–	+	–	52	72	+	–	+	F	+	$+_{g+}$	–	+	+	+	+	+	+	+	+	+	–	+	–							124
Raoultella planticola	–	–	+	–	–	–	–	+	–	+	+	+	+	+	+	+	+	+	F	+	$+_{g+}$	+	+	+	+	+	+	+	+	+	+	–	+	–	–	4–41	0				228
Raoultella terrigena	–	–	+	–	–	–	–	+	–	+	+	+	+	+	+	+	–	+	F	+	$+_{g+}$	+	+	+	+	+	+	+	+	+	+	–	+	–		4–35	0	R	R		228, 402
Similar morphologically to *K. pneumoniae*																																									
Serratia fonticola	–	–	+	+	+	–	+	+	–	+	–	+	–	+	–	+	–	+	F	+	$+_{g+}$	+	+	+	+	+	+	+	$+_s$	+	70	–	+	–	–	4–37					290, 558
Serratia liquefaciens	–	–	+	+	+	–	+	+	–	+	–	+	>	37	+	+	+	+	F	+	$+_{g+}$	64	16	+	+	+	+	+	+	+	+	–	+	–	+						500
Serratia marcescens	–	–	+	+	+	–	+	+	–	+	–	+	40	18	+	73	70	–	F	27	–	77	17	+	+	+	+	–	+		–	–	+	–	+						290
Serratia rubidaea	–	–	+	+	+	–	–	+	–			+		24	+	+	+	+	F	+	–	35	+	+	+	+	+	–	+	+	+	–			+						250, 290
Yersinia aldovae	–	–	+	+	$+_v$	–	+	–	–	+	–	+	+	+	+	–	–	+	F	+	+	+	–	+	+	+	+	+	–	+	+	+	+		–						85
Yersinia bercovieri	–	–	+	+	+	–	+	–	–	+	–	–	+	+	–	$+_s$	–	+	F	+	+	–	–	+	+	+	$+_s$	+	+	+	+	–	+		–						812
Yersinia enterocolitica	–	–	+	–	>	–	>	–	–	+	>	–	+	+	+	15	+	+	F	+	–	–	–	+	+	+	–	+	+	+	36	–			–						250
Yersinia frederiksenii	–	–	+	–	+	–	+	–	–	+	+	–v	+	+	+	+	–	+	F	+	+	44	22	+	+	+	+	+	+	+	20	–			–	25–37	0				423
Motility & VP positive at 25°C, but negative at 37°C																																									
Yersinia kristensenii	–	–	+	–	+	–	+	–	–	+	43	–	+	+	–	–	–	–	F	+	$+_{g-}$	27	60	+	+	+	–	+	–	+	+	–			–	4–41	0				383, 402
Motile at 25°C but not at 35°C. VP –ve at 25°C																																									
Yersinia mollaretii	–	–	+	+	+	–	+	–	–	+	–	+	+	+	–	$+_s$	–	+	F	+	$+_s$	+s	–	+	+	+	$+_s$	+	+	+	+	–	+		–						812
Yersinia pseudotuberculosis	–	–	+	–	–	–	–	–	–	+	–	–	+	+	–	+	–	+	F	+	–	–	–	+	+	+	–	–	–	+	+	–			>						121
Yersinia rohdei	–	–	+	+	+	–	$+_s$	–	–	+	–	$+_s$	$+_s$	+	–	–	–	+	F	+	+	–	+	+	+	+	+	+	+	+	+	–	+		–	25	0	R	R		12
All *Yersinia* and *Hafnia* species should be incubated at 25–28°C																																									

Table 4.11. Gram-negative, oxidase-positive rods.

Test	Gm	Ox	cat	βH	mot	SW	ODC	LDC	ADH	Nit	Ind	Cit	urea	mr	vp	aes	G	onpg	OF	arab	glu	inos	lac	malt	man	mano	sal	sor	suc	tre	Xyl	H₂S	MCA	TCBS	DNase	temp	NaCl	0129 10	0129 150	Amp	Ref
Organism																																									
Pathogenic																																									
Benechea chitinovora	–	?	+	+	+	–		+		+	+	+	–			+			F		+g+								+			+				20–25	1–4				806
Bordetella bronchiseptica	–	+	+	+	+	–	–	–	–	+	–	+	+			–	–	–	O	–	–	–	–	–	–	+	–	–	–	–	–	–	+			25–37	0–4				199, 642
Brucella abortus	–	+	+	v	–		–	–	–	+	–	–	+s			–	–	–	O		v		–	–					–	+	+	–	v			35	0				169
Burkholderia pseudomallei	–	+		v	+	–	–	–	+	+	–	v	–	–	–	v	+	–	O	+	+	+	+	+	+	+	+	+	–	+	+	–	+			20–42					38, 623
API reactions may be neg for urea & Nit, 18% pos for cit																																									
Deleya aquamarinus	–	+	+	–	+	–	+	–	–	+	–	–	+	–	–	–	–	–	O	+s	–	+	–	+	+	+	+	+	+	+	+s	+s			–	4–42	1–10			S	8,45
Previously *Alcaligenes faecalis homari*																																									
Janthinobacterium lividum	–	+	+	+	–	–		+	+	+	–	+	–	+	+	+	+	–	O	–	–	+	–	+	+	+	+	+	+	–	+s					4–30	0–2				496
Purple pigment																																									
Pasteurella multocida	–	+	+	–	–	–	+	–	–	+	+	–	–	–	–	–	–	–	F	v	+	–	–	v	+	–	–	v	+	v	v	–	–			25–37	0				169, 562
Pasteurella skyensis	–	+w	–	–w	–	–	+	+	–	+	+	+	–			–	–	–	F	–	+	–	+	+	+	+	–	–	–	+	–			–		14–32	1.5–2				100
Pasteurella testudinis	–	+	+	+	–	–	–	–	–	+	+	–	–			+	+	+	F	v	+g−	+s	–	85	+	–	–	v	+	v	+		v								709
Plesiomonas shigelloides	–	+	+	90	+	+	+	+	95	+	50	–c	–		v	–	–	+	F	–	+gv	50	+	+	50	+	–	v	–	+	–	–	+	–G	50	10–42	0–3	R	S	S	169
Pseudoalteromonas bacteriolytica	–	+	+w	+	+	–				–		–				+	+		O	+	+							+	–	–	–	–				15–35	3				677
Red pigment																																									
Pseudoalteromonas piscicida	–	+	+	+	+	–	–	–	–	+	–	–				+	+		O	–	+	–	–	+	–	–	–	–	+	–	–		+		+	25–40	5–10				134, 572
Pseudomonas anguilliseptica	–	+	+	–	+	–	–	–	+	–	–	+	+			v	v		I	–	–	–	–	–	–	–	–	–	–	–	–		+	–	v	5–30	0–3	S	R	Pen = R	96, 541
Finnish isolates are gel pos, Japanese isolates are gel neg																																									
Pseudomonas chlororaphis	–	+	+		+	v	v	v	+	+	–	+				+	+		O	+	+		–	+	+	+	–	–	+	+	–					4–35					828
Green colonies. 5 days for growth																																									
Ps. fluorescens biovar I	–	+	+	–	+	–	–	–	+	–	–	+				+	+	–	O	+	+	+	–	+	+	+	–	+	+	+	+		+			4–37					

continued

Table 4.11. Continued.

Test	Gm	Ox	cat	βH	mot	SW	ODC	LDC	ADH	Nit	Ind	Cit	urea	mr	vp	aes	G	onpg	OF	arab	glu	inos	lac	malt	man	mano	sal	sor	suc	tre	Xyl	H2S	MCA	TCBS	DNase	temp	NaCl	0129 10	0129 150	Amp	Ref
Ps. fluorescens biovar II	-	+	+		+	-		-	+	+		+					+	-	O	+	+	+	-	+	+	+	-	+	+		+	+	+			4–37					
Ps. fluorescens biovar III	-	+	+		+	-		-	+	+		+					+	-	O	+	+	-	-	+	>	+	-	>	-		+		+			4–37					
Ps. fluorescens biovar IV	-	+	+		+	-		-	+	+		+					+	-	O	+	+		-	+	+	+	-	+	+		+	+	+			4–37					
Ps. fluorescens biovar V	-	+	+		+	-		-	-	-		+					+	-	O	+	+			+	>	+	-	>	>	>	+		+			4v–37					
Pseudomonas fluorescens	-	+	+	+	+	-		-	+	>	-	+	>	-	-	-	+	-	O	+	+	>	-	+	+	+	-	>	>	>	+	-	+			4–30	0–5	R	R		623
Pseudomonas plecoglossicida	-	+	+	+	+	-		-	+	+	-	+	-			-	-	-	O	-		>					-					-	+			10–30	0–5	R	R		582
Closely related to Pseudomonas putida																																									
Pseudomonas pseudoalcaligenes	-	+	+	-	+	-	>	-	+	>	-	+	-	-	-	-	>	-	OA	+	+	-	-	14	-	-	-	-	-	-	11	-	+		-	15–41	0				297
Pseudomonas putida	-	+	+	>	+	-	-	-	+	-	-	+	44			-	-	-	O	+	+	-	-	21	19	+	-	-v	9	-	+	-	+			4–35	0–5				297, 623
Pseudomonas stutzeri	-	+	+	-	+	-	-	-	-	+	-	+	14			-	>	-	O	v	+	-v	-	+	68	+	-	-	v	-	+	-v	+		-	25–41	0–6				297, 359, 623
Wrinkled colonies, slightly yellow																																									
Roseobacter strain CVSP	-	+	w	-	+	-													-																	23	2				104
Shewanella putrefaciens	-	+	+	-	+	-	+	-	-	+	-	+	-	-	-	-	+	-	O	50	+s	-	-	70	-	9	-	-	35	-	-	+	+	-	+	4–37	0–3				433
Varracalbimi	-	+	-	-α	-	-	-	+	+	-	-			-	-	-	-	-	F	-	+g	-v	+	+	+	+	-	+	v	-	+	-v	-			4–22	2	S	S		771
Environmental																																									
Actinobacillus delphinicola _(Strain from Mesoploden bidens. 10% CO₂ required for growth)_	-	+	-	-w	-	-	+	+	-	+	-		-	-	+	-	-	-	F	-	+		-	-	-	+		-	-	-	-	-	-			30–42					263
Actinobacillus delphinicola _(Strains from Phocoena phocoena. 10% CO₂ required for growth)_	-	+	-	-w	-	-	+	70	+	21	-		-	-	+	-	-	-	F	-	+		-	-	-	+		-	-	-	-	-	-			30–42					263
Actinobacillus delphinicola _(Strains from Stenella coeruleoalba. CO₂ required for growth)_	-	+	-	-w	-	-	+	+	+	+	-		-	-	+	-	-	-	F	-	+		-	-	+	+		-	-	-	-	-	-			30–42					263
Actinobacillus scotiae _(10% CO₂ required for growth. Nit = neg using Rosco tablets)_	-	+	-	-w	-	-	+	70	-	+	-	-	+	-	+	+	-	+	F	-	+	-	+	-	-	+	-	-	-	-	-	-	-		-	25–37	0				265

Identification table (continued). Column headers appear on a preceding page; data read left-to-right per row, followed by growth-temperature range, NaCl/growth range, R/S column and reference number(s).

Organism	Notes	Test results (left → right)	Temp (°C)	Range	R/S	Ref.
Achromobacter xylosoxidans denitrificans	Previously Alcaligenes denitrificans	− + + v + − − − − + − + − − − − − − − − − − O − − + − − −	10–37	0–4	−	297
Allomonas enterica		+ + + +v + − 70 80 − + + + + + + + + + + + +g − F + + +g − + −	20–37	3–5		
Alteromonas macleodii		− + + + + − − − − − − − + + + − + + 20 + + − O + + + − + −	35–40	2	R	286, 815
Aquaspirillum spp.		− + + + + + − − + − − − − + − − + − − − − − − I + − − + −	25	0–2	R	497
Brevundimonas diminuta		− + + α + + − − − − − − − − − − + − − − − − O − − − −v − +14	30–37	0–5	R	
Brevundimonas (Pseudomonas) vesicularis	Yellow colonies	− + + α + − − − − − + 58 52 O + − − 46 − − − − − + − − + 20 −	25	0–5		297
Burkholderia (Pseudomonas) cepacia	Freshwater, yellow pigment	− v+ + + − 66 − + 42 + + O +v + + + + + 81 + − + − − + −	25–40	0	R R	297
Chromobacterium violaceum	Violet pigment. Some non-pigmented strains	− + + +v + + − − 37 + − F +g + − − + − − + − − +	25–37	0	R	482
Colwellia maris		− + + + − − − − − + − O + − − − + − − − + + +	0–22	3–4	S	
Halomonas aquamarina	Previously Alcaligenes faecalis	− + + − + − − − − + − O − − − − − − − − + − −	25–37	0–4		169, 297
Halomonas venusta		− + + + − − − + − + O − v + − − v v + − +	4–37	0–6.5	R S	66, 310
Hydrogenophaga (Pseudomonas) flava		+ w − − − + O +g-d + + + + + + − −				39
Hydrogenophaga (Pseudomonas) palleronii	Yellow pigment	− + w − + − − − − O − − − +	30			39, 834
Hydrogenophaga pseudoflava	Yellow pigment. Nitrate pos with yeast extract	− + w + − − − − − O + +g-d + + + + + + + +	35–41			39, 834
Iodobacter fluviatilis	Violet on MCA. Spreading colonies on 1/4 NA	− + +w + 85 + − − F +g − + + − + − + +	4–25	0–1	R R	502
Mannheimia haemolytica		− + + − − − −v + + − − F + − + +w −	20–37		R	

continued

Table 4.11. Continued.

156

Test	Gm	Ox	cat	βH	mot	SW	ODC	LDC	ADH	Nit	Ind	Cit	urea	mr	vp	aes	G	onpg	OF	arab	glu	inos	lac	malt	man	mano	sal	sor	suc	tre	Xyl	H2S	MCA	TCBS	DNase	temp	NaCl	0129 10	0129 150	Amp	Ref
Marinomonas communis	–	+		–	–	–			–			+					–		O		+	–	–	+	+	+	–	+	–	25	+					35–40					286
Marinomonas vaga	–	–		–	+	–			–	–	–	+					–		O		+	–	–	40	+	+	–	+	–	20	60					35					286
Oceanimonas baumannii	–	+		+	+	–			–	+	–	+	–		–	–	–	–	O		–															10–41	1–7				130
Oceanimonas doudoroffii	–	+		+	+	–			–	+	–	+	–		–	–	–	–	O		–															10–41	2				69, 130
Nitrate may be neg by tube test, pos by API 20NE																																									
Phococenobacter uteri	–	+	–	–	–	–	–	–	–	+	–	–	–	–	+	–	–	+	F	–	+	–	–	–	–	–	–	–	–	–	–	–	–			22–37					266
Pseudoalteromonas antarctica	–	+	+	+	+	–		–	–	–	–	–	–	+	+	–	+	–	–	–	+	–	–	+	+	+	–	–	–	–	–	–	+		–	4–30	0.1–9				115
Pseudoalteromonas citrea	–	+	+	+	+	–	–	–	–	–	–	–	–	–	–	–	+	–	O	–	+	–	–	–	–	+	–	–	–	+	–	–			+	10–30	1–10	R	R		285
Pseudoalteromonas denitrificans	–	+	–	–	+	–	–	–	–	+	–	–	–	–	–	–	+	–	O																+	4–22	1.5–5	R	R		239
Red pigment																																									
Pseudoalteromonas distincta	–	+	+																																						
Pseudoalteromonas elyakovii	–	+	+	+	+	–				–		+					+		O		+							–	–	+						10–37	2				679
Pseudoalteromonas espejiana	–	+	+	+	+					–		–					+		O							+		–	–	+						35	2				150
Pseudoalteromonas flavipulchra	–	+	+	+	+	–	–		–	–	–	–	–	–	–	–	+	–	O		+		–	+	–	+	–	–	–	+	–	–			+	10–44	0.5–10	R	R		286, 398
Pseudoalteromonas haloplanktis haloplanktis	–	+	+w	+	+	–			–	–	+	+	+				+		O		+			+	50	+	–	–	+	+	–				+	4–15	2				285
Pseudoalteromonas haloplanktis tetraodonis	–	+	+	+	+					–		–	–				+		O		+															4–15	2				285
Pseudoalteromonas luteoviolacea	–	+	–						–	–	–	–	–			+	+	–	O		+v		–	+	–	–		–	–	+	–	–			+	10–30	2–4	R	R		287
Produces a violet pigment																																									
Pseudoalteromonas maricaloris	–	+	+	+	+	–				–		–				+	+		O																	10–37	0.5–10	R	R		398

Organism	Temp (°C)	NaCl (%)			Ref.
Pseudoalteromonas nigrifaciens	4–28	2			395
Produces black pigment					
Pseudoalteromonas rubra	25–35	2	R	R	283
Red pigment					
Pseudoalteromonas ulvae	4–25	1–2	S	S	231
Pseudoalteromonas undina	25–35	2			150
Pseudomonas acidovorans	25–35	0			297
Pseudomonas aeruginosa	25–41	0–5	R	R	623
Green pigment. May have brick-red pigment					
Pseudomonas alcaligenes	25–41	0–5			169
Rod or filament					
Pseudomonas aureofaciens	4–37				
Orange pigment					
Pseudomonas mendocina	20–41	0–6			297
Yellow pigment					
Pseudomonas mesophilica	25–35	0			
Fat cells, pleomorphic, vacuolated. Coral-pink pigment					
Pseudomonas perfectomarina	40	2			
Pseudomonas stanieri	40	2			
Roseobacter gallaeciensis	15–37	2			662
Ovoid rods, brown diffusible pigment					
Salinivibrio costicola ssp. *costicola*	5–45	0.5–20	R	S	279, 371
Previously *Vibrio costicola*. Found in hypersaline habitats					
Salinivibrio costicola NCMB 701^T	5–45	0.5–20	R	S	279, 371
Shewanella algae	25–42	0–10			433, 588, 851
Shewanella amazonensis	4–30	0–3			
Antarctic marine organism					

continued

Table 4.11. *Continued.*

Test	Gm	Ox	cat	βH	mot	SW	ODC	LDC	ADH	Nit	Ind	Cit	urea	mr	vp	aes	G	onpg	OF	arab	glu	inos	lac	malt	man	mano	sal	sor	suc	tre	Xyl	H₂S	MCA	TCBS	DNase	temp	NaCl	0129 10	0129 150	Amp	Ref
Shewanella baltica	–	+	+	–	+	–	+	–	–	+	–	+	–	–	–	–	–	–	O														+	+	+	4–30					851
Shewanella benthica	–	+	+	+	+	–	–	–	–	+	–		–	–	–	–	–	–	O	–	+								+			+			+	4–15	0–2				
Shewanella colwelliana	–	+	+	–	+	–	–	–	–	+	+		–		–	–	+	–	O	–	–	–	–	–	–	–v	–	–	–	–	–	–				8–30	1–5	R	R		112, 814
Adheres to surfaces. Sugars go alkaline. Older broth cultures have red-brown pigment																																									815
Shewanella frigidimarina	–	+	+	+	+	–	–v	–	–v	+	–	+v	–	–	–	v	+	–v	O	–	+	–	–	+	+	–v	–	–	+	+v	–v	+v			+	0–28	0–8				112
Antarctic marine organism																																									
Shewanella gelidimarina	–	+	+	+	+	–	–	–	–	+	–	–	–	–	–	–	+	+	O	–	–	–	–	–	–	–	–	–	–	–	–	+			+v	4–15	1–6				112
Antarctic marine organism																																									
Shewanella hanedai	–	+	+	+	+	–	–	–	–	+	–	–		–		–	+	–	O	–	–	–	–	–	–			–	–	–	–	+				4–25	2				409
Shewanella japonica	–	+	+	+	+	+	–		–								+		F	–	+														+	10–37	0–3				397
Shewanella oneidensis	–	+	+	+	+	–	–		–	+	–	–			–	–	+	–	F	–	+			–								+				15–40	0–3				782
Shewanella pealeana	–	+	+	–	+	–				+	–	+			–	–	–	–	O	–	+			–	–			–	–	–	–	+				4–30	0.1–0.7				492
Shewanella woodyi	–	+	+	+	+	–	–	–	–	+	–	–				–	+	–	O	–	–		–	–	–			–	–	–	–	+			+	4–25	2				112
Degrades agar																																									
Sphingomonas paucimobilis	–	+	+	–	+	–	–	–	+	–	–	–				+	–	+	O	–	–	+	+	+	+	+		+	+	+	+	–	–		+	30–37					297
Yellow pigment																																									
Stappia stellulata-like M1	–	+	+	+	+	–				+		+	+		+		+	+	F																	25	1.5				105
Stenotrophomonas (Pseudomonas) maltophilia	–	–	+	α	+	–	+	+	–	–	–	+	–v		v	v	+	+	O	–	+w	+v	+v	+	+		+	–	+v	+	54	+	+		+	35	0–5				623
Yellow colonies																																									

See also Table 4.5 for *Brucella* species.

–c = negative in Chistensen's citrate method; OA = Oxidative; I = Inert reaction, may show an alkaline reaction; –α = negative or alpha haemolysis; numbers refer to percentage of positive strains; numbers in 'temp' and 'NaCl' columns refer to range (°C or % NaCl) at which the organism will grow.

Table 4.12. *Helicobacter* spp.

Test Organism	Gm	Ox	cat	mot	Nal	Kf	Gly	IA	1.5 NaCl	25°C	37°C	42°C	urea	Nit	hip	GGT	AP	H₂S	Ref
H. acinonychis	–		+		R	S	–	–				–	+	–		+	+		327
H. acinonyx	–		+		R	S		–				–	+	–		+	+		540
H. bilis	–	+	+		R	R	+	–	–		+	+	+	+	–			+	540
H. canis	–		–		S	–		+		–		+	–	–			+		540
H. cetorum	–	+	+	+	I	S	–	–		–	+	+	+	–		+	–		327,328,329
H. cholecystus	–	+	+		I	R	+	–		–	+	+	–	+	–	–	+	–	540
H. cinnaedi	–		+		S	I	+	–		–	+	–	–	+	–	–	–	–	540
H. felis	–	+	+	+	R	S	–	–		–	+	+	+	+	–	+	+	–	327,540
H. fenneliae	–		+		S	S	+	+	+		+	–	–	–		–	+		540
H. hepaticus	–	+	+		R	R	+	+	+	–	+	–	+	+	–			+	540
H. muridarum	–	+	+		R	R	–	+	–	–	+	–	+	–	–	+	+	+	540
H. mustelae	–	+	+		S	R	–	+	–	–	+	+	+	+	–	+	+	–	327,540
H. nemestrinae	–		+		R	S	+	–				+	+	–			+		327,540
H. pametensis	–	+	+		S	S	+	–	–		+	+	–	+	–	–	+	–	540
H. pullorum	–		+		S	R	–				+	+	–	+	–				540
H. pylori	–	+	+		R	S			–	–	+	–	+	–	–	+	+	–	327,540
H. trogontum	–	+	+	+	R	R					+	+	+	+	–	+	–	+	540

AP = alkaline phosphatase hydrolysis; cat = catalase; GGT = gammaglutamyl transpeptidase activity; Gly = growth in 1% glycine; Gm = Gram reaction; Hip = hippurate hydrolysis; H₂S = production of hydrogen sulphide; I = intermediate susceptibility; IA = indoxyl acetate hydrolysis; Kf = sensitivity to cephalothin 30 μg; Mot = motility; Nal = susceptibility to naladixic acid 30 μg; Nal = growth in the presence of 1.5% NaCl; Nit = nitrate reduction; Ox = oxidase; S = sensitive; R = resistant;
Urea = urease activity; 25°C, 37°C, 42°C = growth at these temperatures.
Tests for alkaline phosphatase and hydrolysis of indoxyl acetate were done using an ANI-Ident disc from bioMerieux, GGT, AP, hip, Nit, H₂S were done using the API Campy identification system (bioMerieux). Urea hydrolysis was tested using a rapid urea test from Remel. Growth on 1% glycine can be tested on a growth medium such as blood agar (BA), containing 1% glycine. Sensitivity to Na and Kf were done on BA using standard antibiotic sensitivity test methods.

Table 4.13. Gram-positive cocci.

Test / Organism	Gm	Ox	cat	βH	mot	SW	ODC	LDC	ADH	Nit	Ind	Cit	urea	mr	vp	aes	G	onpg	OF	arab	glu	inos	lac	malt	man	mano	sal	sor	suc	tre	Xyl	H₂S	MCA	Coag	DNase	temp	NaCl	Man/An	RAA	Hip	Ref
Pathogenic																																									
Aerococcus viridans	+	–	–	α	–	–				–					–						+																				755
Aerococcus viridans var. *homari*	+	–	w	α		–			–	–				–	–	+	–		F	50	+		+	+	50	+	60	50	+	+				–		10–37	0–10			+	299, 827
Enterococcus faecalis var. *liquefaciens*	+c	–	–							+	–	+		–	–	+	+	+													–			+							
Granulicatella balaenopterae	+c	–	–	–	–				+	–			w+		–	+	–	–	F	–	+		–	+	–			–	–	+	–								–	–	179, 478
Lactococcus garvieae	+	–	–	α	–			–	+	–	–	–	–	+	+	+	–	–	F	–	+	–	–	+	+	+	+	–	V	+	–	–	V			10–45	0–6.5		–	–	464, 731
Growth in 6.5% NaCl, optimum 0% NaCl, 37°C. Resistant to clindamycin																																									638, 780
biotype 1, 2, 12	+	–		α	–				+						+	+							+	+	+	+		–	+	+	–		–				10–45	0–6.5				780
biotype 3, 4, 5, 6, 7, 10, 11, 13	+	–		α	–				+						+	+							+	+	+	+		–	–	+	–		–				10–45	0–6.5				780
biotype 8, 9	+	–		α	–				+						+	+			F	–	+	–	+	+	–	+	+	–	–	+	+	–	–				10–45	0–6.5				780
Lactococcus piscium	+c	–		–	–				–	–	–		–		+	+			F	+	+	–	+	+	+	+	+	–	+	+	+	+	+		–		5–30				–	835
Micrococcus luteus	+	+	+	–	–		–	–	V	–	–	–	V		–		+	–	O	–	–	–	–	–	V		–	–	V	+	–	–	–				10–45	0–10				43
Planococcus species	+	–	+	–	+		–	–	–	+	–		–		+		–	+	O	–	+	–	–	+	–		–	–	–	+	–	–	–				37	2				43
Staphylococcus aureus	+	–	+	+	–		–	+	+	+	–	+	+		+	+			F	–	+		+	+	+	+	–	+	+	+	–			+	+	37–45		+			296
Staphylococcus delphini	+	–	+	+	–		–	+	+	+		+	+		–				F	–	+		+	+	+s	+		+	+	–	–			+	+w	37–45	0–15	–			264, 778
Staphylococcus epidermidis	+	–	+	–	–		–	+	+	80			+		+				F	–	+		+	+	–	+		–	+	–	–			–	–w	25–45	0–7				296
Staphylococcus lutrae	+c	–	+	+	–		–	–	–	+			+		–	+	+	+	F	+	+		+	+	V	+		+	+	+	+	+		+	+w	25–42	0–10				264

Organism	Growth temp (°C)	NaCl/pH range	Ref.
Staphylococcus warneri	25–40	0–10	296
Streptococcus agalactiae	25–37	0–3	71, 242
Group B. No growth at 10°C or 45°C. VP may be negative			
Streptococcus agalactiae (S. difficile)	30	0–1	135, 776
Group B, Type Ib. Hip +ve at 30°C, –ve at 37°C			
Streptococcus dysgalactiae ssp. *dysgalactiae*		0–4	727, 790
Positive to Lancefield group L antisera. Bacitracin = S			
Streptococcus dysgalactiae ssp. *dysgalactiae*		0–4	135, 776
Group C			
Streptococcus iniae ATCC 29177	10–35	2–4	574
No growth at 6.5% NaCl, or 10°C or 45°C. Isolates from fish			
Streptococcus iniae ATCC 29178	10–37	2–4	223, 625
No growth at 6.5% NaCl, or 10°C or 45°C. Isolates from fish			
Streptococcus iniae	37	2–4	223
Streptococcus milleri			87
Strains from humans			
Streptococcus parauberis	10–37	0–4	175, 224
No growth at 6.5% NaCl, or 4°C or 45°C. Neg for strep group D			
Streptococcus phocae	25–37	0	700
Sensitive to bacitracin disc, some strains react with strep group C antisera			
Vagococcus fessus			369
Vagococcus lutrae	37	0	477
Vagococcus salmoninarum	5–30	0–6.5	542, 682; 732, 807
May grow on TCBS, weak delayed reaction with strep group D antisera. Hip & VP pos with API 20 STREP			
Vagococcus describes motile Lancefield group N. *V. salmoninarum* has been reported as negative for Lancefield group N			

continued

Table 4.13. *Continued.*

Test	Gm	Ox	cat	βH	mot	SW	ODC	LDC	ADH	Nit	Ind	Cit	urea	mr	vp	aes	G	onpg	OF	arab	glu	inos	lac	malt	man	mano	sal	sor	suc	tre	Xyl	H₂S	MCA	Coag	DNase	temp	NaCl	Man An	RAA	Hip	Ref
Environmental																																									
Abiotrophia defectiva	+	–	–	α	–	–			–				–		–		+	+	F				+						+	+						20–42				–	421
Abiotrophia para-adiacens	+	–	–	α	–	–			–								–	–	F										+	–											421
Enterococcus faecalis	+	–	–	α	–	–	–	–	+	–	–	–	–		+	+	>	–	F	–	+	–	+	+	+	+	+	>	+	+	–		+			10–45	0–6.5			+	638
Enterococcus faecium	+	–	–	α	v	–			+	–					+	+	–	+	F	+	+	–	+	+	+	+	+	–	+	+	–					10–50	0–6.5			+	
Facklamia miroungae	+	–	–	α	–	–	+		+	–			+		–	–	–	–	F	–			–	–	–			–	–	+						25–42	0–5			–	368
Granulicatella adiacens	+	–	–	α	–	–			–				–		–		–	–	F				–						+	–						20–42				–	421
Granulicatella elegans	+	–	–	α	–	–			+				v		–	–	–	–	F				–	+	–		–	–	+	–						27–37				+	653
Lactococcus lactis Serogroup N	+	–	–	α	–				+	–					+	+	+	–	F	–		+	+	+	–			–	–	+						10–45	0–2			–	731

Phenotypically similar to *E. seriolicida*. Sensitive to clindamycin

Test	Gm	Ox	cat	βH	mot	SW	ODC	LDC	ADH	Nit	Ind	Cit	urea	mr	vp	aes	G	onpg	OF	arab	glu	inos	lac	malt	man	mano	sal	sor	suc	tre	Xyl	H₂S	MCA	Coag	DNase	temp	NaCl	Man An	RAA	Hip	Ref
Planococcus citreus	+	–	+	–	+	–	–	–		–	–	v	–	–	–	–	+		F	–	+	–	–	–	–	–	–	–	–	–	–	–				5–30	0–15			>	326
Planococcus kocurii	+	–	+	–	+	–	–	–		–	–	v	–	–	–	–	>		F	–	>	>	–	–	–	–	–	–	–	–	–	–				5–30	0–10			>	326
Planomicrobium okeanokoites	–v	w	+	–	+	–	+	+	+	–	–		–	–		–	+		O	–	–		–	+	–	–			–		–					20–37	3–5				566
Staphylococcus aureus anaerobius (Anaerobic growth only)				+											–									+										+	+						264
Staphylococcus capitis	+	–	+w	–w	–					+					>	–	–		F	–	+	–	–	–	+	+	–	–	+s	–	–		–	–	w	30–45	0–10				
Staphylococcus carnosus	+	–	+	–	–					+			>		+	–	+		F	–	+	v	v	–	+	+			–	v	–		–		–	15–45	0–15				
Staphylococcus chromogenes			+						+						–				F					>	80				+						–			–			
Staphylococcus cohnii	+	–	+	–w	–					–					+w		+		F	–	+	–	+	80	80	80	–	–	–	+	–		–	–	20	15–45	0–10			–	

Species													Coag																Ref
Staphylococcus cohnii urealyticum	+	–	+	–		–	28	15		+	25	28	+	F	–	+	–	22	+	–	+	–	+	–			–			
Staphylococcus haemolyticus	+	+	+w	–		–		+		+			+	F	–	+	50	+	50	–	+	+	+	–			–		20–45 0–10	
Staphylococcus hominus	+	+	–	–		–		+		w			F	F	–	+	60	+	–	–	+	+	+	–			W		20–45 0–7	264
Staphylococcus hyicus	+	+	–				+	–					F	F	–	–	–	+		+	+	+	+	–			>			264
Staphylococcus intermedius	+	+	+					+		–			F	F	–w	+	–w	+	+	+	+	+	+	–			+			264
Staphylococcus saprophyticus	+	+	–	–				–		+			+	F	–	+	+	+	+	+	+	+	+	–			–		15–40 0–15	
Staphylococcus schleiferi coagulans	+	+								+			F	F	–			+	–	+	–	–	–	–			+			264
Staphylococcus schleiferi schleiferi	+	+								+			F	F	–					+	v	v	–	–			+			264
Staphylococcus simulans	+	+	–w	–				+		–			F	F	–	+	–	+	77	+	+	77	–	–			v		15–45 0–10	
Staphylococcus warneri	+	+w	–	–				–		+			F	F	–	+	+s	+	–	–	+	+	+	–			–w		25–40 0–10	
Staphylococcus xylosus	+	+					80	80					80	F	+	+	80	+	+	–	80	80	+	+			–		15–40 0–10	790
Streptococcus dysgalactiae equisimilis	+	–	+	–				+		–			F	F	–			+	–	–	+	+	–							–
Streptococcus porcinus	+	+						+		+	+		F	F	–	+	+	+	–	+	+	+	+	–			+		10–37 0–6.5	175, 625
Streptococcus uberis	+	–	–	–				+		+	+		F	F	–	+	+	+	–	+	+	+	+	–			+		10–37 0–6.5	175, 625
Vagococcus fluvialis	+cr	–	α	–				–		–	+	–	>	F	+g–	–	+	+	–	+	–	+	–	–			–		5–40 0–6.5	177, 732
ATCC 49515																														629

Weak delayed reaction with strep group D antisera

The list of Streptococci and Staphylococci is not comprehensive, and readers are urged to use a text such as Bergey's Manual, if a *Streptococcus* species does not fit with the species listed here.

Coag = coagulase reaction; Hip = hydrolysis of hippurate; Man An = fermentation of mannitol under anaerobic conditions; RAA = growth on Rogosa Acetate agar.

Table 4.14. Gram-positive rods.

Test / Organism	Gm	Ox	cat	βH	mot	SW	ODC	LDC	ADH	Nit	Ind	Cit	urea	mr	vp	aes	G	onpg	OF	arab	glu	inos	lac	malt	man	mano	sal	sor	suc	tre	Xyl	H₂S	MCA	DNase	temp	NaCl	ZN	MAF	RAA	Hip	Ref
Pathogenic																																									
Arcanobacterium phocae	+		10	+	–	–				–			–		–	–	–		F	–	+g⁻	+	+	+	40	60	–	–	+	+	10						–	–			636
Bacillus cereus	+	+	+	+	+					–	–				+	+	+		O		+														25–45		–	–			74, 307
Bacillus mycoides	+	+	+	+	–						–				+	+	+		O		+														25–37		–	–			307
Carnobacterium piscicola	+sr	–	–	–α	–	–		–	+	–	–	–	–	+	+	+	–	+	F	–	+g⁻	–	40*	+	+	+	+	–*	+	+	–	–		–	0–37	0–6	–	–	–	–	73, 176
*Weak reactions in phenol red medium, negative in purple base medium																																									
Carnobacterium spp.	+r	–	–	–	–	–		–	+	–	–	–	+	+	+	+	–	v	F	+g⁻	+g⁻	+v	+v	+	+	+	+	v	+	+	–	–	–	–	10–37	0–6	–		–		73
Corynebacterium aquaticum	+	–	+	+	+	–		–	–	+	–	–	–	–	+	+	v	+	–	–	–	–	–	–	–		–	–	–	+	–	–	–		4–42	0–5	–		–		75
Corynebacterium testudinoris	+		+	–	–	–				+			–		+	+	–		F	+	+	–	+	+	–		+	+	–					37	0					180
Dermatophilus chelonae	+		+	+	+	–				+			–		–	–	+		F	+	+		–	v	–		–	–	–	–	–	–		25		–				529
Dermatophilus congolensis	+		+	+	+	–				–			+		–	–	+		F	+	+		–	+v	–		–	–	–	–	–	–		37		–				308
Erysipelothrix rhusiopathiae	+r	–	–	α	–	–				–	–		–		–	–	–	–	F	+	+	–	+	+	–w		–	–	–	–	–w	+		5–37		–		–		292
Renibacterium salmoninarum	+	–	+	+	–	–			–			–			–	–	–	–	–	–	–	–	–	–	–		–	–	–	–			–	15–18	0	–				671
Rhodococcus spp.	+r	–	–	–	–	–							+						F	–	–	+	–	–					+		+	–		25	0–5	–				53
No growth 37°C. Growth in 5 % NaCl																																									
Streptomyces salmonaris	+		–														+		O	–	+	+			–				w	+	–	–		+	12–37			–			41
Brick-red to orange pigmented mycelia																																									
Environmental																																									
Actinomyces marimammalium	+	–							–				–		–	–	–		F	–	+	+	+	+	–				–	–	–	–							–	
Arcanobacterium bernardiae	+	–	–	v	–				–				–			–	–		F	–	+		+	+					–	–	–									636
Arcanobacterium haemolyticum	+																																								

Species																					Ref.
Arcanobacterium pluranimalium	+	+	α	–	–									–	F	–			–	–		+	
Arcanobacterium pyogenes	+	–	+	–	–									–	F	$+_g$	–		–	v–		+,–	636
	Acid, clot, reduction in litmus milk													20–40									
Arthrobacter agilis	+c	+	+	+v	–	–	–			+	O	–	–		–			v 20–30 0–1	–				
Arthrobacter nasiphocae	+	+	–					+w	+	–	O	–	+			25–40 5				–		+	8
Arthrobacter rhombi	+c	+	+	–	–			+	–	+	O	–	+	+	+	+	4–30 1–10		–		–		600
Atopobacter phocae	+	–	–			+		–	–	–	F	–	>	v	v	25			–		–		479
Carnobacterium alterfunditum	+r	–	–	–			nt	–	–	w	F	+	w	–	–	0–20 1–6		nt		–		–	412
Carnobacterium divergens	+r	–	–		+			–	+	+	F	+	+	+	+	0–40 0–7		–		–		–	176
Carnobacterium divergens 6251	+r	–	–						–		F					0–35 0–6		–		–		–	649
Carnobacterium funditum	+r	–	+	–		nt		–	–	+	F	+	+	+	+	0–20 1–6		nt		–		–	412
Carnobacterium gallinarum	+r	–	–	–	+			+	+	+	F	+	+	+	+	0–35		–		–		–	176
Carnobacterium inhibens	+r	–	+	–	–			+	w	+	F	$+_g$	+	+	+	0–30 1–6		–		–		+	412
Carnobacterium mobile	+r	–	+	–	–			+	–	+	F	+g	+	+	+	0–35		–		–		–	176
Corynebacterium phocae	+r	+	–		>			–	+	+	F	+	v	–	–			–		–			613
Corynebacterium pseudodiphtheriticum	+r	+	+	–		+		–	–	–	–	–				20–42 0–5		–		–			133
Corynebacterium pseudotuberculosis	+r	+	+	–		+		–	–	–	F	$+_g$				20–42 0–3		–		–			133
Corynebacterium xerosis	+r	+	+	–		+		–	–	–	F	$+_g$	–	–	–	20–35 0–3		–		–			133
Dietzia maris	+	+		–	–	30		–	+	–	O	+	–	–	–	25 0–7		–		–			573
Rhodococcus equi	+			–	+			–		–		–	+	+	+	0–7		–		–			573
Rhodococcus fascians	+	+				+		36	+	+	O	+	+	+	+	25 0–5	40	+		+			573
Weissella strain DS-12	+cr																					

Carnobacterium species are very similar to *Lactobacillus* spp. *Vagococcus* describes motile Lancefield group N strains.
+c = gram positive coccus; +cb = gram positive cocco-bacillus; +cr = gram positive coccoid rod; +r = gram positive rod; Hip = Hippurate hydrolysis; MAF = Modified Acid Fast stain reaction; RAA = Rogosa Acetate Agar; v = variable reactions reported; w = weak reaction; ZN = Ziehl-Neilson.
Also see tables for *Nocardia* and *Mycobacteria*.

Table 4.15. *Mycobacterium* and *Nocardia* spp.

Test / Organism	Gm	Ox	cat	βH	mot	ODC	LDC	ADH	Nit	Ind	Cit	urea	mr	vp	aes	G	onpg	arab	glu	inos	lac	malt	man	mano	sal	sor	suc	tre	Xyl	H2S	temp	NaCl	NaCl 4	50°C	Hip	Ref
Mycobacterium abscessus	AFB		+						−			+																				6.5	+			736
Mycobacterium avium	AFB		+		−				+			−																			37–45	5−				737
Mycobacterium chelonae	AFB	−	+		−				−		−	+							+	−			−	+		−	−	−	−			15–25	0–3	−		−	133
Mycobacterium fortuitum	AFB		+		−				+			+																			28–37	0–5	+			569
Mycobacterium marinum	AFB		+		−				−			+																			28–30	0				111
Rough & smooth colonies. Cells may be long and cross-banded																																				
Mycobacterium neoaurum	AFB																														25	5−				53
Mycobacterium peregrinum	AFB		+								−													+				+	+		25–37	+				551
Mycobacterium poriferae	AFB								−			+																			25–37	5	+			756
Mycobacterium scrofulaceum	AFB								−			+																								569
Mycobacterium simiae	AFB								−			+																								569
Mycobacterium species	AFB	−			−				−			+																			28					337
Mycobacterium triplex-like	AFB																														25–30					345
Nocardia asteroides	+	−	+		−				+	−	+	+w		−	+	−	+	−	+	−	−	−	−	−	−	−	−	+	−	−	20–37	0–4	+	+	v	270,455,457
Nocardia brasiliensis	+		+						+	−	+	+		−	+	+	+	−	+	+	−	−	+	−	−	−	−	−	−		30–45	0–2	+	−	−	270,457
Nocardia brevicatena	+								−	−	−	−		−	+	+		−	−	−	−	−	+	−	−	−	+	+	−							165
Nocardia carnea	+								+		−	−			+			−	+	>			+			+										165
Nocardia caviae	+	−	+		−				+	>		+w		−	+	+		−	+	+	−	+	+	−	−	−	−	v	−			0–4	+	+	v	457
Nocardia crassostreae	+	−	+		−					−		+		−	+	−		−	+			−	−	−	−		+	+	−	−	22–30	0–2	−	+		270

Organism																		20–45	0–4			Ref
Nocardia farcinica	+	–	+			+	–	+	–	–	–			–		>		–	20–45	0–4	+		455
Nocardia flavorosea	+				–	–	–		–	–	–		–	–		–		–			+	–	165
Nocardia nova	+				–	+	–	+		–	–			–	>	–	>	–	20–40	0–4	+		455,805
Nocardia otitidiscaviarum	+				–	+	>	+	+	+	–			–		–		–					165
Nocardia pseudobrasiliensis	+	+			+	+	+	–	+	–	–		–	+	+		+	–	20–37				165,661
Nocardia salmonicida	+	–	+		+	+	+	+	+	–	–		–	–	–		–	–	10–30	0–4	–		455,391
Nocardia seriolae	+	–	–	–	+	–	+	–	+	–	>		–	–	–	+	–	–	20–30	0–2	+	+	455
Nocardia transvalensis					+	+	>	+	>	+				>	>								165
Nocardia vaccinii					+	+	>	+	>	–	+			>	–		+						165
Nocardia spp. (Aust)	+				–	+		+	–														117

5– = no growth in 5% NaCl.

Table 4.16. *Mycobacterium* spp. – additional tests.

Test	niacin	Nitrate	Tween 80	cat	aryl	urease	pyr	Fe	thio	5% NaCl	MCA	Ref
Organism												
Mycobacterium abscessus	–	–	+	+	+	+	+	–	+	+	+	736
Mycobacterium avium	–	–	–	v	–	–	+		+	–		737
Mycobacterium chelonae	v	–	–	+	+	+				–	+	133,737
Mycobacterium fortuitum	–	+	v	+	+	+				+	+	116,737
Mycobacterium marinum	–	–	+	–	–	+	+		+	–	–	737,111
Mycobacterium neoaurum		+			+						–	53
Mycobacterium peregrinum	–	+	–	+	+	+	+		+	+	+	551
Mycobacterium poriferae	–	–	+	+	–	+	+		+	+	–	756
Mycobacterium scrofulaceum	–	–	–	+	–	+	+		+	–		737
Mycobacterium simiae	v	–	–	+	+	+	+		+	–		737
Mycobacterium spp. (new)		–	–	–	–	+	+			–		337

aryl = arylsulfatase; pyr = pyrazinamidase; thio = 2-thiophenecarboxylic acid hydrazide.

Table 4.17. *Mycoplasma* spp.

Test	Gm	Ox	cat	βH	mot	PO₄	ODC	LDC	ADH	Nit	Ind	Cit	urea	mr	vp	aes	G	TTC	OF	arab	glu	inos	lac	malt	man	mano	sal	C	S	tre	Xyl	F&S	temp	NaCl	Pen	Ref
Organism																																				
Pathogenic																																				
Mycoplasma alligatoris						+			–				–		–	–					+		+			+						–	30–34			128
Mycoplasma crocodyli			+	+		+			–				–		–			+			+		+	+		+		+	+			–	25–42		R	441
Mycoplasma mobile		+	+	+	+	+v			–			–	–		–	–		+		+	+		+	+		+		+	+		+	+	4–30		R	440

Cells adhere strongly to glass or plastic. Weak growth on BA. F&S with horse serum, not bovine serum. Cells glide.

Test	Gm	Ox	cat	βH	mot	PO₄	ODC	LDC	ADH	Nit	Ind	Cit	urea	mr	vp	aes	G	TTC	OF	arab	glu	inos	lac	malt	man	mano	sal	C	S	tre	Xyl	F&S	temp	NaCl	Pen	Ref
Mycoplasma phocicerebrale				+		+			+				–			–		–										+	+			–	37			295
Mycoplasma phocidae				α		+			+				–								–															660
Mycoplasma phocirhinis				+		+			–				–					–			–							+	+		+	+	37			295
Mycoplasma testudinis				+		–			–				–			–		–		–	+	–	–	+	+	+	–			–	–	F				350

F&S = film and spots; C = cholesterol requirement for growth; S = serum required for growth; Pen = Sensitivity to penicillin 10 Units. See Chapter 7 for specific biochemical tests for *Mycoplasma*.

Table 4.18. *Listonella* spp.

Test	Gm	Ox	cat	βH	mot	SW	ODC	LDC	ADH	Nit	Ind	Cit	urea	mr	vp	aes	G	onpg	OF	arab	glu	inos	lac	malt	man	mano	sal	sor	suc	tre	Xyl	H₂S	MCA	TCBS	DNase	temp	NaCl	0129 10	0129 150	Amp	Ref
Organism																																									
Pathogenic																																									
Listonella anguillarum ATCC 14181	–	+	+	+	+	–		–	+	+	–	+	–	–	+	–			F	–	+g–	–	–	+	+	+	–	–	+	+	–	–	–	Y	+	10–37	0.5–7	S	S	R	506

Previously *Vibrio anguillarum*

Test	Gm	Ox	cat	βH	mot	SW	ODC	LDC	ADH	Nit	Ind	Cit	urea	mr	vp	aes	G	onpg	OF	arab	glu	inos	lac	malt	man	mano	sal	sor	suc	tre	Xyl	H₂S	MCA	TCBS	DNase	temp	NaCl	0129 10	0129 150	Amp	Ref
Listonella anguillarum NCIMB 6	–	+	+	+	+		–	–	+	+	+	–	–	+	+	–	F	+	F	+	+g–	–	–	+	+	+	–	–	+	+	–	–	–	wG	+	4–30	0.5–8	S	S		81, 563

MR may be negative or weak. Citrate is positive or negative

Serotype 01 strains tend to be positive for arabinose

Test	Gm	Ox	cat	βH	mot	SW	ODC	LDC	ADH	Nit	Ind	Cit	urea	mr	vp	aes	G	onpg	OF	arab	glu	inos	lac	malt	man	mano	sal	sor	suc	tre	Xyl	H₂S	MCA	TCBS	DNase	temp	NaCl	0129 10	0129 150	Amp	Ref
Listonella pelagia NCMB 2253	–	+	+	v	+		–	–	–	–	–	+	–	+	–	–	F		F	–	+g–	–	–	+	+	+	–	–	+	+	–	–	–	v	–	10–37	0.5–7	S	S	S	506, 563

Previously *Vibrio pelagius* II

wG = weak growth of green-coloured colonies; v = variable reaction.

Also check **Table 4.19** (*Moritella* spp.), **Table 4.20** (*Photobacterium* spp.), **Table 4.21** (*Vibrio* – pathogenic species), **Table 4.22** (*Vibrio* – environmental species).

Table 4.19. *Moritella* spp.

Test	Gm	Ox	cat	βH	mot	SW	ODC	LDC	ADH	Nit	Ind	Cit	urea	mr	vp	aes	G onpg	OF	arab	glu	inos	lac	malt	man	mano	sal	sor	suc	tre	Xyl	H₂S	MCA	TCBS	DNase	temp	NaCl	0129 10	0129 150	Amp	Ref
Organism																																								
Pathogenic																																								
Moritella marina ATCC 15381, NCMB 1144ᵀ	–	+	+	+	+	+	–	–	–	+	–	–	+	–	–	+	+	F	–	+g–	–	–	–	–	–	–	–	–	–	–	–		–	+	0–20	3–5	R	S	S	506, 766
																																								Mano = pos, DNase = neg (Ref 506)
Moritella marina ATCC 15381, NCMB 1144ᵀ	–	+					+	+		+	–		+	+	–	+	+	F	–	+g–	–	–	–	–	–	–	–	–	–					+	4–20	3	S	S	S	81, 82
Moritella viscosa NCIMB 13584	–	+	+	+	+	–	+	+	+	+	–		+	5	–	–	+	–	–	+g–	–	–	–	–	–	–	–	–	–	–	–		–	+	4–21	1–4	R	S	S	506
																																								Isolates from Scotland, Norway. May need 1% peptone added to media for growth. No growth at 25°C
Moritella viscosa	–	+	+				–	+									+	–	–	+	–	–	+	+	+	–	–	–	–	–	–		–		4–21	2–3	R	S		82
																																								Isolates from south-west Iceland only. No growth at 25°C
Moritella viscosa	–	+	+				–	–									+	–	–	+	–	–	+	+	+	–	–	–	–	–	–		–		4–21	2–3	R	S		82
																																								Isolates from North Iceland
Environmental																																								
Moritella japonica	–	+	+																																		R	R	S	585

Also check **Table 4.18** (*Listonella* spp.), **Table 4.20** (*Photobacterium* spp.), **Table 4.21** (*Vibrio* – pathogenic species), **Table 4.22** (*Vibrio* – environmental species).

Table 4.20. *Photobacterium* spp.

Test	Gm	Ox	cat	βH	mot	SW	ODC	LDC	ADH	Nit	Ind	Cit	urea	mr	vp	aes	G onpg	OF	arab	glu	inos	lac	malt	man	mano	sal	sor	suc	tre	Xyl	H₂S	MCA	TCBS	DNase	temp	NaCl	0129 10	0129 150	Amp	Ref
Organism																																								
Pathogenic																																								
P. damselae ssp. *damselae* ATCC NCMB 2184ᵀ	–	+	w	+	+w	–	–	+	+	+	–	–	+	+	+	–	–	F	–	+g+	–	–	+	–	+	–	–	–	–	–	–	G	+	+	10–37	0.5–5	S	S	S	268, 506, 618
																																								80% of *P. damselae* strains may be positive for trehalose (ref 745)
P. damselae ssp. damselae biotypes																																								
biotype 1	–	+	+	+	+	–	–	+	+	+	–	–	+	+	+	–	–	F	–	+g+	–	–	+				–				–	G				0.5–7	S	S	R	618

continued

Table 4.20. Continued.

Test	Gm	Ox	cat	βH	mot	SW	ODC	LDC	ADH	Nit	Ind	Cit	urea	mr	vp	aes	G	onpg	OF	arab	glu	inos	lac	malt	man	mano	sal	sor	suc	tre	Xyl	H2S	MCA	TCBS	DNase	temp	NaCl	0129/10	0129/150	Amp	Ref
biotype 2	–	+	+	+	+	–	–	–	+	+	–	–	+	+	+	–	–	–	F	–	+g–	–	–	+			–	–						G			0.5–7	S	S		618
biotype 3	–	+	+	+	+	–	–	–	+	+	–	–	+	+	+	–	–	–	F	–	+g–	–	–	+			–	–				–		G			0.5–5	S	S		618
biotype 4	–	+	+	+	+	–	–	–	+	+	–	–	+	+	–	–	–	–	F	–	+g–	–	–	+			–	–				–		G			0.5–7	S	S		618
biotype 5	–	+	+	+	+	–	–	+	+	+	–	–	+	+	+	–	–	–	F	–	+g–	–	–	+			–	–				–		G			0.5–7	S	S		618
biotype 6	–	+	+	+	+	–	–	–	+	+	–	–	+	+	+	–	–	–	F	–	+g–	–	–	+			–	–				–		G			0.5–7	S	S		618
biotype 7	–	+	+	+	+	–	–	+	+	+	–	–	+	+	+	–	+	+	F	–	+g–	–	–	+			–	–				–		G			0.5–5	R	R		618
biotype 8	–	+	+	+	+	–	–	–	+	+	–	–	+	+	+	–	+	–	F	–	+g+	–	–	+			–	–				–		G			0.5–7	R	R		618
biotype 9	–	+	+	+	+	–	–	+	+	+	–	–	+	+	+	–	+	–	F	–	+g+	–	–	+			–	–				–		G			0.5–7	S	S		618

Haemolysis may be more apparent on BA than MSA-B

Test	Gm	Ox	cat	βH	mot	SW	ODC	LDC	ADH	Nit	Ind	Cit	urea	mr	vp	aes	G	onpg	OF	arab	glu	inos	lac	malt	man	mano	sal	sor	suc	tre	Xyl	H2S	MCA	TCBS	DNase	temp	NaCl	0129/10	0129/150	Amp	Ref
P. damselae ssp. piscicida ATCC 17911	–cb	+	+	–	–	–	–	+	+	–	–	–	–	w	+	–	–	–	F	–	+g–	–	–	–	–	+	–	–	+w	–	–	–	–	–	73	15–35	0.5–3	S	S	S	518, 745

Short rods, bipolar staining. VP results variable or neg by API 20E (289, 459)

Environmental

Test	Gm	Ox	cat	βH	mot	SW	ODC	LDC	ADH	Nit	Ind	Cit	urea	mr	vp	aes	G	onpg	OF	arab	glu	inos	lac	malt	man	mano	sal	sor	suc	tre	Xyl	H2S	MCA	TCBS	DNase	temp	NaCl	0129/10	0129/150	Amp	Ref
P. angustum NCIMB 1895	–	+	+	–	–	–	–	–	–	+	–	–	–	+	+	+	+	+	F	–	+g–	–	–	+	–	+	–	–	+	v	+	–	–	–	+	4–37	0.5–6	S	S	S	745
P. iliopiscarium ATCC 51760	–	+	+	–	+	–	–	+	+	+	–	–	–	+	+	+	–	–	F	–	+g+	–	–	+	–	+	–	–	+	v	–	–	–	G	+	4–25	0.5–2	S	S		81
P. leiognathi ATCC 25521	–	–	+	–	+	–	–	+	–	v	–	–	–	+	+	+s	+s	–	F	–	+g–	–	–	–	–	+	–	–	–	–	–	–	–	–		20–30	0.5–5				340, 745
P. leiognathi LMG 4228	–	–	–	–	–	–	–	–	+	–	–	–	–	–	+	–	+	+	F	–	+g–	–	v	–	–	+	–	–	–	–	–	–	+	–	–	20–35	0.5–6	S	S	S	745
P. phosphoreum NCIMB 844	–	–	+	–	v	–	–	+	+	+	–	–	–	+	+	v	–	+	F	–	+g+	–	–	+	+	+	–	–	–	–	–	–			+	4–30	0.5–8	R	S	S	81
P. phosphoreum NCIMB 1282	–	–	+	–	v	–	–	–	+	+	–	–	–	+	+	v	+	+	F	–	+g+	–	–	+	–	+	–	–	–	–	–	–			–	4–21	1–8	S	S		81
P. profundum	–	–					–	+	+	–	–				+		–		F		+g+			+	–	+	–	–	–	–					–	5–20					586

Also check **Table 4.18** (*Listonella* spp.), **Table 4.19** (*Moritella* spp.), **Table 4.21** (*Vibrio* – pathogenic species), **Table 4.22** (*Vibrio* – environmental species).

Table 4.21. *Vibrio* – pathogenic species.

Test / Organism	Gm	Ox	cat	βH	mot	SW	ODC	LDC	ADH	Nit	Ind	Cit	urea	mr	vp	aes	G onpg	OF	arab	glu	inos	lac	malt	man	mano	sal	sor	suc	tre	Xyl	H₂S	MCA	TCBS	DNase	temp	NaCl	0129 10	0129 150	Amp	Ref
V. alginolyticus	–	+	+	+v	+	+	53	+	–	+	+	60	–	+	83	–	+ –	F	–	$+_g^-$	–	–	+	+	+	33	–	+	+	–	–	+	Y	+	15–42	1–10	R	S	R	552, 821
V. alginolyticus NCIMB 1903	–	+					–	+	–	+				+	+	+	+	F	–	$+_g^-$	–	–	+	+	–	–	–	+	–					+	15–42	1–10	S	S		81
V. cholerae 01	–	+	+	+	+	–	+	+	–	+	+	+v	–	–	–	–ᵃ	+ +	F	–	$+_g^-$	–	–	+	+	80	+v	–	+	+	–	–	+	Y	+	4–42	0–6v	S	S	R	507, 821
V. cholerae non 01	–	+	+	+	+	–	+	+	–	+	+	+	–	–	+	–ᵃ	+ +	F	–	$+_g^-$	–	–	+	+	70	–	–	+	+	–	–	+	Y	+	10–42	0–3	S	S	Sv	507
V. cholerae 0139	–	+		+	+	–	+	+	–		+				75		+	F	–	$+_g^-$	–	–	+		+		+	+					Y			0–3	R	R	S	9, 392
V. mimicus	–	+	+	+	+	–	+	+	–	+	+	95	–	–	–	–	+ +	F	–	$+_g^-$	–	–	+	+	+	–	–	–	+	–	–	+	G	+	4–42	0–6	S	S	R	210, 230, 507
V. parahaemolyticus	–	+	+	+	+	+	+	+	–	62	+	63	–	+	–	–	+ –	F	80	$+_g^-$	–	–	+	+	+	53	–	–	+	–	–	+	G	+	20–40	3–8	R	S	V	272, 552
V. parahaemolyticus ATCC 43996	–	+	+	+	+	+	+	+	–	+	+	–	+s	+	–	–	+ –	F	+	$+_g^-$	–	–	+	+	+	–	–	–	+	–	–	+	G	w	20–40	0–3	R	PS		135
V. parahaemolyticus ATCC 27969	–	+	+		+	+	+	+	–	+	–	+	–			+	+ –	F	–	$+_g^-$	–	–	+	+	+	–	–	–			–		G					S		499
V. (carchariae) harveyi ATCC 35084	–	+	+	–	+	+s	+	+	–	+	+	+s	+	+	–	–	v –	F	–	$+_g^-$	–	–	+	+	+	–	+s	–	+	–	–	+	Y/G	+	10–40	0.5–8	R	S		11, 135, 847
Colonies of *V. harveyi* slowly spread across the plate. Cit may take 3–5 days																																								70,81
V. harveyi ATCC 14126, ATCC 14129	–	+	+	w	+	–	+	+	–	+	+	+w	–	–	–	+	+ +	F	–	$+_g^-$	–	+	+	+	+	–	–	+	–	–	–	–	Y	–	12–40	3–6	R	S		11
Haemolysis against sheep RBC																																								
V. vulnificus ATCC 27562 biotype 1, serovar non-E	–	+	+	+	+	+	+	+	–	+	+	+	–	+	–	–	+ +	F	–	$+_g^-$	–	+	+	–	–	–	–	–v	+	–	–	+	G	–	20–42	3–6	S	S		98
Human strain, USA – non-virulent for eels. Cit = neg in API 20E but pos in Simmons tube citrate																																								
V. vulnificus ATCC 27562 biotype 1	–	+	+	+	+	–	89	+	–	+	+	+	–	+	–	–	+ +	F	–	$+_g^-$	–	+	+	44	+	+	44	–	+	–	–	+	G	–	37	0.5–7	S	S		746
As originally described by Tison *et al.* (1982) using conventional media																																								

continued

Table 4.21. *Continued.*

Test	Gm	Ox	cat	βH	mot	SW	ODC	LDC	ADH	Nit	Ind	Cit	urea	mr	vp	aes	G	onpg	OF	arab	glu	inos	lac	malt	man	mano	sal	sor	suc	tre	Xyl	H₂S	MCA	TCBS	DNase	temp	NaCl	0129 10	0129 150	Amp	Ref
V. vulnificus ATCC 33184 biotype 1, serovar non-E	–	+	+	+	+	–	+	+	–	+	+	+	–	+	–	–	+	+	F	–	$+_{g^-}$	–	+	+	+	+	–	–	–	+	–	–	+	G		37	0.5–6	S	S	S	98
Human strain – USA																																									
V. vulnificus biotype 1, serovar non-E	–							+			+														v																201
Isolated from diseased eels. Belgium (man = neg), Sweden (man = pos)																																									
V. vulnificus ATCC 33187 biotype 2, serovar E	–	+	+	+	+	–	–	+	+	+	–	+	–	+	+	–	+	+	F	–	$+_{g^-}$	–	+	+	–	–	–	–	–	+	–	–	+	G	+	20–35	3–6	S	S	S	98
Human strain, USA – virulent for eels																																									
V. vulnificus ATCC 33149 biotype 2, serovar E, 04	–	+	+	+	+	–	–	+	–	+	–	+	–	+	–	–	+	+	F	–	$+_{g^-}$	–	+	+	–	–	–	–	–	+	–	–	+	G	+	20–35	3–6	S	S	S	98
Virulent for eels – Japanese strain																																									
V. vulnificus biotype 2 serovar E	–	+	+	+	+	–	20	+	–	+	–	+	–	+	–	–	+	+	F	–	$+_{g^-}$	–	+	+	+	–	–	–	–	+	–	–	+	G	+	20–35	3–6	S	S	S	26
Taiwanese strains – avirulent for eels																																									
V. vulnificus ATCC 33148 biotype 2	–	+	+	+	+	–	+	+	–	+	–	+	–	+	–	–	+	+	F	–	$+_{g^-}$	–	+	+	+	+	+	–	–	+	–	–	–	G		37	0.5–7	S	S	S	746
As originally described by Tison *et al.* (1982) using conventional media																																									
V. vulnificus biogroup 3	–	+	+	+	+	–	+	+	–	+	+	–	–	+	–	–	+	–s	F	–	$+_{g^-}$	–	–	+	–	–	–	–	–	+	–	–	–	G				S	S		101, 565
Human strain – Israel																																									
V. trachuri	–	+	+	+	+	–	–	+	–	+	+	–	–	–	–	–	–	–	F	–	+	–	+	+				+	+		–			Y		30	3–7	S	S		400
V. shilonii	–	+	+	+	+	+	–	+	–	+	+	–	–				+	+	F	–	$+_{g^+}$	–	–						+				Y		16–37	2–4	S	S		59, 458
V. fischeri ATCC 7744	–	+	+	–v	+	–	–	+	–	+		v	+	+	+	–	–	+	F	–	$+_{g^-}$	–	–	+	+	–	60	–	40	–	–			–, G	+	10–30	0.5–6	S	S	S	
Yellow-orange pigment. West description = mano +ve																																									
V. fischeri NCMB 1281, ATCC 7744	–	+	+	+	+	–	+	+	–	+	–	–	+				+			–	$+_{g^-}$	–	–	v	+v	+	+v	+	–	+v	–			v	+		0.5–5	S	S		81, 342
V. fischeri ATCC 25918	–	+	+	–	–	–	+	+	–	+	–	–	–	–	–	–	–	–	F	–	+	–	–	+	–	–	+	–	–	–	–		–	–	+	10–30	1–6	R	S	R	745
VP +ve in API 20E, –ve in MRVP medium																																									

V. fischeri NCMB 1281	−	+	+	+	+	−	+	−	−	−	−	−	F	−	+g−	−	−	+	−	+	−	−	+	−	−	v	+	10–30	0.5–7	S	S	S	506
Grey colonies																															
V. logei ATCC 29985^T	−	+	+	+	−	+	+	−	+	+	−	−	F	−	+g−	−	+	+	−	+	−	+	+	−	−		+	4–22	0.5–5	S	S	S	506
V. logei ATCC 15382	−	+	+	−	−	+	+	+	+	+	−	−	F	−	+g−	+	+	+	+	+	−	+	+	+	−		+	4–25	0.5–5	S	S	S	506
V. logei NCMB 1143	−	+	+	−	+	+	+	+	+	+	−	>	F	−	+g−	−	+	+	>	+	+	+	+	−	−		+	4–30	0.5–5	S	S	S	81, 745
V. logei NCIMB 2252	−	+		+	+	−	+	+	+	+	+	−	F	+	+g−	−	−	+	−	+	−	−	−	−	−		+	4–22	0.5–5	S			81
V. logei	Yellow-orange pigment. Strains are citrate negative																															
V. proteolyticus	−	+	+	+	+	33	+	−	+	−	−	−	F	−	+g−	−	−	+	−	+	−	+	50	76	+	Y/G	+	20–40	1–10	R	S	S	67, 788
V. proteolyticus AHLDA 1735	−	+	+	+	+	+	+	+	+	+	−	w	F	−	+g−	w	+	+	+	+	+	+	+	−	+	G	−	42–	1–10	PS	S		135
V. fluvialis ATCC 33809	−	+	>	+	+	−	−	72	+	−	F	+	+g−	−	+	+	−	+	−	+	−	+	+	Y	+	10–35	1–6	R	S	V	81, 620
V. fluvialis NCTC 11327	−	+	>	+	+	−	−	−	+s	F	+	+g−	−	+	+	−	+	−	+	−	+	+	Y	+	10–35	1–6	R	S	V	123, 485
API 20 E pos for VP and indole																															
V. furnissii ATCC 35016^T	−	+	>	+	+	−	−	−	+	F	+	+g+	−	+	+	−	+	−	+	−	+	+	Y	+	20–37	1–8	R	S	R	123, 485
API 20 E pos for VP and indole																															
V. furnissii ATCC 11218	−	+	+	+	−	+	+	−	+	F	+	+g+	−	+	+	−	+	−	+	−	+	+	Y	w	20–37	1–8	R	S		135
ADH = neg at 0% NaCl. Pos ADH at 2% NaCl																															
V. splendidus I ATCC 33125	−	+	+	90	+	43	+	−	+	+	−	F	−	+g−	−	−	+	−	+	−	+	−	G	−	4–37	1–6	S	S	S	281, 506, 620
Type strain is citrate negative																															
V. splendidus II NCMB 2251	−	+	+	v	+	+	+	−	+	+	−	F	−	+g−	−	−	+	+	−	+	−	+	G	+	4–30	0.5–7	S	S	R	81, 254 / 81, 506, 819
V. tubiashii ATCC 19109	−	+	+	83	+	30	+	−	+	>	+	F	−	+g−	−	−	+	+	−	+	−	+	Y	+	10–37	0.5–6	S	S	S	108, 321

continued

Table 4.21. *Continued.*

Test	Gm	Ox	cat	βH	mot	SW	ODC	LDC	ADH	Nit	Ind	Cit	urea	mr	vp	aes	G	onpg	OF	arab	glu	inos	lac	malt	man	mano	sal	sor	suc	tre	Xyl	H₂S	MCA	TCBS	DNase	temp	NaCl	0129 10	0129 150	Amp	Ref
V. tubiashii NCMB 1340ᵀ	−	+	+	w	+	−	−	−	+s	+	+	−	−	+	−	−	+	+	F	−	+g−	−	−	+	+	+	−	−	+	+	−	−	−	Yw	+	10–30	0.5–7	S	S	S	135, 506
V. ichthyoenteri	−	+	+		+	−	−	−	−	+	−	−	−		−		−	−	F	−	+g−	−	−	+	+	+	−	−	+	+	−	−		Yw		15–30	1–6	S	S	S	389
V. tapetis	−cb	+	+	+	+	−	−	−	−	+	+	−	−		−	−	+	+	F	−	+g−	−	−	+	+	+		−	−	+	−	−		G		4–22	1–5	S	S	S	108, 146
Aesculin reported positive (ref 610)																																									587
V. ordalii ATCC 33934	−cv	+	+	+	+	−	−	−	−	−	−	−v	−	−	−	−	+	−	F	−	+g−	−	−	+	+	−	−	−	+	−	−	−		−	+	4–30	0.5–7	S	S	Rv	506, 680, 819
V. ordalii DF-3K	−cv	+	+	+	+	−	−	−	−	−	−	−	−	−	−	−	+	−	F	−	+g−	−	−	+	−	−	−	−	+	−	−	−		−	+	4–30	0.5–7	S	S	Rv	198, 680
V. ordalii NCIMB 2167	−	+																	F		+g−			+	+	+	−	−	+	+		−			+	21–37	0.5–5	S	S		81
V. pectenicida	−	+	+	+	+	+	−	−	−	+	−	+	−	−	−	−	+	−	F	−	+g−	−	−	+	−	−	−	−	−	−	−	−		−	+	4–30	1–6	S	S	S	470
V. penaeicida	−	+	+	+	+	−	−	−	−	+	50	67	+	+	−	−	+	+	F	−	+g−	−	50	+	−	+	−	−	+	+	−	−		G		20–30	1–3	S	S	S	187, 388
Type strain = indole neg																																									
V. salmonicida	−	+	+	−	+	−	−	−	−	−	−	−	−	−	−	−	−	−	F	−	+gv	−	−	−	+	+	−	−	−	+	−	−		G	+	1–22	0.5–4	S	S	S	198, 232, 506
V. salmonicida NCIMB 2262	−	+					−	−		−	−	−	−		−		+		F	−	+g−	−	−	+	−	−	−	−	−	−					+	1–22	1–4	S	S		81
V. corallilyticus YB1ᵀ	−	+	+		+		+	+	+	+	+	+	−				+	+	F	+	+								+					Y		25–30	1–7	S	S	R	83,84
4 of 6 strains are ADH negative																																									

ᵃ *V. cholerae* is negative for aesculin, but some strains cause blackening of the medium due to production of melanin. To determine a true negative the loss of fluorescence must be measured at 354 nm. See under 'aesculin' in interpretation of tests.

In this table, Vibrios are grouped according to their ODC, LDC, ADH reactions. Use this as a starting point in identifications. 42− = negative growth at 42°C.

Also check **Table 4.18** (*Listonella* spp.), **Table 4.19** (*Moritella* spp.), **Table 4.20** (*Photobacterium* spp.), **Table 4.22** (*Vibrio* – environmental species).

Table 4.22. Vibrio – environmental species.

Test / Organism	Gm	Ox	cat	βH	mot	SW	ODC	LDC	ADH	Nit	Ind	Cit	urea	mr	vp	aes	G	onpg	OF	arab	glu	inos	lac	malt	man	mano	sal	sor	suc	tre	Xyl	H2S	MCA	TCBS	DNase	temp	NaCl	0129 10	0129 150	Amp	Ref
Environmental																																									
V. diabolicus	–	+	+		+	+	+	+	–	+	+	–	–		–	–	–	–	F	–	+	–	–	+	+	+	–	–	+	+	–	–				20–45	2–5	S	S		635
V. rotiferianus	–	+	+		+	–	+	+	–	+	+	–	>		–	–	+	+	F	+	+g+	–		+	+	+		–	+	+	–	–		Y		28–40	2–6	S	S		305
V. cincinnatiensis	–	+	+		+	–	–	+	–	+	+	+	–		+		+	+	F	+	+g+	+	–	–	–	+	+	–	+	+	–	–		Y	+	22–37	1–6	R	S	S	120
V. mediterranei	–	+	+	+	+	–	–	+	–	+	+	–	–	+	–	75	+	+	F	–	+g–	–	–	+	+	+	25	+	+	+	–	–	–	Y	+	20–30	3–6	S	S	S	631
V. metschnikovii	–cr	–	+	+	+	–	–	–	+	–	20	30	–	+	+	40	+	+	F	–	+g–,v	–,v	59	+	+	+	–	30	+	+	–	–	V	Y,–	+	4–40	0.5–7	S	S	V	483, 819
V. brasiliensis LMG 20546^T	–cr	+	+		+	–	–	–	+	+	+		–	+	+		+	+	F	–	+	–		+	+	+		–	+	+	–	–		Y		20–40	2–6	S	S		740
V. neptunius LMG 20536^T	–cr	+	+		+	–	–	–	+	+	+	+	–	+	+		+	–	F	–	+	–		–		+		–	+	+	–	–		Y		20–35	2–6	S	S		740
V. pacinii LMG 19999^T	–	+	+		+	–	–	–	+	+	–	w	–	+	66		>	–	F	+	+	–	+	–					+	+	–	–		Y		4–35	1.5–6	S	S		147
V. xuii LMG 21346^T	–cr	+	+		+	–	–	–	+	+	+	–	–	+	+		–	–	F	+	+	–		+	+	+		–	+	+	–	–		Y		20–40	2–8	S	S		740
V. orientalis NCMB 2195^–	–	+	+	–	+	–	–	–	+	+	+	+	–	+	–	–	+	+	F	–	+g–	–	–	+	+	+	–	–	+	+	–	–		Y	+	4–35	0.5–8	S	S		506
Some LDC positive strains																																									
V. aerogenes	–	–	+		+	–	–	–	+	+	+	+	–		–	+	+	+	F	–	+g+	+	–	+	+	+	–	–	+	+	+	–	–	Y	+	20–35	1–7	R	R	S	692
V. aestuarianus	–	+	+	–	+	–	–	–	+	+	+	+	–	+	–	+	+	+	F	–	+g–	–	+	+	+	+	–	v+	+	+	–	–	+	Y	+	4–37	0.5–6	S	S	S	747
V. diazotrophicus	–	+	+		+	–	–	–	+	+	+	+	16	+	–	+	+	+	F	+	+g–	–	+	+	+	–	50	–	+	+	+	–		Y	–	10–35	0.5–6	R	S	S	319, 819, 820
V. lentus	–	+	+		+	–	–	–	+	+	+	+	+		–		+	+	F	–	+g–	–	–	+	+	+	–	–	–	+	–	–		G	+	4–30	2–6	R	R		513
Thornley's ADH = pos, Møller ADH = neg. Type strain = man neg																																									
V. mytili	–	+	+		+	–	–	–	+	+	–	–	–	–	–	–	–	–	F	+	+g+	–	–	+	+	–	+	–	+	+	+	–	–	Y	–	10–37	1–10	R	S		632
V. nereis	–	+	–	+	+	–	–	–	+	+	+	+	–	+	–	–	+	–	F	–	+g–	–	–	+	–	+	–	–	+	+	–	–		Y	+	20–35	3–6	R	S	S	819
Enterovibrio norvegicus	–	+	+		+	–	–	–	+	–	+	–	–	–	–	+	+	+	F	–	+	–	–	+	+				+	+	–	–		G	–	20–28	1.5–6	R	R		741

continued

Table 4.22. *Continued.*

Test	Gm	Ox	cat	βH	mot	SW	ODC	LDC	ADH	Nit	Ind	Cit	urea	mr	vp	aes	G ongg	OF	arab	glu	inos	lac	malt	man	mano	sal	sor	suc	tre	Xyl	H₂S	MCA	TCBS	DNase	temp	NaCl	0129 10	0129 150	Amp	Ref
V. scophthalmi	–	+	+	–	+	–	–	–	90	+	–	–	–	+	–	+	–	F		+g–	–	–	+	–	+	–	–	+	–	–	–	–	Y	–	22–35	0.5–3	S	S		149, 254
V. cyclitrophicus	–	+	+		+		–	–	+w	–	–	–	–		–	+	–	F	–	+g–	–	–	+	+	–	–	–	+	+	+	–				4–37	2–10			S	338
V. agarivorans	–	+	+	–	+	–	–	–	–	+	–	+	–	+	–	–	+	F	–	+g–	–	–	w	w	–	–	–	–	+	+	–	–	G	–	20–37	1–6	S	S	S	135, 514
Hydrolyses agar. Gelatin neg in API 20E																																								
V. campbellii	–	+	+	–	+	–	–	–	–	+	+	58	–	50	–	50	–	F	–	+g–	–	–	+	+	50	82	–	–	+	–	–		G	+	20–35	3–6	R	S	R	820
V. gazogenes	–cr	–	+	+	+	–	–	–	–	–	–	+	–		–	+	+	F	+	+g+	–	–	+	+	+	+	+	+	+	+	–		Y	–	20–42	3–6	S	S		67
Red to orange colonies																																								
V. halioticoli	–	+	+	–	–	–	–	–	–	+	+v	–	–	+	–	–	–	F	–	+g–	–	–	+	+	+v	–	–	–v	–	–	–	–	G	–	15–30	2–3	S	S		678
V. hollisae	–cv	+	+	w	+w	–	–	–	–	+	+	–	–	–	–	–	–	F	+	+g+	–	–	–	–	+	–	–	–	+	–	–	–	–	–	30–37	1–6	R	S	S	346, 580
Haemolysis weak at 24 h, positive after 7 days																																								
V. natriegens	–	+	+	–	+	–	–	–	–	+	–	+	62	+	–	+	67	F	+	+g–	60	–	+	+	30	+	–	+	+	–	–		–Y	+	10–40	3–6	R	S	S	44, 820
V. natriegens NCMB 1900	–	+	+	–	+	–	–	–	–	+	–	+	–	–	–	–	+	F	+	+g–	–	–	+	+	+	+	–	+	+	–	–		Y	+	4–37	0.5–7	S	S	S	506, 563
V. navarrensis	–	+	+	–	+	–	–	–	–	+	+	+	–	–	–	+	+	F	–	+g–	–	–	+	+	+	–	–	+	–	–	–		Y	–	10–40	0.5–7	S	S	S	767
V. nigripulchritudo	–	+		–		–	–	–	–	+	+	+	–	50	–	50	+	F	–	+g–	+	+	+	+	–	–	v+	–	+	–	–		G	+	20–30	3–5	R	S	S	277
V. wodanis *Colonies blue-black*	–	+	+	+	+	–	–	–	–	–	89	+	20	+	–	+	+	F	–	+g–	–	–	+	43	+	–	–	83	+	–			30	+	4–25	1–4	S	S	S	506
NCIMB 13582 *Colonies blue-black*																																								
V. rumoiensis	–	+	+						–	+	–	+	+	+	–	–	–	F	+	+	–		+	+	+			+	+	–				+	2–34	3–6	S	S	S	850
V. calviensis	–	+	+	+	+				–	+	+	+	+	+	–	+	+	F		+g–			+	+									G		4–30	1.5–12	S	S	S	216

V. natriegens strain NCMB 1900 was initially recorded as the type strain for *Listonella pelagia* I (Macián *et al.*, 2000).

In this table, Vibrios are grouped according to their ODC, LDC, ADH reactions. Use this as a starting point in identifications.

Also check **Table 4.18** (*Listonella* spp.), **Table 4.19** (*Moritella* spp.), **Table 4.20** (*Photobacterium* spp.), **Table 4.21** (*Vibrio* – pathogenic species).

Laboratory Worksheet

Case Number ....

Test / Isolate No.	Gm	Ox	cat	βH	mot	SW	ODC	LDC	ADH	Nit	Ind	Cit	urea	mr	vp	aes	G	onpg	OF	arab	glu	inos	lac	malt	man	mano	sal	sor	suc	tre	Xyl	H₂S	MCA	TCBS	DNase	temp	NaCl 0/3	0129 10	0129 150	Amp	Identification

Gm = Gram; Ox = oxidase; cat = catalase; βH = β haemolysis; mot = motility; SW = swarming; ODC = ornithine decarboxylase; LDC = lysine decarboxylase; ADH = arginine dihydrolase; Nit = nitrate; Ind = indole; Cit = citrate; urea = urea; mr = methyl red; vp = Voges-Proskauer; aes = aesculin; G = gelatin; onpg = o-nitrophenyl-β-galactopyranoside (β-galactosidase test); OF = Oxidative fermentative; arab = arabinose; glu = glucose; inos = inositol; lac = lactose; malt = maltose; man = mannitol; mano = manrose; sal = salicin; sor = sorbitol; suc = sucrose; tre = trehalose; xyl = xylose; H₂S = hydrogen sulphide; MCA = MacConkey agar; TCBS = thiosulphate-citrate-bile-sucrose agar; DNase = deoxyribonuclease agar; temp = temperature; NaCl 0/3 = growth in 0% and 3% salt; 0129 10 = sensitivity to vibrio static disc 10 µg; 0129 150 = sensitivity to vibrio static disc at 150 µg; Amp = sensitivity to ampicillin 10 µg.

Table 4.23. API 20E database biochemical results.

Organism	onpg	ADH	LDC	ODC	cit	H2S	ure	TDA	ind	vp	G	glu	man	inos	sor	rha	suc	mel	amy	arab	ox	NO2	N2	mot	MCA	O	F	Temp	Time	Inoc %NaCl	Ref	Strain
Acinetobacter/Moraxella	–	–	–	–	–	–	–	–	–	–	–	–	–	–	–	–	–	–	–	–	+										41	
Actinobacillus delphinicola	–	+	–	–	–	–	–	–	–	+	–	+	–	–	–	–	–	–	–	–	+	+	–	–	–	+	+	37	48	0.85	263	Dolphin
Actinobacillus delphinicola	–	21	+	70	–	–	–	–	–	+	–	+	–	–	–	–	–	–	–	–	+	+	–	–	–	+	+	37	48	0.85	263	Porpoise
Actinobacillus delphinicola	–	+	–	–	–	–	–	–	–	+	–	+	–	–	–	–	–	–	–	–	+	+	–	–	–	+	+	37	48	0.85	263	Whale
Actinobacillus scotiae	+	–	–	+	–	–	+	–	–	+	–	+	–	–	–	–	–	–	–	–	+	+	–	–	–	+	+	37	48	0.85	265	
Aeromonas allosaccharophila	+	v	+	v	v	–	–	–	+	–	+	+	+	–	–	v	+	v	–	v	+	+	+	+	+	+	+	27	24		41	
Aeromonas bestiarum	+	+	+	–	–	–	–	–	+	–	+	+	+	–	–	+	+	–	+	+	+	+	+	+	+	+	+	27	24	0.85	452	Carp
Aeromonas caviae	+	+	–	–	10	–	–	–	+	8	80	+	+	–	–	–	+	–	–	+	+	+	+	+	+	+	+				21	
Aeromonas caviae	+	+	+	–	+	–	–	–	+	–	+	+	+	–	–	–	+	–	–	+	+	+	+	+	+	+	+	25	48	0.85	240	ATCC 15468^T
Aeromonas culicicola	+	+	+	–	+	–	–	–	+	+	+	+	+	–	–	–	+	–	–	–	+	+	+	+	–	+	+	30	24	0.85	624	NCIM 5147^T
Aeromonas encheleia	+	+	–	–	–	–	–	–	+	–	+	+	+	–	–	75	75	+	–	–	+	+	+	+	–	+	+	27	24	0.85	241,452	
Aeromonas eucrenophila	+	+	–	–	–	–	–	–	+	–	+	+	+	–	–	–	–	–	+	–	+	+	+	+	–	+	+					
Aeromonas hydrophila	+	+	+	–	–	–	–	–	+	+	+	+	+	–	–	–	+	–	–	+	+	+	+	+	+	+	+	25	48	0.85	240	ATCC 7966^T
Aeromonas hydrophila	+	+	+	–	v	–	–	–	+	+	–	+	+	–	–	–	+	–	–	+	+	+	+	+	+	+	+				674b	ATCC 7966^T
Aeromonas hydrophila	+	+	+	–	v	–	–	–	+	+	+	+	+	–	–	+	+	–	+	+	+	+	+	+	+	+	+				322	ATCC 14715
Aeromonas hydrophila	+	+	–	–	+	–	–	–	+	–	+	+	+	–	–	–	+	–	+	+	+	+	+	+	+	+	+			2	509	WFM 504
Aeromonas hydrophila	+	+	+	–	–	–	–	–	+	+	+	+	+	–	+	–	+	–	+	+	+	+	+	+	+	+	+	23	48	2	227a	
Aeromonas janadaei	+	+	+	–	+	–	–	–	+	+	+	+	+	–	–	–	–	50	–	–	+	+	+	+	+	+	+				143	
Aeromonas janadaei	+	+	+	–	+	–	–	–	+	+	+	+	+	–	–	–	–	+	–	–	+	+	+	+	+	+	+	25	48	0.85	135	AHLDA 1718
Aeromonas popoffii	+	+	–	–	+	50	–	–	+	–	+	+	+	–	–	–	–	–	–	–	+	+	+	+	+	+	+				450	
A. salmonicida salmonicida	–	+	+	–	–	–	–	–	–	–	–	+	+	–	–	–	+	–	+	–	+	+	–	–	+	+	+				450	
A. salmonicida salmonicida	+	+	+	–	–	–	–	–	+	–	+	+	+	–	–	–	+	–	+	–	+	+	–	–	+	+	+				468	NCIMB 1102
A. salmonicida achromogenes	–	–	–	–	–	–	–	–	–v	–	–	+	+	–	–	–	+	–	–	–	+	+	–	–	+	+	+				322	
A. salmonicida achromogenes	–	–	–	–	–	–	–	–	+	+v	–	+	+	–	–	–	+	–	–	–	+	+	–	–	+	+	+				468	NCIMB 1110
A. salmonicida masoucida	+	+	–	–	–	–	–	–	+	–	+	+	+	–	–	–	+	–	+	–	+	+	–	–	+	+	+				322	
A. salmonicida salmonicida	+	+	+	–	–	–	–	–	–	–	+	+	+	–	–	–	+	–	+	–	+	+	–	–	+	+	+				322	
A. salmonicida salmonicida	–	+	+	–	v	–	–	–	–	–	+	+	+	–	–	–	+	–	+	–	+	+	–	–	+	+	+				674	
A. salmonicida salmonicida	–	+	+	–	–	–	–	–	–	–	+	+	–	–	–	–	+	–	–	–	+	+	–	–	+	+	+				674	
A. salmonicida atypical	–	–	–	–	–	–	–	–	+w	–	–	–	–	–	–	–	–	–	–	–	+			–	+	w	w					
A. salmonicida atypical	–	–	–	–	–	–	–	–	+	–	–	+	+	–	–	–	w	–	–	–	+	+	–	–	w	w	w				761	Australian strain
A. salmonicida atypical	–	+w	–	–	–	–	–	–	+	–	–	+	+	–	–	–	w	–	–	–	+	+	–	–	w	w	w	25	72	0.85	135	AHLDA 1334
Aeromonas sobria	+	+	+	–	–	–	–	–	+	+	+	+	+	–	–	–	+	–	–	–	+	+	+	+	+	+	w	25	48	0.85	240	CIP 74.33^T
Aeromonas sobria	+	+	+	+	50	–	–	–	+	+	+	+	+	–	–	–	+	–	–	28	+	+	+	+	+	+	+				21	

Aeromonas trota	+	+	+	–	+	–	–	–	+	–	+	+	+	–	–	–	–	–	+	–	+	+		+	+	+	+	37	24	0.85	142	ATCC 49657[T]
Aeromonas veronii veronii	+	–	+	+	+	–	–	–	+	+	+	+	+	–	–	–	+	–	+	–	+	+		+	+	+	+	27	24	0.85	347,2	
Allomonas spp.	+	+	–	–	–	–	–	–	+	–	+	+	+	–	–	–	+	–	+	–										2	509	WFM401
Burkholderia pseudomallei	–	+	–	–	63	–	–	–	–	–	+	+	+	+	+	–	43	–	60	+	+	+		+	+	+	–	37	48	0.85	38	
Carnobacterium divergens	+	+	–	–		–	–	–	–	+			–																			
Citrobacter freundii	+	v	–	–	–	+	–	–	–	–	–	–	–	–	+	+	+	+	–	+	–										41	
Edwardsiella ictaluri	–	–	+	+	–	–	–	–	–	–	–	+	–	–	–	–	–	–	–	–	–	+		+		+	+				334,41	
Edwardsiella tarda	–	–	+	w	+	+	–	–	+	–	–	+	–	–	–	–	–	–	–	–	–	+			+	+	+	25	48	0.85	135	AHLDA 135
Edwardsiella tarda	–	–	+	+	–	+	–	–	+	–	–	+	–	–	–	–	–	–	–	–	–										41	
Edwardsiella tarda – atypical strains	–	–	+	–	+	+	–	–	+	–	–	+	+	–	–	–	–	–	–	–	–							25	48	0.85	845	
Enterovibrio norvegicus	+	+	–	–	–	–	–	–	+	–	–	+	–	–	–	–	–	–	–	–	+	–		+		+	+	28	48	2	741	LMG 19839[T]
Flavobacterium branchiophilum	–	–	–	–	–	–	–	–	–	–	+	+	–	–	–	–	+	+	–	–	–										41	
Flavobacterium columnare	–	–	–	–	–	+	–	–	–	–	+	–	–	–	–	–	–	–	–	–	+										41	
Flavobacterium columnare	+	–	–	–		+	–	–	–	–	+	–	–	–	–	–	–	–	–	–	+	+		–	–	–	–	25	48	0.5	689	
Flavobacterium gillisiae	–	+	–	–	–	–	–	–	–	–	–	+	+	–	–	–	+	–		–	–	–						20	48	1.5	533	
Flavobacterium hibernum	+	–	–	–	–	–	–	–	–	–	+	+	–	–	–	+	+	–	–	+	–	+						25	48	1.5	532	
Flavobacterium hydatis	–	–	–	–	–	–	–	–	–	–	+	+	+	–	–	–	+	–	–	+	–										41	
Flavobacterium johnsoniae	+	–	–	–	+	–	–	–	–	+	+	–	–	–	–	–	–	–	–	–	+	+						25	72	0.85	135	AHLDA 1714
Flavobacterium psychrophilum	–	–	–	–	–	–	–	–	–	–	+	–	–	–	–	–	–	–	–	–	+										41	
Flavobacterium tegetincola	–	–	–	–	–	–	–	–	–	–	–	+	+	–	–	–	–	–	–	–	–	–						20	48	1.5	533	
Flavobacterium xantham	–	–	–	–	–	+	–		–	–	+	+	+	–	–	–	+	–	–	–	+	+						20	48	1.5	533	
Hafnia alvei	–	–	+	+	+	–	–	–	–	–	–	+	+	–	–	–	–	–	–	+	–	+									41	
Hafnia alvei	+	–	+	+	–	–	–	–	–	–	–	+	+	–	–	+	–	–	–	+	–	+		+	+	+	+				652	
Hafnia alvei	+	–	+	+	+	–	–	–	–	+	–	+	+	–	–	+	–	–	+	+	–	+		+	+	+	+	25	48	0.85	135	AHLDA 1729
Hafnia alvei	+	–	+	+	+	–	–	–	–	–	–	+	+	–	–	+	–	–	–	+	–	+		+	+	+	+	25	48	0.85	203	ATCC 51873
Halomonas cupida	–	+	+	+	–	–	–	–	–	–	–	–	+	–	+	+	+	–	–	–	–										41	
Janthinobacterium lividum	–	+	–	–	+	–	–	–	–	+	+	+	+	–	–	–	–	–	–	–	+										41	
Klebsiella pneumoniae	+	–	v	–	+	–	v	–	–	+	–	+	+	+	+	+	+	+	+	+	–	+		+	+	+	+				41	
Listonella anguillarum	+	+	–	–	+	–	–	–	+	+	+	+	+	–	+	–	+	–	–	–	+	+		+	v	+	+	25	48	0.85	240	NCMB 6
Listonella anguillarum	+	+v	–	–	+	–	–	–	+	+	+	+	+	–	+	–	+	–	+	+v	+	+				+	+				428,41	
Listonella anguillarum	+	+	–	–	–	–	–	–	+	+	+	+	w	–	+	–	w	–	–	–	+	+			–	+	+	25	48	2		
Listonella anguillarum	+	+	–	–	+	–	–	–	+	+	+	+	+	–	+	–	+	–	–	+	+	+			–	+	+	25	48	2	135	AHLDA 1730

continued

Table 4.23. *Continued.*

Tests	onpg	ADH	LDC	ODC	cit	H2S	ure	TDA	ind	vp	G	glu	man	inos	sor	rha	suc	mel	amy	arab	ox	NO2	N2	mot	MCA	O	F	Temp	Time	Inoc %NaCl	Ref	Strain
Moritella marina	–	–	–	–	–	–	–	–	–	–	+	+	–	–	–	–	–	–	–	–	+	+	+	+	–	+	+				766	
Moritella viscosa	–	–	+	–	–	–	–	–	–	–	>	+	–	–	–	–	–	–	–	–	+	+	+	+	–	+	+				132	
Pantoea agglomerans	+	–	–	–	+	–	–	–	–	+	+s	+	+	–	–	+	+	–	–	+	–	+	–	+	–	+	+				291	
Pantoea agglomerans	+	–	–	–	+	–	–	–	–	+	+s	+	+	–	–	–	+	–	–	+	–	+									41	
Pantoea dispersa	+	–	–	–	+	–	–	–	–	+	+s	+	+	–	–	+	+	v	+	+	–	+	–	+	–	+	+				291	
Pasteurella multocida	–	–	–	–v	–	–	–	–	+	–	–	+	+	–	+	–	+	–	–	–	+	+	–	–	–	+	+				428	
Pasteurella testudinis	+	–	–	–	–	–	–	–	+	–	–	+	+	70	–	60	+	30	–	30	+	+	–	–	+	+	+				709	
Phocoenobacter uteri	–	–	–	–	–	–	–	–	–	+	–	+w	–	–	–	–	–	–	–	–	+	+	–	–	–	+	+				266	
Photobacterium angustum	–	+	–	–	–	–	–	–	–	+	+	+	–	–	–	–	+	–	–	–	+	–	–	–	–	+	+	26	72	1.5	745	NCIMB 1895
P. damselae ssp. *damselae*	–	+	–	–	–	–	+	–	–	+	–	+	+	–	–	–	–	–	–	–	+	+	+	+	+	+	+	26	72	1.5	745	ATCC 33539ᵀ
P. damselae ssp. *damselae*	–	+	+	–	+	–	–	–	–	+	–	+	+	–	–	–	–	–	–	–	–	+	+	+	+	+	+	25	48	0.85	240	ATCC 33539ᵀ
P. damselae ssp. *damselae*	–	+	–	–	–	–	–	–	–	–	–	+	+	–	–	–	–	–	+	–	–	+	+	+	+	+	+	26	72	1.5	745	NCIMB 2184
P. damselae ssp. *damselae*	–	+	55	–	15	–	85	–	–	+	20	+	–	–	–	–	–	–	15	–	75	+	+	+	+	+	+	26	72	1.5	745	ATCC 17911
P. damselae ssp. *piscicida*	–	+	–	–	20	–	–	–	–	+	–	+	–	–	–	–	–	–	–	–	+	–	–	–	46	+	+	26	72	1.5	745	Aberrant strain
P. damselae ssp. *piscicida*	–	–	–	–	–	–	+	–	–	+	–	+	–	–	–	–	–	–	–	–	+	–	–	–	+	+	+	26	72	1.5	745	ATCC 17911
P. damselae ssp. *piscicida*	–	+	–	–	–	–	–	–	–	–	–	+w	–	–	–	–	–	–	–	–	+	–	–	–	–	+	+				428	ATCC 17911, 29687
P. damselae ssp. *piscicida*	–	+	–	–	–	–	–	–	–	+	–	+	–	–	–	–	–	–	–	–	+	+	–	–	+	+	+	22	48	1.5	#	ATCC 29690, 17911
Photobacterium iliopiscarium	–	+	+	–	–	–	–	–	–	–	–	+	–	–	+	–	–	–	–	–	+	+	+	+	+	+	+				599	
Photobacterium leiognathi	–	+	–	–	–	–	–	–	–	+	–	+	–	–	–	–	–	–	–	–	+	+	–	–	–	+	+	26	72	1.5	745	LMG 4228
Photobacterium phosphoreum	–	+	w	–	–	–	–	–	+	+	+	+	–	–	–	–	+	–	+	–	–	+	–	–	+	+	+	26	72	2	509	IB39
Plesiomonas shigelloides	+	+	+	+	–	–	–	–	+	–	–	+	–	+	–	–	–	–	+	–	+	–	–	+	–	+	+				41	
Providencia (Proteus) rettgeri	–	–	–	–	–	–	+	+	+	–	–	+	+	+	–	+	–	–	–	–	–	+	–	+	+	–	+				41	
Providencia rustigianii	–	–	–	–	+w	–	+	+	+	–	–	+	+	+	–	–	+w	–	–	–	–	+	–	+	+	+	+				559	
Pseudoalteromonas ulvae	–	–	–	–	+	–	–	–	–	–	+	–	–	–	–	–	–	–	–	–	+	+	–	+	+	+	–	23	48	1.5	231	
Pseudomonas anguilliseptica	–	–	–	–	–	–	–	–	–	–	+	–	–	–	–	–	–	–	–	–	+	+	–	+	+	+	–				41	
Pseudomonas fluorescens	+	+	–	–	+	–	–	–	–	–	+	+	+	+	+	–	+	–	–	+	+	–	–	+	–	+	–				41	
Pseudomonas fluorescens/putida	–	75	–	–	75	–	–	–	–	10	27	25	–	–	–	–	–	25	1	20	+	26	–	+	–	+	–				*	
Pseudomonas plecoglossicida	–	–	–	–	–	–	–	–	–	–	–	+	–	–	–	–	+	–	–	–	+	–	–	+	–	+	+				428	
Pseudomonas putida	–	–	–	–	–	–	–	–	–	–	–v	–	–	–	–	–	–	–	–	–	+	–	–	–	–	+	+					
Rahnella aquatilis	+	–	–	–	+	–	–	–	–	+	–	+	–	+	+	+	+	+	+	+	–	+	+	+	+	+	+				151	

Salegentibacter salegens																						−		20	48	1.5	533	
Salmonella cholerasuis arizonae	+	−	+	−	+	+	−	−	−	−	+	+	−	+	−	+	−	−	−	−							41	
Serratia fonticola	+	−	+	+	+	−	−	−	−	+	−	+	+	−	+	+	−	+		+	−	+	+	+	+	+			558	
Serratia fonticola	+	−	+	+	+	−	−	−	−	−	−	+	+	+	+	+	20	+		+	−	+	+	+	+	+			290	
Serratia liquefaciens	+	−	+	+	+	−	v	−	−	v	+	+	+	v	+	−	+	v	+	−	−	+							41	
Serratia plymuthica	+	−	−	−	+	−	−	−	−	+	+	+	+	+	−	−	+	−	−	+	−	+							41	
Shewanella algae	−	−	−	+	−	+	−	−	−	−	+	−	−	−	−	−	−	−	+	+	+	+	+	−				433	
Shewanella frigidimarina	−	−v	−	−	+v	+	−	−	−	−	+	+	+	−	−	+	−	−	−	+	+	+		+	+			112	
Shewanella gelidimarina	−	−	−	−	−	+	−	−	−	−	+	−	−	−	−	−	−	−	−	+	+	+						112	
Shewanella putrefaciens	−	−	−	+	+	+	−	−	−	−	+	−	−	−	−	35	−	−	49	+	+	+	+	+	−			433	
Shewanella woodyi	−	−	−	−	−	−	−	−	−	−	+	−	−	−	−	−	−	−	−	+	+	+		+	−			112	
Tenacibaculum maritimum	−	−	−	−	−	−	−	−	−	−	+	−	−	−	−	−	−	−	−	+							41	
Vibrio aestuarianus	+	+	v	−	+				+		+	+	+				+		+											
Vibrio agarivorans	+	−	−	−	−	−	−	−	−	−	+	+	−	−	−	−	−	+	−	+	+		−	+	+	25	48	2	135	AHLDA 1732
Vibrio alginolyticus	−	−	+	53	v	−	−	−	+	83	v	+	+	−	−	+	−	67	−	+	+	+		+	+	552				
Vibrio alginolyticus	−	−	+	−	−	−	−	−	+	w	+	+	+	−	−	+	−	−	−	+	+	+	+	−	+	+			135	
Vibrio alginolyticus	−	−	+	+	−	−	−	−	+	+	+	+	+	−	−	+	v	v	−	+	+	+		+	+	650				
Vibrio brasiliensis	+	+	−	−	+	−	−	−	+	+	+	+	+	−	−	+	−	+	−	+	+	+	nt	+	+	25	48	2	740	LMG 20546[T]
Vibrio cholerae non-01	+	−	+	+	v	−	−	−	+	+	+	+	+	−	−	+	−	−	+		+	+	+	+				740		
Vibrio cholerae 01	+	−	+	+	v	−	−	−	+	−	+	+	+	−	−	+	−	−	+		+	+	+	+						
Vibrio cholerae	+	−	+	+	w	−	−	−	+	+	w	+	+	−	−	+	−	−	+	+	+	+	+	+	+			2	509,552	WF110r
Vibrio diabolicus	−	−	+	+	−	−	+	+	+	−	+	+	+	−	−	+	−	+	−	+	+	+	+	+					635	
Vibrio fischeri	−	−	+	−	−	−	−	−	−	+	−	+	−	−	−	−	−	+	−	+	+	+	−	+	+	26	72	1.5	745	ATCC 25918
Vibrio fluvialis	+	+	−	−	+	−	−	−	+	−	+	+	+	−	−	+	−	−	+	+	+	+	+	+	+	25	48	0.85	240	MEJ 311
Vibrio fluvialis	+	+	−	−	81	−	−	−	+	56	+	+	+	−	−	+	−	−	+	+	+	+	+	+	+	37	48	0.85	485,687	
Vibrio furnissii	+	+	−	−	63	−	−	−	88	75	+	+	+	−	−	+	−	−	+	+	+	+	+	+	+	37	48	0.85	687	
Vibrio furnissii	+	+	−	−	w	−	−	−	+	w	−	+	+	−	−	+	−	−	+	+	+	+	+	+	+	25	48	0.85	135	ATCC 11218
Vibrio furnissii	+	+	−	−	+	−	−	−	+	−	+	+	+	−	−	+	−	−	+	+	+	+	+	+	+	25	48	0.85	240	ATCC 35016[T]
Vibrio halioticoli	−	−	−	−	−	−	−	−	−	−	w	−	−	−	−	−	−	−	+	+		−	w	w	25	48	2	135	AHLDA 1734
Vibrio harveyi	−	−	+	+	+w	−	−	−	+	−	+	+	+	−	−	+	−	+	−	+	+		+	+	25	48	2	581	ATCC 14129
Vibrio (carchariae) harveyi	−	−	+	+	+	−	+	−	+	−	+	+	+	−	−	+	−	+	−	+	+	+	+	+	+	25	48	2	135,581, 847	ATCC 35084
Vibrio ichthyoenteri	−	−	−	−	−	−	−	−	−	−	+	−	−	−	−	+w	−	−	−	+	+	+						389	

continued

Table 4.23. Continued.

Tests	onpg	ADH	LDC	ODC	cit	H₂S	ure	TDA	ind	vp	G	glu	man	inos	sor	rha	suc	mel	amy	arab	ox	NO₂	N₂	mot	MCA	O	F	Temp	Time	Inoc %NaCl	Ref	Strain
Vibrio mediterranei	+	+	+	–	–	–	–	–	+	–	+	+	+	+	+	–	+	–	+	+	+	+	–	+	–	+	+	25	48	2	135	AHLDA 1733
Vibrio mimicus	+	–	+	+	+	–	–	–	+	–	+	+	+	–	–	–	–	–	–	–	+	+	–	+	+	+	+				161,210	
Vibrio mimicus	+	–	+	+	–	–	–	–	+	–	+	+	+	–	–	–	–	–	–	–	+	+	–	+	+	+	+				161	
Vibrio mytili	+	–	–	–	–	–	–	+	+	–	–	+	+	–	–	–	+	–	+	–	+	+	–	+	+	+	+				635	
Vibrio navarrensis	–	–	–	–	–	–	–	–	+	–	+	+	+	–	–	–	+	–	+	–	+	+	+	+	+	+	+				768	
Vibrio neptunius	–	+	–	–	+	–	–	+	+	+	+	+	+	–	–	–	+	–	–	–	+	+	–	+	nt	+	+	25	48	2	740	LMG 20536ᵀ
Vibrio nereis	–	–	–	–	–	–	–	+	+	–	–	+	–	–	–	–	+	–	–	–	+	+	–	+	–	+	+				635	
Vibrio ordalii	–	–	–	–	–	–	–	–	–	–	+	+	+	–	–	–	+	–	–	–	+	–	–	–		+	+				428	DF 3K
Vibrio pacinii	v	+	–	–	w	–	–	+	–	66	v	+	+	–	–	–	+	66	+	–	+	+	–	+		+	+	28	48	1.5	306	LMG 1999ᵀ
Vibrio parahaemolyticus	21	–	+	+	+	–	–	–	+	–	+	+	+	–	–	–	–	–	+	77	+	+	–	+	+	+	+		48		*552	
Vibrio parahaemolyticus	–	–	+	+	–	–	–	–	+	–	+	+	+	–	–	–	–	–	–	w	+	+	–	+	+w	+	+		48	1.5	135	ATCC 43996
Vibrio parahaemolyticus	–	–	+	+	+	–	–	–	+	–	+	+	+	–	–	–	–	–	–	+	+	+	–	+	+w	+	+				428	
Vibrio parahaemolyticus	–	–	+	+	–	–	+	–	+	–	+	+	+	–	–	–	–	–	–	+	+	+	–	+	+	+	+				10	ATCC 17802
Vibrio pectenicida	–	–	–	–	+	–	–	–	–	–	+	+	–	–	–	–	–	+	+	–	+	+	–	+	nt	+	+				470	
Vibrio penaeicida	+	–	–	–	+	–	nt	–	–	–	+	+	–	–	–	–	–	–	+	–	+	+	–	+	nt	+	+	25		2	388	
Vibrio proteolyticus	–	+	+	–	+	–	–	–	+	+	+	+	+	–	–	–	–	–	–	–	+	+	–	+	w	+	+	25	48	2	135	AHLDA 1735
Vibrio rotiferianus	+	–	+	+	–	–	83	+	+	–	+	+	+	–	–	–	+	+	+	+	+	nr	–	+		+	+	25	48	2	305	LMG 21460ᵀ
Vibrio salmonicida	–	–	–	–	–	–	–	–	–	–	–	+	–	–	–	–	–	–	–	–	+	–	–	–		+	+				41	
Vibrio scophthalmi	–	90	–	–	–	–	–	–	+	–	+	+	+	–	–	–	+	–	–	–	+	+	–	+	nt	+	+	25	48	1.5	149,254	CECT 4638
Vibrio splendidus biovar I	+	+	–	–	–v	–	–	–	+	–	+	+	+	–	–	–	+	–	–	–	+	+	–	+	nt	+	+	22	48	0.85	254	
Vibrio splendidus	+	+	–	–	–	–	–	–	+	–	+	+	+	–	–	–	v	+	+	–	+	+	–	+		+	+	25	48	2	281	
Vibrio splendidus	–	–	–	–	–	–	–	–	+	–	–	+	+	–	–	–	+	+	+	–	+	+	–	–	+	+	+	25	48	2	466	
Vibrio tapetis	+	–	–	–	–	–	–	–	+	–	+	+	+	–	–	–	+	–	+	–	+	+	–	+	+	+	+	25		2	108,146	
Vibrio tubiashii	+	+s	–	–	–	–	+	+	+	–	+	+	+	–	–	–	+	–	+	–	+	+	–	+	+	+	+				321,635	
Vibrio tubiashii	+	–	–	–	–	–	–	–	+	–	+	+	+	–	–	–	+	–	+	–	+	+	–	+	nt	+	+	25	48	2	135	
Vibrio vulnificus biotype 1	+	–	+	+	–	–	–	–	+	–	+	+	+	–	–	–	–	–	+	–	+	+	–	+	+	+	+				26	f
V. vulnificus biotype 2	+	–	+	+	+	–	–	–	–	–	+	+	+	–	–	–	–	–	+	–	+		–	+		+					26	
V. vulnificus biotype 2	+	–	+	+	+	–	–	–	–	–	+	+	+	–	–	–	–	–	+	+	+		–	+			+				26	

Organism	1	2	3	4	5	6	7	8	9	10	11	12	13	14	Temp	Time	%NaCl	Code	Strain	Ref
V. vulnificus biotype 2 sero E	+	−	+	+	−	−	+	+	−	−	−	+	+	+	25	48		g		26
V. vulnificus biotype 2 sero E	+	−	+	+	−	−	+	+	−	−	−	+	+	+				a	ATCC 33187	98
V. vulnificus biotype 2 sero E	+	−	+	−	−	−	+	+	−	−	−	+	+	+				b	ATCC 33149	98
V. vulnificus biotype 2 sero E	+	−	+	+	−	−	+	+	−	−	−	+	+	+				c		98
V. vulnificus biotype 2 sero E	+	−	+	+	−	−	+	+	−	−	−	−	+	+				d		98
V. vulnificus biotype 2 sero E	−	−	+	−	−	−	+	+	−	−	−	+	+	+				e		98
V. vulnificus	+	−	+	+	−	+	+	+	−	+	w	−	+	+					WF8A1110	509
V. xuii	−	+	−	+	+	+	+	−	+	+	+	+	+	+	25	48	2		LMG 21346ᵀ	740
Yersinia intermedia	+	−	+	+	+	+	+	−	−	+	+	+	−	−						41
Yersinia ruckeri	+	−	+	−	−	−	+	+	−	−	−	−	−	−						41
Yersinia ruckeri	+	−	+	−	+	−	+	+	−	−	−	−	+	−						674,717
Yersinia ruckeri	+	−	+	−	−	−	+	+	+	−	−	−	−	−						717
Yersinia ruckeri	+	−	+	−	−	−	+	+	+	−	−	−	−	−						717
Yersinia ruckeri	+	−	+	−	+	−	+	+	−	−	−	−	−	+	25	48	0.85		AHLDA 1313	135,203
Yersinia ruckeri	+	−	+	−	+	−	+	+	−	−	−	−	−	+						674
Yersinia ruckeri	+	−	+	−	−	+	+	+	−	+	−	−	−	+	25	48	0.85			184
Yersinia ruckeri	+	−	+	−	+	−	+	+	−	+	−	−	−	+	25	48	0.85			184
Yersinia ruckeri	+	−	+	−	−	−	+	+	−	−	−	−	−	+	25	48	0.85			184
Zobellia galactanivorans	+	−	−	−	−	+	+	+	+	+	+	−	+	+	30	7d	1.5			61
Zobellia uliginosa	+	−	−	−	−	+	−	+	−	+	+	−	+	+	30	7d	1.5			61

= References: 149, 289, 518, 674, 751, 855.

ongp = β-galactosidase; ADH = arginine dihyrolase; LDC = lysine decarboxylase; ODC = ornithine decarboxylase; cit = citrate; H₂S = production of hydrogen sulphide; ure = urease; TDA = tryptophane deaminase; ind = indole; vp = Voges-Proskauer; G = gelatin; glu = glucose; man = mannitol; inos = inositol; sor = sorbitol; rha = rhamnose; suc = sucrose; mel = melibiose; amy = amygdalin; arab = arabinose; ox = oxidase; NO₂ = reduction to nitrite gas; mot = motility; MCA = growth on MacConkey; O = oxidative fermentation; F = fermentation; Temp = temperature of incubation; Time = time of incubation; h = hours; d = days; Inoc = inoculum; %NaCl = final NaCl concentration in inoculum; w = weak; + = positive reaction; − − = negative reaction; s = slow reaction; numbers indicate percentage of positive strains. ᵃ = note that A. hydrophila strains are considered to be negative for sorbitol. ᵇ = The incubation time was not stated for this result. It is important to incubate for 48 h to ensure correct LDC result; *from API database; Ref = reference number.

Table 4.24. API 20E database numbers (organisms listed alphabetically).

Organism	Strain	API 20 E number	No of strains	Notes	Ref
Acinetobacter/Moraxella		0000004	1		41
Actinobacillus delphinicola		200500416	2	Dolphin, whale	263
Actinobacillus scotiae	NCTC 12922	111500416	1		265
Aeromonas allosaccharophila		724613657	1		41
Aeromonas allosaccharophila		724616657	1		41
Aeromonas allosaccharophila		724617657	1		41
Aeromonas bestiarum		704713757	5	Carp	452
Aeromonas caviae	ATCC 15468^T	324612657			21,240
Aeromonas encheleia		304613457			241
Aeromonas hydrophila	ATCC 7966^T	704712657	1		240
Aeromonas hydrophila	ATCC 7966^T	304512657	1	Important to incubate for 48 h for correct LDC result	674
Aeromonas hydrophila	ATCC 14715	324713757	1		322
Aeromonas janadaei		724714457	2		135,143
Aeromonas salmonicida		200610417			450
Aeromonas salmonicida	AHLDA 1334	204402417	1	'Atypical' strain, imported goldfish (Western Australia)	135
Aeromonas salmonicida achromogenes		004412417			322
Aeromonas salmonicida achromogenes	NCIMB 1110	004512417			468
Aeromonas salmonicida masoucida		304512517			322
Aeromonas salmonicida salmonicida		200610417			674
Aeromonas salmonicida salmonicida		300610517			322
Aeromonas salmonicida salmonicida		600610417			674
Aeromonas salmonicida salmonicida	NCIMB 1102	700610517			468
Aeromonas sobria	CIP 74.33^T	704712457			240
Aeromonas sobria		724712457			21
Aeromonas trota	ATCC 49657^T	724610557			142,560
Aeromonas veronii veronii	ATCC 35623	114612557	1		347
Aeromonas veronii veronii	ATCC 35604	114712457	1	Human, USA	347
Aeromonas veronii veronii	ATCC 35606	134612557	1		347
Aeromonas veronii veronii	ATCC 35605	134712557	1	Wound isolate, human, USA	347
Aeromonas veronii veronii	ATCC 35622	514712557	1		347
Aeromonas veronii veronii		534712557	12	Various strains	2,347,452
Burkholderia cepacia		0004004	3/10		38
Burkholderia cepacia		0206006	1/10		38
Burkholderia cepacia		4304004	4/10		38
Burkholderia cepacia		5304004	1/10		38
Burkholderia pseudomallei		0006727	1/91		38
Burkholderia pseudomallei		0206706	1/91		38
Burkholderia pseudomallei		2006704	1/91		38
Burkholderia pseudomallei		2006706	4/91		38
Burkholderia pseudomallei	NCTC8016	2206707	1/91		38
Burkholderia pseudomallei		2006707	5/91		38
Burkholderia pseudomallei		2006726	1/91		38
Burkholderia pseudomallei		2006727	22/91		38
Burkholderia pseudomallei		2202704	1/91		38
Burkholderia pseudomallei		2202706	3/91		38
Burkholderia pseudomallei		2206704	4/91		38
Burkholderia pseudomallei		2206706	21/91		38
Burkholderia pseudomallei		2206707	12/91		38
Burkholderia pseudomallei		2206727	15/91		38
Edwardsiella hoshinae		454412057			*
Edwardsiella hoshinae		474412057			*
Edwardsiella ictaluri		410400057			41,334
Edwardsiella tarda		454400057			41
Edwardsiella tarda	AHLDA 135	474400057			135

Table 4.24. *Continued.*

Organism	Strain	API 20 E number	No of strains	Notes	Ref
Edwardsiella tarda – atypical		4644100..	4	Atypical strains from red sea bream, Japan	845
Enterobacter agglomerans		120712057			41
Enterovibrio norvegicus		30440044.			741
Flavobacterium branchiophilum		0006060..			41
Flavobacterium columnare		0402004..			41
Flavobacterium columnare		140200410			689
Flavobacterium gillisiae		200412000			533
Flavobacterium hibernum		100603210			532
Flavobacterium hydatis		000612210			41
Flavobacterium psychrophilum		0002004..			41
Flavobacterium tegetincola		00041000.			533
Flavobacterium xantham		04061241.			533
Hafnia alvei		430410257			41
Hafnia alvei		510411257			652
Hafnia alvei		530511357			135
Hafnia alvei	ATCC 51873	5304112..	1		203
Halomonas cupida		6100530..			41
Janthinobacterium lividum		2207104..			41
Klebsiella pneumoniae		521577357			41
Listonella anguillarum	V10	304452456	1	Japan	49
Listonella anguillarum	V239	304572557	1	Japan	49
Listonella anguillarum	PT-87050	304652456	2	Japan	49
Listonella anguillarum		304752456	1		
Listonella anguillarum	V244 to V246	304752476	5	Japan	49
Listonella anguillarum	HT-77003	304752657	1	Japan	49
Listonella anguillarum	V240	324472757	1	Japan	49
Listonella anguillarum	V241	324562757	1	Japan	49
Listonella anguillarum	ET-78063	324632657	1	Japan	49
Listonella anguillarum	UB 4346, 434	324712677	2	Spain	49
Listonella anguillarum	NCMB 6	3247524..	1		240
Listonella anguillarum	UB (ET-1)	324752557	1	Japan	49
Listonella anguillarum	V318, AHLDA 1730	324752656	2	Japan, Australia	49,135
Listonella anguillarum	RH-8101, AVL	324752657	6	France, Japan	49
Listonella anguillarum		3247527..			41
Listonella anguillarum		324752756			428
Listonella anguillarum		324752756			49
Listonella anguillarum	LMG 3347	324752757	4	Japan, Norway, Spain	49
Listonella anguillarum	UB A078	324752777	1	Spain	49
Listonella anguillarum	V320	324772656	1	Spain	49
Listonella anguillarum	UB A054–56	324772657	4	Japan, Spain	49
Listonella pelagia		100412456			
Listonella pelagia		104612456			
Listonella pelagia		124612456			
Moritella marina		000600456			766
Moritella viscosa		400400456			132
Moritella viscosa		400600456			132
Pantoea agglomerans	CDC 1429–71	100577357			291
Pantoea agglomerans	ATCC 14589	120516357			291
Pantoea agglomerans	NCPPB 2285	120517257			291
Pantoea agglomerans	DNA HG 14589	120517357			291
Pantoea agglomerans		120713257			291
Pantoea dispersa		120713357			291
Pantoea dispersa		120717357			291

continued

Table 4.24. *Continued.*

Organism	Strain	API 20 E number	No of strains	Notes	Ref
Pasteurella multocida		014452456			428
Phocoenobacter uteri		000100416			266
Phocoenobacter uteri		000500416			266
Photobacterium angustum	NCIMB 1895	200702406			745
Photobacterium damselae damselae	LMG 7892	200500457			49
Photobacterium damselae damselae		201500457			41
Photobacterium damselae damselae	ATCC 33539[T]	201500457			745
Photobacterium damselae damselae	NCIMB 2184	220400157			745
Photobacterium damselae damselae	ATCC 33539[T]	600400457			240
Photobacterium damselae damselae	LMG 13639	600500457	6	Various sources	49
Photobacterium damselae piscicida	P90029	001100407	1	Atypical strain	745
Photobacterium damselae piscicida	ATCC 17911	200400406		ATCC 29687	428
Photobacterium damselae piscicida	ATCC 17911	200500406		ATCC 29690, and other strains. Refs: 149, 518, 855, 751	289,674,745
Photobacterium damselae piscicida	ATCC 17911	200500407	Majority	NCIMB 2058	745
Photobacterium damselae piscicida		220500407			745
Photobacterium iliopiscarium		600400457	1		599
Photobacterium leiognathi	LMG 4228	200500017			745
Photobacterium phosporeum	IB39	6046021..			509
Plesiomonas shigelloides		714420457			41
Providencia rettgeri		007431057			41
Providencia rustigianii		026402057			559
Pseudomonas aeruginosa		0206006	1/18		38
Pseudomonas aeruginosa		2002004	1/18		38
Pseudomonas aeruginosa		2006004	5/18		38
Pseudomonas aeruginosa		2206004	9/18		38
Pseudomonas aeruginosa		2206006	1/18		38
Pseudomonas anguilliseptica		000200440			41
Pseudomonas anguilliseptica		020000440			96,541,828
Pseudomonas anguilliseptica		020200440			96,541,828
Pseudomonas fluorescens		0204004..	2/8		38
Pseudomonas fluorescens		2000004..	5/8		38
Pseudomonas fluorescens		220000443	1/8		38
Pseudomonas fluorescens		220000453			
Pseudomonas stutzeri		0000004..	1/9		38
Pseudomonas stutzeri		0004004..	5/9		38
Pseudomonas stutzeri		0004104..	3/9		38
Rahnella aquatilis		100557317			*
Rahnella aquatilis		120547257			151
Rahnella aquatilis		120557317			*
Salmonella cholerasuis arizonae		560621057			41
Serratia fonticola		530475257			290
Serratia fonticola		530477257			290
Serratia fonticola		530555257			558
Serratia plymuthica		120732257			41
Shewanella algae		050200453			433
Shewanella frigidimarina		06061245.			112
Shewanella frigidimarina		26061245.			112
Shewanella gelidimarina		04020045.			112
Shewanella putrefaciens		070200453			433
Shewanella putrefaciens		070200653			433
Shewanella putrefaciens		070202653			433
Shewanella woodyi		000200452			112
Tenacibaculum maritimum		000200410			41
Vibrio agarivorans	AHLDA 1732	100410556	2	From abalone. Pathogenicity unknown	135

Table 4.24. *Continued.*

Organism	Strain	API 20 E number	No of strains	Notes	Ref
Vibrio alginolyticus		404712456			135
Vibrio alginolyticus		404712457			135
Vibrio alginolyticus		414712457			650
Vibrio alginolyticus		414712557			552
Vibrio alginolyticus		414716557			650
Vibrio alginolyticus		434712457			552
Vibrio brasiliensis	LMG 20546[T]	32471255.			740
Vibrio cholerae	LMG 16741	204612457	1	Shrimp, Thailand	49
Vibrio cholerae	LMG 16742	324602557	1	*Pe. orientalis*, PRC, Thailand	49
Vibrio cholerae 01		514612457			
Vibrio cholerae non 01		514712457			
Vibrio cholerae	PS-7701	524712457	1	*Pl. altivelis*, Japan	49
Vibrio cholerae	PS-7705	524712476	1	*Pl. altivelis*, Japan	49
Vibrio cholerae	91/1198	534612457	1	*Carassius auratus*, Australia	49
Vibrio cholerae 01		534612457			
Vibrio cholerae non 01	LMG 16743	534712457	1		49
Vibrio cholerae	WF 110r	534712557			509,552
Vibrio diabolicus		416612557			635
Vibrio fischeri	ATCC 25918	400500556			745
Vibrio fluvialis		304612657			485,687
Vibrio fluvialis		304712657			485,687
Vibrio fluvialis		324412757			485,687
Vibrio fluvialis	and MEJ 311	324612657	2		240,485,687
Vibrio fluvialis		324712657			485,687
Vibrio furnissii	ATCC 35016[T]	324612657			240
Vibrio furnissii		304712657			687
Vibrio furnissii		324712657			687
Vibrio furnissii	ATCC 11218	324512657		Citrate, indole, VP, weak reaction	135
Vibrio (carchariae) harveyi		415412557			135,581,847
Vibrio (carchariae) harveyi		415612557			135,581,847
Vibrio harveyi	ATCC 14129	434612557			581
Vibrio (carchariae) harveyi		435412557			135,581,847
Vibrio (carchariae) harveyi	ATCC 35084	435612557			135,581,847
Vibrio halioticoli	AHLDA 1734	000410456	1	Abalone	135
Vibrio ichthyoenteri		00040245.			389
Vibrio mediterranei	AHLDA 1733	704672556	1	Abalone	135
Vibrio mimicus		514610457	1		161
Vibrio mimicus		534610457	1		161,210
Vibrio mytili		106412556			635
Vibrio neptunius	LMG 20536[T]	22670245.			740
Vibrio nereis		00640245.			635
Vibrio ordalii	LMG 13544	000402446	1		49
Vibrio ordalii	LMG 10951	000402476	3	Salmon	49
Vibrio ordalii	V-306	000402556	1	Amago trout, Japan	49
Vibrio ordalii	RF-2, PT-81025	000402576	8	*O. mykiss*, *Pl. altivelis*, Japan	49
Vibrio ordalii	V-11	000412446	1	*O. mykiss*, Japan	49
Vibrio ordalii	F378, F380, V-250	000602546	4	Salmon, Australia; *O. mykiss*, Japan	49
Vibrio ordalii	F379, F381	000612446	2	Salmon, Australia	49
Vibrio ordalii	DF 3K	000612446	1	Denmark	428
Vibrio ordalii	V-302	304752476	4	*O. mykiss*, Japan	49
Vibrio parahaemolyticus		410610657			552
Vibrio parahaemolyticus		410610757			552
Vibrio parahaemolyticus	ATCC 43996	414610657	1		135
Vibrio parahaemolyticus		430610657			552

continued

Table 4.24. *Continued.*

Organism	Strain	API 20 E number	No of strains	Notes	Ref
Vibrio parahaemolyticus		434610657			428
Vibrio parahaemolyticus	ATCC 17802	434610657			10
Vibrio pectenicida		020600456			470
Vibrio penaeicida		12060455.			388
Vibrio penaeicida		12460455.			388
Vibrio proteolyticus	AHLDA 1735	624750456	1	*Artemia*	135
Vibrio rotiferianus	LMG 21460^T	5176167..			305
Vibrio salmonicida	RVAU 890206–1/12	000410040	1	*Salmo salar*, Faroe Islands	49
Vibrio salmonicida	RVAU 881129–1/2	000410440	5	*Salmo salar*, Faroe Islands	49
Vibrio salmonicida	RVAU 881129–1/1	000410446	1	*Salmo salar*, Faroe Islands	49
Vibrio salmonicida	RVAU 890206–1/10	000412640	1	*Salmo salar*, Faroe Islands	49
Vibrio salmonicida	RVAU 890206–1/7	520600556	1	*Salmo salar*, Faroe Islands	49
Vibrio scophthalmi		20040245.			149
Vibrio splendidus		004416516			466
Vibrio splendidus	UB S292	204610456	1	Seawater, Sweden	49
Vibrio splendidus	90–0652	224614446	1	Striped trumpeter, Australia	49
Vibrio splendidus	RVAU 88–12–686	304610456	3	*Sa. Salar*, Norway	49
Vibrio splendidus	UB S236	304610556	1	Seawater, Sweden	49
Vibrio splendidus	RVAU 88–12–711	304612456	1	*Sp. auratus*, Norway	49
Vibrio splendidus	89–1638	304612556	1	*O. mykiss*, Australia	49
Vibrio splendidus		304614556			281
Vibrio splendidus		304616556			281
Vibrio splendidus	RVAU 88–12–717	324410556	1	*Sc. maximus*, Norway	49
Vibrio splendidus	LMG 16752	324414557	1	Oyster, Spain	49
Vibrio splendidus	RVAU 88–12–712	324610456	1	*Solea solea*, Norway	49
Vibrio splendidus	UB S308, LMG 16747	324610556	2	Seawater, Sweden; oyster, Spain	49
Vibrio splendidus	LMG 16745	324610557	1	*Sp. auratus*, Greece	49
Vibrio splendidus	LMG 16744, 16750	324612557	2	*Sp. auratus*, Greece; oyster, Spain	49
Vibrio splendidus	LMG 16748	324614556	1	Oyster, Spain	49
Vibrio splendidus	LMG 16749	324616556	1	Oyster, Spain	49
Vibrio splendidus	LMG 16746	324712556	1	Oyster, Spain	49
Vibrio tapetis		10460055.			108,146
Vibrio tubiashii		106612556			321,635
Vibrio tubiashii		304412556			321,635
Vibrio vulnificus biovar 1 non-serovar E	M06–24	434610557	1	Human blood, USA	98
V. vulnificus biovar 1 non-serovar E	532	504600557	2	Diseased eel, Belgium, avirulent for eels	33,98
V. vulnificus biovar 1 non-serovar E	628	514412757	1	Paguara fish, Venezuela	98
V. vulnificus biovar 1 non-serovar E	530	514602557	7	Diseased eel, Belgium, virulent	33,98
V. vulnificus biovar 1 non-serovar E	ATCC 27562	514610557	1	Human strain, USA	26,98
V. vulnificus biovar 1 non-serovar E		514610557	23	Human (USA), eels (Spain, Sweden), shrimp (Thailand)	33,98
V. vulnificus biovar 1 non-serovar E	B9629, C7184	514610557		Human wound infection, USA	33,98
V. vulnificus biovar 1 non-serovar E	UMH1, 374	514610557		Human wound infection, USA	33,98
V. vulnificus biovar 1 non-serovar E	ATCC 27562	534600557		Human strain, USA	33
V. vulnificus biovar 1 non-serovar E	E109	534600557		Healthy eel, Spain, avirulent for eels	98
V. vulnificus biovar 1 non-serovar E	C7184	534610557	1	Human blood, USA, avirulent for eels	98
V. vulnificus biovar 1 serovar E	534	514410557	2	Diseased eel, Sweden, avirulent for eels	33,98
V. vulnificus biovar 1 serovar E	ATCC 33186	514610557		Human blood, USA, avirulent for eels	98
V. vulnificus biovar 1 serovar E	521	534600557		Unknown, Australia, avirulent for eels	98
V. vulnificus biovar 1	LMG 12092	104602557	1	Eel, Belgium	33

Table 4.24. *Continued.*

Organism	Strain	API 20 E number	No of strains	Notes	Ref
V. vulnificus biovar 1	169	104610557	1	Eel, Belgium	33
V. vulnificus biovar 1	E109	114000557	1	Eel, Spain	33
V. vulnificus biovar 1	160	114610557	1	Eel, Belgium	33
V. vulnificus biovar 1	M626	410610457	1	Eel, Spain	33
V. vulnificus biovar 1	Vv1	504600557		Human wound infection, USA	33
V. vulnificus biovar 1	UMH1	504610557	1	Human wound infection, USA	33
V. vulnificus biovar 1	M631	510610557	1	Eel, Spain	33
V. vulnificus biovar 1	167	514650557	1	Eel, Belgium	33
V. vulnificus biovar 1	VIB 521	516600557	1	Unknown	33
V. vulnificus biovar 1		524610557			97
V. vulnificus biovar 2 serovar E	171	100600557	1	Diseased eel, Sweden	98
V. vulnificus biovar 2	171	104600557	1	Eel, Belgium	33
V. vulnificus biovar 2 serovar E	NCIMB 2136	400600557	1	Diseased eel, Japan	98
V. vulnificus biovar 2		420600557			97
V. vulnificus biovar 2 serovar E	520	500400557	1	Shrimp, Taiwan, virulent for eels	98
V. vulnificus biovar 2 serovar E	NCIMB 2137	500600457	1	Diseased eel, Japan	33,98
V. vulnificus biovar 2 serovar E	ATCC 33149	500600557	1	Diseased eel, Japan	33,98,356
V. vulnificus biovar 2	NCIMB 2138	500600557	1	Diseased eel, Japan	98
V. vulnificus biovar 2 serovar E	121	510200457	1	Diseased eel, Sweden	98
V. vulnificus biovar 2 serovar E	526	510200557	1	Diseased eel, Sweden	98
V. vulnificus biovar 2 serovar E	524	510600557		Diseased eel, Norway	98
V. vulnificus biovar 2	NCIMB 2136	510600557	1	Diseased eel, Japan	33
V. vulnificus biovar 2		510600557	14	Eels (Japan, Norway – serovar E, Spain, Sweden), shrimp, human	33,98
V. vulnificus biovar 2 serovar E	523	514600557	5	Eels (Belgium – serovar E, Spain, Sweden), sea bream (Spain)	33,98
V. vulnificus biovar 2 serovar E	NCIMB 2138	520600557	1	Diseased eel, Japan	98
V. vulnificus biovar 2 serovar E	E86	520600557	2	Diseased eel, (Japan, Spain)	26,98
V. vulnificus biovar 2 serovar E	525	530600457	1	Diseased eel, Sweden	98
V. vulnificus biovar 2 serovar E	ATCC 33187	530600557	1	Human strain, USA	98
V. vulnificus biovar 2 serovar E	E105	530600557		Diseased eel, Spain	26,98
V. vulnificus biovar 2 serovar E		530600557	2	Human (USA), eel (Spain)	26,98
V. vulnificus biovar 2 serovar E		530610557	1	Taiwanese strains, avirulent for eels	26
V. vulnificus	818	414412557	1	Unknown, France	98
V. vulnificus	822	504410557	1	Shrimp, Senegal	98
V. vulnificus	WF8A1110	534610357			509
Vibrio xuii	LMG 21346[T]	20651275.			740
Yersinia intermedia		115457157			41
Yersinia intermedia		115457357			41
Yersinia intermedia		115477357			41
Yersinia intermedia		115477357			41
Yersinia ruckeri		110410057			674,717
Yersinia ruckeri		110450057			717
Yersinia ruckeri		110550057			717
Yersinia ruckeri	ATCC 29473	510410057	2		184,674
Yersinia ruckeri	AHLDA 1313	510510057	1		135
Yersinia ruckeri		5105500	20	Environmental strains	184
Yersinia ruckeri		5107500	1	Environmental strains	184
Yersinia ruckeri		530610057			41
Zobellia galactanovorans		10061361.			61
Zobellia uliginosa		100.0..5.			61

*from API database.
To be used in conjunction with the API database supplied by bioMerieux.

Table 4.25. API 20E database numbers (numbers in ascending order).

API 20 E number	Organism	Strain	No of strains	Notes	Ref
0000004..	Acinetobacter/Moraxella		1		41
0000004..	Pseudomonas stutzeri		1/9		38
000100416	Phocoenobacter uteri				266
0002004..	Flavobacterium psychrophilum				41
000200410	Tenacibaculum maritimum				41
000200440	Pseudomonas anguilliseptica				41
000200452	Shewanella woodyi				112
0004004..	Pseudomonas stutzeri		5/9		38
0004004..	Burkholderia cepacia		3/10		38
000410456	Vibrio halioticoli	AHLDA 1734	1	Abalone	135
000402446	Vibrio ordalii	LMG 13544			49
00040245.	Vibrio ichthyoenteri				389
000402476	Vibrio ordalii	LMG 10951		Salmon	49
000402576	Vibrio ordalii	RF-2, PT-81025		O. mykiss, Pl. altivelis, Japan	49
00041000.	Flavobacterium tegetincola				533
000410040	Vibrio salmonicida	RVAU 890206–1/12		Salmo salar, Faroe Islands	49
0004104..	Pseudomonas stutzeri		3/9		38
000410440	Vibrio salmonicida	RVAU 881129–1/2		Salmo salar, Faroe Islands	49
000410446	Vibrio salmonicida	RVAU 881129–1/1		Salmo salar, Faroe Islands	49
000412640	Vibrio salmonicida	RVAU 890206–1/10		Salmo salar, Faroe Islands	49
000500416	Phocoenobacter uteri				266
000600456	Moritella marina				766
000602546	Vibrio ordalii	F378, F380		Salmon, Australia	49
0006060..	Flavobacterium branchiophilum				41
000612210	Flavobacterium hydatis				41
000612446	Vibrio ordalii	F379, F381		Salmon, Australia	49
000612446	Vibrio ordalii	DF 3K			428
0006727..	Burkholderia pseudomallei		1/91		38
001100407	Photobacterium damselae piscicida	P90029	1	Atypical strain	745
004412417	Aeromonas salmonicida achromogenes				322
004416516	Vibrio splendidus				466
004512417	Aeromonas salmonicida achromogenes	NCIMB 1110			468
00640245.	Vibrio nereis				635
007431057	Providencia rettgeri				41
014452456	Pasteurella multocida				428
020000440	Pseudomonas anguilliseptica				96,541,828
020200440	Pseudomonas anguilliseptica				96,541,828
0204004..	Pseudomonas fluorescens		2/8		38
020600456	Vibrio pectenicida				470
0206006..	Burkholderia cepacia		1/10		38
0206006..	Pseudomonas aeruginosa		1/18		38
0206706..	Burkholderia pseudomallei		1/91		38
026402057	Providencia rustigianii				559
0402004..	Flavobacterium columnare				41
04020045.	Shewanella gelidimarina				112
04061241.	Flavobacterium xantham				533
050200453	Shewanella algae				433
06061245.	Shewanella frigidimarina				112
070200453	Shewanella putrefaciens				433
070200653	Shewanella putrefaciens				433

Table 4.25. *Continued.*

API 20 E number	Organism	Strain	No of strains	Notes	Ref
070202653	*Shewanella putrefaciens*				433
100.0..5.	*Zobellia uliginosa*				61
100410556	*Vibrio agarivorans*	AHLDA 1732	2	From abalone, pathogenicity unknown	135
100412456	*Listonella pelagia*				
100577357	*Pantoea agglomerans*	CDC 1429–71			291
100600557	*Vibrio vulnificus* biovar 2 serovar E	171		Diseased eel, Sweden	98
100603210	*Flavobacterium hibernum*				532
10061361.	*Zobellia galactanovorans*				61
10460055.	*Vibrio tapetis*				108,146
104600557	*Vibrio vulnificus* biovar 2	171		Eel, Belgium	33
104602557	*Vibrio vulnificus* biovar 1	LMG 12092		Eel, Belgium	33
104610557	*Vibrio vulnificus* biovar 1	169		Eel, Belgium	33
104612456	*Listonella pelagia*				
106412556	*Vibrio mytili*				635
106612556	*Vibrio tubiashii*				321,635
110410057	*Yersinia ruckeri*				674,717
110450057	*Yersinia ruckeri*				717
110550057	*Yersinia ruckeri*				717
111500416	*Actinobacillus scotiae*	NCTC 12922	1		265
114000557	*Vibrio vulnificus* biovar 1	E109		Eel, Spain	33
114610557	*Vibrio vulnificus* biovar 1	160		Eel, Belgium	33
114612557	*Aeromonas veronii veronii*	ATCC 35623	1		347
114712457	*Aeromonas veronii veronii*	ATCC 35604	1	Human, USA	347
115457157	*Yersinia intermedia*				41
115457357	*Yersinia intermedia*				41
115477357	*Yersinia intermedia*				41
120516357	*Pantoea agglomerans*	ATCC 14589			291
120517257	*Pantoea agglomerans*	NCPPB 2285			291
120517357	*Pantoea agglomerans*	DNA HG 14589			291
120547257	*Rahnella aquatilis*				151
12060455.	*Vibrio penaeicida*				388
120712057	*Enterobacter agglomerans*				41
120713257	*Pantoea agglomerans*				291
120713357	*Pantoea dispersa*				291
120717357	*Pantoea dispersa*				291
120732257	*Serratia plymuthica*				41
12460455.	*Vibrio penaeicida*				388
124612456	*Listonella pelagia*				
134612557	*Aeromonas veronii veronii*	ATCC 35606	1		347
134712557	*Aeromonas veronii veronii*	ATCC 35605	1	Wound isolate, human, USA	347
140200410	*Flavobacterium columnare*				689
2000004..	*Pseudomonas fluorescens*		5/8		38
2002004..	*Pseudomonas aeruginosa*		1/18		38
200400406	*Photobacterium damselae piscicida*	ATCC 17911		ATCC 29687	428
20040245.	*Vibrio scophthalmi*				149
200412000	*Flavobacterium gillisiae*				533
200500017	*Photobacterium leiognathi*	LMG 4228			745

continued

Table 4.25. *Continued.*

API 20 E number	Organism	Strain	No of strains	Notes	Ref
200500406	*Photobacterium damselae piscicida*				149,289, 751,855
200500406	*Photobacterium damselae piscicida*	ATCC 17911		ATCC 29690	674,745
200500407	*Photobacterium damselae piscicida*	ATCC 17911	Majority	NCIMB 2058	745
200500416	*Actinobacillus delphinicola*		2	Dolphin, whale	263
200500457	*Photobacterium damselae damselae*	LMG 7892			49
2006004..	*Pseudomonas aeruginosa*		5/18		38
200610417	*Aeromonas salmonicida*				450
200610417	*Aeromonas salmonicida salmonicida*				674
2006704..	*Burkholderia pseudomallei*		1/91		38
2006706..	*Burkholderia pseudomallei*		4/91		38
2006707..	*Burkholderia pseudomallei*		5/91		38
2006726..	*Burkholderia pseudomallei*		1/91		38
2006727..	*Burkholderia pseudomallei*		22/91		38
200702406	*Photobacterium angustum*	NCIMB 1895			745
201500457	*Photobacterium damselae damselae*				41
201500457	*Photobacterium damselae damselae*	ATCC 33539			745
204402417	*Aeromonas salmonicida*	AHLDA 1334	1	'Atypical' strain, goldfish	135
204610456	*Vibrio splendidus*	UB S292		Seawater, Sweden	49
204612457	*Vibrio cholerae*	LMG 16741		Shrimp, Thailand	49
20651275.	*Vibrio xuii*	LMG 21346[T]			740
2200004..	*Pseudomonas fluorescens*		1/8		38
2202704..	*Burkholderia pseudomallei*		1/91		38
2202706..	*Burkholderia pseudomallei*		3/91		38
220400157	*Photobacterium damselae damselae*	NCIMB 2184			745
220500406	*Photobacterium damselae piscicida*				745
2206004..	*Pseudomonas aeruginosa*		9/18		38
2206006..	*Pseudomonas aeruginosa*		1/18		38
2206704..	*Burkholderia pseudomallei*		4/91		38
2206706..	*Burkholderia pseudomallei*		21/91		38
2206707..	*Burkholderia pseudomallei*	NCTC8016	1/91		38
2206707..	*Burkholderia pseudomallei*		12/91		38
2206727..	*Burkholderia pseudomallei*		15/91		38
2207104..	*Janthinobacterium lividum*				41
224614446	*Vibrio splendidus*	90–0652		Striped trumpeter, Australia	49
22670245.	*Vibrio neptunius*	LMG 20536[T]		Rotifers, bivalve larvae	740
26061245.	*Shewanella frigidimarina*				112
300610517	*Aeromonas salmonicida salmonicida*				322
324612657	*Aeromonas caviae*	ATCC 15468[T]			240
30440044.	*Enterovibrio norvegicus*	LMG 19839[T]			741
304412556	*Vibrio tubiashii*				321,635
304452456	*Listonella anguillarum*				49
304512517	*Aeromonas salmonicida masoucida*				322
304512657	*Aeromonas hydrophila*	ATCC 7966	1		674
304512657	*Vibrio furnissii*	ATCC 11218	1	Citrate, indole, VP weak reaction	135
304572557	*Listonella anguillarum*				49
304610456	*Vibrio splendidus*	RVAU 88–12–686		*Sa. Salar*, Norway	49
304610556	*Vibrio splendidus*	UB S236		Seawater, Sweden	49
304612456	*Vibrio splendidus*	RVAU 88–12–711		*Sp. auratus*, Norway	49
304612556	*Vibrio splendidus*	89–1638		*O. mykiss*, Australia	49

Table 4.25. *Continued.*

API 20 E number	Organism	Strain	No of strains	Notes	Ref
304612657	*Vibrio fluvialis*				485,687
304613457	*Aeromonas encheleia*				241
304614556	*Vibrio splendidus*				281
304616556	*Vibrio splendidus*				281
304652456	*Listonella anguillarum*				49
304712657	*Vibrio fluvialis*				485,687
304712657	*Vibrio furnissii*				687
304752476	*Listonella anguillarum*				
304752476	*Listonella anguillarum*				49
304752657	*Listonella anguillarum*				49
324410556	*Vibrio splendidus*	RVAU 88–12–717		*Sc. maximus*, Norway	49
324412757	*Vibrio fluvialis*				485,687
324414557	*Vibrio splendidus*	LMG 16752		Oyster, Spain	49
324472757	*Listonella anguillarum*				49
324512657	*Vibrio furnissii*	ATCC 11218	1	Citrate, indole, VP weak reaction	135
324562757	*Listonella anguillarum*				49
324632657	*Listonella anguillarum*				49
324602557	*Vibrio cholerae*	LMG 16742		*Pe. orientalis*, PRC	49
324610456	*Vibrio splendidus*	RVAU 88–12–712		*Solea solea*, Norway	49
324610556	*Vibrio splendidus*	UB S308		Seawater, Sweden	49
324610557	*Vibrio splendidus*	LMG 16745		*Sp. auratus*, Greece	49
324612557	*Vibrio splendidus*	LMG 16744		*Sp. auratus*, Greece	49
324612657	*Aeromonas caviae*				21
324612657	*Vibrio fluvialis*	and MEJ 311	3		485,687,240
324612657	*Vibrio furnissii*	ATCC 35016[T]			240
324614556	*Vibrio splendidus*	LMG 16748		Oyster, Spain	49
324616556	*Vibrio splendidus*	LMG 16749		Oyster, Spain	49
32471255.	*Vibrio brasiliensis*	LMG 20546[T]		Bivalve larvae	740
324712556	*Vibrio splendidus*	LMG 16746		Oyster, Spain	49
324712657	*Vibrio fluvialis*				485,687
324712657	*Vibrio furnissii*	ATCC 11218			135,687
324712677	*Listonella anguillarum*				49
324713757	*Aeromonas hydrophila*	ATCC 14715	1		322
324752557	*Listonella anguillarum*				49
324752656	*Listonella anguillarum*	and AHLDA 1730	2	Japan, Australia (freshwater fish)	49,135
324752756	*Listonella anguillarum*				41,49,428
324752757	*Listonella anguillarum*				49
324752777	*Listonella anguillarum*				49
324772656	*Listonella anguillarum*				49
324772657	*Listonella anguillarum*				49
400400456	*Moritella viscosa*				132
400500556	*Vibrio fischeri*	ATCC 25918			745
400600456	*Moritella viscosa*				132
400600557	*V. vulnificus* biovar 2 serovar E	NCIMB 2136		Diseased eel, Japan	98
404712456	*Vibrio alginolyticus*				135
404712457	*Vibrio alginolyticus*				135
410400057	*Edwardsiella ictaluri*				41,334
410610457	*V. vulnificus* biovar 1	M626		Eel, Spain	33

continued

Table 4.25. *Continued.*

API 20 E number	Organism	Strain	No of strains	Notes	Ref
410610657	*Vibrio parahaemolyticus*				552
410610757	*Vibrio parahaemolyticus*				552
414412557	*V. vulnificus*	818		Unknown, France	98
414610657	*Vibrio parahaemolyticus*	ATCC 43996			135
414712457	*Vibrio alginolyticus*				650
414712557	*Vibrio alginolyticus*				552
414716557	*Vibrio alginolyticus*				650
415412557	*Vibrio (carchariae) harveyi*				135,581,847
415612557	*Vibrio (carchariae) harveyi*				135,581,847
416612557	*Vibrio diabolicus*				635
420600557	*V. vulnificus* biovar 2				97
4304004..	*Burkholderia cepacia*		4/10		38
430410257	*Hafnia alvei*				41
430610657	*Vibrio parahaemolyticus*				552
434610557	*V. vulnificus* biovar 1 non-serovar E	M06–24		Human blood, USA	98
434610657	*Vibrio parahaemolyticus*				428
434610657	*Vibrio parahaemolyticus*	ATCC 17802			10
434612557	*Vibrio harveyi*	ATCC 14129			581
434712457	*Vibrio alginolyticus*				552
435412557	*Vibrio (carchariae) harveyi*				135,581,847
435612557	*Vibrio (carchariae) harveyi*	ATCC 35084			135,581,847
454400057	*Edwardsiella tarda*				41
4644100..	Atypical *Edwardsiella tarda*		4	Atypical strains from red sea bream, Japan	845
474400057	*Edwardsiella tarda*	AHLDA 135		ATCC 15947[T]	135,845
500400557	*V. vulnificus* biovar 2 serovar E	520		Shrimp, Taiwan, virulent for eels	98
500600457	*V. vulnificus* biovar 2 serovar E	NCIMB 2137		Diseased eel, Japan	33,98
500600557	*V. vulnificus* biovar 2 serovar E	ATCC 33149		Diseased eel, Japan	33,98,356
500600557	*V. vulnificus* biovar 2	NCIMB 2138		Diseased eel, Japan	98
504410557	*V. vulnificus*	822		Shrimp, Senegal	98
504600557	*V. vulnificus* biovar 1 non-serovar E	532		Diseased eel, Belgium, avirulent for eels	98
504600557	*V. vulnificus* biovar 1	Vv1		Human wound infection, USA	33
504610557	*V. vulnificus* biovar 1	UMH1		Human wound infection, USA	33
510200457	*V. vulnificus* biovar 2 serovar E	121		Diseased eel, Sweden	98
510200557	*V. vulnificus* biovar 2 serovar E	526		Diseased eel, Sweden	98
510410057	*Yersinia ruckeri*	ATCC 29473			184,674
510411257	*Hafnia alvei*				652
510510057	*Yersinia ruckeri*	AHLDA 1313	2	Channel catfish	135,203
510550057	*Yersinia ruckeri*		20	Environmental isolates	184
510600557	*V. vulnificus* biovar 2	NCIMB 2136		Diseased eel, Japan	33
510600557	*V. vulnificus* biovar 2 serovar E	524		Diseased eel, Norway	98
510610557	*V. vulnificus* biovar 1	M631		Eel, Spain	33
510710057	*Yersinia ruckeri*	YR55, YR80	2	Environmental isolates	184
510750057	*Yersinia ruckeri*		1	Environmental isolates	184

Table 4.25. *Continued.*

API 20 E number	Organism	Strain	No of strains	Notes	Ref
514410557	*V. vulnificus* biovar 1 serovar E	534		Diseased eel, Sweden, avirulent for eels	98
514412757	*V. vulnificus* biovar 1 non-serovar E	628		Paguara fish, Venezuela	98
514600557	*V. vulnificus* biovar 2 serovar E	523		Diseased eel	98
514602557	*V. vulnificus* biovar 1 non-serovar E	530		Diseased eel, Belgium, virulent	98
514610457	*Vibrio mimicus*				161
514610557	*V. vulnificus* biovar 1 non-serovar E	ATCC 27562		Human strain, USA	26,98
514610557	*V. vulnificus* biovar 1 non-serovar E	B9629, C7184		Human wound infection, USA	33,98
514610557	*V. vulnificus* biovar 1 non-serovar E	UMH1, 374		Human wound infection, USA	33,98
514610557	*V. vulnificus* biovar 1 serovar E	ATCC 33186		Human blood, USA, avirulent for eels	98
514612457	*Vibrio cholerae* 01				
514650557	*V. vulnificus* biovar 1	167		Eel, Belgium	33
514712457	*Vibrio cholerae* non 01				
514712557	*Aeromonas veronii veronii*	ATCC 35622	1		347
516600557	*V. vulnificus* biovar 1	VIB 521		Unknown	33
5176167..	*Vibrio rotiferianus*	LMG 21460^T		Rotifer	305
520600556	*Vibrio salmonicida*	RVAU 890206–1/7		*Salmo salar*, Faroe Islands	49
520600557	*V. vulnificus* biovar 2 serovar E	NCIMB 2138		Diseased eel, Japan	98
520600557	*V. vulnificus* biovar 2 serovar E	E86		Diseased eel, Spain	26,98
521577357	*Klebsiella pneumoniae*				41
524610557	*V. vulnificus* biovar 1				97
524712457	*Vibrio cholerae*	PS-7701		*Pl. altivelis*, Japan	49
524712476	*Vibrio cholerae*	PS-7705		*Pl. altivelis*, Japan	49
5304004..	*Burkholderia cepacia*		1/10		38
5304112..	*Hafnia alvei*	ATCC 51873	1		203
530475257	*Serratia fonticola*				290
530477257	*Serratia fonticola*				290
530510057	*Yersinia ruckeri*	AHLDA 1313			135
530511357	*Hafnia alvei*	AHLDA 1729			135
530555257	*Serratia fonticola*				558
530600457	*V. vulnificus* biovar 2 serovar E	525		Diseased eel, Sweden	98
530600557	*V. vulnificus* biovar 2 serovar E	ATCC 33187		Human strain, USA	98
530600557	*V. vulnificus* biovar 2 serovar E	E105		Diseased eel, Spain	26,98
530610057	*Yersinia ruckeri*				41
530610557	*V. vulnificus* biovar 2 serovar E			Taiwanese strains, avirulent for eels	26
534600557	*V. vulnificus* biovar 1 non-serovar E	ATCC 27562		Human strain, USA	33
534600557	*V. vulnificus* biovar 1 non-serovar E	E109		Healthy eel, Spain, avirulent for eels	98

continued

Table 4.25. *Continued.*

API 20 E number	Organism	Strain	No of strains	Notes	Ref
534600557	*V. vulnificus* biovar 1 serovar E	521		Unknown, Australia, avirulent for eels	98
534610357	*V. vulnificus*	WF8A1110			509
534610457	*Vibrio mimicus*				161,210
534610557	*V. vulnificus* biovar 1 non-serovar E	C7184		Human blood, USA, avirulent for eels	98
534612457	*Vibrio cholerae*	91/1198		*Carassius auratus*, Australia	49
534612457	*Vibrio cholerae* 01				
534712457	*Vibrio cholerae* non 01				
534712557	*Aeromonas veronii veronii*				2,347
534712557	*Vibrio cholerae*	WF 110r			509,552
560621057	*Salmonella cholerasuis arizonae*				41
600400457	*Photobacterium iliopiscarium*				599
600400457	*Photobacterium damselae damselae*	ATCC 33539ᵀ		Usually VP positive	240
600500457	*Photobacterium damselae damselae*	LMG 13639	6	Various sources	49
600610417	*Aeromonas salmonicida salmonicida*				674
6046021..	*Photobacterium phosphoreum*	IB39			509
6100530..	*Halomonas cupida*				41
624750456	*Vibrio proteolyticus*	AHLDA 1735	1	*Artemia*	135
700610517	*Aeromonas salmonicida salmonicida*	NCIMB 1102			468
704672556	*Vibrio mediterranei*	AHLDA 1733	1	Abalone	135
704712457	*Aeromonas sobria*	CIP 74.33ᵀ			240
704712657	*Aeromonas hydrophila*	ATCC 7966ᵀ	1		240
704713757	*Aeromonas bestiarum*		5	Carp	452
714420457	*Plesiomonas shigelloides*				41
724610557	*Aeromonas trota*	ATCC 49657ᵀ	1		142,560
724613657	*Aeromonas allosaccharophila*		1		41
724616657	*Aeromonas allosaccharophila*		1		41
724617657	*Aeromonas allosaccharophila*		1		41
724712457	*Aeromonas sobria*				21
724714457	*Aeromonas janadaei*		2		135,143

Not all references reported the 9-digit profile number, hence some species only have a 7- or 8-digit profile number with no results for growth on MCA or OF results.

Table 4.26. API 20NE database biochemical results.

Organism	NO3	TRP	Glu	ADH	Ure	Esc	Gel	Png	Glu	Ara	Mne	Man	Nag	Mal	Gnt	Cap	Adi	Mlt	Cit	Pac	Ox	Temp	Time	Inoc	Ref	Strain
Acinetobacter calcoaceticus	–	–	–	–	–	–	–	–	–	–	–	–	–	–	–	+	–	–	+	+	–		24		41	
Actinobacillus delphinicola	+	–	–	–	–	–	–	–	–	–	–	–	–	–	–	–	–	–	–	–	+				263	Dolphin
Actinobacillus delphinicola	+	–	–	+	–	–	–	–	–	–	–	–	–	–	–	–	–	–	–	–	+				263	Porpoise
Actinobacillus delphinicola	+	–	–	–	–	–	–	–	–	–	–	–	–	–	–	–	–	–	–	–	+				263	Whale
Aeromonas hydrophila	+	v	+	+	–	+	+	+	+	v	+	+	+	+	+	+	–	+	v	–	+				41	
A. salmonicida achromogenes/masoucida	+	v	–	–	–	v	v	–	v	–	v	v	–	–	–	–	–	–	–	–	+				41	
A. salmonicida salmonicida	+	v	v	v	–	+	+	–	+	–	v	+	+	+	+	+	–	+	–	–	+				41	
Aeromonas sobria	+	v	+	+	–	–	+	+	+	–	+	+	+	+	+	+	–	+	+	–	+				41	
Alteromonas nigrifaciens	–			–	–	–	+	–	+	–	+	–	+	+	+	+	+	+	+	+	+				395	
Bordetella bronchiseptica	–	–	–	–	+	–	–	–	–	–	–	–	–	–	–	–	+	–	+	+	+	37		0.85	102	Type strain
Bordetella bronchiseptica	+	–	–	–	+	–	–	–	–	–	–	–	–	–	–	–	+	–	+	+	+	37		0.85	102	Type strain
Bordetella bronchiseptica	+	–	–	–	+	–	–	–	–	–	–	–	–	–	–	–	+	–	+	–	+	37		0.85	102	Type strain
Brevundimonas diminuta	–	–	–	–	–	–	–	–	–	–	–	–	–	–	–	–	–	–	–	–	+				685	
Brucella spp.	+	–	–	–	+	–	–	–	–	–	–	–	–	–	–	–	–	–	–	–	+	37		0.85	262	
Chromobacterium violaceum	–	–	+	+	–	–	+	–	+	–	–	–	+	–	+	+	–	+	+	–	+				482	Pigmented
Chromobacterium violaceum	+	–	+	+	–	–	+	–	+	–	–	–	+	–	+	+	–	+	–	–	+				482	Non-pigmented
Chryseobacterium balustinum	+	+	–	–	+	+	+	–	–	–	w	–	–	w	–	–	–	–	w	–	+	37	48	0.85	773	
Chryseobacterium gleum	+	v	–	–	+	+	+	–	+	+	w	–	–	+	–	–	–	–	–	–	+	37	48	0.85	773	
Chryseobacterium indologenes	–	+	–	–	+	+	+	–	w	–	w	–	–	w	–	–	–	–	w	–	+	37	48	0.85	773	
Chryseobacterium indoltheticum	–	+	+	–	–	+	+	–	w	–	w	–	–	w	–	–	–	–	–	w	+	37	48	0.85	773	
Chryseobacterium meningosepticum	–	+	+	–	18	+	+	–	+	–	+	+	+	–	–	–	–	w	27	–	+	37	48	0.85	773	
Chryseobacterium scophthalmum	–	–	w	–	+	+	+	–	+	–	+	–	+	+	+	–	–	–	–	–	+				557	
Flavobacterium hibernum	+	–	+	–	–	+	+	–	+	+	+	–	+	+	+	+	–	+	–	–	–				532	
Halomonas cupida	+	–	–	–	–	–	–	–	+	+	+	–	+	+	+	+	–	+	+	+	–				69	
Halomonas venusta	+	–	–	–	+	+	–	+	+	–	+	+	+	+	+	+	v	+	+	+	+				310	
Janthinobacterium lividum	+	–	–	–	–	+	–	–	+	–	+	+	+	+	–	v	v	+	+	+	+				496	

continued

Table 4.26. *Continued.*

	NO₃	TRP	Glu	ADH	Ure	Esc	Gel	Png	Glu	Ara	Mne	Man	Nag	Mal	Gnt	Cap	Adi	Mlt	Cit	Pac	Ox	Temp	Time	Inoc	Ref	Strain
Listonella anguillarum	+	+	+	+	–	–	+		+	+	+	+	+	+	+		–	+	+	+	+				108	
Myroides odoratimimus	–	–	–	–	+	–	+		–	–	–	–	–	–	–	–	–	+w	–	–	+				774	
Myroides odoratus	–	–	–	–	+	–	+		–	–	–	–	–	–	–	–	–	+w	–	–	+		24		774	
Oceanomonas baumannii	+	–	–	–	–	–	–	–	–	–	–	–	–	–	–	+	–	+	+	–	+	25	48	1.5	130	
Oceanomonas doudoroffii	+	–	–	–	–	–	–	–	–	–	–	–	–	–	–	+	–	+	+	–	+	25	48	1.5	130	
Phocoenobacter uteri	+	–	+w	–	–	–	–	–	–	–	–	–	–	–	–	–	–	–	–	–	+				266	
Photobacterium damselae	+	–	+	+	+	–	–	–	+	–	+	–	–	–	–	–	–	v	–	–	+				41	
P. damsela damsela	+	–	+	+	+	v	–	–	+	–	+	–	+	+	–	–	+	–	–	+	+				289	
P. damsela piscicida	–	–	+	+	–	–	–	–	+	–	+	–	+	+	–	–	+	–	–	–	+				289	
Plesiomonas shigelloides	+	+	+	+	–	–	–	+	v	–	–	–	v	v	v	v	–	v	–	–	+				41	
Pseudoalteromonas antarctica	–	–	+	–	–	–	+	–	+	–	+	+	–	+	+	–	+	–	+	–	+	5	5d		115	
Pseudoalteromonas bacteriolytica	–		+		+		+		+		+	–	–	–	–			–	–	+	+				677	
Pseudoalteromonas haloplanktis							+		+				–		–			–							288	
Pseudomonas anguilliseptica	–w	–	–	–	–	–	+	–	–	–	–	–	v	+	–	–	+	+	–	–	+				541	
Pseudomonas chlororaphis	–	–	–	–	–	v	–	–	–	+	+	+	–	–	+	v	–	+	+	–	v				41	
Pseudomonas fluorescens	v	–	–	v	–	–	v	–	+	v	+	+	v	–	+	+	–	+	+	–	+				41	
Pseudomonas plecoglossicida	+	–	–	+	–	–	–	–	+	–	–	–	–	–	+	+	–	+	+	+	+	25	48		582	
Pseudomonas putida	+	–	–	+	–	–	–	–	+	v	v	v	–	–	+	+	–	v	+	+	+				582	
Shewanella algae LMG 2265	+	–	–	–	–	–	+	–	–	–	–	–	+	–	–	+	–	+	–	–	+				851	
Shewanella baltica NCTC 10735	+	–	+	–	–	–	+	–	+	–	–	–	+	+	+	–	–	+	+	–	+				851	
Shewanella pealeana		–	–					–	+	–				+				+	+		+				492	
Shewanella putrefaciens	+	–	+	–	–	–	+	–	–	+	–	–	+	+	–	+	–	+	+	–	+				433	
Sphingomonas subarctica	–	–	–	–	–	+	+	+	+	+	+	–	+	+	–	–	–	+	+	–	+					

Species	Strain	NO₃	TRP	Glu	ADH	Ure	Esc	Gel	Png	Glu	Ara	Mne	Man	Nag	Mal	Gnt	Cap	Adi	Mlt	Cit	Pac	Ox				Ref
Stappia stellulata-like strain M1		+		–	+	+	–	+	+	+	+	+	+	+	+	+	+	+	+	+	–	+				105
Vagococcus salmoninarum		–	–	+	–	+	–	+														–				542
Vibrio alginolyticus		+	+	+	–	+	–	+	+	v	v	v	v	v	+	+	v	–	+	+	+	+				41
Vibrio alginolyticus		+	+	+	–	+	–	+	+	+	+	+	+	+	+	+	nt	nt	+	nt	–	+				108
Vibrio alginolyticus		+	+	+	+	+	+	+	+	+	+	+	+	+	+	+	–	–	+	–	+				
Vibrio coralliilyticus	YB	+	–	+	–	+	–	+	83	+	+	+	+	+	83	83	+	–	+	83	+	30	48	3	83,84
Vibrio (carchariae) harveyi	ATCC 35084	+	+	+	–	+	+	+	+	+	+	+	+	+	+	+	+	+	+	+	+				11
Vibrio cholerae		+	+	+	–	+	–	+	–	v	v	v	+	+	v	+	+	–	+	+	+				41
Vibrio diabolicus		+	+	+	–	+	–	+	+	+	+	+	+	–	–	+	+	–	+	+	+				635
Vibrio harveyi	ATCC 14126	+	+	+	–	+	+	+	+	+	+	+	+	+	+	+	+	+	+	+	+				11
Vibrio harveyi		+	+	+	–	+	–	+	+	+	+	+	+	+	+	+	+	+	+	+	+				11
Vibrio ichthyoenteri		+	–	+	–	+			33	–	–	50	50		–					–	+				389
Vibrio mimicus		+	+	+	–	+	–	+	+	+	+	+	+		–	+			+	+	+				230
Vibrio mytili		+	+	+	+	+	+	+	+	+	+	+	+		–	+	+	+		+					635
Vibrio navarrensis		+	+	+	–	+	+	+	+	+	+	+	+	–	+	60	–	+	+						768
Vibrio nereis		+	+	+	–	+								–	+	+	–	–	+						635
Vibrio penaeicida		+	–	+	–	+			+	+	+	+	+	+	+	+			+						388
Vibrio shilonii		+	+	+	–	+	–	+	+	+	+	+	+						+	+	+				458
Vibrio tapetis		+	+	+	–	+	–	+	–	–	+	–	–	–	–	–	–	–							108
Vibrio tubiashii		+	+	+	–	+	+	+	+	+	+	+	+	–	+	–	+	–	–	–	–				635
Zobellia galactanivorans		+	–	+	–	+	+	+	+	+	+	+	+	–	w	–	–	–	–	–	+	30	7d	2.5	61
Zobellia uliginosa		+	–	+	–	+	+	+	+	+	+	+	w	–	–	–	–	–	–	–	+	30	7d	2.5	61

NO₃ = nitrate; TRP = indole; Glu = glucose fermentation; ADH = arginine dihydrolase; Ure = urease; Esc = aesculin; Gel = gelatin hydrolysis; Png = *p*-nitrophenyl-β-D-galactopyranoside; Glu = glucose assimilation; Ara = L-arabinose assimilation; Mne = mannose assimilation; Man = mannitol assimilation; Nag = *N*-acetylglucosamine; Mal = maltose assimilation; Gnt = gluconate assimilation; Cap = caprate assimilation; Adi = adipate assimilation; Mlt = malate assimilation; Cit = citrate assimilation; Pac = phenylacetate; Ox = oxidase; *API database.

Table 4.27. API 50CH database biochemical results.

	1	2	3	4	5	6	7	8	9	10	11	12	13	14	15	16	17	18	19	20	21	22	23	24	25	26	27	28	29
Test	Gly	Ery	Dara	Lara	rib	dXyl	lXyl	ado	mdx	gal	glu	fru	mne	Lsor	rha	dul	ino	man	sor	mdm	mdg	nag	amy	arb	esc	sal	cel	mal	lac
Organism																													
Aerococcus viridans var. *homari*					w					+	+	+	+					+	v		v	+	+	+	+	+	+	+	+
Aeromonas bestiarum HG2	+	–		+	+	–	–	–	–	+	+	+	+	–	–	–	–	+	+	–	+	+	–	+	+	+	67	67	67
Aeromonas caviae HG 4	+	–		+	+	–	–	–	–	+	+	+	31	–	–	–	–	+	7	–	–	+	–	+	+	+	90	+	87
Aeromonas culicicola	+	–	–	–	+	–	–	–	–	+	+	+	+	–	–	–	–	+	–	–	–	+	–	–	–	–	–	+	–
Aeromonas encheleia	+	–	–			–	–	–	–	+	+	+	+		–	–	–	+	–	–	–					+	–	+	–
Aeromonas eucrenophila HG6	+	–	–	+	+	–	–	–	–	+	+	+	+	–	–	–	–	+	–	–	+	+	–	+	+	+	+	+	+
Aeromonas hydrophila HG1	+	–		+	+	–	–	–	–	+	+	+	+	–	–	–	–	+	–	–	–	+	–	+	+	+	–	+	–
Aeromonas hydrophila HG3	+	–		+	+	–	–	–	–	+	+	+	+	–	–	–	–	+	+	–	+	+	–	+	+	+	62	+	54
Aeromonas media HG 5	+	–		+	+	–	–	–	–	+	+	+	+	–	–	–	–	+	–	–	–	+	–	+	+	+	+	+	+
Aeromonas popoffii	+	–						–		+	+		+				–	+								–	–	+	–
A. salmonicida 'atypical'																													
A. salmonicida achromogenes	+			–	+					+	+	+	+					+			–	–		–			–	–	
Baltic sea turbot	–			–						–	+	+						–			–	–		–	v		–	–	
Baltic sea flounder	–			–	v					–	+	+	v					v			–	–		–	+	v		+	
Baltic sea dab	+			–	–					v	+	+	v					–			–	+w		–	+	+			
Baltic sea blenny	+			–	+					+	+	+	+					+			–	+		–	–			+	
'Atypical' Pigment pos	+			–	+					+	+	+						+			–	17							
'Atypical' Pigment neg	44			–	+					44	+	+						+			–	88							
'Atypical' Pigment neg	44			–	+					44	+	+						–			–	88							
Australian strain	–			–						–								–			–			–	–				
A. salmonicida typical	+			+	+					+		+	+					+	–		+	+		+	+	+	–	+	–
A. salmonicida salmonicida	+			+	+					+	+	+	–					+			+	+		+	+	+	–		

201

Test	30 mel	31 suc	32 tre	33 inu	34 mlz	35 raf	36 amd	37 glyg	38 xylt	39 gen	40 tur	41 lyx	42 tag	43 Dfuc	44 Lfuc	45 Darl	46 Larl	47 gnt	48 2ket	49 5ket	Temp	Time	Inoc	Ref	Strain	Strain
Organism																										
Aerococcus viridans var. homari	v	+	+	+		+				+	v							+		v				827		
Aeromonas bestiarum HG2	–	+	+	–	–	+	+	+	–	–		–	–	–	–	–	–	+	–	–	30	48	0.85	427	CDC 9533–76	
Aeromonas caviae HG 4	–	+	+	–	–	–	+	+	–	36	–	–	–	–	–	–	–	+	–	–	30	48	0.85	427,21	ATCC 15468[T]	
Aeromonas culicicola	+	+	+	–	–	–	–	+	–	–	–	–	–	–	–	–	–	+	–	–	30	24	0.85	624	NCIMB 5147[T]	
Aeromonas encheleia			+			–										–										
Aeromonas eucrenophila HG6	–	+	+	–	–	–	+	+	–	–			–	–	–	–	–	–	–	–	30	48	0.85	427	ATCC 23309[T]	
Aeromonas hydrophila HG1	–	+	+	–	–	–	+	+	–	–		–	–	–	–	–	–	+	–	–	30	48	0.85	427	ATCC 7966[T]	
Aeromonas hydrophila HG3	–	+	+	–	–	–	+	+	–	–		–	–	–	–	–	–	–	–	–	30	48	0.85	427	CDC0434–84	
Aeromonas media HG 5	–	+	+	–	–	–	+	+	–	–	–	–	–	–	–	–	–	–	–	–	30	48	0.85	427	CDC 0862–83	CDC 9072–83
Aeromonas popoffii	–	–	+	–	–	–										–										
A. salmonicida 'atypical'																										
A. salmonicida achromogenes	–	+	+		–		+	+	–		–	–	–	–	–	–	–	–	–	+s	20	7d	0.5	831		
Baltic sea turbot	–	–	–	–	–		–	–	–	–	–	–	–	–	–	–	–	–	–	+	20	14d		832		
Baltic sea flounder	–	v	+	–	–		v	+	–	–	–	–	–	–	–	–	–	–	–	+	20	14d		832		
Baltic sea dab	–	+	–	–	–		–	–	–	–	–	–	–	–	–	–	–	–	–	+	20	14d		832		
Baltic sea blenny	–	+	+	–	–		+	+	–	–	–	–	–	–	–	–	–	–	–	+w	20	14d		832		
'Atypical' Pigment pos		+	68				100	3													20	7d		323,352		
'Atypical' Pigment neg		+	27				63	75													20	7d		323		
'Atypical' Pigment neg		+					63	75													20	7d		323		
Australian strain	+	+	–															–						761		
A. salmonicida typical	–	–	–	–			+	+										+		–	20–22	7d		352		
A. salmonicida salmonicida	–	–	–		–		+	+			–							+		+s	20	7d	0.5	831		

continued

Table 4.27. *Continued.*

Test	1 Gly	2 Ery	3 Dara	4 Lara	5 rib	6 dXyl	7 lXyl	8 ado	9 mdx	10 gal	11 glu	12 fru	13 mne	14 Lsor	15 rha	16 dul	17 ino	18 man	19 sor	20 mdm	21 mdg	22 nag	23 amy	24 arb	25 esc	26 sal	27 cel	28 mal	29 lac
Aeromonas sobria	+	–	–	28	+	–	–	–	–	+	+	+	+	–	–	–	–	+	–	–	24	+	–	–	16	–	52	+	16
Aeromonas trota	+	–	–	–	+	–	–	–	–	+	+	+	+	–	–	–	–	+	–	–	–	+	+	–	–	–	+	+	–
Aeromonas veronii sobria HG8/10	+	–	–	14	+	–	–	–	–	+	+	+	+	–	–	–	–	+	–	–	29	+	–	–	10	–	57	+	10
Aeromonas veronii veronii	–	–								+	+		+					+			+				+	+	+	+	–
Arcanobacterium bernardiae	+	v			v			+		+	+	+						–			+					+		+	–
Arcanobacterium phocae	+	–	–	–	+					+	+	+	70				+	30				+						+	+
Arcanobacterium phocae	–				–					v	+	–	+				+	–				+						+	>
Arcanobacterium pyogenes	–	+			+	+		+		+	+	+	+				+	–	+			–	–		–	–	+	+	+
Arthrobacter rhombi	+	–	–	–	+	+	–	–	–	+	+	+	+	–	v	–	–	+	–	–	–	–	+	+	+	+	+	+	+
Brevundimonas diminuta	–	–	–	–	–	–	–	–	–	–	–	–	–	–	–	–	–	+	–	–	–	–	–	–	–	–	–	–	–
Brevundimonas vesicularis	–	–	–	–	–	+	–	–	–	–	+	–	–	–	–	–	–	–	–	–	–	+	–	–	–	–	–	+	–
Carnobacterium divergens					+						+	+	+									+	+			+	+	+	
Carnobacterium gallinarum					+	+					+	+	+								+	+	+		+	+	+	+	+
Carnobacterium mobile					+						+	+	+									+				+	+	+	
Carnobacterium piscicola	+	–	–	–	+	–	–	–	–	+w	+	+	+	–	–	–	–	+	v	+	+	+	+	+	+	+	+	+	+w
Chryseobacterium scophthalmum																											–		
Enterococcus faecium	v			+														+	+										
Flavobacterium columnare	–	–	–	–	–	–	–	–	–	–	–	–	–	–	–	–	–	–	–	–	–	–	–	–	–	–	–	–	–
Flavobacterium psychrophilum	–	–	–	–	–	–	–	–	–	–	–	–	–	–	–	–	–	–	–	–	–	–	–	–	–	–	–	–	–
Flexibacter litoralis	–	–	–	–	–	–	–	–	–	–	–	–	–	–	–	–	–	–	–	–	–	–	–	–	–	–	–	–	–
Flexibacter polymorphus	–	–	–	–	–	–	–	–	–	–	–	–	–	–	–	–	–	–	–	–	–	–	–	–	–	–	–	–	–
Flexibacter roseolus	–	–	–	–	–	–	–	–	–	–	–	–	–	–	–	–	–	–	–	–	–	–	–	–	–	–	–	–	–
Flexibacter ruber	–	–	–	–	–	–	–	–	–	–	–	–	–	–	–	–	–	–	–	–	–	–	–	–	–	–	–	–	–
Janthinobacterium lividum	+	–	+	+	+	+	–	–	–	+	+	+	+	+	v	–	+	+	+	+	–	v	–	+	–	+	+	+	+
Klebsiella planticola	+	–	–	+	+	+	–	+	–	+	+	+	+	+	+	–	+	+	+	+	+	+	+	+	+	+	+	+	+
Lactococcus garvieae	–	–	–	–	+	–	–	–	–	+	+	+	+	–	+	–	–	+			–	+	+	+	+	+	+	+	–

continued

Test	30 mel	31 suc	32 tre	33 inu	34 mlz	35 raf	36 amd	37 glyg	38 xylt	39 gen	40 tur	41 lyx	42 tag	43 Dfuc	44 Lfuc	45 Darl	46 Larl	47 gnt	48 2ket	49 5ket	Temp	Time	Inoc	Ref	Strain
Aeromonas sobria	12	+	+	-	-	-	+	+	-	-	-	-	-	-	-	-	-	+	-	-	29	48		21	
Aeromonas trota	-	-	+	-	-	-	+	+	-	-	-	-	-	-	-	-	-	-	-	-	37	24	0.85	20	ATCC 49657^T
Aeromonas veronii sobria HG8/10	-	+	+	-	-	-	+	+	-	-		-	-	-	-	-	-	-	-	-	30	48	0.85	427	CDC 0437–84
Aeromonas veronii veronii	-	-	+			-			+							-								274	
Arcanobacterium bernardiae	-	+	+	-	70	-	+	v	+	-	+	-	-			+	v		-	+	37	48		636	
Arcanobacterium phocae	-	+	-	-		-	+	+	-	-	+	-	30	-	-	-	-	-	-	+	37	24		613	
Arcanobacterium phocae	-	v	-	-	-	v	-	-	-	-	-	-	-	-	-	-	-	-	-	-	37	48		641	
Arcanobacterium pyogenes	-	-	+	-	+	-	+	+	-	+	+	-	-	-	-	+	-	+	-	-	37	48		600	
Arthrobacter rhombi	+	+	+	+	-	-	-	-	-	+	+	-	-	-	-	+	-	+	-	-	25	48		685	
Brevundimonas diminuta	-	-	-	-	-	-	-	-	-	-	-	-	-	-	-	-	-	-	-	-	28	4		685	
Brevundimonas vesicularis	-	-	-	-	-	-	-	-	-	-	-	-	-	-	-	-	-	-	-	-	28	4		685	
Carnobacterium divergens	-	+	+	-	+	-	-	-	-	+	+	-	-	-	-	-	-	+	-	-	25	7d		176	
Carnobacterium gallinarum	-	+	+	-	+	-	-	-	-	-	+	-	+	-	-	-	-	+	-	-				176	
Carnobacterium mobile	+	+	+	+	-	-	-	-	-	-	-	-	–v	-	-	-	-	-	-	-				176	
Carnobacterium piscicola	+	+	+	+	+	v	-	-	-	+	+w	-	-	-	-	-	-	+	-	-	25	72		176,73	
Chryseobacterium scophthalmum	v	v																v							
Enterococcus faecium	v	v	-	-	-	-	-	-	-	-	-	-	-	-	-	-	-	>	-	-					
Flavobacterium columnare	-	-	-	-	-	-	-	-	-	-	-	-	-	-	-	-	-	-	-	-	22	12		88,89,211	
Flavobacterium psychrophilum	-	-	-	-	-	-	-	-	-	-	-	-	-	-	-	-	-	-	-	-	22	12		88,89,211	
Flexibacter litoralis	-	-	-	-	-	-	-	-	-	-	-	-	-	-	-	-	-	-	-	-				89	
Flexibacter polymorphus	-	-	-	-	-	-	-	-	-	-	-	-	-	-	-	-	-	-	-	-				89	
Flexibacter roseolus	-	-	-	-	-	-	-	-	-	-	-	-	-	-	-	-	-	-	-	-				89	
Flexibacter ruber	-	-	-	-	-	-	-	-	-	-	-	-	-	-	-	-	-	-	-	-				89	
Janthinobacterium lividum	-	+	-	v	-	v	-	-	+	-	-	+	-	-	+	+	-	-	+	-				496	
Klebsiella planticola	+	+	+	-	-	+	-	-	+	-	-	+	-	-	+	-	-	-	-	-					
Lactococcus garvieae	-	-	+	-	-	-	-	-	-	+	-	-	-	-	+	-	-	-	-	-	28	24–96		174,236	ATCC 49156

Table 4.27. *Continued.*

Test	1 Gly	2 Ery	3 Dara	4 Lara	5 rib	6 dXyl	7 lXyl	8 ado	9 mdx	10 gal	11 glu	12 fru	13 mne	14 Lsor	15 rha	16 dul	17 ino	18 man	19 sor	20 mdm	21 mdg	22 nag	23 amy	24 arb	25 esc	26 sal	27 cel	28 mal	29 lac
Lactococcus garvieae	–	–	–	–	+	–	–	–	–	+	+	+	+	–	–	–	–	+	–		–	+	+	+	+	+	+	+	–
Biotypes 1, 2, 12																		+											
Biotypes 3, 7, 11, 13																		+											
Biotypes 4, 5, 6, 10,																		+											
Biotypes 8, 9																		–											
Lactococcus piscium	–	–	–	+	+	+	–	–	–	+	+	+	+	–	–	–	–	+	–	–	–	+	+	+	+	+	+	+	+
Moritella viscosa	–	–	–	–	+	–	–	–	–	–	+	+	–	–	–	–	–	–	–	–	–	+	–	–	–	–	–	+	–
Pedobacter heparinus	–	–	–	+	–	+	–	–	–	+	+	+	+	–	–	–	–	–	+	–	–	–	–	–	+	–	+	+	+
Pedobacter piscium	–	–	–	–		v	–	–	–	+	+	+	+	–	–	–	–	+	–	+	+	–	+	+	+	+	+	+	+
Photobacterium angustum	–	–	–	–	+	+	–	–	–	+	+	+	+	–	–	–	–	–	–	–	–	+	–	–	–	–	–	+	–
P. damselae damselae	+	–	–	–	+	–	–	–	–	+	+	+	+	–	–	–	–	–	–	–	–	+	–	–	–	–	+	+	–
P. damselae piscicida	–	–	–	–	+	–	–	–	–	+w	+	+	+	–	–	–	–	–	–	–	–	+	–	–	–	–	–	–	–
P. damselae piscicida	–	–	–	–	+	+	–	–	–	+	+	+	+	–	–	–	–	+	–	–	–	+	–	–	–	–	–	+	+
Photobacterium iliopiscarium	v	–	–	–	+	–	–	–	–	+	+	+	+	–	–	–	–	–	–	–	–	+	–	–	–	–	–	+	–
Photobacterium leiognathi	+	–	–	–	+	–	–	–	–	+	+	+	+	–	–	–	–	–	–	–	–	+	–	–	–	–	–	–	–
Streptococcus (difficile) agalactiae	–	–	–	–	+	–	–	–	–	–	+	+	+	–	–	–	–	–	–	–	–	+	–	–	–	–	–	+	–
Streptococcus (difficile) agalactiae	–	–	–	–	+	–	–	–	–	–	+	+	+	–	–	–	–	–	–	–	+	+	–	–	–	–	–	+	–
Streptococcus (difficile) agalactiae	–	–	–	–	+	–	–	–	–	–	+	+	+	–	–	–	–	+	–	–	+	+	–	–	–	–	–	+	–
Streptococcus iniae	–	–	–	–	+	–	–	–	–	?	+	+	+	–	–	–	–	+	–	+	–	+	–	+	+	+	+	+	–
Streptococcus iniae	–	–	–	–	+	–	–	–	–	–	+	+	+	–	–	–	–	+	–	–	+	+	–	+	+	+	+	+	–
Streptococcus iniae	–	–	–	–	+	–	–	–	–	–	+	+	+	–	–	–	–	+	–	–	–	+	–	+	+	+	+	+	–
Streptococcus iniae	–	–	–	–	+	–	–	–	–	+	+	+	+	–	–	–	–	+	+	–	–	+	+	+	+	+	+	+	+
Streptococcus parauberis	–	–	–	–	+	–	–	–	–	+	+	+	+	–	–	–	–	+	+	–	–	+	+	+	+	+	+	+	+
Streptococcus phocae	–	–	–	–	+	–	–	–	–	–	+	+	+	–	–	v	–	+	–	–	–	+	–	–	–	–	–	+	–
Streptococcus porcinus	+	–	–	–	+	–	–	–	–	+	+	+	+	–	–	–	–	+	+	–	–	+	v		+	+	+	+	v
Streptococcus uberis	–	–	–	–	+	–	–	–	–	+	+	+	+	–	–	v	–	+	+	–	–	+	+	+	+	+	+	+	+

Test	30 mel	31 suc	32 tre	33 inu	34 mlz	35 raf	36 amd	37 glyg	38 xylt	39 gen	40 tur	41 lyx	42 tag	43 Dfuc	44 Lfuc	45 Darl	46 Larl	47 gnt	48 2ket	49 5ket	Temp	Time	Inoc	Ref	Strain	Strain
Lactococcus garvieae	–	–	+	–	–	–	–	–	–	+	–	–	–	–	–	–	–	–	–	–	28	24–96		174,211, 236,237, 780	ATCC 49156	
Biotypes 1, 2, 12		+											+											780		
Biotypes 3, 7, 11, 13		–											+											780		
Biotypes 4, 5, 6, 10,		–											–											780		
Biotypes 8, 9		–																						780		
Lactococcus piscium	+	+	+	–	+	+	+w	–	–	+	+	–	–	–	–	–	–	+	–	–	25	48	0.85	835	NCFB 2778	
Moritella viscosa	–	–	–	–	–	–	+	+	–	–	–	–	–	–	–	–	–	–	+	–	15	72		132	NCIMB 13484	
Pedobacter heparinus	–	+	–	–	–	–	+	–	–	–	+	–	–	–	+	–	–	–	–	+	28	48	0.5	728		
Pedobacter piscium	+	+	+	–	–	+	–	–	–	+	+	–	–	–	–	+	–	–	–	–	28	48	0.5	728		
Photobacterium angustum	–	–	–	–	–	–	–	–	–	–	–	–	–	–	–	–	–	–	–	–	26	72	1.5	745	NCIMB 1895	
P. damselae damselae	–	–	+	–	–	–	–	–	–	–	–	–	–	–	–	–	–	–	–	–	26	72	1.5	745	ATCC 35083	ATCC 33539
P. damselae piscicida	–	+w	–	–	–	+w	>	–	–	–	–	–	–	–	–	–	–	–	–	–	26	72	1.5	288,745	NCIMB 2058	NCIMB 25918, ATCC 17911
P. damselae piscicida	–	–	–	–	–	–	–	–	–	–	–	–	–	–	–	–	–	–	–	–	26	72	1.5	745	P 90029	Atypical strain
Photobacterium iliopiscarium	–	–	v	–	–	–	–	–	–	–	–	–	–	–	–	–	–	v	–	–	22		2	599		
Photobacterium leiognathi	–	–	–	–	–	–	–	–	–	–	–	–	–	–	–	–	–	+	–	+	26	72	1.5	745	LMG 4228	
Streptococcus (difficile) agalactiae	–	+	+	–	+	–	+	+	–	–	–	–	–	–	–	–	–	–	–	–	30					
Streptococcus (difficile) agalactiae	–	+	+	–	+	–	+	+	–	+	–	–	–	–	–	–	–	–	–	–	30	72		233	ND 2–22	
Streptococcus (difficile) agalactiae	–	+	+	–	+	–	+	+	–	+	–	–	–	–	–	–	–	–	–	–	30	72		233	ND 2–22	
Streptococcus iniae	–	+	+	–	+	–	+	+	–	+	–	–	–	–	–	–	–	–	–	–	30					
Streptococcus iniae	–	+	+	–	+	–	+	+	–	+	–	–	–	–	–	–	–	–	–	–	30	72		233	ND 2–16	
Streptococcus iniae	–	+	+	–	+	–	+	+	–	+	–	–	–	–	–	–	–	–	–	–	30	72		233,235	ND 2–16	
Streptococcus iniae	–	+	+	–	+	–	+	+	–	+	–	–	–	–	–	–	–	–	–	–	24	72	0.85	183		
Streptococcus parauberis	–	+	+	v	–	–v	–v	–	–	+	–	–	+	–	–	–	–	–	–	–	37	24		224		
Streptococcus phocae	–	–	+	–	–	–	–	–	–	–	–	–	–	–	–	–	–	–	–	–	37	24		700		
Streptococcus porcinus	–	+	+	–	–	–	>	–	–	–	–	–	–	–	–	–	–	–	–	–	37	24–48	0.5	175		
Streptococcus uberis	–	+	+	+	–	–	–	–	–	+	–	–	+	–	–	–	–	–	–	–	37	24		224		

continued

Table 4.27. *Continued.*

Test	1 Gly	2 Ery	3 Dara	4 Lara	5 rib	6 dXyl	7 lXyl	8 ado	9 mdx	10 gal	11 glu	12 fru	13 mne	14 Lsor	15 rha	16 dul	17 ino	18 man	19 sor	20 mdm	21 mdg	22 nag	23 amy	24 arb	25 esc	26 sal	27 cel	28 mal	29 lac
Tenacibaculum maritimum	–	–	–	–	–	–	–	–	–	–	–	–	–	–	–	–	–	–	–	–	–	–	–	–	–	–	–	–	–
Vagococcus fluvialis	–	–	–	–	+	–	–	–	–	–	+	+	+	–	–	–	–	+	+	–	+	+	+	+	+	+	+	+	–
Vagococcus fluvialis	–	–	–	–	+	–	–	–	–	–	+	+	+	–	–	–	–	+	–	–	+	–	+	+	+	+	+	+	–
Vagococcus salmoninarum	–	–	–	–	60	+w	–	–	–	–	+	+	+	90	–	–	–	–	70	–	–	+	+	+	+w	+	60	70	–
Vibrio carchariae ATCC 35084	–	–	–	–	+	–	–	–	–	+	+	+	+	–	–	–	–	+	–	–	–	+	–	–	+w	–	+	+	–
Vibrio diabolicus	+	–	–	–	+	–	–	–	–	+	+	+	+	–	–	–	–	+	–	–	–	+	+	–	–	–	–	+	–
Vibrio fischeri	–	–	–	–	+	–	–	–	–	+	+	+	+	–	–	–	–	–	–	–	–	+	–	–	–	+	+	+	–
Vibrio harveyi ATCC 14129	–	–	–	–	+	–	–	–	–	+	+	+	+	–	–	–	–	+	–	–	–	+	–	–	+w	–	+	+	–
Vibrio mytili	+	–	–	+	+	+	–	–	–	+	+	+	+	–	–	–	–	+	–	–	–	+	+	+	+	+	+	+	–
Vibrio nereis	–	–	–	–	+	–	–	–	–	–	+	+	+	–	–	–	–	–	–	–	–	+	–	–	–	–	–	+	–
Vibrio penaeicida	+sl	–	–	–	–	–	–	–	–	–	+	+	+	–	–	–	–	–	–	–	–	–	–	–	–	–	+	+	–
Vibrio salmonicida	v	–	–	–	+	–	–	–	–	+	+	+	–	–	–	–	–	+	–	–	+	+	–	–	–	–	–	+	–
Vibrio splendidus	v	–	–	–	+	–	–	–	–	+	+	+	+	–	–	–	–	+	–	–	+	+	–	–	+	–	+	+	>
Vibrio tubiashii	–	–	–	–	+	–	–	–	–	+	+	+	+	–	–	–	–	+	–	–	+	+	+	–	–	–	+	+	–
Yersinia kristensenii	+	–	–	+	+	+	–	–	–	+	+	+	+	+	+	–	–	+	+	–	–	+	>	+	–	–	+	+	+
Zobellia galactanovorans	–	–	–	+	+	+	–			+	+	+	+	–	–	–	–	+	–				+		+			+	+
Zobellia uliginosa	–									+	+				–		–	–							+			+	

Test	30 mel	31 suc	32 tre	33 inu	34 mlz	35 raf	36 amd	37 glyg	38 xylt	39 gen	40 tur	41 lyx	42 tag	43 Dfuc	44 Lfuc	45 Darl	46 Larl	47 gnt	48 2ket	49 5ket	Temp	Time	Inoc	Ref	Strain
Tenacibaculum maritimum	–	–	–	–	–	–	–	–	–	–	–	–	–	–	–	–	–	–	–	–	22	12	ASW	88,89,211	
Vagococcus fluvialis	–	+	+	–	–	–	–	–	–	+	+	–	–	–	–	–	–	–	–	–				177	Species
Vagococcus fluvialis	–	–	+	–	–	–	–	–	–	+	v	–	–	–	–	–	–	–	–	–				177,498	NCDO 2497
Vagococcus salmoninarum	–	+	+	–	–	–	–	–	–	50	–	–	+	–	60	–	–	–	+w	–	25	2–7d		542,807	NCFB 2777
Vibrio carchariae ATCC 35084	–	+	+	–	–	–	+	+	–	–	–	–	–	–	–	–	–	+	–	–				581	
Vibrio diabolicus	–	+	+	–	–	–	+	+	–	–	–	–	–	–	–	–	–	+	–	–	25	48	2	635	
Vibrio fischeri	–	–	–	–	–	–	+	–	–	–	–	–	–	–	–	–	–	–	–	–	26	72	1.5	745	ATCC 25918
Vibrio harveyi ATCC 14129	–	+	+	–	–	–	+	+	–	–	–	–	–	–	–	–	–	+	–	–	25	24	2	581	
Vibrio mytili	–	+	+	–	+	–	+	+	–	+	v	–	–	–	–	–	–	+	–	–	25	48	2	635	
Vibrio nereis	–	+	+	–	–	–	+	–	–	–	v	–	–	–	–	–	–	+	–	–	25	48	2	635	
Vibrio penaeicida	–	–	+	–	–	–	–	+	–	–	–	–	–	–	–	–	–	–			25	7d		388	JMC 9123
Vibrio salmonicida	–	–	+	–	–	–	–	–	–	–	–	–	–	–	–	–	–	+						232	
Vibrio splendidus	+	–	+	–	–	–	+	+	–	–	–	–	–	–	10	–	–	+			25	48	2	281	
Vibrio tubiashii	–	+	+	–	–	–	+	+	–	–	–	–	–	–	–	–	–	+			25	48	2	635	
Yersinia kristensenii	–	–	+	–	–	–	+	–													28	48	0.85	86	
Zobellia galactanovorans		+										–	–								30	7d	2.5	23	DSM 12802^{T}
Zobellia uliginosa		+										w	+						+		30	7d	2.5	61	DSM 2061^{T}

Gly = glycerol; Ery = erythritol; Dara = D-arabinose; Lara = L-arabinose; rib = ribose; dXyl = D-xylose; ado = adonitol; mdx = β-methyl-D-xyloside; gal = galactose; glu = glucose; fru = fructose; mne = mannose; Lsor = L-sorbose; rha = rhamnose; dul = dulcitol; ino = inositol; man = mannitol; sor = sorbitol; mdm = α-methyl-D-mannoside; mdg = α-methyl-D-glucoside; nag = N-acetylglucosamine; amy = amygdalin; arb = arbutin; esc = aesculin; sal = salicin; cel = cellobiose; mal = maltose; lac = lactose; mel = melibiose; suc = sucrose; tre = trehalose; inu = inulin; mlz = melizitose; raf = raffinose; amd = starch; glyg = glycogen; xylt = xylitol; gen = gentiobiose; tur = D-turanose; lyx = D-lyxose; tag = D-tagatose; Dfuc = D-fucose; Lfuc = L-fucose; Darl = D-arabitol; Larl = L-arabitol; gnt = gluconate; 2ket = 2-ketogluconate; 5ket = 5-ketogluconate. Numbers indicate the percentage of strains with a positive result.

Table 4.28. API Coryne database biochemical results.

	Nit	Pyz	Pyra	Pal	β gur	β gal	α glu	βNAG	esc	ure	gel	glu	rib	xyl	man	mal	lac	sac	gly	cat	Ref
Actinomyces marimammalium	–	–	–	v	–	+		+			–	–				+				–	370
Arcanobacterium bernardiae					–	–			–		–		+	–			–		+	–	480
Arcanobacterium haemolyticum					–				–		–		–	–			+		–	v	480
Arcanobacterium phocae	–	–	+	+	–	+	+	–	–	–	–		+	v	30	+	v	+	+	v	480,636
Arcanobacterium phocae	–	–	+	+	–	–	+	–	–	v	–		–	–	–	+	v	v	–	+	613
Arcanobacterium pluranimalium	–	–	v	–	+	–	–	–	w+	–	+	–	+	–	–	+	–	–	–	+	480
Arthrobacter nasiphocae	–	–	+	+	–	+	+	–	+	–	+	–	–	–	–	–	+	–	–	+	182
Arthrobacter rhombi	–	–	–	–	–	+	+	–	+	–	+	–	–	–	+	+	+	+	–	+	600
Corynebacterium aquaticum	–	+	+	+	–	+	+	–	+	–	+v				+	–					133
Corynebacterium pseudodiphtheriticum	+	+	+	+	–	–	–	–	–	+	–					–					133
Corynebacterium pseudotuberculosis	–	–	–	+	–	+	–	–	–	+	–					–					133
Corynebacterium testudinoris	+	–	–	–	–	–	v	–	+	–	–		+		+	+	–	+	–	+	180
Corynebacterium xerosis	+	–	+	+	–	–	–	–	–	–	–					–					133

Nit = nitrate; Pyz = pyrrolidonyl acrylamidase; Pyra = pyrazinamidase; Pal = alkaline phosphatase; β gur = β glucuronidase; β gal = β galactosidase; α glu = α glucosidase; βNAG = N-acetyl-β glucosaminidase; esc = aesculin; ure = urease; gel = gelatin; glu = glucose; rib = ribose; xyl = xylose; man = mannitol; mal = maltose; lac = lactose; sac = sucrose; gly = glycogen; cat = catalase. Ref = reference; + = positive reaction; – = negative reaction; numbers indicate percentage of positive strains; w = weak reaction.

Table 4.29. API 20 Strep database biochemical results.

	vp	hip	aes	pyra	α-gal	β-gur	β-gal	Pal	Lap	ADH	rib	ara	man	sor	lac	tre	inu	raf	amd	glyg	Hem	Ref
Abiotrophia defectiva	–	–	–	+	+	–	+	–	+	–	–	–	–	–	+	+	–	+	+	–		
Carnobacterium inhibens	–	+	+	–	–	–	+	–	–	–	+	–	+	–	+w	+	+w	–	+	–		412
Carnobacterium piscicola	+	–	+	+	–	–	–	nt	–	+	+	–	+	+	v	+	+	v	–	+		542
Enterococcus avium	+	–	+	+	–	–	–	–	–	–	+	+	+	+	+	+	+	+	+	–	α	156
Enterococcus durans	+	–	+	+	+	–	+	–	+	+	+	–	–	–	+	+	–	–	+	–	α	156
Enterococcus faecalis	+	–	+	+	–	–	+	–	+	+	+	–	+	+	+	+	–	–	+	–	γ	156
Enterococcus faecium		+			–	–	+	–	+	+	+	+	+	–	–	–	+	–	–	v		
Granulicatella adiacens	–	–	–	+	–	+	–	–	+	–	–	–	–	–	–	–	+	–	–	–		
Granulicatella balaenopterae	–	–	+	–	–	–	–	–	+	+	–	–	–	–	–	+	–	–	–	–	α	478,773
Granulicatella elegans	–	+	–	+	–	–	–	–	+	+	–	–	–	–	–	–	–	+	–	–	α	653
Lactococcus garvieae	+	–	+	+	–	–	–	–	+	–	+	–	+	+	–	+	–	–	–	–	α	156
Lactococcus garvieae	+	–	+	+	–	–	–	–	+	+	+	–	+	–	–	+	–	–	–	–	α	237
Lactococcus lactis ssp. cremoris	+	–	+	–	–	–	+	–	+	+	+	–	–	–	–	+	–	–	–	–	γ	156
Lactococcus lactis ssp. lactis	+	–	+	+	–	–	–	–	+	+	+	–	–	–	–	+	–	–	+	–	γ	156
Lactococcus raffinolactis	+	–	+	–	+	–	–	–	+	–	–	–	–	–	+	+	–	+	+	–	γ	156
Streptococcus iniae Fish strains	–	–	+	+	–	+	–	+	+	70	+	–	46	–	–	+	–	+	+	+	αβ	183,223
Streptococcus iniae Human strains	–	–	+	–	–	–	–	+	+	–	+	–	+	–	–	+	–	–	+	–	αβ	223
Streptococcus iniae	–	–	+	+	–	+	–	+	+	+	+	–	+	–	–	+	–	–	+	+	β	235,848
Streptococcus (difficile) agalactiae	+	–	–	–	–	+	–	+	+	+	+	–	–	–	–	–	–	–	–	+		233
Streptococcus (difficile) agalactiae	+	+	–	–	–	+	–	+	+	–	+s	–	–	–	–	–	–	–	–	+		776
Streptococcus phocae	–	–	–	–	–	–	–	+	+	+	+	–	–	–	–	–	–	–	–	–		700
Streptococcus porcinus	+	–	+	+	–	+	–	+	+	+	+	–	+	+	–	+	–	–	v	–	β	175
Vagococcus fluvialis	+v	–	+	+	–	–	–	–v	+v	–	+	–	+	+	–	+	–	+	+	+	α	629
Vagococcus salmoninarum	+	+	+	+	–	–	–	+	+	–	70	–	–	70	–	+	–	–	+	–	α	542

VP = Voges-Proskauer; hip = hippurate; aes = aesculin; pyra = pyrrolidonylarylamidase; α-gal = α galactosidase; βgur = β glucuronidase; β-gal = β galactosidase; Pal = alkaline phosphatase; Lap = leucine arylamidase; ADH = arginine dihydrolase; rib = ribose; ara = arabinose; man = mannitol; sor = sorbitol; lac = lactose; tre = trehalose; inu = inulin; raf = raffinose; amd = amygdalin; glyg = glycogen; Hem = haemolysis; Ref = reference; nt = not tested. Numbers show percentage of positive strains. + = positive result; – = negative result; w = weak reaction; s = slow reaction. α = α haemolysis; β = β haemolysis; γ = no haemolysis.

Table 4.30. API Rapid ID32 Strep database biochemical results.

Organism	1 ADH	1.1 βglu	1.2 βgar	1.3 βgur	1.4 αgal	1.5 pal	1.6 rib	1.7 man	1.8 sor	1.9 lac	1A tre	1B raf	0 vp	0.1 appa	0.2 βgal	0.3 pyrA	0.4 βnag	0.5 gta	0.6 hip	0.7 glyg	0.8 pul	0.9 mal	0A mel	0B mlz	1C suc	1D lara	1E darl	0C mbdg	0D tag	0E βman	1F cdex	0F ure	Ref
Abiotrophia defectiva	–	–	50	–	+	–	–	–	–	50	+	50	–	50	+	75	–	–	–	–	+	+	–	–	+	–	–	–	v	–	–	–	*,421
Abiotrophia para-adiacens	–	+	–	50	–	–	–	–	–	–	–	–	–	+	–	–	50	–	–	–	–	+	–	–	+	–	–	–	–	–	–	–	421
Actinomyces marimammalium	–	–	+	–	–	v	–	–	–	+	–	–	–	+	+	–	+	+	–	v	–	+	–	–	–	–	–	–	–	–	–	–	370
Aerococcus viridans	–	70	3	30	60	–	28	75	25	79	91	42	–	–	10	83	–	–	92	10	10	95	–	–	+	–	–	65	–	–	–	–	*
Arcanobacterium bernardiae			–	–	–	–	+			–					–	–	–	–	–	+													480
Arcanobacterium haemolyticum			–	v			–			+					–	v				–													480
Arcanobacterium phocae			+	–	+w		+			v					+	–				+													480
Arcanobacterium pluranimalium	–	–	–	+	–	–	+	–	–	–	–	–	–	+	–	+	–	v	+	–	–	–	–	–	–	–	–	–	–	–	–	–	480
Arcanobacterium pyogenes			+	+	–		+			+					+	–		v	v	v													480
Atopobacter phocae	+	–	–	–	–	+	+	–	–	v	v	–	–	w–	–	w	–	–	–	+	+	+	–	–	v	–	–	–	–	–	+	–	479
Carnobacterium alterfundium	–					–	–	–					–			–	–	–				+			+								181
Carnobacterium divergens	+					–	+	–					+			+	+	+				+			+								181
Carnobacterium funditum	–					–	–	–					–			–	–	+				+			+								181
Carnobacterium gallinarum	+					–	+	–					+			+	+	+				+			+								181
Carnobacterium inhibens	–	+		+	–	–	+	+	–	+w	+	–	–	–	–	+	+	–	+	–	–	+	–	–	+	+	+	+	–	–	–	–	412
Carnobacterium mobile	+	–				–	–	–					–			+	–	–				–	w	w	+								181
Carnobacterium piscicola	+	–		+	–	–	+	+	v	–	+	w	+	+	–	+	+	+	–	–	–	+	w	w	+	–	–	w	+	–	+	–	479
Enterococcus faecalis	+	+	–	–	–	–	+	+	–	+	+	–	+	64	–	+	80	85	–	–	–	+	w	w	+	–	–	+	+	–	+	–	638
Erysipelothrix rhusiopathiae	42	–	+	–	–	28	–	–	–	75	–	–	–	64	–	+	80	85	–	–	–	+	–	–	+	–	–	–	+	–	+	–	*
Facklamia miroungae	+	–	–	–	–	–w	nr	–	–	–	+	–	–	+	–	+	–w	w	–	–	–	+	–	–	–	–	–	–	–	–	–	+	368
Granulicatella adiacens	–	+	–	30	–	–	–	–	–	–	+	–	–	+	–	70	25	–	–	–	–	+	–	–	+	–	–	–	+	–	–	–	*,421
Granulicatella balaenopterae	+	–	+	–	–	–	–	–	–	–	+	–	+	+	+	+w	+	–	–	–	+	+	–	–	+	–	–	+	+	–	–	+w	478,479
Granulicatella elegans	+	–	–	–	–	–	–	–	–	–	+	+	–	–	–	+	–	–	+	–	–	+	–	–	+	–	–	–	–	–	+	–v	421
Lactococcus garvieae	+	+	+	–	–	–	+	+	–	–	+	–	+	+	–	+	+	+	+	–	–	+	–	–	+	–	–	–	+	–	+	–	638
Lactococcus garvieae	+	+	–	–	–	–	35	75	–	50	+	–	+	+	–	74	10	–	–	–	–	75	–	50	50	–	–	85	50	–	50	–	*
Biotype 1								+								+	–								+				+		–		*,780
Biotype 2								+								+	–								+				+		+		*,780
Biotype 3								+								+	–								–				+		–		*,780
Biotype 4								+								–	–								–				–		–		*,780
Biotype 5								+								–	+								–				–		–		*,780

211

Organism	Ref
Biotype 6	*,780
Biotype 7	*,780
Biotype 8	*,780
Biotype 9	*,780
Biotype 10	*,780
Biotype 11	*,780
Biotype 12	*,780
Biotype 13	*,780
Lactococcus lactis lactis	*
Streptococcus agalactiae	*
Streptococcus agalactiae	γ242
Streptococcus agalactiae (difficile)	135
Streptococcus dysgalactiae ssp. dysgalactiae	727
Streptococcus iniae – fish	
Streptococcus iniae – human	
Streptococcus iniae – fish	135σ
Streptococcus iniae – fish	135#a
Streptococcus iniae – fish	135#b
Streptococcus parauberis	*,224, 745
Streptococcus uberis	*
Vagococcus fessus	369
Vagococcus fluvialis	177,369, 629
Vagococcus lutrae	477
Vagococcus salmoninarum	682,732
Vagococcus salmoninarum	682,732

Lancefield group L = positive. Streptex groups A–G = negative

ADH = arginine dihydrolase; βglu = β glucosidase; βgar = β galactosidase; βgur = β glucuronidase; αgal = α galactosidase; pal = alkaline phosphatase; rib = ribose; man = mannitol; sor = sorbitol; lac = lactose; tre = trehalose; raf = raffinose; vp = acetoin production (Voges-Proskauer); appa = alanine-phenylalanine-proline arylamidase; pyrA = pyroglutamic acid arylamidase; βgal = β galactosidase; βnag = N-acetyl-β glucosaminidase; gta = glycyl-tryptophane arylamidase; hip = hippurate; glyg = glycogen; pul = pullulan; mal = maltose; mel = melibiose; mlz = melezitose; suc = sucrose; lara = L-arabinose; darl = D-arabinose; tag = tagatose; βman = β mannosidase; cdex = cyclodextrin; ure = urease. Ref = reference; w = weak reaction; + = positive reaction; – = negative reaction; s = slow reaction. Numbers indicate percentage of positive strains. * = from API database; # = AHLDA 1722 strain from imported aquarium fish (^{a} = results at 37°C incubation; ^{b} = results from 25°C incubation); σ = strain from Queensland; γ = strains from mullet and seabream from Kuwait.

Table 4.31. API ZYM database results.

Well / Enzyme	1 con	2 alk	3 C4 est	4 C8 est lip	5 C14 lip	6 aryl	7 val	8 cys	9 try	10 chr	11 acp	12 np	13 α-gal	14 β-gal	15 β-glucr	16 α-glu	17 β-glu	18 N-aβglu	19 man	20 fuc	Temp °C	Time H	Inoc % NaCl	Strain	Ref
Organism																									
Actinobacillus delphinicola	–	5	1	3	–	3	–	–	–	–	5	0–2	–	–	–	–	–	–	–	–					263
Actinomyces viscosus	–	–	–	+	–	+	+	–	–	–	+	+	+	+	–	–	+	–	–	+					41
Aequorivita antarctica	–	+	+	+	–	+	–	–	–	–	–	+	–	+	–	–	–	–	–	–	20				113
Aerococcus viridans var. homari	–	+	+	+	–	+	–	–	–	–	–	+	–	–	–	–	–	–	–	–					41
Aeromonas hydrophila	–	5	–	4	2	4	1	–	3	–	5	1	–	3	–	1	–	4	–	–	25	4	0.85	NCTC 7810	135
Aeromonas janadaei	–	1	2	3	–	2	w	–	–	–	2	2	–	2	–	w	–	2	–	–	25	4	0.85	AHLDA 1718	135
Aeromonas salmonicida	–	3	–	1	–	2	w	–	1	–	5	1	–	–	–	–	–	2	–	–	25	20	0.85	AHLDA 1334	135
Aeromonas veronii sobria	–	5	–	4	1	3	w	–	3	–	5	1	1	4	–	1	–	5	–	–	25	4	0.85	AHLDA 1684	135
Alteromonas macleodii	–	4	–	2	–	2	2	–	4	–	4	1	–	–	–	–	–	–	–	–					283
Arcanobacterium phocae	–	4	2	2	–	3	–	1	v	–	5	–	3	1	–	5	–	–	–	–					636
Arthrobacter aurescens	–	–	–	+	–	+	–	+	+	–	+	+	–	+	+	+	+	–	+	–					41
Arthrobacter nasiphocae	–	+	1	–	–	–w	–	–	+	–	+	+	–	+	–	+	+	–	–	–	MI	MI	MI	CCUG 42953	182
Atopobacter phocae	–	+	1	1	–	–	–	–	–	–	+		–	v	–	–	–	–	–	–					479
Bordetella bronchiseptica	–	w–1	1	1	–	3–4	w	–	–	–	2	w–2	–	–	–	–	–	–	–	–	7	4	0.85	Various	135
Brevundimonas diminuta	–	+	+	+	–	+	–	–	+	+	+		–	–	–	–	–	–	–	–					685
Brevundimonas vesicularis	–	+	+	+	–	+	v	–	+	–	+		–	–	–	+	–	–	–	–					685
Carnobacterium alterfunditum	–	+	+	–	–	–	–	–	–	–	+	+	–	–	–	–	+	–	–	–	25		0.85	CCUG 34643	649
Carnobacterium divergens	–	+	+	–	–	–	–	–	–	–	w	+	–	–	–	–	–	–	–	–	25		0.85	CCUG 30094	649
Carnobacterium divergens 6251	–	–	+	w	–	–	–	–	–	–	–	w	–	–	+	–	–	–	–	–	25		0.85	6251	649
Carnobacterium funditum	–	+	+	–	–	–	–	–	–	–	+	+	–	–	–	–	+	–	–	–	25		0.85	CCUG 34644	649
Carnobacterium gallinarum	–	+	+	–	–	–	–	–	–	–	–	w	–	–	–	–	–	–	–	–	25		0.85	CCUG 30095	649
Carnobacterium inhibens	–	–	–	–	–	–	–	–	–	–	–	–	–	–	–	–	+	–	–	–					412
Carnobacterium inhibens	–	–	–	–	–	–	–	–	–	–	–	+	–	–	–	+	+	–	–	–	25		0.85	CCUG 31728	649
Carnobacterium mobile	–	+	+	–	–	–	–	–	–	–	–	w	–	–	–	–	–	–	–	–	25		0.85	CCUG 30096	649
Carnobacterium piscicola	–	–	–	–	–	–	–	–	–	–	–		–	–	+	–	–	–	–	–					479
Cellulophaga lytica	–	5	3	4	1	5	4	2	3	1	5	5	1	1	–	3	2	5	2	–	22	12	ASS		89
Chryseobacterium balustinum	–	5	1	2	2	5	5	3	2	1	5	5	–	–	–	3	3	2	–	–	25	12	ASS		89,557

Species	1	2	3	4	5	6	7	8	9	10	11	12	13	14	15	16	17	18	19	20	21				
Chryseobacterium gleum	–	5	2	3	1	5	5	2	2	2	5	5	5	–	–	–	4	5	4	–	–	22	12		557
Chryseobacterium indologenes	–	5	2	3	1	5	4	2	2	2	5	5	5	–	–	–	5	–	4	–	–	22	12		557
Chryseobacterium indoltheticum	–	5	2	3	1	5	4	1	–	–	4	4	4	–	–	–	4	–	3	–	–	22	12		557
Chryseobacterium meningosepticum	–	5	4	4	3	5	5	5	3	4	4	5	5	3	–	–	4	2	5	2	2	37	4	ASS	89
Chryseobacterium scophthalmum	–	5	3	4	1	5	4	2	4	1	5	5	5	–	–	–	3	1	2	–	–				557
Corynebacterium phocae	–	+	–	+	–	v	–	–	–	–	+w	–	+	–	–	–	+	+	–	–	–				613
Corynebacterium testudinoris	–	–	w	w	–	+	–	–	–	–	+w	–	–	–	–	–	v	+	–	–	–				180
Cytophaga allerginae	–	5	2	4	1	5	5	2	1	3	5	5	5	3	–	2	5	5	4	–	–	22	12	ASS	89
Cytophaga arvensicola	–	5	2	3	1	5	5	2	4	1	4	4	4	4	–	1	4	3	5	2	3	22	12	ASS	89
Cytophaga fermentans	–	5	1	3	–	5	2	–	–	–	5	5	5	5	–	–	2	5	5	–	–	22	12	ASS	89
Cytophaga hutchinsonii	–	4	2	4	1	4	4	3	–	–	3	2	2	–	–	–	–	–	–	–	–	22	12	ASS	89
Cytophaga latercula	–	5	2	4	1	5	5	3	5	5	4	4	4	–	–	–	–	–	3	–	–	22	12	ASS	89
Edwardsiella tarda	–	5	–	–	w	1	1	–	–	–	4	2	4	–	–	–	–	–	3	–	–	25	20	0.85	135 (AHLDA 135)
Empedobacter brevis	–	5	3	4	–	5	5	3	3	2	5	5	5	–	–	–	4	–	–	–	–	22	12	ASS	89
Enterovibrio norvegicus	–	+	–	–	+	+	w	–	–	–	+	w	w	–	nr	–	nr	nr	nr	–w	nr	28	48	2	741 (LMG 19839^T)
Facklamia miroungae	–	–w	–	–	–	w	5	2	–	–	w	w	w	–	–	–	5	–	–	–	–				368
Flavobacterium aquatile	–	+	+	+	1	+	5	+	+	2	1	2	2	–	–	–	5	–	–	–	–	22	12	ASS	89,92
Flavobacterium aquatile	–	+	+	+	–	+	4	2	+	–	+	+	+	–	–	–	+	–	+	–	–	18	48		603 (ATCC 11947)
Flavobacterium branchiophilum	–	5	2	3	–	5	5	2	2	–	4	4	4	–	–	–	1	–	–	–	–	25	12	ASS	92
Flavobacterium branchiophilum	–	+	+	+	–	+	+	+	+v	–	+	+	+	v	–	–	+	+	–	–	–	18	48	0.85	603
Flavobacterium branchiophilum	–	+	+	+	–	+	+	+	+	–	+	+	+	–	–	–	+	–	–	–	–	25	12	ASS	802
Flavobacterium columnare	–	5	2	3	–	4	4	1	3	1	3	3	3	–	–	–	3	–	1	–	–	25	12	ASS	88,92
Flavobacterium columnare	–	5	–	2	–	5	5	3	–	–	5	4	5	–	–	–	–	–	–	–	–				376
Flavobacterium columnare	–	+	+	+	–	+	+	+	–	–	+	+	+	–	–	–	–	–	–	–	–	18	48		603 (NCMB 2248)
Flavobacterium columnare	–	5	–	2	–	5	2	2	–	–	1	1	1	–	–	–	3	–	4	–	–	25	4	0.85	135 (AHLDA 1468)
Flavobacterium flevense	–	5	1	2	1	5	1	1	–	–	3	3	3	1	5	–	3	–	4	–	–	22	12	ASS	89,92
Flavobacterium frigidarium	–	5	–	2	–	5	5	1	1	–	5	5	5	–	–	–	–	–	–	–	–				376
Flavobacterium hydatis	–	5	2	4	1	4	5	2	2	1	4	5	4	–	–	–	4	–	2	–	–	22	12	ASS	89,92
Flavobacterium hydatis	–	5	–	1	–	5	5	3	3	–	5	5	5	–	–	–	5	–	5	–	–				376
Flavobacterium johnsoniae	–	5	1	3	1	5	5	2	2	1	4	5	4	3	–	–	4	1	4	1	–	25	12	ASS	89,92
Flavobacterium johnsoniae	–	5	3	4	1	4	5	3	3	2	5	4	5	5	5	1	5	2	5	–	–	25	12	ASS	89

continued

Table 4.31. *Continued.*

Well	1	2	3	4	5	6	7	8	9	10	11	12	13	14	15	16	17	18	19	20	Temp	Time	Inoc	Strain	Ref
Enzyme	con	alk	C4 est	C8 est lip	C14 lip	aryl	val	cys	try	chr	acp	np	α-gal	β-gal	β-glucr	α-glu	β-glu	N-aβglu	man	fuc	°C	H	% NaCl		
Organism																									
Flavobacterium johnsoniae	–	5	2	3	1	5	5	3	1	2	5	4	3	3	1	5	2	5	–	–	25	12	ASS		89
Flavobacterium johnsoniae	–	5	–	1	–	3	1	1	–	–	4	3	2	1	–	3	3	2	–	–	25		ASS		376
Flavobacterium johnsoniae	–	+	+	+	+	+	+	+	+	+	+	+	+	+	–	+	+	+	–	–	18	48		ATCC 17061	603
Flavobacterium pectinovorum	–	5	3	3	–	4	4	2	+	+	4	4	–	2	1	4	5	3	–	–	22	12		NCIMB 9059	92
Flavobacterium psychrophilum	–	5	2	3	1	5	1	–	–	–	3	3	–	–	–	–	–	–	–	–	20	12		NCIMB 1947	89,92
Flavobacterium psychrophilum	–	+	–	+	–	+	+	–	–	–	+	+	–	–	–	–	–	–	–	–	18	48		NCMB 1947	603
Flavobacterium saccharophilum	–	5	3	4	–	4	4	2	–	–	5	3	2	4	–	5	–	4	–	–	22	12	ASS	NCIMB 2072	92
Flavobacterium saccharophilum	–	5	–	1	–	2	1	2	–	–	5	2	–	1	–	1	–	1	–	–	22	12		NCMB 2072	376
Flavobacterium succinicans	–	5	3	3	–	4	4	2	1	–	5	5	–	–	–	4	2	4	–	–	22	12	ASS	NCIMB 2277	92
Flexibacter aggregans	–	5	3	4	1	5	4	3	3	–	5	5	1	3	–	4	2	5	3	3	22	12	ASS	NCIMB 1443	89
Flexibacter aurantiacus*	–	5	2	4	1	5	5	3	–	–	5	3	–	1	–	5	3	3	–	–	22	12	ASS	NCIMB 1382	89
Flexibacter aurantiacus*	–	5	2	4	1	5	5	3	–	–	5	4	–	–	–	5	3	4	–	–	22	12	ASS	NCIMB 1455	89
Flexibacter canadensis	–	5	2	3	1	5	5	3	4	1	5	5	4	1	–	5	3	4	–	3	22	12	ASS	ATCC 29591	89
Flexibacter flexilis	–	5	3	4	–	5	4	3	1	2	3	1	–	–	–	3	3	–	–	–	22	12	ASS	NCMB 1377	89
Flexibacter flexilis	–	+	–	+	–	+	+	+	–	+	+	+	–	–	–	+	+	–	–	–				ATCC 23079	603
Flexibacter litoralis	–	5	2	4	1	5	5	3	5	1	5	5	–	–	–	2	–	–	–	–	22	12	ASS	NCMB 1366	89
Flexibacter polymorphus	–	5	2	3	1	5	5	3	4	1	2	2	–	–	–	2	–	–	–	–	30	12	ASS	ATCC 27820	89
Flexibacter roseolus	–	4	2	3	1	3	3	2	1	3	2	2	–	–	–	–	–	–	–	–	25	12	ASS	NCIMB 1433	89
Flexibacter ruber	–	5	3	4	2	3	3	2	1	3	3	3	–	–	–	–	–	–	–	–	25	12	ASS	NCIMB 1436	89
Flexibacter sancti	–	5	–	2	–	4	1	–	4	4	4	5	4	4	–	4	3	4	–	2	22	12	ASS	NCIMB 1379	89
Flexibacter tractuosus	–	5	2	3	1	5	4	3	1	4	5	4	–	–	–	5	1	–	–	–	22	12	ASS	NCIMB 1408	89
Granulicatella balaenopterae	–	–	–	+	–	+	–	–	–	–	–	–	–	–	–	–	–	+	–	–					478
Hafnia alvei	–	5	–	w	1	3	2	–	w	–	4	1	–	2	–	2	1	w	–	–	25	4	0.85	AHLDA 1729	135
Listonella anguillarum	–	+	+	+	–	+	+	+	–	–	+	+	–	–	–	–	–	+	–	–	22	24		ATCC 14181	506
Listonella anguillarum	–	4	–	2	–w	1	–	–	–	3	–	–	–	–	–	–	–	–	–	–	25	24	2	NCMB 6	81
Listonella anguillarum	–	3	1	2	2	2	1	1	–	3	–	w	–	w	–	–	2	–	–	–	25	24	2	AHLDA 1730	135
Listonella pelagia II	–	+	+	+	–	+	–	–	+	–	+	+	–	–	–	–	–	–	–	–	22	24		NCMB 2253	506
Mannheimia haemolytica	–	5	1	1–w	–	1–3	0–1	–	–	–	5	1–2	–	1–2	–	–	–	–	–	0–2	37	4	0.85	Various	135

Species																								(25)	(12)	ASS	Strain	Ref
Marinilabilia salmonicolor	−	5	1	2	−	−	−	−	−	−	4	2	1	3	−	−	1	−	5	5	5	5	−	−	25	12	ASS	NCMB 2216	89
Moritella marina	−	+	+	+	−	−	−	−	−	−	+	+	+	−	−	−	−	−	−	−	−	−	−	−	22	24		NCMB 1144	506
Moritella viscosa	−	+	+	+	+	−	−	−	−	−	+	+	+	−	−	−	−	−	−	−	−	−	−	−	22	24		NCMB 13584	506
Myroides odoratimimus	−	+	+	+	−w	3	−w	−	−	−	−	−	−	−	−	−	−	−	−	−	−	−	−	−				NCTC 11180	774
Myroides odoratus	−	+	+	+	−w	1	−w	−	−	−	−w	+	+	−	−	−	−	−	−	−	−	−	−	−				NCTC 11036	774
Myroides odoratus	−	5	3	2	−	2	−	1	−	5	5	5	5	−	−	−	−	−	−	−	−	−	−	−	30	12		NCTC 11036	89
Pasteurella multocida	−	5	1−w	1−2	2	2	0−w	−	−	−	5	5	1	−	−	−	−	−	−	−	−	−	−	−	37	4	0.85	Various	135
Pasteurella skyensis	−	5	1	2	2	2	−	−	−	−	5	5	5	−	−	−	−	−	−	−	−	−	−	−	25	4d	1.5	NCTC 13204	100
Pedobacter heparinus	−	+	+	+	+	+	−	−	−	−	+	+	+	+	+	+	+	+	−	−	−	−	−	−	28	4	0.85	NCIB 9290	728
Pedobacter heparinus	−	5	2	4	−	4	−	2	1	−	4	2	2	−	−	−	−	−	−	−	−	−	−	−	22	12		NCIB 9290	89
Pedobacter piscium	−	+	+	+	+	+	+	+	+	+	+	+	+	+	+	+	+	+	−	−	−	−	−	−	28	4	0.85		728
Phocoenobacter uteri	−	5	1	1	−	+	−	−	−	−	5	5	5	−	−	−	−	−	−	−	−	−	−	−				NCMB 2184	266
Photobacterium damselae	−	+	+	+	−	+	−	−	−	−	+	+	+	−	−	+	−	−	−	−	−	−	−	−	22	24		AHLDA 1683	506
Photobacterium damselae	−	5	−	2	−	4	−	−	1	5	5	5	1	4	4	−	−	−	−	−	−	−	−	−	25	20	2		135
Photobacterium piscicida	−	+	+	+	−	+	−	−	−	w	+	+	+	−	−	−	−	−	−	−	−	−	−	−				AHLDA 192	855
Plesiomonas shigelloides	−	4	1	4	−	1	−	−	−	−	5	5	1	w	2	5	5	−	−	−	−	−	−	−	25	4			135
Pseudoalteromonas antarctica	−	+	+	+	+	+	+	+	+	+	+	+	+	+	+	+	+	w	−	−	−	−	−	−					115
Pseudoalteromonas citrea	−	−	−	−	+	+	+	+	−	−	−	−	−	−	w	w	−	−	−	−	−	−	−	−					285
Pseudoalteromonas rubra	−	4	2	2	1	1	4	1	2	−	4	+	2	2	−	−	−	−	−	−	−	−	−	−					283
Pseudomonas anguilliseptica	−	+	+	+	−	−	−	−	−	−	+	+	+	−	−	−	−	−	−	−	−	−	−	−				NCIMB 1949[T]	96
Pseudomonas plecoglossicida	−	+	+	+	+	+	+	+	−	−	+	+	+	−	−	−	−	−	−	−	−	−	−	−					582
Renibacterium salmoninarum	−	+	+	+	−	+	−	−	−	−	+	+	+	+	−	+	−	−	−	−	−	−	−	−					41
Shewanella algae	−	+	+	13	+	+	+	+	−	−	+	+	+	+	+	+	−	−	−	−	−	−	−	−					433
Shewanella baltica	−	−	−	−	−	w	w	−	−	−	−	−	−	−	w	w	−	−	−	−	−	−	−	−					851
Shewanella putrefaciens	−	+	+	+	67	78	89	−	+	+	+	+	+	+	+	+	−	−	−	−	−	−	−	−					433
Sphingobacterium multivorum	−	+	+	+	+	−	−	−	−	−	+	+	+	+	+	+	−	−	−	−	−	−	−	−					364
Sphingobacterium multivorum	−	5	4	4	1	5	1	1	−	2	4	5	5	4	3	5	3	−	−	−	−	−	−	−	37	4	ASS	NCTC 11343	89
Sphingobacterium spiritivorum	−	+	−	−	−	+	−	−	1	−	+	1	+	+	+	+	−	−	−	−	−	−	−	−				NCTC 11386	365
Sphingobacterium spiritivorum	−	5	2	4	1	1	5	5	5	2	4	5	5	3	2	5	3	4	−	−	−	−	−	−	30	12	ASS	NCTC 11386	89
Staphylococcus lutrae	−	5	5	5	−	5	−	5	5	−	5	5	5	1	−	−	−	−	−	−	−	−	−	−					264
Tenacibaculum maritimum	−	5	1	3−4	2−5	5	5	5	5	5	5	5	5	0−1	−	−	−	−	−	−	−	−	−	1	18	7d	2		605

continued

Table 4.31. *Continued.*

Organism	1 con	2 alk	3 C4 est	4 C8 est lip	5 C14 lip	6 aryl	7 val	8 cys	9 try	10 chr	11 acp	12 np	13 α-gal	14 β-gal	15 β-glucr	16 α-glu	17 β-glu	18 N-aβglu	19 man	20 fuc	Temp °C	Time H	Inoc % NaCl	Strain	Ref
Tenacibaculum maritimum	–	5	3	4	1	5	5	3	1	2	5	5	–	–	–	–	–	–	–	–	25	12	2		89
Tenacibaculum maritimum	–	5	2	3	1	4	5	2	3	1	5	5	–	–	–	–	–	–	–	–	22	12	0.85	ATCC 43398	743
Tenacibaculum ovolyticum	–	+	+	+	–	+	+	–	–	–	+	+	–	–	–	–	–	–	–	–					324
Vagococcus fessus	–	v	+w	+w	–	+	–	v	–	+	–	–	+	v	–	–	v	–	–	–					369
Vagococcus lutrae	–	–w	–	+	–	+	–	–	–	+	+	–	+	–w	–	+	+	+	–	–					477
Vibrio agarivorans	–	1	w	1	–	1	1	–	2	–	5	w	–	2	–	–	–	1	–	–	25	24	2	AHLDA 1732	135
Vibrio alginolyticus	–	5	–	2	w	4	–	–	–	–	5	1	–	–	–	–	–	1	–	–	25	4			135
Vibrio brasiliensis	–	+	+	+	+	+	+	–	–	–	–	+	–	+	–	–	–	–	–	–	25	24	1.5	LMG 20546^T	740
Vibrio calviensis	–	+	+	+	+	+	+	–	–	–	+	+	–	+	–	–	–	–	–					DSM 14347^T	216
Vibrio cholerae non-01	–	3	–	3	1	2	1	w	–	–	1	1	–	2	–	1	–	4	–	–	25	24	0.85	AHLDA 996	135
Vibrio diabolicus	–	3	2	3	1	3	–	–	2	2	3	–	–	–	–	–	–	–	–	–					635
Vibrio fischeri	–	+	–	+	–	+	–	–	–	–	+	+	–	–	–	–	–	+	–	–	22	24		NCMB 1281	506
Vibrio furnissii	–	3	–	2	1	4	2	2	–	–	1	1	–	1	–	–	–	4	–	–	25	24	0.85	ATCC 11218	135
Vibrio halioticoli	–	1	w	1	–	1	w	–	–	–	–	1	–	–	–	w	–	3	–	–	25	24	2	AHLDA 1734	135
Vibrio harveyi	–	3	2	2	–	1	w	–	–	1	5	1	–	–	–	–	–	1	–	–	25	4	0.85	ATCC 35084	135
Vibrio harveyi	–	4	2	2	–	1	w	–	–	2	3	1	–	–	–	–	–	2	–	–	25	4	2	ATCC 35084	135
Vibrio harveyi	–	+	+	+	–	–	–	–	–	–	+	+	–	–	–	–	–	–	–	–				ATCC 35084	847
Vibrio logei	–	+	+	+	–	+	–	–	–	–	+	+	+	+	–	–	–	4	–	–	22	24		ATCC 15382	506
Vibrio logei	–	+	+	+	+	+	+	–	–	–	+	+	+	+	–	–	–	–	–	–	22	24		NCMB 1143	506
Vibrio logei	–	+	+	+	–	+	–	–	–	–	+	+	–	–	–	–	–	–	–	–	22	24		ATCC 29985	506
Vibrio mediterranei	–	5	2	2	2	1	–	–	–	–	4	2	–	–	–	–	–	2	–	–	25	24	2	AHLDA 1733	135
Vibrio mimicus	–	5	–	2	–	4	1	–	–	w	5	1	–	1	–	–	–	3	–	–	25	4	2	AHLDA 1654	135
Vibrio mytili	–	–	2	3	1	3	–	–	1	–	2	–	–	–	–	–	–	2	–	–				CECT 632	635
Vibrio natriegens	–	+	+	+	–	+	–	–	–	–	+	+	+	–	–	–	–	–	–	–	22	24		NCMB 1900	506
Vibrio neptunius	–	+	+	+	+	+	+	–	–	–	+	+	–	–	–	–	–	+	–	–	22	24	1.5	LMG 20536^T	740
Vibrio nereis	–	–	1	1	–	2	–	–	–	–	–	–	–	–	–	–	–	–	–	–				LMG 3895	635
Vibrio ordalii	–	+	+	+	–	+	–	–	–	–	+	+	–	–	–	–	–	+	–	–	22	24		NCMB 2167	506

	1	2	3	4	5	6	7	8	9	10	11	12	13	14	15	16	17	18	19	20	Temp	Time	%NaCl		
Vibrio ordalii	–	5	–	w	–	4	w	–	–	–	w	w	–	w	–	–	–	–	–	–	25	4	2	DF 3K	428
Vibrio orientalis	–	+	+	+	+	+	+	–	–	–	+	+	+	+	–	+	–	–	–	–	22	24		NCMB 2195	506
Vibrio pacinii	–	+	+	+	–	+	+	w	–	–	+	+	66	+	+	+	–	–	–	–	28	4	1.5	LMG 1999^T	306
Vibrio parahaemolyticus	–	5	–	3	3	5	5	–	1	–	5	1	–	+	–	–	–	–	–	–	25	20	2	ATCC 43996	135
Vibrio parahaemolyticus	*	+	–	w	–	–	–	–	–	–	+	+	–	+	–	–	–	–	–	–				ATCC 17802	499
Vibrio parahaemolyticus	–	+	–	–	–	+	+	–	+	–	+	+	+	+	–	–	–	–	–	–				ATCC 27969	499
Vibrio proteolyticus	–	2	1	2	2	–	–	–	–	–	w	w	–	w	–	–	–	–	–	–	25	24	2	AHLDA 1735	135
Vibrio rotiferianus	–	+	+	+	+	+	+	+	+	–	+	+	+	–	–	–	–	–	–	–	25	24	2	LMG 21460^T	305
Vibrio salmonicida	–	+	+v	+	–	+	+	–	–	–	+	+	–	–	–	–	–	–	–	–	22	24		NCMB 2262	506
Vibrio splendidus I	–	+	+	+	+	+	+	+	+	–	+	+	+	+	+	+	+	–	–	–	22	24		NCMB 1^T	506
Vibrio splendidus I	–	+	+	+	-v	+	+	–	+	–	+	V	–	+	+v	+v	+	–	–	–					281
Vibrio splendidus II	–	+	+	+	+	+	+	+	+	–	+	+	–	+	–	–	–	–	–	–	22	24		NCMB 2251	506
Vibrio tapetis	–	+	+	+	+	+	+	–	–	–	+	+	–	+	–	–	–	–	–	–	22	18	2	B1090^T	587
Vibrio tubiashii	–	2	1	3	2	2	2	–	–	–	–	–	–	–	–	–	–	–	–	–				LMG 10936	635
Vibrio tubiashii	–	+	+	+	+	+	+	+	+	–	+	+	–	+	–	–	–	–	–	–	22	24		NCMB 1340^T	506
Vibrio tubiashii	–	+	+	+	+	+	+	+	+	–	+	+	–	+	–	–	–	–	–	–	22	24		NCMB 1340	506
Vibrio vulnificus biovar I	–	5	w	1	w	w	1	–	–	–	w	1	–	w	–	–	–	–	–	–	25	4	0.85	AHLDA 1716	135
Vibrio vulnificus biovar I	–	5	1	1	w	w	w	w	–	–	–	w	–	w	–	–	–	–	–	–	25	4	2	AHLDA 1716	135
Vibrio wodanis	–	+	83	+	86	–	–	–	–	–	69	+	–	77	–	–	–	–	–	–	22	24		NCIMB 13582	506
Vibrio xuii	–	+	+	+	+	1	1	–	–	–	+	+	–	+	–	–	–	–	–	–	25	24		LMG 21346^T	740
Yersinia ruckeri	–	1	–	w	1	–	–	–	–	–	2	1	–	3	–	1	–	–	3	–	25	20	1.5	AHLDA 1313	135
Zobellia galactanivorans						3	3				3	3		2						3	30	7d	2.5		61
Zobellia uliginosa						1	1				1	1		1						1	30	7d	2.5		61

*Isolates NCMB 1382 and NCMB 1455 previously identified as *Flexibacter aurantiacus* are now thought to be *Flavobacterium johnsoniae*.

1 = control well; 2 = 2-naphthyl-phosphate; 3 = 2-naphthyl-butyrate; 4 = 2-naphthyl-caprylate; 5 = 2-naphthyl-myristate; 6 = L-leucyl-2-naphthylamide; 7 = L-valyl-2-naphthylamide; 8 = L-cystyl-2-naphthylamide;

9 = N-benzoyl-DL-arginine-2-naphthylamide; 10 = N-glutaryl-phenylalanine-2-naphthylamide; 11 = 2-naphthyl-phosphate; 12 = naphthol-AS-BI-phosphate; 13 = 6-Br-2-naphthyl-α-D-galactopyranoside; 14 = 2-naphthyl-β-D-galactopyranoside;

15=naphthol-AS-BI-β-D-glucuronide; 16 = 2-naphthyl-α-D-glucopyranoside; 17 = 6-Br-2-naphthyl-β-D-glucopyranoside; 18 = 1-naphthyl-N-acetyl-β-D-glucosaminide; 19 = 6-Br-2-naphthyl-α-D-mannopyranoside; 20 = 2-naphthyl-α-L-fucopyranoside.

Temp = temperature of incubation; Time = time of incubation; H = hours; d = days; Inoc = inoculum; %NaCl indicates final NaCl concentration in the inoculum. MI = Manufacturer's Instructions; nr = not reported; Ref = reference number.

Ref 603 classified a reaction of 2 or less as negative.

Table 4.32. Strains quoted in references.

Organism	Strain number	Reference number
Actinobacillus scotiae	NCTC 12922	265
Actinomyces marimammalium	CCUG 41710^T = CIP 106509^T	370
Aequorivita antarctica	ACAM 640^T, DSM 14231^T	113
Aequorivita crocea	ACAM 642^T, DSM 14239^T	113
Aequorivita lipolytica	ACAM 641^T, DSM 14236^T	113
Aequorivita sublithincola	ACAM 643^T, DSM 14238^T	113
Aeromonas allosaccharophila HG15	CECT 4199	427
Aeromonas bestiarum HG2	CDC 9533-76	1,21,427
Aeromonas caviae HG4	ATCC 15468	427,21
Aeromonas encheleia HG16	CECT 4342 = LMG 16330	241,379,427
Aeromonas eucrenophila HG6	ATCC 23309, CDC 0859-83	427,21
Aeromonas eucrenophila	LMG 3774 = NCMB 74	379
Aeromonas eucrenophila	LMG 13057	379
Aeromonas hydrophila dhakensis	P21^T = LMG 19562^T = CCUG 45377^T	383
Aeromonas hydrophila hydrophila	ATCC 7966^T = LMG 2844^T	383
Aeromonas hydrophila HG1	ATCC 7966^T, CDC 9079-79	21,427,818
Aeromonas hydrophila HG3	CDC 0434-84	427
A. hydrophila anaerogenes = *A. caviae*	ATCC 15468	818
A. hydrophila anaerogenes	ATCC 15467	Taxonomy 2000
Aeromonas janadaei HG 9	CDC 0787-80, ATCC 49568	143,427
Aeromonas media HG 5B	CDC 9072-83, CDC 0435-84	427,21
Aeromonas media HG 5A	CDC 9072-83, CDC 0862-83	427,21
Aeromonas popoffii	LMG 17541^T	380
Aeromonas salmonicida achromogenes	NCMB 1110	450,475
Aeromonas salmonicida pectinolytica	DSM 12609^T = 34 mel	615
A. salmonicida salmonicida	ATCC 14174	450,818
A. salmonicida salmonicida	SVLT -1, -2, -5, -6, Non-pigmented strains	450
A. salmonicida salmonicida	NCMB 1102	475
Aeromonas salmonicida HG3	CDC 0434-84	21
Aeromonas schubertii HG12	CDC 2446-81, ATCC 43700, formerly Enteric group 501	348,427
Aeromonas sobria HG7	CIP 7433, CDC 9538-76	21,427,818
Aeromonas veronii sobria HG8/10	CDC 0437-84	427
Aeromonas spp. HG11	CDC 1306-83	427
Aeromonas trota	ATCC 49657^T = LMG 12223^T HG 14	142,427
Aeromonas veronii	ATCC 35604^T = CDC 1169-83	21,347
Aeromonas veronii	HG 8, CDC-0437-84	21
Alteromonas aurantia	ATCC 33046, NCMB 2052	286
Alteromonas citrea	NCMB 1889	285
Alteromonas rubra	NCMB 1890	283
Arcanobacterium phocae	DSM 10002^T, M1590/94/3^T	636
Arcanobacterium pluranimalium	CCUG 42575^T = CIP 106442	480
Arthrobacter rhombi	CCUG 38813^T	600
Atopobacter phocae	CCUG 42358^T = CIP 106392	479
Bordetella bronchiseptica	ATCC 19395, ATCC 4617, NCTC 8344	102
Brevundimonas diminuta	ATCC 11568, LMG 2089 = CCUG 1427	685
Brevundimonas vesicularis	ATCC 11426 = CCUG 2032 = LMG 2350	685

219

Table 4.32. *Continued.*

Organism	Strain number	Reference number
Brucella maris biovar I	NCTC 12890, 2/94	404
Brucella maris biovar II	NCTC 12891, 1/94	404
Brucella maris biovar III		247,404
Carnobacterium inhibens	CCUG 31728^T, strain K1	411,412
Carnobacterium piscicola	ATCC 35586	73,353,682
Chromobacterium violaceum		482
Corynebacterium aquaticum	RB 968 BA	73,133
Edwardsiella ictaluri	CDC 1976-78, ACC 33202	334
Edwardsiella tarda	ATCC 15947^T	374,640
Enterococcus faecalis	ATCC 19433	638
Enterovibrio norvegicus	LMG 19839^T, CAIM 430^T	741
Flavobacterium aquatile	NCIB 8694^T, LMG 4008^T	89,92
Flavobacterium branchiophilum	ATCC 35035, BGD-7721	802
Flavobacterium frigidarium	ATCC 700810 = NCIMB 13737	376
Flavobacterium gillisiae	ACAM 601^T	533
Flavobacterium hibernum	ATCC 51468 = ACAM 376^T	532
Flavobacterium hydatis (*C. aquatilis*)	ATCC 29551	720
Flavobacterium tegetincola	ACAM 602^T	533
Flexibacter polymorphus	ATCC 27820	494
Granulicatella balaenopterae	CCUG 37380^T, M1975/96/1	478
Iodobacter fluviatilis	NCTC 11159^T	502
Janthinobacterium lividum	NCIMB 9230, NCIMB 9414, DSM 1522	496
Lactococcus garvieae	NCDO 2155	224,638
Lactococcus garvieae	ATCC 49156^T	236,464,682
Lactococcus lactis	ATCC 19435	731
Lactococcus piscium	NCFB 2778	835
Listonella anguillarum	ATCC 14181	506
Listonella anguillarum	NCMB 6 = ATCC 19264 (Bagge and Bagge strain)	341
Listonella anguillarum	NCMB 407 = PL 1	341
Listonella anguillarum	NCMB 571 (Hoshina)	341
Listonella anguillarum	NCMB 828 = ATCC 14181 (strain 4063, Smith)	341
Listonella pelagia I (said to be *V. natriegens*)	NCMB 1900^T	506
Listonella pelagia II	NCMB 2253	506
Mesophilobacter marinus	IAM 13185	583
Moritella marina	NCMB 1144^T = ATCC 15381	82,506,766
Moritella viscosa	NCIMB 13584^T = NVI 88/478^T	82,506
Mycoplasma crocodyli	ATCC 51981	441
Mycoplasma mobile	163K	439
Myroides odoratimimus	NCTC 11180, LMG 4029	565
Myroides odoratus	NCTC 11036, LMG 1233	565
Nocardia seriolae	JCM 3360	455
Nocardia spp., Australian strain	98/1655	117
Pantoea agglomerans	ATCC 27155^T, NCTC 9381	249,291
Pantoea agglomerans	ATCC 27155^T, ATCC 12287	325
Pantoea dispersa	ATCC 14589^T	291
Pasteurella skyensis	NCTC 13204^T, NCIMB 13593^T	100
Phocoenobacter uteri	NCTC 12872	266
Photobacterium angustum	NCIMB 1895	745
Photobacterium damselae ssp. *damselae*	ATCC 33539 = NCIMB 2184^T	289,504,506,705,745
Photobacterium damselae ssp. *damselae*	ATCC 35083	268,745
Photobacterium damselae ssp. *piscicida*	ATCC 17911 & NCIMB 2058	289,745
Photobacterium damselae ssp. *piscicida*		751
Photobacterium damselae ssp. *piscicida*	ATCC 29690, ATCC 17911	518
Photobacterium fischeri	NCMB 1281^T = ATCC 7744	506
Photobacterium fischeri	ATCC 25918	745
Photobacterium fischeri	ATCC 7744	340,818
Photobacterium iliopiscarium	ATCC 51760	599,767

continued

Table 4.32. *Continued.*

Organism	Strain number	Reference number
Photobacterium leiognathi	LMG 4228, NCIMB 1895	745
Photobacterium logei	ATCC 15382 = NCMB 1143 = PS 207	506
Photobacterium logei	NCMB 1143	506
Photobacterium logei	ATCC 29985^T	506
Photobacterium logei	ATCC 15382	82
Photobacterium logei	NCIMB 2252, ATCC 29985	506
Providencia friedericiana	DSM 2620	559
Pseudoalteromonas antarctica	CECT 4664^T, NF3	115
Pseudoalteromonas citrea	ATCC 29719^T, NCMB 1889	285
Pseudoalteromonas elyakovii	ATCC 700519^T, KMM 162^T	679
Raoultella planticola	ATCC 33531	228
Raoultella terrigena	ATCC 33257	228
Renibacterium salmoninarum	ATCC 33209	671
Serratia fonticola	ATCC 29844	290
Shewanella algae	LMG 2265, IAM 14159	433,792,851
Shewanella baltica	NCTC 10735, DSM 9439, CECT 323, IAM 1477, LMG 2250	851
Shewanella colwelliana	ATCC 39565	815
Shewanella frigidimarina	ACAM 591^T	112
Shewanella gelidimarina	ACAM 456^T	112
Shewanella (Alteromonas) hanedai	ATCC 33224	409
Shewanella japonica	KMM 3299, LMG 19691 = CIP 106860	397
Shewanella oneidensis	ATCC 700550^T	782
Shewanella pealeana	ANG-SQ1^T	492
Shewanella putrefaciens	ATCC 8071	433,792
Shewanella woodyi	ATCC 51908^T, MS32	112
Sphingobacterium spiritivorum	NCTC 11386	365
Staphylococcus delphini	DSM 20771^T	778
Staphylococcus lutrae	DSM 10244, M340/94/1	264
Streptococcus (difficile) agalactiae	ND 2-22, CIP 103768	233
Streptococcus (difficile) agalactiae	LMG 15977	776
Streptococcus iniae	ND 2-16, CIP 103769	233
Streptococcus iniae	ATCC 29177	626
Streptococcus parauberis	NCDO 2020	224
Streptococcus phocae	NCTC 12719, 8399 HI	700
Streptococcus uberis	NCDO 2038	224
Tenacibaculum (Flexibacter) maritimum	NCMB 2154^T	801
Tenacibaculum (Flexibacter) ovolyticum	NCIMB 13127 = EKD002	324
Vagococcus fessus	CCUG 41755	369
Vagococcus fluvialis	NCDO 2497, NCFB 2497	177,629
Vagococcus lutrae	CCUG 39187	477
Vagococcus salmoninarum	NCFB 2777	682,807
Vibrio aerogenes	ATCC 700797 = CCRC 17041, FG1	692
Vibrio aestuarianus	ATCC 35048, LMG 7909	149,747
Vibrio agarivorans	CECT 5084, CECT 5085^T = DSM 13756	514
Vibrio agarivorans	AHLDA 1732	135
Vibrio brasiliensis	LMG 20546^T	740
Vibrio calviensis	RE35/F12^T = CIP 107077^T = DSM 14347^T	216
Vibrio (carchariae) harveyi	ATCC 35084^T	11,135
Vibrio cincinnatiensis	ATCC 35912	120
Vibrio diabolicus	CNCM I-1629 = HE800	635
Vibrio diazotrophicus	ATCC 33466	319
Vibrio fluvialis	ATCC 49515, NCDO 2497 = NCFB 2497	177,732
Vibrio fluvialis	NCTC 11327	123,485
Vibrio furnissii	ATCC 35016 = CDC B3215	123
Vibrio furnissii	ATCC 11218	135
Vibrio furnissii	Group F	687

Table 4.32. *Continued.*

Organism	Strain number	Reference number
Vibrio gazogenes	ATCC 29988	818
Vibrio halioticoli	IAM 14596^T	678
Vibrio harveyi	ATCC 14126^T	11,818
Vibrio hollisae	KUMA871, ATCC 33564^T	346,580
Vibrio lentus	CECT 5110^T = DSM 13757 = 40MA	513
Vibrio mediterranei	CECT 621^T, LMG11258	149,631
Vibrio metschnikovii	NCTC 8563	819
Vibrio metschnikovii	NCTC 8443	818
Vibrio mimicus	ATCC 33653^T	210,507,818
Vibrio mytili	CECT 632T	635
Vibrio natriegens	ATCC 14048	44
Vibrio navarrensis	CIP 1397-6	767,768
Vibrio neptunius	LMG 20536^T	740
Vibrio nereis	LMG 3895^T	635
Vibrio ordalii	NCMB 2167^T, ATCC 33509 = DF$_3$K = Dom F$_3$	506,680
Vibrio orientalis	NCMB 2195^T	506
Vibrio pacinii	LMG 1999^T	306
Vibrio pectenicida	CIP 105190^T, A365	470
Vibrio penaeicida	JCM 9123, KH-1, IFO 15640	388
Vibrio proteolyticus	AHLDA 1735	135
Vibrio proteolyticus	CW8T2	788
Vibrio rotiferianus	LMG 21460^T	305
Vibrio salmonicida	NCMB 2262^T	232,506
Vibrio salmonicida	NCMB 2245	506
Vibrio salmonicida	90/1667-10c	506
Vibrio scophthalmi	A089, CECT 4638^T	254,149
Vibrio shilonii (possibily *V. mediterranei*)	ATCC BAA-91^T = DSM 13774 = AK-1	59,458
Vibrio splendidus I	NCMB 1^T (= ATCC 33125^T)	149,506
Vibrio splendidus I	ATCC 33125	281,620
Vibrio splendidus II	NCMB 2251	149,506
Vibrio tapetis	B1090^T, CECT 4600	108,587
Vibrio tubiashii	NCMB 1340^T	506
Vibrio tubiashii	ATCC 19109^T	321
Vibrio tubiashii	LMG 10936T	635
Vibrio vulnificus		149
V. vulnificus biovar I non-serovar E f	ATCC 27562	26,746
V. vulnificus biovar 2, serovar E a	ATCC 33187	98
V. vulnificus biovar 2, serovar E b	ATCC 33149	98
V. vulnificus biovar 2, serovar 04 b	ATCC 33149	356
V. vulnificus biovar 2, serovar E c	NCIMB 2138	98
V. vulnificus biovar 2 serovar E d	NCIMB 2137	98
V. vulnificus biovar 2, serovar E e	NCIMB 2136	98
V. vulnificus biovar 2, serovar E g	Taiwanese strains	26
V. vulnificus biovar 2	ATCC 33148	746
Vibrio wodanis	NCIMB 13582 = NVI 88/441^T	506
Vibrio xuii	LMG 21346^T	
Yersinia rohdei	ATCC 43380, CDC 3022-85	12
Yersinia aldovae	ATCC 35236, CDC 669-83	85

5

Technical Methods

_____

5.1 Total Bacterial Count (TBC)

Diagnostic, research, or industry-based laboratories often monitor pond or tank water for bacterial load as an indicator of water quality, and hence as a disease indicator. This is done by a test known as the TBC.

There are a number of methods for TBCs and these may be found in most microbiology textbooks. A basic method for TBCs is described here.

Generally, TBCs are performed using a good general-purpose medium that will support the growth of the majority of the organisms that are expected to be cultured. A selective medium for *Vibrio* species such as TCBS is often used as a culture medium for TBCs. However, this medium has a tendency to produce a lower count than an equivalent sample inoculated to MSA-B plate (or BA in the case of freshwater samples) even when the majority of bacteria cultured are *Vibrio* spp. Therefore, it is not the best primary culture medium.

The time between collection of the water sample and testing is important. Bacteria in the water sample will multiply within a few hours at room temperature. Thus, if there is a delay of more than 1–2 h between collection and testing, an inaccurate assessment of the bacterial load of the original water sample will be obtained. The water sample must be stored at 4°C either during transport or while waiting to be tested in the laboratory. There is no effect on bacterial count due to the size of the collection container (Simon and Oppenheimer, 1968).

There is no predetermined optimal bacterial load of pond or tank water, and this will vary anyway according to the number of fish in the water, that is, the stocking density. It is suggested that laboratories regularly monitor a particular tank or set of tanks and keep a record of their results. The bacterial load or TBC is then related back to the health of the fish. If a daily log or graph is kept of the TBC against the health of the fish, then the laboratory can eventually determine the optimal bacterial load for healthy fish, as opposed to the load seen when the fish appear diseased.

Suggested sites for monitoring bacterial load are the intake water supply, the filtered water supply, fish tanks and the feed supply. The sediment at the bottom of the tank will have a greater bacterial count than the water body, particularly when there are dead or diseased larvae or fish.

Research in an oyster hatchery indicated that four main genera were involved in the bacterial load; these were *Vibrio* species, *Alteromonas* species, *Pseudomonas* species and *Flavobacterium* species. *Vibrio* species were the predominant organism. Virulence testing showed that a third of these *Vibrio* species, which included *Listonella anguillarum*, and two of the ten *Alteromonas* species were pathogenic. The isolates were not identified to species level. Infection was not seen in the larvae when they were dosed with the non-pathogenic bacteria. However, fatal infection was produced when the bacterial load of the pathogenic bacteria was 1×10^7–5×10^7. At a lower dose of 1×10^5–5×10^5, infection still occurred, but took 2–3 days longer to develop and for clinical signs to be seen. Thus, infection rates are related to the bacterial load of pathogenic bacteria (Garland *et al.*, 1983).

Bacteria are part of the diet of marine filter feeders. In another study on healthy oysters (*Crassostrea gigas*) and horse mussel (*Modiolus modiolus*), the total bacterial load of normal flora in the haemolymph and soft tissue was 2.6×10^4 colonies/ml, and 2.9×10^4 colonies/g, respectively. The predominant flora and percentage of the total flora were *Pseudomonas* spp. (61.3%), *Vibrio* spp. (27%) and *Aeromonas* spp. (11.7%) in the haemolymph, with the predominant bacteria in the soft tissues being *Vibrio* spp. (38.5%), *Pseudomonas* spp. (33%) and *Aeromonas* spp. (28.5%). When the oysters and horse mussel were challenged in virulence studies using *Vibrio salmonicida*, the TBC increased to 10^5 colonies/ml in the haemolymph and 6×10^7 colonies/g in the soft tissue (Olafsen *et al.*, 1993).

Table 5.1 indicates the bacterial counts obtained in both studies.

The microflora in the intestines of healthy jack mackerel (*Trachurus japonicus*) was found to be 4.6×10^6 colonies/g and consisted of *Vibrio* species. The total bacterial load in the stomach was 2.6×10^5 colonies/g, with the predominant flora being *Vibrio* spp., *Achromobacter* spp., with smaller counts of *Pseudomonas* spp., *Flavobacterium* spp., *Corynebacterium* spp., *Bacillus* spp. and *Sarcina* spp. (Aiso *et al.*, 1968).

Equipment

A pipette or pipettes capable of measuring 100 µl to 1000 µl
1.5 ml microfuge tubes

An agar medium that supports the growth of the target bacteria, such as BA for freshwater or MSA-B, or MA 2216 for salt water
Inoculating loop
Bunsen burner
Incubator set to 25°C

Water

Collect a volume of water
Do dilutions at neat, 10^{-1}, 10^{-2}, 10^{-3}

Dilutions

Neat = 100 µl to a plate and lawn inoculate (spread evenly over the plate)
10^{-1} = 900 µl of sterile distilled water + 100 µl of sample
10^{-2} = 900 µl of sterile distilled water + 100 µl of dilution 10^{-1}
10^{-3} = 900 µl of sterile distilled water + 100 µl of dilution 10^{-2}

Method

Place 100 µl (or 10 µl) of each dilution into the centre of an agar plate. Use a separate plate for each dilution. Spread each inoculum evenly over each plate using a sterile bent glass rod, or a flame-sterilized inoculating loop. It is important to ensure that the inoculum is spread evenly over the plate so that individual colonies are obtained for counting. Clumps of colonies will lead to erroneous results.

Table 5.1. Total bacterial counts at sites in oyster hatcheries: indication of healthy and diseased states.

Sample site	Total bacterial count (cells/ml)
Seawater (filtered, no larvae)[a]	10^3–10^{3} [a]
Tank water[a]	10^4–10^5 [a]
Sediment[a]	10^7–10^8 cells/g[a]
Haemolymph healthy oyster[b]	2.6×10^4 [b]
Soft tissue healthy oyster[b]	2.9×10^4 [b]
Load of pathogenic bacteria at which mortality is seen. Rapid onset of infection[a]	1×10^7–5×10^7 [a]
Load of pathogenic bacteria at which mortality is seen. Slower onset of infection[a]	1×10^5–5×10^5 [a]
Haemolymph in diseased oyster[b]	10^5 [b]
Soft tissue in diseased oyster[b]	6×10^7 [b]

[a]Data from Garland *et al.* (1983); [b]Data from Olafsen *et al.* (1993).

Incubation

Place plates in a sealed plastic container and incubate at room temperature or an incubator set to 25°C.

Counting the bacterial colonies

Examine the plates at 24 and 48 h.
Count the number of colonies on a plate: using a felt pen, mark each colony from the back of the Petri dish to assist with counting
Count = N

Calculation to obtain CFU/ml

If 100 μl is placed on to a plate and then lawn inoculated, this is a plate dilution of 10^{-1}. If 10 μl were placed on a plate, this would be a plate dilution of 10^{-2}.

CFU/ml = $N \times$ dilution $\times$ plate dilution.

Thus if 268 colonies are counted from the 10^{-2} dilution and 100 μl was inoculated to the plate, the count = $268 \times 10^2 \times 10 = 2.68 \times 10^5$ CFU/ml.

5.2 Microscopy

Most laboratory staff will be familiar with the use of a light microscope. However, for those who are new to the use of a microscope, some basics are explained here.

Before use, a microscope should be set up for Koehler illumination. This ensures that the light entering the microscope is focused so as to produce an evenly illuminated field.

- Place a stained smear on to the stage. Focus using the ×10 objective.
- Close the field diaphragm – this is the diaphragm at the base of the microscope.
- Close the aperture of the iris diaphragm. This is the aperture just under the stage that the glass slide rests on.
- There will be a small circle of light. Centre this circle of light using the condenser centring screws.

- Next, adjust the height of the condenser under the stage so that the edge of the circle of light is in sharp focus.
- Now open the field diaphragm so that it just disappears from view.
- Open the aperture diaphragm to suit the contrast required.

Staff who are new to the use of a microscope should remember that examination of a Gram-stained smear of bacterial cells is best observed using the oil immersion lens, which together with the ×10 objective gives a magnification of 1000 times. The aperture on the substage condenser should be open to half its setting with the light intensity set so that it is comfortable for the eyes. For the examination of wet preparations this aperture will need to be almost closed so that a good contrast is obtained and cellular structures are defined.

Only use lens tissue paper for cleaning the objective lens. Always remove the oil from the lens after use, as the oil can damage the structures within the objective. If oil does get on to any of the lenses, it can be removed by soaking some lens tissue with petroleum spirit and gently wiping across the lens until the oil has been removed.

5.3 Storage of Isolates

Isolates may be stored at −80°C in Lab Lemco broth supplemented with 10% glycerol. For marine organisms use marine broth 2216 supplemented with 30% glycerol (Bowman and Nichols, 2002).

There are a number of different media for freeze-drying of bacteria. One method is to suspend organisms in 1 ml of inositol horse serum in Wheaton serum bottles or appropriate container depending on freeze-drying equipment. Snap freeze in liquid nitrogen and follow instructions for freeze-drying equipment (see Chapter 7).

An analysis of different cryoprotectants tested for *Flavobacterium columnare*, *F. psychrophilum* and *Tenacibaculum maritimum* recommends the use of a medium containing two-thirds Brucella broth (Difco) and one-third horse or fetal calf serum (Desolme and Bernardet, 1996).

6

Techniques for the Molecular Identification of Bacteria

The polymerase chain reaction (Saiki *et al.*, 1985; Mullis and Faloona, 1987) has wide applications in both diagnostic and research laboratories. It is routinely being used in the diagnosis and identification of bacteria. Likewise, 16S rDNA sequencing is being used more widely and is a useful tool for the identification of bacteria that are difficult to identify by biochemical tests. This chapter covers the basics of both these techniques.

6.1 Molecular Identification by PCR Using Specific Primers

A number of specific primers for the detection of aquatic organisms by PCR have been reported. However, most, if not all, of these primers will undergo a long period of validation before they are routine in a diagnostic laboratory (Bader *et al.*, 2003). It is suggested they be used in tandem with biochemical identification methods.

The PCRs that have been reported for aquatic organisms are summarized in Table 6.1. Each forward and reverse primer is listed together with the recommended annealing temperature, number of cycles and expected product length in base pairs. The concentration of reagents is not given here because of the wide variety used for the individual PCRs. The reader is advised to either refer to the reference or optimize the particular PCR for their laboratory conditions and equipment.

DNA extraction techniques, basic PCR protocols and preparation of reagents are covered in this chapter.

Aeromonas *spp. PCR*

The primers used for *Aeromonas hydrophila* (Nielsen *et al.*, 2001) also produce a band at 685 bp with an Australian strain of atypical *A. salmonicida* (AHLDA 1334). *Aeromonas hydrophila* strains that are positive using these primers are negative for the phenotypic tests of LDC and cellobiose fermentation. This PCR may assist in determining that isolates are *A. hydrophila*, as phenotypic tests are not always reliable (Nielsen *et al.*, 2001). However, phenotypic tests would need to be carried out to differentiate from atypical *A. salmonicida*. *A. hydrophila*, ATCC 7810, which is positive for LDC, does not produce an amplified product by this PCR.

A PCR that detects the presence of the aerolysin gene, detected DNA from all haemolytic strains of *A. hydrophila* that were cytotoxic to vero and CHO cells, and produced enterotoxin as detected by suckling-mouse assays. No aerolysin gene was detected in non-haemolytic strains of *A. hydrophila*, non-haemolytic strains of *A. caviae*, or strains of *A. sobria*. Extraneous bands were found with some strains of *A. caviae* and *Plesiomonas shigelloides*, but not at the correct molecular weight (Pollard *et al.*, 1990). The aerolysin gene is considered to be a useful virulence marker for detecting virulent pathogenic *Aeromonas* species (Kong *et al.*, 2002). Performing PCRs using the primers of Nielsen *et al.* (2001), and the primers of Pollard *et al.* (1990), may be useful in identifying virulent strains of *A. hydrophila*.

©N.B. Buller 2004. *Bacteria from Fish and Other Aquatic Animals: a Practical Identification Manual* (N.B. Buller)

Table 6.1. List of specific primers available for PCR detection of aquatic organisms.

Organism	Forward primer 5'–3'	Reverse primer 5'–3'	AT	C	Bp	Reference
Aeromonas caviae	AER8 CTGCTGGCTGTGACGTTACTCGCAG Digest with *Alu*I, 37°C for 1 h. Produces fragments 180 & 80 bp	AER9 TTCGCCACCGGTATTCCTCCAGATC	62	30	260	Khan and Cerniglia, 1997
Aeromonas caviae Primers are specific for haemolytic strains of *A. caviae* only.	CAV1a GAGCCAGTCCTGGGCTCAG	CAV1b GCATTCTTCATGGTGTCGGC	65	30	381	Wang *et al.*, 1996
Aeromonas caviae and *A. trota*	AER1 AGTTGGAAACGACTGCTAATA Digest with *Alu*I, 37°C for 1 h. No digestion	AER2 ACGCAGCAGATATTAGCTTCAG	68	30	316	Khan and Cerniglia, 1997
Aeromonas hydrophila Does not differentiate from *A. encheleia*	AH-F GAAAGGTTGATGCCTAATACGTA	AH-R CGTGCTGGCAACAAAGGACAG	60	28	685	Nielsen *et al.*, 2001
Aeromonas hydrophila Aerolysin gene	Aero1a CCAAGGGGTCTGTGGCGACA	Aero1b TTTCACCGGTAACAGGATTG	55	60	209	Pollard *et al.*, 1990
Aeromonas hydrophila (specific for HG1) Produces primer dimer bands at base of gel	7e5-F6 GCTCTCATTGCAGAGGCCGTTACTAGG	7e5-R78 TCTGCACGAAACTTCAAGCAGCTTTG	55	35	242	Oakey *et al.*, 1999
Aeromonas salmonicida ssp. *salmonicida*	ASS-F AGCCTCCACGCGCTCACAGC	ASS-R AAGAGGCCCCATAGTGTGGG	60	30	512	Miyata *et al.*, 1996
Aeromonas salmonicida ssp. *salmonicida* Use in nested PCR with Universal 16S rDNA primers	AS1 GGCCTTTCGCGATTGGATGA	AS2 TCACAGTTGACACGTATTAGGCGC	55	30	271	Hoie *et al.*, 1997 Taylor and Winton, 2002
Aeromonas sobria Detects cytotoxin gene	SOBF GCG ACC AAC TAC ACC GAC CTG	SOBB GGACTTGTAGAGGGCAAC CCG				Filler *et al.*, 2000
Aeromonas trota	AL1 TTGCCGCCCAGGCCGGTGCTG	AL2 ACCACTGTGTGGACCAGGGTA	66	30	622	Khan *et al.*, 1999
Edwardsiella spp. Amplifies *sodB* gene from all *Edwardsiella* spp.	E1F ATGTCRTTCGAATTACCTGC	497R TCGATGTARTARGCGTGTTCCCA	42	35	454	Yamada and Wakabayashi, 1999
Flavobacterium species-specific Based on 16Sr DNA	Col-72F GAAGGAGCTTGTTCCTTT	Col-1260R GCCTACTTGCGTAGTG	60	30	?	Triyanto *et al.*, 1999
Flavobacterium columnare within 16S rDNA	FvpF GCCCAGAGAAATTTGGAT	FvpR1 TGCGATTACTAGCGAATCC	59	25	1192	Bader *et al.*, 2003

Target	Primer 1	Primer 2				Reference
Flavobacterium columnare Genomovar 1 Sequence within 16S rDNA	Col-Ta TTCAGATGGCTTCATTTG	Col-Tb CCGTTTACGGGCGGTTGGAATACAG	54	30	?	Triyanto *et al.*, 1999
Flavobacterium columnare Genomovar 2 Sequence within 16S rDNA	Col-T1 ATTAAATGGCATCATTTA	Col-T2 TCGTTTACGGCGTGGGACTACCA	52	30	621	Triyanto *et al.*, 1999
Flavobacterium columnare Genomovar 3 Sequence within 16S rDNA	Col-T11 GATGTGGCCTCACATTGTG	Col-Tb CCGTTTACGGGCGGTTGGAATACAG	56	30	?	Triyanto *et al.*, 1999
Flavobacterium psychrophilum 16S rRNA	PSY1 GTTGGCATCAACACACT	PSY2 CGATCCTACTTGCGTAG	57	30	1089	Wiklund *et al.*, 2000
Flavobacterium psychrophilum 16S rRNA	FP1 GTTAGTTGGCATCAACAC	FP2 TCGATCCTACTTGCGTAG	54	35	1088	Urdaci *et al.*, 1998
Flavobacterium psychrophilum 16S rRNA	Psy1 CGATCCTACTTGCGTAG	Psy2 GTTGGCATCAACACACT	45 50	30 39	1100	Toyama *et al.*, 1994
Lactococcus garvieae Based on the dihydropteroate synthase gene	SA1B10-1-F CATTTACGATGGCGCAG	SA1B10-1-R CGTCGTGTTGCTGCAACA	58	30	709	Aoki *et al.*, 2000
Lactococcus garvieae Based on 16S rDNA	pLG-1 CATAACAATGAGAAATCGC	pLG-2 GCACCCTCGCGGGTTG	55	35	1100	Zlotkin *et al.*, 1998a
Lactococcus garvieae	IRL TTTGAGAGTTTGATCCTGG	LgR AAGTAATTTTCCACTCTACTT	45	35	482	Pu *et al.*, 2002
Lactococcus piscium Does not differentiate from *L. plantarum*, *L. raffinolactis*	IRL TTTGAGAGTTTGATCCTGG	PiplraR CGTCACTGAGGGGCTGGAT	45	35	863	Pu *et al.*, 2002
Listonella anguillarum Haemolysin gene	VAH-P1 ACCGATGCCATCGCTCAAGA	VAH1-P2 GGATATTGACCGAAGAGTCA	55	30	603	Hirono *et al.*, 1996
Listonella anguillarum toxR gene	VA-U2 CACTTCGCAACCCGAAGAGACA	VA-D1 CTGCTTAGGTGCCAGTTCTCCA	62	20	307	Okuda *et al.*, 2001
Mycobacterium genus	T39 GCGAACGGGTGAGTAACACG	T13 TGCACACAGGCCACAAGGGA	50	30	924	Talaat *et al.*, 1997
Mycobacterium chelonae	Restrict with ApaI to give fragments at 812 and 112 bp	Restrict with BanI to give fragments at 562 and 362 bp				Talaat *et al.*, 1997

continued

Table 6.1. *Continued.*

Organism	Forward primer 5'–3'	Reverse primer 5'–3'	AT	C	Bp	Reference
Mycobacterium fortuitum	Restrict with *Apa*I to give fragments at 677, 132, 115 bp	Restrict with *Ban*I to give fragments at 562 and 362 bp				Talaat *et al.*, 1997
Mycobacterium marinum	Restrict with *Apa*I to give fragments at 677, 132, 115 bp	No restriction with *Ban*I				Talaat *et al.*, 1997
Mycobacterium triplex-like Nested PCR	For1 CGAAAGCGTGGGGAGGGAACA For2 GGTGTGGGTTTCCTTCCTT	Rev1 AGACCCGATCCGAACTGAGACC Rev2 ACGGGGCCATTGTAGCAT	55 55	38 55	409	Herbst *et al.*, 2001
Nocardia genus	NG1 ACCGACCACAAGGGG	NG2 GGTTGTAACCTCTTCGA	55	30	596	Laurent *et al.*, 1999
Photobacterium damselae ssp. *damselae* and *piscicida* Utilizes presence or absence of urease gene	Ure-5 TCCGAATAGGTAAAGCGGG Car1 GCTTGAAGAGATTCGAGT	Ure-3 CTTGAATATCCATCTCATCTGC Car2 CACCTCGCGGTCTTGCTG	60 60	30 30	448 267	Osorio *et al.*, 2000
Pseudomonas anguilliseptica	PAF GACCTCGCGCCATTA	PAR CTCAGCAGTTTTGAAAG	46	35	439	Blanco *et al.*, 2002
Pseudomonas plecoglossicida Based on *gyrB* coding region Used TaqMan real time PCR	GBPA-F CCTGCTGAAGGACGAGCGTTCG	GBPA-R AACCAGGTGAGTACCACCGTCG	68	50		Sukenda and Wakabayashi, 2000
Renibacterium salmoninarum Detects p57 gene	CAAGGTGAAGGGAATTCTTCCACT	GACGGCAATGTCCGTTCCCGGTTT				Brown *et al.*, 1994
Renibacterium salmoninarum Detects p57 gene	FL7 CGCAGGAGGACCAGTTGCAG	RL11 GGAGACTTGCGATGCGCCGA	60	35	349	Miriam *et al.*, 1997
Internal probe	FL10 GGTGTAACGATAATGCGCCA	RL11	60	35	149	
Renibacterium salmoninarum	F GATCGTGAAATACATCAAGG	R GGATCGTGTTTTATCCACC	60	30	149	Leon *et al.*, 1994
Salinivibrio (Vibrio) costicola IGS^Glu (cosB)-specific primer	VCOS-F CTGACGCTATTCTTGCGA	VCOS-R GTAATCACATTCGTAAATGC	55	35	186	Lee *et al.*, 2002
Streptococcus agalactiae (β-haemolytic, group B). Semi-nested PCR. Cross-reaction with *S. porcinus*	DSF1 TGCTAGGTGTTAGGCCCTTT DSF2 GGCCTAGAGATAGGCTTTCT	DSR1 CTTGCGACTCGTTGTACCAA DSR1	67 67	30 30	450 265	Ahmet *et al.*, 1999

Streptococcus iniae 16S rRNA	Sin-1 (CTAGAGTACACATGTACT(AGCT)AAG)	Sin-2 GGATTTCCACTCCCATTAC	55	35	300	Zlotkin et al., 1998b
Tenacibaculum (Flexibacter) maritimum 16S rRNA	Mar1 TGTAGCTTGCTACAGATGA	Mar2 AAATACCTACTCGTAGGTACG	58	39	400	Bader and Shotts, 1998; Cepeda et al., 2003
Tenacibaculum (Flexibacter) maritimum 16S rDNA	MAR1 AATGGCATCGTTTTAAA	MAR2 CGCTCTCTGTTGCCAGA	45	30		Toyama et al., 1996
Vibrio cholerae 01 ctxA gene	CTX2 CGGGCAGATTCTAGACCTCCTG	CTX3 CGATGATCTTGGAGCATTCCCAC	60	25	564	Fields et al., 1992
Vibrio cholerae 01 El Tor Positive for rtxA, rtxC and ctxB genes	RtxA-F CTGAATATGAGTGGGTGACTTACG	RtxA-R GTGTATTGTTCGATATCCGCTACG	55	30	417	Chow et al., 2001
Vibrio cholerae non-01 Positive for rtxA and rtxC genes	RtxC-F CGACGAAGATCATTGACGAC	rtxC-R CATCGTCGTTATGTGGTTGC	55	30	263	Chow et al., 2001
Vibrio cholerae 01 classical Positive for ctxB gene only (negative for rtxA and rtxC genes)	ctxB2 GATACACATAATAGAATTAAGGAT	ctxB3 GGTTGCTTCTCATCATCGAACCAC	55	30	460	Chow et al., 2001
Vibrio diazotrophicus IGS^Glu(diaA)-specific primer	VDIA-F AGATTCTCTTGATGAGTGCC	VDIA-R TACCTACATCTCTAAGAGACATAG	55	35	300	Lee et al., 2002
Vibrio fischeri LuxA gene	LuxA-F GTTCTTAGTTGGATTATTGG	LuxA-R TCAGTTCCATTAGCTTCAAATCC	40	40	428	Lee and Ruby, 1995
Vibrio fluvialis IGS^Glu(fluA)-specific primer	VFLU-F ATAAAGTGAAGAGATTCGTACC	VFLU-R GTATTCCTGAATGGAATACAC	55	35	278	Lee et al., 2002
Vibrio hollisae gyrB gene	HG-F1 GCTCTGTCGGAAAAACTTGA	HG-R2 ATGCTCAAAATGGAACACAG	55	30	363	Vuddhakul et al., 2000
Vibrio hollisae toxR gene	HT-F3 CTGCCCAGACACTCCCTCTTC	HT-R2 CTCTTCCTTACCATAGAAACCG	62	24	306	Vuddhakul et al., 2000
Vibrio nigripulchritudo IGS^Glu (nigA)-specific primer	VNIG-F CATTTCTTTGAAACAGAAAGT	VNIG-R TAGATAAGGGGATTGTTGCTA	55	35	114	Lee et al., 2002
Vibrio parahaemolyticus gyrB gene. This gene may also be present in V. alginolyticus	VP1 CGGGCGTGGGTGTTTCGGTAGT	VP2r TCCGCTTCGGCGCTCATCAATA	60	30	285	Venkateswaran et al., 1998; Kim et al., 1999
Vibrio parahaemolyticus tl gene. This gene may also be present in V. alginolyticus	L-tl AAAGCGGATTATGCAGAAGCACTG	R-tl GCTACTTTCTAGCATTTTCTGC	58.6	30	450	Bej et al., 1999

continued

Table 6.1. *Continued.*

Organism	Forward primer 5'–3'	Reverse primer 5'–3'	AT	C	Bp	Reference
Vibrio parahaemolyticus toxR gene May get weak non-specific bands with *V. alginolyticus*, *V. vulnificus*	ToxR1 GTCTTCTGACGCAATCGTTG	ToxR2 ATACGAGTGGTTGCTGTCATG	63	20	368	Kim *et al.*, 1999
Vibrio parahaemolyticus Cloned fragment pR72H specific for this species	VP33 TGCGAATTCGATAGGGTGTTAACC	VP32 CGAATCCTTGAACATACGCAGC	60	35	387 or 320	Lee *et al.*, 1995 Robert-Pillot *et al.*, 2002
Vibrio penaeicida 16S rDNA	VpF GTGTGAAGTTAATAGCTTCATATC	VR CGCATCTGAGTGTCAGTATCT	62	35	310	Saulnier *et al.*, 2000
Vibrio proteolyticus IGS[IA] (proC)-specific primer	VPRO-C GCATTCTTACGAGTGTG	VPRO-R ATTAGTTGTATTCAAATA	55	35	133	Lee *et al.*, 2002
Vibrio salmonicida IGS[9] (salA)-specific primer	VSAM-F TGCGATTTATGAGTGTTCA	VSAM-R ACTCTTCATTGAGAGTTCTG	55	35	275	Lee *et al.*, 2002
Vibrio splendidus IGS[9] (spnA; spnD)-specific primer	VSPN-F GATTTAGTTAAAGCCAGAGC	VSPN-R CCTGATAACTGTTTGCCG	55	35	240, 294	Lee *et al.*, 2002
Vibrio trachuri (junior synonym of *V. harveyi*)	Pstf-1a TGCGCTGACGTGTCTGAATT	Pstf-1b AAGCAGCGATGACAAGCAGT	60	35	417	Iwamoto *et al.*, 1995b
Vibrio tubiashii IGS[IA] (tubA)-specific primer	VTUB-F TGGGTCTTTCAGGCCCG	VTUB-R CGACGAATGACCGTTGTC	55	35	394	Lee *et al.*, 2002
Vibrio vulnificus Nested PCR	P1 GACTATCGCATCAACAACCG P3 GCTATTTCACCGCCGCTCAC	P2 AGGTAGCGAGTATTACTGCC P4 CCGCAGAGCCGTAAACCGAA	57 59	50 50	704 222	Lee *et al.*, 1998
Vibrio vulnificus Cytolysin-haemolysin gene	F CGCCGCTCACTGGGCAGTGGCTG	R GCGGGTGGTTCGGTTAACGGCTGG	64.5	30–50		Coleman *et al.*, 1996
Yersinia ruckeri	YER 8 GCGAGGAGGAAGGGTTAAGTG	YER 10 GAAGGCACCAAGGCATCTCTG	60	25	575	Gibello *et al.*, 1999
Yersinia ruckeri 16S rDNA	Ruck1 CAGCGGAAAGTAGCTTG	Ruck2 TGTTCAGTGCTATTAACACTTAA	55	30	409	LeJeune and Rurangirwa, 2000
Yersinia ruckeri yruR/yruI quorum sensing gene	IF2 GAGCGCTACGACAGTCCCAGATAT	IR2 CATACCTTTAACGCTCAGTTCGAC	65	40	1000	Temprano *et al.*, 2001

Multiplex PCR						
Virulent *Aeromonas* species	Aero-F TGTCGGGSGATGACATGGAYGTG	Aero-R CCAGTTCCAGTCCCACCACTTCA	62	35	720	Kong *et al.*, 2002
Vibrio cholerae	Esp-F GAATTATTGGCTCCTGTGCAGG	Esp-R ATCGCTTGGCGCATCACTGCCC				
Vibrio parahaemolyticus	Vpara-F GCTGACAAAACAACAATTTATTGTT	Vpara-R GGAGTTTCGAGTTGATGAAC				Kong *et al.*, 2002

For multiplex reaction, primers were used at 0.2 µM for Aero, 0.1 µM for Esp, and 1.0 µM for Vpara

The Aero primer pairs detect virulent species of *A. hydrophila, A. eucrenophila, A. popoffii, A. sobria* and *A. trota*

E. coli (uidA gene)	L-UIDA TGGTAATTACCGACGAAAACGGC	R-UIDA ACGCGTGGTTACAGTCTTGCG	55–60	30	147	Brasher *et al.*, 1998
Vibrio cholerae (ctx gene)	L-CTX CTCAGACGGGATTTGTTAGGCACG	R-CTX TCTATCTCTGTAGCCGGTATTACG			302	
Vibro parahaemolyticus (tl gene) (may amplify *V. alginolyticus*)	L-TL AAAGCGGATTATGCAGAAGCACTG	R-TL GCTACTTTCTAGCATTTTCTGC			450	
Vibrio vulnificus (cth gene)	L-CTH TTCCAACTTCAAACCGAACTATGAC	R-CTH GCTACTTTCTAGCATTTTCTGC			205	
Salmonella typhimurium (invA)	L-INVA CTCTACTTAACAGTGCTCGTTTAC	R-INVA TTGATAAACTTCATCGCACCGTCA			273	
Universal protocol						
Vibrio cholerae (toxin gene)	VC-1 GGCAGATTCTAGACCTCCT	VC-2 TCGATGATCTTGGAGCATTC			563	Wang *et al.*, 1997 (quotes Fields *et al.*, 1992)
Vibrio vulnificus (cytolysin gene)	VV-1 CTCACTGGGGCAGTGGCT	VV-2 CCAGCCGTTAACCGAACCA			383	Wang *et al.*, 1997 (quotes Brauns *et al.*, 1991)
Vibrio parahaemolyticus (genomic DNA)	VIP-1 GAATTCGATAGGGTGTTAACC	VIP-2 ATCCTTGAACATACGCAGC			381	Wang *et al.*, 1997 (quotes Lee *et al.*, 1995)

AT = annealing temperature; C = number of cycles; Bp = product size in base pairs; IGS = intergenic spacers.

Edwardsiella *spp. PCR*

The *sodB* gene (coding for iron-cofactored super-oxide dismutase) from *Edwardsiella* spp. was amplified (454 bp for all species) and sequenced. Differences in nucleotide sequence divided the species of *Edwardsiella* into a pathogenic and non-pathogenic cluster. Cluster I comprises pathogenic strains of *E. tarda* from Japanese eel, Japanese flounder (*Paralichthys olivaceus*), Nile tilapia (*Oreochromis niloticus*) and ayu (*Plecoglossus altivelis*), atypical *E. tarda* from red sea bream, *Edwardsiella* species from Japanese eel, and *E. ictaluri*. Non-pathogenic *E. tarda* and *E. hoshinae* were found in cluster II (Yamada and Wakabayashi, 1999).

Flavobacterium columnare *PCR*

Primers FvpF and FvpR2, located within the 16S rDNA are said to be specific for *F. columnare* (Bader *et al.*, 2003). Three primer sets, also located within the 16S rDNA gene, differentiate the three genomovars of *F. columnare*. Three separate forward primers are available, with genomovars 1 and 3 having the same reverse primer (see Table 6.1). Primers for amplifying genomovar 2 produce a product band at 621 bp, but also produce bands at 800 and 1000 bp in all *Flavobacterium* species (Triyanto *et al.*, 1999). The sensitivity of the PCRs can be increased from tissue samples by using universal primers (Bader *et al.*, 2003) or universal primers followed by the *Flavobacterium* species-specific set of Col-72F and Col-1260F (Table 6.1) (Triyanto *et al.*, 1999).

Pseudomonas plecoglossicida *PCR*

This PCR was conducted using the TaqMan methodology for quantitative real-time PCR. The internal control and target DNA probes were TGT-P 5′-(FAM) AGATGGCGTGGGCGTTGAAGTAGCGC (TAMRA)-3′, and ISD-P 5′-(VIC) CCTTCACCACCACGGCCGAGCGTGAG (TAMRA)-3′ (Sukenda and Wakabayashi, 2000).

Tenacibaculum *(Flexibacter) maritimum PCR*

A nested PCR using universal primers of 20F and 1500R in the first PCR reaction, followed by a nested PCR using *T. maritimum*-specific primers Mar1 and Mar2 in the nested reaction allows

detection of this organism direct from fish tissue (Cepeda *et al.*, 2003). Ready-to-Go PCR beads (Amersham Pharmacia Biotech) were used, with 1 pmol of each primer in a 25 μl reaction. Cycling conditions for the first PCR reaction used a preheating step of 95°C for 5 min, 30 cycles of 95°C for 30 s, 57°C for 30 s, 72°C for 60 s with a final cycle that used an extension step at 72°C for 5 min. The second PCR reaction used a preheating step of 94°C for 2 min, followed by 40 cycles of 94°C for 2 s, 54°C for 2 s, 72°C for 10 s with a final cycle that used an extension step of 4 min.

PCR for Vibrio *spp.*

Many of these PCRs will need to undergo a period of validation, and as more information is gathered and research done, the specificity of these PCRs will also be assessed. Such a case has been found with PCRs for some of the *Vibrio* species. There is some doubt as to the specificity of some of the primers for *Vibrio parahaemolyticus* and *V. alginolyticus*. The *gyrB* gene has been suggested to differentiate *V. parahaemolyticus* from *V. alginolyticus* (Venkateswaran *et al.*, 1998); however, the primers also detected the gene in *V. alginolyticus* when the suggested annealing temperature of 58°C was used. Specificity for *V. parahaemolyticus* was improved at an annealing temperature of 60°C (Kim *et al.*, 1999). The *tl* gene that encodes for a thermolabile haemolysin and used in a PCR for detecting *V. parahaemolyticus* (Bej *et al.*, 1999) was found to occur not only in *V. parahaemolyticus* but was also detected in species of *V. alginolyticus* (Robert-Pillot *et al.*, 2002). The *toxR* gene, which is involved in the regulation of many genes, is conserved amongst the *Vibrio* species; however, there is a low degree of homology that allows for selection of species-specific primers. A PCR that targets the *toxR* gene in *V. parahaemolyticus* may amplify non-specific amplicons from *V. algino-lyticus* and *V. vulnificus* (Kim *et al.*, 1999). Studies suggested that to date the PCR that appears to be the most specific for *V. parahaemolyticus* is a PCR that detects a fragment termed R72H. This section of DNA is composed of a non-coding region and a phosphatidylserine synthetase gene and the primers amplify an amplicon of either 387 or 320 bp, both of which are considered specific for *V. parahaemolyticus* (Lee *et al.*, 1995; Robert-Pillot *et al.*, 2002).

Multiplex PCR

A multiplex PCR enabled simultaneous detection of the human pathogens *E. coli, Salmonella typhimurium, V. cholerae, V. parahaemolyticus* and *V. vulnificus* from shellfish samples. An optimized PCR was used with 2.5 mM MgCl₂, and an annealing temperature of 55°C (Brasher *et al.*, 1998).

Nested PCR

Any of the above primers in Table 6.1 that are designed within the 16S rRNA can be used in a nested PCR reaction to improve detection sensitivity. The first PCR reaction is done with eubacterial primers, which amplify the entire 16S rDNA, and the specific primers are then used for a nested PCR reaction. The detection of *Flavobacterium columnare* was improved in tissue samples when specific primers were used in a nested PCR with eubacterial universal primers (Triyanto *et al.*, 1999; Bader *et al.*, 2003). For details on universal primers see section 6.3.

An optimized nested PCR was reported for *Aeromonas salmonicida, Flavobacterium psychrophilum* and *Yersinia ruckeri*. The first round PCR used universal primers for the 16S rDNA gene and the second round of primers were species-specific for the three organisms tested. The PCR conditions were optimized so that the same conditions could be used for all primers (Taylor and Winton, 2002). The universal primers of Weisburg *et al.* (1991) with forward primer fD2 5′-AGAGTTTGATCATGGCTCAG-3′ and reverse primer rP2 5′-GTTTACCTTGTTACGACTT-3′ were used to generate a 1500 bp fragment of the 16S rDNA gene. This template was then used in second round PCR with primers for *Aeromonas salmonicida* AS1 and AS2 (Høie *et al.*, 1997), *Flavobacterium psychrophilum* PSY1 and PSY2 (Toyama *et al.*, 1994), and primers YER8 and YER10 for *Yersinia ruckeri* (Gibello *et al.*, 1999). In the PCR reaction mix, the final concentration of MgCl₂ was 2.0 mM, dNTP was 200 µM, with 100 pmol final concentration for each primer and 1.25 U/50 µl of Taq. Thermocycling conditions were also standardized with an initial denaturation step of 95°C for 4 min, followed by 30 cycles of 95°C for 45 s, annealing for 45°C for universal primers and 55°C for specific primers, with an extension of 72°C for 90 s. A final cycle with an extension step of 72°C for 4 min was used. A final cycle of 4°C infinity was used as a holding cycle (Taylor and Winton, 2002).

Outline of steps for PCR using specific primers

Step	Method
1	Extract DNA from bacterial colonies
2	Amplify DNA using specific primers
3	Visualize DNA on agarose gel

DNA extraction from bacterial cells

There are many methods for the extraction of DNA from bacterial cells and these include both manual methods and commercially available kits. Such kits include Instagene (Bio-Rad), AquaPure™ genomic DNA kits (Bio-Rad), Chelex-based resin (Bio-Rad), Puregene (Gentra Systems), PrepMan™ (Applied Biosystems), MasterPure (Astral Scientific – Epicentre), Wizard® Genomic (Promega), and the Dneasy Tissue system and QiAamp® system from Qiagen.

MANUAL METHOD 1.   Suspend bacterial cells in 100 µl of a solution of 1 mM EDTA-0.5% Triton-X-100. Boil in a microwave on HIGH setting for 5 min (Lee *et al.*, 1998).

MANUAL METHOD 2.   Make a suspension of bacterial cells to 0.5 McFarland density. Centrifuge at 13,000 rpm for 5 min. Suspend pellet of cells in a digestion buffer of 50 mM Tris-HCl (pH 8.5), 1 mM EDTA, 0.5% SDS and 200 µg/ml of proteinase K. Incubate for 3 h at 55°C with agitation. Heat inactivate the proteinase K for 10 min at 95°C. Cool to 4°C, then centrifuge at 13,000 rpm for 10 min. Use supernatant in the PCR reaction. However, the PCR master mix must contain a final concentration of Tween 20 to neutralize the effect of SDS, which is inhibitory to Taq polymerase at concentrations as low as 0.01% (Goldenberger *et al.*, 1995).

This method is also suitable for tissue samples with an overnight incubation in digestion buffer followed by sonication (Goldenberger *et al.*, 1995).

MANUAL METHOD 2 FOR GRAM-POSITIVE BACTERIA. For Gram-positive bacteria treat cells with lysozyme (1 mg lysozyme/ml TE, pH 8.0).

Extraction of DNA from tissue

As for extraction of DNA from cultured bacterial cells, bacterial DNA can also be extracted from tissue samples. There are many methods available for both manual and commercially available methods. The following are some suggestions.

MANUAL METHOD 1.   To 100 µl of homogenized tissue add 100 µl of Chelex-100 resin (Bio-Rad Laboratories or Sigma-Aldrich). Heat at 56°C for 10 min. Add 200 µl of 0.1% Triton-X-100 and boil for 10 min. Cool on ice, then centrifuge at 12,000 rpm for 3 min. Use 5 µl in PCR reaction (Khan and Cerniglia, 1997).

MANUAL METHOD 2.   Culture 10 g of tissue into 90 ml of TSBYE medium (30 g tryptic soy broth powder with dextrose, (Difco); 6 g yeast extract; 1 l of water). Incubate at 25°C overnight with shaking. Take 0.5 ml of the upper phase of the sample and mix with 1 ml of sterile phosphate buffered saline (PBS, 0.05 mol/l, pH 7.4). Centrifuge at 9000 g for 3 min. Wash the pellet three times in PBS and once with sterile water. Resuspend pellet in 50 µl of water, then dilute 1:10 with 1% Triton-X-100 and place in boiling waterbath for 5 min. Place immediately on ice. Use 2 µl in PCR mixture (Wang et al., 1997).

6.2 PCR Protocols

Any PCR that is introduced into the laboratory needs to be optimized for the Taq enzyme that is used by that laboratory and the primers that are prepared. Differences may occur between different batches of primers and this may be seen with the concentration of the dNTPs. For example, a pair of primers that may be optimal at a final concentration of 100 µM dNTP mix in the PCR reaction; when re-ordered a concentration of 200 µM may be optimal.

Method 1 – standard protocol

Table 6.2 suggests a standard PCR protocol that may be used to amplify from most target sequences. It is also a basic starting point from which to do optimizations. When optimizing the concentrations in the PCR mix, the suggested ranges for testing are: dNTP concentration (100–200 µM), primer concentration (0.1–1 µM), magnesium chloride concentration (1–4 mM) and Taq enzyme (1–2.5 Units). Too much or too little of any of these reagents will cause nonspecific background, mispriming or insufficient product generated. In particular, high concentrations of dNTP, primer or $MgCl_2$ will lead to non-specific product (Saiki, 1989; Innis and Gelfand, 1990). A negative and positive control should be included with each PCR run.

Sometimes bovine serum albumin, gelatin, Tween 20 or DMSO can be added to the PCR reaction mix to help stabilize the Taq enzyme. DMSO may assist in preventing the inhibition of the function of the Taq enzyme caused by inhibitors present in tissue or from contamination by agar and agar-containing media such as Stuarts transport medium. However, not more than 2% should be used, as greater than this concentration is inhibitory to Taq.

It is suggested that the reagent volumes for the PCR master mix be set up in an Excel spreadsheet with the appropriate formulae in each cell. When a different number of samples are tested then the master mix is easily calculated.

Aliquot PCR master mix into 20 µl volumes. Add 30 µl of sterile paraffin oil to each tube if

Table 6.2.   Standard PCR protocol.

Reagent	×1 (µl)	×10 (µl)	Final concentration
Water	11.8	= 25 minus the volume of reagents	
PCR buffer (x10)	2.5	= 2.5 × 10	1
dNTP mix (2 mM of each)	2.5	= 2.5 × 10	200 µM (0.2 mM)
$MgCl_2$ (25 mM)	2	= 2 × 10	2 mM
Primer forward (20 pmol = 5 µM)	0.5	= 0.5 × 10	0.1 µM
Primer reverse (20 pmol = 5 µM)	0.5	= 0.5 × 10	0.1 µM
Taq enzyme (5 U/µl)	0.2	= 0.2 × 10	1 Unit
Total volume including volume of DNA to be added	25	= 25 × 10	

thermocycler does not have a hot lid. Tubes can be stored at −20°C until required.

Add 5 μl of template (DNA) and place tubes on thermocycler.

Method 2 – commercial PCR reaction mixture

Many companies offer master mix that contains buffer, MgCl₂, and dNTPs. The user adds the DNA template, primers and water. An example is Ready-To-Go® PCR beads from Amersham Pharmacia Biotech. This is used as a 25 μl reaction mix to which primers and 5 μl of DNA are added. BIOMIX ready-to-go from Astral Scientific (Bioline), PCR Master Mix (Promega), and IQ supermix (Bio-Rad laboratories) are other examples of commercially available PCR Master Mix.

Thermocycling conditions

The cycling conditions will depend on the type of thermocycler, the annealing temperature of the primers and the type of *Taq* enzyme used, whether it is a hot-start enzyme or not.

A standard set of cycling conditions begins with one cycle at 95°C for 1–5 min. Then, 25–35 cycles of denaturation at 95°C for 30 s, annealing at 55–68°C 30 s, extension at 72°C for 30 s. A final cycle is suggested as for the last cycle but with an extension time of 10 min. The thermocycler can be set to do a hold cycle at 4°C.

The primer annealing temperature depends upon the length of the primer, the GC content and concentration. A recommended annealing temperature is 5°C below the true T_m of the primers. The T_m is usually provided on the data sheet sent with the primer. The T_m can be estimated by calculating 2°C for A or T, and 4°C for G or C in the primer sequence. The annealing temperature affects the specificity of the reaction. A lower than optimal annealing temperature may lead to mispriming of non-target sequence or the mis-extension of incorrect nucleotides at the 3′ end of the primers. A low extension temperature together with high dNTPs also favours mispriming and non-specificity (Saiki, 1989; Innis and Gelfand, 1990).

The optimal number of cycles is between 25 and 35. Increasing the number of cycles may lead to problems with the PCR, as a plateau effect is reached where product is no longer amplified and a result is that non-specific product may be amplified preferentially. This plateau is reached according to the number of target copies initially present in the sample and the amount of DNA synthesized. Also, reagent exhaustion occurs with an extended number of amplification cycles. The recommended number of cycles per starting material is 25–30 (for 3×10^5 target molecules), 30–35 (1.5×10^4 molecules), 35–40 (1×10^3 molecules), 40–45 (50 molecules) (Saiki, 1989; Innis and Gelfand, 1990).

Gel electrophoresis

A standard gel for the detection of amplified product is as follows. Prepare a 2% agarose gel using chromosomal grade agarose (Bio-Rad). Load 5 μl of amplification product and 5 μl of 2 × loading buffer into the wells. Run gel at 5 v/cm for 2 h. On a Bio-Rad PowerPac 300, 80 v for 90 min produces well-separated bands. Always include a molecular weight marker such as a 100 bp marker.

The gel is stained for 1 h in 1 litre of distilled water containing 50 μl of ethidium bromide. Ethidium bromide is a carcinogen and therefore appropriate precautions need to be taken, such as wearing lab coats and gloves.

The gel is photographed over a UV trans-illuminator. Safety note: Do not expose the skin or eyes to UV light. Laboratory coats, gloves and face shields must be worn when operating UV transilluminators that are not housed within a closed instrument.

Reagent preparation of stock and working solutions

dNTP concentration for PCR amplification

For dNTPs supplied as 100 mM. The final concentration in the PCR is required at either 100 μM or 200 μM depending on results in optimization tests.

Therefore, prepare each dNTP to 10 mM by diluting 1:10. From this prepare a working master mix of a dNTP solution by adding equal volumes of each dNTP plus an equal volume of water; that is, a 1:5 dilution for each dNTP to give a final concentration of each dNTP of 2 mM. For a final concentration of 100 mM in the PCR master mix, use 1.25 μl. For 200 mM use 2.5 μl in a master mix of 25 μl total volume.

The nucleotides can also be purchased as a mixture containing all four dNTPs. An example is PCR nucleotide mix (Promega) with each nucleotide at 10 mM concentration.

Preparation of primers

A data sheet is usually sent with the primer, which details the concentration, melting temperature etc of the primer. Prepare a stock solution of 100 pmol in TE buffer (pH 7.0). Therefore, if the data sheet states a primer is 3500 pmol, add 350 µl of TE buffer to the dehydrated primer to obtain a stock solution of 100 pmol.

Prepare a working solution in distilled water of the required concentration, usually 20 pmol.

Primer quantification by spectrophotometry

An accurate quantification can be done by spectrophotometer. Dilute the stock solution of primer 1:20 (i.e. 50 µl + 950 µl of water).

The UV absorbance is read at 260 nm. The concentration of the primer or oligonucleotide in µg/ml = absorbance × dilution × weight per OD.

Single-stranded DNA (ssDNA) at a concentration of 33 µg/ml has an absorbance of 1. Therefore 33 µg of ssDNA = 1 OD unit.

Therefore, in the above example, the concentration of ssDNA µg/ml = absorbance × 20×33.

Double-stranded DNA (dsDNA): A_{260} = OD_{260} = 1 for a 50 µg/ml solution.

Oligomer – quantitation

For a 20-mer, a stock solution with A_{260} = 1, contains 5 nmol
5 nmol = 33 µg/(20×325)
For a 40-mer, a stock solution with A_{260} = 1 contains 2.5 nmol
2.5 nmol = 33 µg/(40×325)

Oligomer – conversion of pmol of primer to µg of primer

Multiply pmol by (length × 325)/1,000,000
Example: 51809.88 pmol of a 20-mer (from data sheet)
$(51809.88 \times 20 \times 325)/1,000,000 = 336.7$ µg primer

Conversion of µg of primer to pmol of primer

Multiply by 1,000,000/(length × 325).
Example: 365.73 µg of a 20-mer (from data sheet)
$(365.73 \times 1,000,000)/(20 \times 325) = 56,266.15$ pmol of primer

Primer concentration for PCR amplification

Micromolar concentrations of primer = pmol/µl
Thus 20 pmol of primer in 100 µl PCR mixture = 20 micromolar (20 µM)

Equation for estimating volume required from a stock concentration

A basic formula for estimating the required volume from a stock solution is as follows:
(initial concentration) × (volume needed) = (final concentration) × (volume of sample)
Therefore, if you have a stock solution of 5 µM of primer with a final concentration required of 0.1 µM in a 25 µl reaction volume the equation is:
$5 \mu M \times \chi = 0.1 \mu M \times 25 \mu l$
$\chi = (0.1 \mu M \times 25)/5 \mu M$
$\chi = 0.5 \mu l$. Therefore add 0.5 µl of the 5 µM stock (or working solution) to a 25 µl reaction volume.

Storage of oligonucleotides

Oligonucleotides should be stored at –20°C. Store in aliquots to prevent multiple freeze and thaw of oligonucleotides.

6.3 Molecular Identification by 16S rDNA Sequencing

Sequencing the 16S rRNA has been used extensively to study bacterial evolution and phylogeny. With a vast number of 16S rRNA sequences available in the National Centre for Biotechnology Information (NCBI) and the Ribosomal Database Project (RDP) databases, sequencing the 16S rDNA is an essential tool in bacterial systematics and the identification of new species.

It is also a useful tool when used with biochemical tests for bacterial identification in the diagnostic laboratory. Ribosomal RNA contains variable and highly conserved regions that evolve very slowly and, therefore, can be specific to a genus.

The 16S rRNA is found in the ribosomes, which consist of proteins and RNA. In prokaryotes the ribosomes measure 70S, and consist of two subunits, 50S and 30S. S is the sedimentation rate, or Svedberg unit, during centrifugation. The 16S rRNA is found in the 30 S subunit, whereas the 50S subunit contains the 5S and 23S rRNA molecules. 16S is approximately 1600 nucleotides, 23S, 3000 nucleotides and 5S is approximately 120 nucleotides.

Within the gene encoding for the 16S rRNA are areas of sequence that are conserved across the different genera, and it is these areas that have been used for the design of the universal eubacterial primers. Not all primers will bind to the DNA of the bacterium of interest; however, if a combination of primers is used then a section of DNA will be amplified, which can then be sequenced for identification.

Outline of steps for sequencing

Step Method
1   Extract DNA from bacterial colonies
2   Amplify 16S rDNA using universal eubacterial primers
3   Visualize DNA on agarose gel
4   Clean up amplified product
5   Quantify DNA concentration
6   Perform sequencing PCR reaction
7   Sequencing (send to a sequencing facility)
8   Analysis of sequence information using BioEdit
9   BLAST search for sequence identification

DNA extraction

DNA is extracted from bacterial colonies as described previously.

Amplification of 16S rDNA using universal eubacterial primers

The rDNA contains a number of sites that are conserved across genera, and as such there are a number of primers that can be used to amplify a part or all of the 16S rDNA for many genera. The universal primers A, B and C amplify regions that are universally conserved across prokaryotes and eukaryotes and which were initially proposed by

Lane *et al.* (1985). Since then, a number of different variations have been proposed. Some primers are universal for bacteria only, some amplify the entire 16S rDNA, whereas others amplify different regions. The convention for identifying the primers is to name them according to the number of the *E. coli* position at which the 3' end of the primer anneals.

Some of the primers that are available to either amplify all or part of the 16S rDNA are detailed in Table 6.3. Not all primer combinations will be appropriate for all bacteria. Table 6.4 indicates primer combinations that have proved useful. The primers may be ordered from a company such as Qiagen or Invitrogen.

Primers A, B, and C are universal for organisms from the three primary kingdoms, archaebacteria, eubacteria and eukaryotes (Lane *et al.*, 1985). Primer EUBB (7–26) of Suzuki and Giovannoni (1996) is the same as primer 27f of Lane (1991), except for a degeneracy (**M**) that possibly makes the primer more universal.

Primer 27f was modified because complementarity at the 3' ends led to self-priming followed by primer depletion. The modified primer was referred to as POmod (Wilson *et al.*, 1990). Complementarity in primers and between primer pairs can lead to primer dimers, which is the product that is amplified in preference to the desired sequence on the DNA template (Watson, 1989).

The suggested pairs of primers of 63f and 1387r (Table 6.4) were found to improve upon the primer pair of 27f and 1392r (Table 6.3) (Lane *et al.*, 1991), particularly with difficult DNA templates (Marchesi *et al.*, 1998).

The entire rDNA can be amplified with two sets of primer pairs. The primer pairs POmod and PC3mod amplify 789 base pairs of the rDNA and the pairs P3mod and PC5 amplify the remainder of the rDNA of 721 base pairs (Wilson *et al.*, 1990). Alternatively, the entire 16S rRNA can be amplified with 27f or EUBB as the forward primer and 1525 or EUBA as the reverse primer (Weisburg *et al.*, 1991).

The universal primers 16/23S-F and 16/23S-R were used to amplify the intergenic spacer (IGS) region between the 16S rRNA and the 23S rRNA genes of the *Vibrio* species and this region was subsequently used to design specific primers for the detection of eight *Vibrio* species. A standard PCR reaction mix, and thermocycling conditions with an annealing temperature of

Table 6.3. Universal eubacterial primers for 16S rDNA sequencing.

Primer (E. coli numbering)	Primer sequence 5' to 3'	Target	Reference
27f (EUBB) (7–26)	AGAGTTTGATCMTGGCTCAG	Most Eubacteria	Lane, 1991; Weisburg et al., 1991; Suzuki and Giovannoni, 1996
20F	AGAGTTTGATCATGGCTCAG	Eubacterial	Weisburg et al., 1991
POmod (8–22) modified from 27f	AGAGTTTGATCMTGG	Eubacterial kingdom only	Wilson et al., 1990
63f (43–63)	CAGGCCTAACACATGCAAGTC		Marchesi et al., 1998
357r (339–357)	CTCCTACGGGAGGCAGCAG	Most Eubacteria	Lane, 1991; Weisburg et al., 1991
530f (515–530)	GTGCCAGCMGCCGCGG	Most Eubacteria, Eukaryotes, Archaebacteria	Lane, 1991
P3mod f (787–806)	ATTAGATACCCTDTAGTCC	Eubacterial kingdom only	Wilson et al., 1990
519r (519–536)	GWATTACCGCGGCKGCTG	Primer A. Universal for all kingdoms	Lane et al., 1985; Lane, 1991
PC3mod r (787–806)	GGACTAHAGGGTATCTAAT	Eubacterial kingdom only	Wilson et al., 1990
685r3	TCTRCGCATTYCACCGCTAC	Most Gram-positives, cyanobacteria	Lane, 1991
907r (907–926)	CCGTCAATTCMTTTRAGTTT	Primer B. Universal for all kingdoms	Lane et al., 1985; Lane, 1991
926f	AAACTYAAAKGAATTGACGG	Most Eubacteria, Eukaryotes, Archaebacteria	Lane, 1991
1100r (1100–1114)	GGGTTGCGCTCGTTG	Most Eubacteria	Lane, 1991; Weisburg et al., 1991
1114f	GCAACGAGCGCAACCC	Most Eubacteria	Lane, 1991
1387r (1387–1404)	GGGCGGWGTGTACAAGGC		Marchesi et al., 1998
1392r (1392–1406)	ACGGGCGGTGTGTRC	Primer C. Universal for all kingdoms	Lane et al., 1985; Lane, 1991
1406f	TGYACACACCTCCCGT	Most Eubacteria, Eukaryotes, Archaebacteria	Lane, 1991
PC5 r (1492–1507) Modified from 1492r	TACCTTGTTACGACTT	Eubacterial kingdom only	Wilson et al., 1990; Lane, 1991
1492r (1492–1512)	TACGGYTACCTTGTTACGACTT	Most Eubacteria, Archaebacteria	Lane, 1991
1500R	GGTTACCTTGTTACGACTT	Eubacterial	Weisburg et al., 1991
1525r (1525–1541)	AAGGAGGTGWTCCARCC	Universal for all kingdoms	Lane, 1991; Weisburg et al., 1991
1525r EUB A modified from 1525r	AAGGAGGTGATCCANCCRCA	Eubacterial kingdom only	Suzuki and Giovannoni, 1996

M = C:A; R = A:G; K = G:T; W = A:T (Lane et al., 1985). Other mixtures of nucleotides (known as degeneracies or wobbles) are S = C:G; Y = C:T; V = A:G:C; H = A:C:T; D = A:G:T; B = C:G:T; N = A:G:C:T. F = same sequence as rRNA. R = means the complement of rRNA sequence. The primer positions correspond to the nucleotide numbering system of E. coli 16S rRNA (e.g. 9–27). The C in the name of the primers of Wilson et al. (1990) refers to the complementary sequence.

Table 6.4. Suggested primer pairs.

Forward primer 5'–3'	Reverse primer 5'–3'	Reference
27f GAGTTTGATCCTGGCTCAG	1392R ACGGGCGGTGTGTRC	Lane, 1991
63f CAGGCCTAACACATGCAAGTC	1387R GGGCGGWGTGTACAAGGC	Marchesi et al., 1998
530F GTGCCAGCMGCCGCGG	1100R GGGTTGCGCTCGTTG	Lane, 1991
POmod. AGAGTTTGATCMTGG	PC3mod. GGACTAHAGGGTATCTAAT	Wilson et al., 1990
P3mod. ATTAGATACCCTDTAGTCC	PC5. TACCTTGTTACGACTT	Wilson et al., 1990
16/23S-F. (1390–1407) TTGTACACACCGCCCGTC	16/23S-R. (474–456) CCTTTCCCTCACGGTACTG	Lee et al., 2002

55°C, was used. (Lee *et al.*, 2002). See Table 6.1 for specific primers.

Primers for sequencing the 16S rDNA gene for the identification of Vibrio spp.

There is high sequence homology for the 16S rRNA of all *Vibrio* species and therefore sequencing part of the 16S rDNA will not be of much use when trying to identify to species level. The entire 16S rDNA needs to be sequenced for identification to species level. This can be achieved by amplifying the entire 16S rDNA with universal primers (Table 6.5) and then using this amplified product as a template for eight sequencing primers (Table 6.6) (Thompson *et al.*, 2001a).

For the Universal PCR mix use a standard master mix. Use standard cycling parameters with an annealing temperature of 55°C. A product of approximately 1.5 kilo base pairs (kbp) is amplified.

Once the entire 16S rDNA is amplified and purified, the sequencing primers are used in individual sequencing PCR reactions to sequence the 1.5 kbp product. For purification of the amplified product and for quantification of DNA see page 240. Table 6.7 details the sequencing master mix.

PCR master mix for universal primers

Use the standard PCR protocol as a guide. DMSO at no more than a 2% final concentration can be added to improve the specificity of hybridization of the primers (Wilson *et al.*, 1990). Prepare PCR master mix in a single tube. Aliquot 20 µl volumes to 0.6 ml or 0.2 ml microfuge tubes depending upon the requirements of the thermocycler. Add 20–30 µl of paraffin oil to each tube (if hot lid is not available on the thermocycler). Tubes can be labelled and stored at –20°C until required for use.

For the PCR add 5 µl of template DNA to a 20 µl reaction tube and place in thermocycler.

Thermocycling conditions for universal primers

The thermocycling conditions are much the same as for the universal PCR, except that a relaxed annealing temperature is used. Therefore, the cycling conditions begin with one cycle at 95°C for 1–5 min. Then, 25–35 cycles of denaturation at 95°C for 30 s, annealing at 45–55°C 30 s, extension at 72°C for 30 s. A final cycle is suggested as for the last cycle but an extension time of 10 min. The thermocycler can be set to do a hold cycle at 4°C. The annealing temperature can be between 45 and 55°C.

Table 6.5. Universal primers for amplification of 16S rDNA from *Vibrio* spp.

Forward primer 5′–3′	Reverse primer 5′–3′	Reference
EUBB (7-26) AGAGTTTGATCMTGGCTCAG	EUB-A AAGGAGGTGATCCANCCRCA	Weisburg *et al.*, 1991; Suzuki and Giovannoni, 1996; Thompson *et al.*, 2001a
MH1 AGTTTGATCMTGGCTCAG	MH2 TACCTTGTTACGACTFCACCCCA	Thompson *et al.*, 2001a

Table 6.6. Sequencing primers for the 16S rDNA from *Vibrio* spp.

Primer name	*E. coli* position	Sequence 5′–3′	Reference
16F358	339–358	CTCCTACGGGAGGCAGT	Modified from primer 357r from Lane, 1991
16F536	519–536	CAGCAGCCGCGGTAATAC	Thompson *et al.*, 2001a
16F926	908–926	AACTCAAAGGAATTGACGG	Modified from primer 926f from Lane, 1991
16F1112	1093–1112	AGTCCCGCAACGAGCGCAAC	Thompson *et al.*, 2001a
16F1241	1222–1241	GCTACACACGTGCTACAATG	Thompson *et al.*, 2001a
16R339	358–339	ACTGCTGCCTCCCGTAGGAG	Modified from primer 342r from Lane, 1991
16R519	536–519	GTATTACCGCGGCTGCTG	Modified from primer 519r from Lane, 1991
16R1093	1112–1093	GTTGCGCTCGTTGCGGGACT	Thompson *et al.*, 2001a. Similar to 1100r from Lane, 1991

Thompson *et al.*, 2001a.

Visualize DNA on agarose gel

To check that the DNA has been amplified, visualize the product on a 2% agarose gel.

Clean-up amplified product

Before sequencing, the amplified product needs to be purified from the other components in the PCR reaction. Commercial kits such as QIAquick PCR kit or QIAprep Spin kit (Qiagen), PCR Kleen™ spin columns (Bio-Rad), Wizard® PCR Preps (Promega) can be used for post-PCR clean-up.

Quantify DNA concentration

Once the DNA is free of other contaminants it needs to be quantified for the PCR sequencing reaction. This can be done spectrophotometrically by reading the absorbance at 260 nm. The concentration of double-stranded DNA in μg/ml = absorbance × dilution × weight per OD. Weight per OD for dsDNA is 50 μg/ml.

Alternatively, the concentration of DNA can be estimated by gel electrophoresis in a 2% agarose gel and estimating the amount of DNA against a quantitative DNA molecular weight marker, which has known molecular weights in nanograms per band. Examples are Hyper-Ladder IV from Bioline, GeneRuler™ from MBI Fermentas.

For the sequencing PCR, the concentration of DNA needs to be 10–30 ng for a reaction volume of 10 μl.

Concentrating DNA, ethanol precipitation

If the DNA needs concentrating to obtain the desired concentration for sequencing, then an ethanol precipitation is used. If oil has been used as an overlay, remove as much as possible before the post-PCR clean-up or ethanol precipitation. This method is from Sambrook *et al.* (1989).

1. Add a 1/10 volume of 3 M sodium acetate pH 4.6
2. Add an equal volume of 95% ethanol (non-denatured alcohol)
3. Centrifuge 13,000 rpm for 15 min at room temperature
4. Carefully remove the supernatant
5. Wash twice with 70% ethanol
6. Wash once with 95% ethanol

7. Air dry, or vacuum dry
8. Add the desired volume of water or TE buffer to the dried pellet.

3 M sodium acetate = 24.6 gm of sodium acetate (molecular weight = 82.03) in 100 ml of distilled water.

Perform PCR sequencing reaction

The sequencing PCR is done using the BigDye™ Terminator v3.0 kit from Applied Biosystems.

There is only one primer per sequencing reaction. Therefore, the universal forward primer used in the amplification PCR is used in one sequencing tube, and the reverse primer in a second tube. It is advisable to do each primer in duplicate or triplicate to check for errors due to mispriming.

The sequencing PCR master mix (Table 6.7) is the recommended protocol from Applied Biosystems for use with their BigDye V3.0 Cycle Sequencing kit.

The full reaction volume is that recommended by the supplier. The volume of water is used to adjust for the concentration of DNA. Sequencing results will not be obtained if the DNA is too dilute. For a thermocycler without a hot lid, overlay tubes with 30 μl mineral oil.

Thermocycling conditions for the sequencing PCR

Perform one cycle at 96°C for 2 min. Run 25 cycles of 96°C for 30 s, 50°C annealing for 30 s, 60°C for 4 min. Hold at 4°C. Times may be shorter, depending upon the thermocycler used.

Product purification before sequencing

The product from the sequencing PCR reaction needs to be purified to remove salts and unincorporated dye terminators otherwise the first 100

Table 6.7. Sequencing PCR master mix.

Reagent	Half reaction (μl)	Full reaction (μl)
Water (ultra-pure)	0	0
Terminator ready mix	4	8
Primer F	0.5	1
DNA (10-30 ng/μl)	5.5	11
Volume μl (including DNA)	10	20

bases of the sequence will not be readable. Clean-up can be done by using Centrisep columns from Applied Biosystems, Micro Bio-Spin™ 30 columns from Bio-Rad or DyeEx Dye-Terminator Removal system from Qiagen.

A manual protocol also gives good results and is based on an ethanol precipitation.

1. Into a 0.6 ml microfuge tube place 25 µl of 95% ethanol (non-denatured, absolute alcohol)
2. Add 1 µl of 3 M sodium acetate, pH 4.6
3. Add entire PCR product (10 µl if a half reaction volume was used)
4. Place on ice for 10–20 min. Some methods recommend a strict 10 min only
5. Centrifuge 13,000 rpm at room temperature in a table-top centrifuge for 30 min. Keep the hinges of the tube to the outside so that the DNA pellet can be more easily located
6. Carefully remove supernatant without disturbing the pellet, which may not be visible
7. Immediately add 125 µl of 70% ethanol and gently roll the tube
8. Centrifuge 13,000 rpm for 5 min
9. Carefully remove the supernatant without disturbing the pellet
10. Air dry or vacuum dry the pellet
11. Send the dried product to a sequencing facility.

Sequencing

The sequencing reaction is not covered in this manual. Usually, this reaction is carried out by a specialized laboratory.

Analysis of sequence information using BIOEDIT

The sequence information can be analysed by a number of different programs. Proprietary software from Applied Biosystems is available for purchase from the company.

However, there are a number of freeware programs available and a very comprehensive one is called BIOEDIT. It is available for download from http://www.mbio.ncsu.edu/Bioedit/bioedit

Citation: Hall, T.A. (1999) BioEdit: a user-friendly biological sequence alignment editor and analysis program for Windows 95/98/NT. *Nucleic Acids Symposium Series* 41 95–98.

USING BIOEDIT. A brief explanation of the use of BIOEDIT for analysing sequence data is explained here, but the reader is referred to the help notes under 'General use of BIOEDIT' contained within the program.

The sequence information is usually e-mailed to the client laboratory. Open the e-mail and save the attachments into a folder on the hard drive.

Open the BIOEDIT program. Select OPEN SEQUENCE SET. This opens an 'open file' window. The sequence files will have the ending .ab1. Select the sequence to be examined and double click. This places the title of the sequence in the left-hand side of the BIOEDIT window, and the DNA sequence in the right-hand side window. A chromatogram window is also seen. Position the two windows so that both can be seen and check the sequence for inaccuracies. In many cases where a no base is called (N), checking the chromatograph can interpret the correct base. Do the same with the duplicate sequence. Once both are checked a consensus sequence can be obtained.

First select the duplicate sequence by clicking in the left-hand window to highlight the sequence name. Go to the menu bar and EDIT, COPY SEQUENCE. Then click on the window containing the first sequence. Go to the menu bar and EDIT, PASTE SEQUENCE. The primary and duplicate sequences will now be in the one window.

While holding down the shift key, click on both sequence names in the left-hand window to highlight both sequences. Go to menu bar SEQUENCE, PAIRWISE ALIGNMENT, ALIGN TWO SEQUENCES (optimal global alignment). A new window appears with the alignment result for the two sequences. From this window go to the menu bar and select ALIGNMENT, CREATE CONSENSUS SEQUENCE. A window now appears with the consensus sequence. To save the consensus sequence, highlight the consensus sequence in the left-hand window, go to the menu bar EDIT, COPY SEQUENCE. Open a new word document and select PASTE in the Word menu bar. The consensus sequence will appear thus:

>Consensus
GGTACTGACC etc.

Re-name the sequence as appropriate. This format is known as 'fasta' format, and is the format required for a BLAST search.

ALIGNING TWO SEQUENCES USING CLUSTAL W. BIOEDIT also contains the CLUSTAL program. To use CLUSTAL within BIOEDIT to align two sequences, select the NEW ALIGNMENT icon. Go to the document with the sequence to be aligned. Make sure the sequence is in fasta format. Go back to the BIOEDIT NEW ALIGNMENT window. Under FILE, select IMPORT FROM CLIPBOARD. The sequence name and sequence will appear in the window. Do this for the second sequence. Under ACCESSORY APPLICATION, select CLUSTAL. Run CLUSTAL. When it is finished, close the CLUSTAL window to see the alignment in the BIOEDIT window.

If the reverse complement is needed, under SEQUENCE go to NUCLEIC ACID, then to REVERSE COMPLEMENT.

CLUSTAL W citation: Thompson, J.D., Higgins, D.G. and Gibson, T.J. (1994) Clustal W: improving the sensitivity of progressive multiple sequence alignment through sequence weighting, position-specific gap penalties and weight matrix choice. *Nucleic Acids Research* 22, 4673–4680.

WEB SOFTWARE FOR MULTIPLE ALIGNMENTS.
http://dot.imgen.bcm.tmc.edu:9331/multi-align/multi-align.html
http://www.ebi.ac.uk/clustalw/Clustal   alignment program
http://www.ncbi.nlm.nih.gov/blast/bl2seq/bl2.html Blast for alignment of two sequences
http://www.mbio.ncsu.edu/Bioedit/bioedit BioEdit program for sequence analysis
http://www.technelysium.com.au/chromas.html or   http://bioinfo.weizmann.ac.il/pub/software/chromas (Chromas is a sequence editor)

BLAST SEARCH FOR SEQUENCE IDENTIFICATION. The National Centre for Biotechnology Information is located at http://www. ncbi.nlm.nih.gov Select the BLAST icon. There is a tutorial there. However, BLAST is very easy to use. Select NUCLEOTIDE BLAST. In the search window paste the sequence in fasta format. Press NOW BLAST IT. Press FORMAT to get the results. The results are placed in a queue and it may take 1 or 2 minutes for the results to be downloaded to the desktop.

Citation for BLAST: Altschul, S.F., Madden, T.L., Schäffer, A.A., Zhang, J., Zhang, Z., Miller, W. and Lipman, D.J. (1997) 'Gapped BLAST and PSI-BLAST: a new generation of protein database search programs'. *Nucleic Acids Research* 25, 3389–3402.

The Ribosomal Database Project contains ribosomal sequence information at http://rdp.cme.msu.edu/html/

6.4 Fluorescence *in Situ* Hybridization (FISH)

FISH is used increasingly in clinical laboratories for the identification of bacteria from clinical samples (DeLong *et al.*, 1989; Hogardt *et al.*, 2000; Jansen *et al.*, 2000).

Using the same principle, oligonucleotides or primers that are species-specific as shown by a BLAST search of experimentation can be used to develop sensitive FISH assays for aquatic bacteria. The primers listed in Table 6.1 would be a good starting point for developing specific FISH assays that could be applied to smears of colonies grown on agar plates. Although none of these primers has been tested as being suitable for specific probes, it was suggested that eight primers that were species-specific for *V. costicola*, *V. diazotrophicus*, *V. fluvialis*, *V. nigripulchritudo*, *V. proteolyticus*, *V. salmonicida*, *V. splendidus* and *V. tubiashii* would be suitable for use as specific probes (Lee *et al.*, 2002).

Outline of steps for FISH

Step	Method
1	Prepare smear from bacterial colonies
2	Hybridization
3	Washing
4	Microscopic examination

Oligonucleotide probes

Primers should be selected from Table 6.1 and a BLAST search conducted to ascertain whether the primers are species-specific. The single-stranded oligonucleotides are synthesized and covalently labelled with fluorescein isothiocyanate at the 5′ end.

A universal positive probe is prepared using EUB primer 5′-GCTGCCTCCCGTAGGAGT-3′. This sequence corresponds to positions 338–355 on the *E. coli* numbering system. A universal negative probe is prepared using primer non-EUB   5′-ACTCCTACGGGAGGCAGC-3′ (Amann *et al.*, 1990; Jansen *et al.*, 2000). Fluorescein isothiocyanate is added to the 5′ end of each primer.

Prior to use, the probes are diluted to a concentration of 10 ng/ml in hybridization buffer.

Preparation of smears

Bacterial cells from a colony are emulsified in a drop (10–15 µl) of sterile saline or sterile distilled water within a marked area on a glass microscope slide. The suspension should not be too thick, but should allow individual cells to be seen under the microscope. Prepare one drop per probe to be tested plus drops for the positive and negative controls. Allow to air dry.

Once dry, the cells are fixed to the slide by using a fixative solution of 4% formaldehyde in 96% ethanol. Gram-positive organisms must be permeabilized before application of the hybridization buffer. The fixed slide is placed into permeabilization buffer of 1 mg/ml of lysozyme for 5 min. Gram-negative organisms do not need the permeabilization step (Jansen *et al.*, 2000).

Hybridization

The fixed slides are hybridized in a hybridization buffer of 20 mM Tris-HCl, 0.9 M NaCl, 0.1% sodium dodecyl sulphate, pH 7.2, which contains the probe at a concentration of 10 ng/ml. The hybridization buffer (10–15 µl) can be placed on top of the fixed cells. Gram-negative organisms are hybridized for 45 min, and Gram-positives for 2 h at a temperature of 50°C (Jansen *et al.*, 2000).

One hybridization method pretreated the slides by placing them in ascending ethanol steps of 50, 80 and 96% for 3 min each step. The hybridization buffer then contained 30–40% formamide with 50 ng of probe. This concentration of formamide improved the specificity of the probe (Hogardt *et al.*, 2000). However, a concentration of 20% (vol/vol) of formamide did not show an increase in probe specificity (Jansen *et al.*, 2000).

Washing

The slides are washed in washing buffer of 20 mM Tris-HCl, 0.9 M NaCl, pH 7.2, at a temperature of 50°C for 10 min (Jansen *et al.*, 2000).

Examination of smears

The slides are mounted with VectaShield (Vector Laboratories, Burlingame, California), or Citifluor (Citifluor Ltd, London, UK).

The slides are examined for cells showing fluorescence (positive result), using a fluorescence microscope and filter set capable of detecting fluorescein. Fluorescence should be seen with the positive control, and no fluorescence with the negative control.

7

Preparation of Media for Culture and Identification

7.1 General Isolation and Selective Media

Acetate agar. See Rogosa medium under 'Test media'

Alkaline peptone water (APW) (May be used as an enrichment medium for *Vibrio* species; Furniss *et al.*, 1978)

Reagent	Amount	Preparation of media	Description of growth characteristics
Peptone	10.0 g	Dissolve reagents in distilled water and pH to 8.6. Dispense in 10 ml aliquots to McCartney bottles and autoclave at 121°C for 20 min.	May be used as an enrichment medium to isolate *Vibrio* species from contaminated samples such as faeces and polluted water. To be effective, the APW cultures should be subcultured at 6 h if incubated at 37°C and overnight if incubated at 18–20°C.
Sodium chloride	10.0 g		
Distilled water	1000 ml		

Amies transport medium (Oxoid)
This transport medium can be purchased ready-made as swabs in tubes of media, or prepared from a powdered medium available from Oxoid.

Formulae: 10.0 g charcoal (pharmaceutical), 3.0 g sodium chloride, 1.15 g sodium hydrogen phosphate, 0.2 g potassium dihydrogen phosphate, 0.2 g potassium chloride, 1.0 g sodium thioglycollate, 0.1 g calcium chloride, 0.1 g magnesium chloride, 4.0 g agar, 1000 ml distilled water, pH 7.2.

Anacker-Ordal agar (AO) (Used for growth of freshwater and marine *Cytophaga*, *Flavobacterium* and *Flexibacter* spp.; Anacker and Ordal, 1955, 1959)

Reagent	Amount	Preparation of media	Description of growth characteristics
Bacto-tryptone (Difco)	0.5 g	Add all reagents to 1000 ml of distilled water, and adjust pH to 7.2–7.4. Autoclave at 121°C for 15 min (15 lb/20 min). Cool to 50°C and pour media into Petri dishes. Store plates in sealed plastic bags at 4°C.	Using a cotton-tipped swab, collect samples from skin lesions and gills of fish and inoculate the plate. Colonies of *F. columnare* appear at 2–5 days as yellow-pigmented, rhizoid colonies with a thin spreading growth. *See photographic section.*
Yeast extract (Difco)	0.5 g		
Sodium acetate	0.2 g		
Beef extract (Difco) or Lab-Lemco powder (Oxoid)	0.2 g		Early colonies may be viewed with the aid of a stereomicroscope. Subculture by cutting out a block of agar containing the colony, and invert on to new plate. A Pasteur pipette prepared in the shape of a 'hockey stick' with the end unsealed can be used to pick out a plug of agar.
Agar (Difco or Oxoid Agar No 1)	9 g/l		
Distilled water	1000 ml		

Anacker-Ordal agar – marine (AO-M) (Used for growth of marine *Cytophaga*, *Flavobacterium* and *Flexibacter* spp. and *Tenacibaculum* (*Flexibacter*) *maritimum*; Anacker and Ordal, 1955, 1959) Add artificial seawater salts (Sigma) at 38 g/l. If using seawater, use sterile at 50–100% final concentration (Ostland *et al.*, 1999b).

Anaerobe plates (ANA). For growth of anaerobic Gram-positive and Gram-negative organisms (Oxoid manual)

Reagent	Amount	Preparation of media	Description of growth characteristics
Oxoid Wilkins-Chalgren Anaerobe agar	21.50 g	Suspend powder in water in a 1000 ml Schott bottle with magnetic flea. Autoclave at 121°C for 20 min then cool to 50°C in waterbath. Aseptically add blood. Pour into Petri dishes.	Inoculated plates are incubated at the appropriate temperature in an anaerobic atmosphere.
Distilled water	500 ml		
Sterile equine blood	30 ml		

Gram-negative anaerobe plates (ANA-GN). For growth of anaerobic Gram-negative organisms (Oxoid manual)

Reagent	Amount	Preparation of media	Description of growth characteristics
Oxoid Wilkins-Chalgren Anaerobe agar	21.50 g	Dissolve agar in water. Autoclave at 121°C for 20 min then cool to 50°C in a waterbath. Reconstitute 1 vial of supplement with 10 ml sterile distilled water and add to base. Aseptically add blood. Pour into Petri dishes.	Inoculated plates are incubated at the appropriate temperature in an anaerobic atmosphere.
Distilled water	500 ml		
Sterile equine blood	15 ml		
Oxoid G-N selective supplement	10 ml		

Artificial seawater – ASW: Sea salts (Sigma product number S 9883)

Reagent	Amount	Preparation of media	Description of growth characteristics
Sea salt (Sigma)	38 g	Add salts to distilled water and pH to 7.6. Autoclave at 121°C for 15 min.	May be added to media for growth of marine organisms. Add to AO medium for isolation of marine *Flavobacterium* and *Tenacibaculum* species.
Distilled water	1000 ml		

Addition of 18.7 g/l makes a 50% seawater concentration of 17.5‰ salinity.

Artificial seawater medium (Lewin, 1974 – Used for isolation of marine *Flexibacter*/ *Flavobacterium* group. Baumann *et al.*, 1971, quoted in MacLeod, 1968, is the same medium but without the trace element mixture – Used for marine *Vibrio* species)

Reagent	Amount	Preparation of media	Description of growth characteristics
NaCl	20.0 g	Dissolve reagents in distilled water. Add 1.0 ml of trace element solution. Adjust pH to 7.5 and autoclave at 121°C for 20 min.	Can be used as an inoculating medium for biochemical identification sets or as an initial isolation broth for marine organisms. Add agar at 15 g/l and pour into Petri dishes if a solid medium is required.
MgSO$_4$.7H$_2$O	5.0 g		
KCl	1.0 g		
CaCl$_2$.2H$_2$O	1.0 g		
Distilled water	1000 ml		
B (soluble salt of element)	0.5 mg/ml	Prepare a stock solution of trace elements so that the final concentration of each element is either 0.5 mg/l or 0.01 mg/l.	Lewin used this medium for marine *Flavobacterium* and *Flexibacter* species. The medium used by MacLeod and Baumann did not contain the trace element solution.
Fe (soluble salt of element)	0.5 mg/ml		
Mn (soluble salt of element)	0.5 mg/ml		
Co (soluble salt of element)	0.01 mg/ml		
Cu (soluble salt of element)	0.01 mg/ml		
Mo (soluble salt of element)	0.01 mg/ml		
Zn (soluble salt of element)	0.01 mg/ml		

Blood agar – BA

Reagent	Amount	Preparation of media	Description of growth characteristics
Oxoid Columbia BA base	19.5 g	Suspend agar base in water. Autoclave at 121°C for 15 min. Cool in waterbath to 50°C. Add blood to cooled agar, mix well and pour into Petri dishes to depth of approximately 3 mm. Store at 4°C in sealed plastic bags. (Plates can be left overnight on the bench and sealed in plastic bags the following day. This prevents too much moisture build-up once stored in the plastic bags.)	Plates are inoculated with a swab of the specimen, and incubated at the appropriate temperature and atmosphere. Plates are examined daily for growth and haemolysis.
Distilled water	500 ml		
Sterile equine blood	15 ml		

Bordetella bronchiseptica selective agar – CFPA medium (Smith and Baskerville, 1979; Rutter, 1981; Hommez *et al.*, 1983)

Reagent	Amount	Preparation of media	Description of growth characteristics
Columbia agar base (Oxoid)	19.5 g	Add agars to distilled water and autoclave at 121°C for 15 min. Cool to 50°C. Add sterile blood to cooled agar mixture. Reconstitute 1 vial of Bordetella supplement with 2 ml of distilled water and add to mixture. Add 5 ml of antibiotic stock. Pour plates and store at 4°C.	Colonies are 1 mm at 48 h and may be haemolytic or non-haemolytic, opaque, smooth and pearl-like, or rough, translucent, raised in the centre with an undulating margin, depending on phase variation.
Agar technical No. 3 (Oxoid)	10.0 g		
Distilled water	500 ml		
Bordetella Pertussis supplement (Oxoid, Code SR082E)	2 ml		
Sterile equine blood	15 ml		
Penicillin	20 mg	Antibiotic stock: Add 20 mg of each antibiotic to 10 ml of normal saline. Store in fridge. Handle furaltadone with caution and use mask and gloves.	
Furaltadone	20 mg		
Normal saline	10 ml		

Brucella agar (Available from Difco or Oxoid media suppliers. See also under Farrell's medium)

Burkholderia pseudomallei selective media – glycerol medium (Thomas *et al.*, 1979)

Reagent	Amount	Preparation of media	Description of growth characteristics
Agar No 3 (Oxoid)	2.4 g	Add all reagents to distilled water (glycerol may be warmed to aid pipetting). Autoclave at 121°C for 15 min. Cool to 50°C and pour into Petri dishes. Store plates at 4°C.	Plates are a mauve colour. Colonies appear at 24 h, smooth, mauve-coloured with slight metallic sheen. As incubation increases, colonies become wrinkled and umbonate. Incubate for 4 days.
Glycerol	6.0 ml		
Crystal violet stock solution	0.5 ml		
Distilled water	194 ml		
Crystal violet – Stock (1/5000 dilution)	0.5 g	Add 0.5 g of crystal violet to distilled water and stir until dissolved. Store at room temperature.	
Distilled water	100 ml		

Burkholderia pseudomallei selective broth (Modified from Thomas *et al.*, 1979)

Reagent	Amount	Preparation of media	Description of growth characteristics
MacConkey broth (purple) (Oxoid)	100 ml	Prepare MacConkey broth and add reagents except antibiotics. Autoclave at 121°C for 15 min. Cool to 50°C and add filter-sterilized antibiotics. Aseptically dispense into sterile McCartney bottles.	Detection of *Burkholderia* can be improved by incubating material in broth for 24 and 48 h, followed by subculture to plates.
Crystal violet	0.001 g		
Gentamycin	0.8 mg		
Streptomycin sulphate	5000 units		

Burkholderia pseudomallei **selective media – Ashdown's medium** (Ashdown, 1979a)

Reagent	Amount	Preparation of media	Description of growth characteristics
Tryptone soy agar	40 g	Add all ingredients except gentamycin to distilled water. Autoclave at 121°C for 15 min. Cool to 50°C. Add filter sterilized gentamycin. Mix well and pour into Petri dishes.	Colonies are flat, rough, wrinkled after 3 days. Colonies of *Burkholderia pseudomallei* absorb the Neutral red after 3 days whereas *Pseudomonas* species have no dye uptake at 3 days.
Glycerol	40 ml		
Crystal violet	5 mg		
Neutral red	50 mg		
Gentamycin	4 mg		
Distilled water	1000 ml		

Carbon dioxide atmosphere. See under 'Test media'

Cellobiose-Colistin agar – For selective isolation of *Vibrio vulnificus* (Massad and Oliver, 1987; Høi *et al.*, 1998a)

Reagent	Amount	Preparation of media	Description of growth characteristics
Solution 1		Adjust pH to 7.6. Autoclave at 121°C for 15 min and cool to 55°C.	The medium is olive green to light brown. Incubate plates at 40°C for 24–48 h. *V. vulnificus* appears as yellow colonies surrounded by a yellow zone due to fermentation of cellobiose. *V. cholerae* appears as purple colonies surrounded by a blue zone.
Bacto-peptone (Difco)	10 g		
Beef extract (Difco)	5 g		
NaCl	20 g		
Bromothymol blue	40 mg		
Cresol Red	40 mg		
Agar	15 g		
Distilled water	900 ml		
Solution 2		Filter sterilize. Add to cooled reagents of solution 1 and mix. Dispense into Petri dishes.	
Cellobiose	15 g		
Colistin	0.03 mg/ml (4×10^5 U/l)		
Distilled water	100 ml		

CFPA media. See under *Bordetella bronchiseptica* medium

Dermatophilus **selective medium – polymyxin plates** (Abu-Samra and Walton, 1977)

Reagent	Amount	Preparation of media	Description of growth characteristics
Columbia agar base (Oxoid)	19.5 g	Add agar base to distilled water and autoclave at 121°C for 15 min. Cool to 50°C and aseptically add blood and polymyxin. Polymyxin B may be dissolved in sterile normal saline first.	Grind scab material using a sterile pestle and mortar. Transfer finely ground material to a bijou bottle and add 2 parts of distilled water. Shake thoroughly then allow to settle for 15 min. Take a loopful of surface material and plate to BA and polymyxin plates. Incubate at 25 or 37°C depending on the habitat of the infected animal. Examine plates at 24 and 48 h for presence of adherent pitted dry or mucoid colonies. This medium is not completely selective and some contaminating bacteria will grow.
Distilled water	500 ml		
Sterile equine blood	50 ml		
Polymyxin B (Use at a ratio of 1000 IU/ml of medium)	62.5 mg		

Dubos medium. For isolation of *Cytophaga hutchinsonii* (Bernardet and Grimont, 1989)

Reagent	Amount	Preparation of media	Description of growth characteristics
NaNO₃	0.5 g	Add all reagents except cellobiose, to distilled water. Adjust pH to 7.2 and autoclave at 121°C for 15 min. Cool to 50°C and add filter-sterilized cellobiose solution. Mix well and pour into Petri dishes.	
K₂HPO₄	1 g		
MgSO₄.7H₂O	0.5 g		
KCl	0.5 g		
FeSO₄.7H₂O	0.01 g		
Distilled water	1000 ml		
Agar	15 g		
D-cellobiose	30% w/v	Prepare cellobiose solution and filter-sterilize.	

Note: For isolation of *Cellulomonas* species add 0.5 g yeast extract.

***Edwardsiella ictaluri* medium (EIM)**. For isolation of *Edwardsiella ictaluri* (Shotts and Waltman, 1990)

Reagent	Amount	Preparation of media	Description of growth characteristics
Bacto-tryptone (Difco)	10 g	Dissolve all reagents in distilled water and adjust the pH to 7.0–7.2. Autoclave at 121°C for 15 min. Cool to 50°C and add the 10 ml filter-sterilized solution 1 containing mannitol, colistin, bile salts and fungizone. Mix well and pour into Petri dishes.	This medium is used for the isolation of *Edwardsiella ictaluri*. Most Gram-negative bacteria are inhibited with the exception of *Proteus* spp., *Serratia marcescens*, *Aeromonas hydrophila* and *Yersinia ruckeri*. Gram-positive bacteria are inhibited with the exception of *Enterococci*. 90% of *E. tarda* isolates grow on this medium.
Yeast extract (Difco)	10 g		
Phenylalanine	1.25 g		
Ferric ammonium citrate	1.2 g		
Sodium chloride	5.0 g		
Bromothymol blue	0.03 g		
Agar (Difco)	17 g		
Distilled water	990 ml		*E. ictaluri* is seen as 0.5–1.0 mm green, translucent colonies at 48 h. Colonies of *E. tarda* have a similar size and appearance.
Solution 1			
Mannitol	3.5 g	Add reagents to 10 ml of distilled water and filter-sterilize.	*Proteus* species are 2–3 mm, brownish-green and may swarm. *S. marcescens* colonies are 2–3 mm and reddish brown. *Y. ruckeri* are 1–2 mm yellowish green, *A. hydrophila* are 2–3 mm yellowish-green opaque colonies and *Enterococci* appear as 0.5 mm yellowish colonies.
Colistin	10 mg		
Bile salts	1 g		
Fungizone	0.5 mg		
Distilled water	10 ml		

Electrolyte supplement (Added to biochemical test media to improve growth of organism when Na$^+$ alone is insufficient; Lee *et al.*, 1979)

Reagent	Amount	Preparation of media	Description of growth characteristics
NaCl	100 g	Add all reagents to distilled water and autoclave at 121°C for 15 min.	If addition of NaCl alone to biochemical identification media does not improve the growth of an organism, this electrolyte supplement may improve growth. It is added at a rate of 0.1 ml of supplement per 1.0 ml of medium.
MgCl$_2$.6H$_2$O	40 g		
KCl	10 g		
Distilled water	1000 ml		

***Erysipelothrix* selective medium**. See 'Wood's' broth, and 'Packer's' plates

Farrell's medium. For the growth of *Brucella* spp. (Farrell, 1974)

Reagent	Amount	Preparation of media	Description of growth characteristics
Oxoid blood agar base	20.0 g	Add agar to distilled water and autoclave at 121°C for 15 min. Cool to 50°C and leave at this temperature while the other reagents are prepared.	Incubate plates at 37°C in 10% CO$_2$ for 14 days.
Distilled water	500 ml		Isolates from sea mammals appear after 4 days incubation, whereas isolates from seals do not appear until day 10 of incubation, or do not grow. Incubation of plates should continue for 14 days.
Normal horse serum	25 ml	Inactivate the horse serum at 50°C for 35 min.	
25% Dextrose (D-Glucose)	20 ml	Add 125 g of D-glucose to 375 ml distilled water and dissolve by gentle heating. Dispense into 20 ml aliquots into McCartney bottles and with lids loose, autoclave at 121°C for 15 min. Store at 4°C.	
Oxoid Brucella selective supplement	10 ml	Dissolve Brucella supplement in 5 ml sterile distilled water and 5 ml methanol. Incubate for 10–15 min at 37°C and shake well to dissolve.	It has been suggested that the isolation of strains from seals may be improved by decreasing the concentration or removing bacitracin and/or nalidixic acid from the medium (Foster *et al.*, 2002).
		For complete Farrell's medium aseptically combine all ingredients while stirring.	Always inoculate a non-selective medium such as blood agar or serum dextrose medium.
		Mix well and pour media into plates to an approximate depth of 3 mm.	
Final media preparation		Combine all prepared reagents (serum, dextrose, supplement) into agar. Mix well and pour into Petri dishes to a depth of 3 mm.	Colonies are 1–2 mm in diameter at 5 days incubation, pale yellow, translucent, convex and round with entire edges.

The original medium uses the following antibiotic concentrations: bacitracin (25 units/ml), vancomycin (20 μg/ml), polymixin B (5 units/ml), nalidixic acid (5 μg/ml), nystatin (100 units/ml) and cycloheximide (100 μg/ml) (Farrell, 1974).

Farrell's medium is available from Oxoid as Brucella medium base, to which Brucella selective supplement (Oxoid code SR83) is added. The selective supplement is based on the Farrell formulation.

***Flavobacterium maritimus* media (FMM)**. May improve primary isolation for *T. maritimum* (Pazos *et al.*, 1996)

Reagent	Amount	Preparation of media	Description of growth characteristics
Peptone	5.0 g	Add reagents and powders to sterile seawater. Adjust pH to 7.2–7.4 and sterilize by autoclaving at 121°C for 20 min.	Colonies pale-yellow, flat, irregular with uneven edges. Reduces growth of *Vibrio* species and *Aeromonas* species that may outgrow *Flexibacter* on media such as MSA-B or MA 2216.
Yeast extract	0.5 g		
Sodium acetate	0.01 g		
Agar	15 g		
Sterile seawater	1000 ml		

***Flavobacterium psychrophilum* medium (FPM)**. May improve isolation rate and colony size for *F. psychrophilum* (Daskalov *et al.*, 1999)

Reagent	Amount	Preparation of media	Description of growth characteristics
Tryptone T (Oxoid)	0.5 g	Add all reagents (except sugars and skimmed milk) to distilled water and autoclave at 121°C for 15 min. Cool in a 50°C waterbath.	Colonies grow after 3–6 days and are an intense yellow colour with a spreading and irregular shape.
Yeast extract (Oxoid)	0.5 g		
Beef extract (Oxoid)	0.2 g		
Sodium acetate trihydrate (Sigma)	0.2 g		
D(+) galactose (Sigma)	0.5 g	Prepare supplements (galactose, glucose, rhamnose and skimmed milk) as separate 10% (w/v) solutions, and filter by sterilization thorough a 0.22 μm pore size Millipore Millex porosity filter. Add filter-sterilized solutions to cooled agar medium. Pour into Petri dishes and store at 4°C.	
D(+) glucose (BDH)	0.5 g		
L-rhamnose (Sigma)	0.5 g		
Skimmed milk (Oxoid)	0.5 g		
Water	1000 ml		
Agar – bacteriological agar no. 1 (Oxoid)	9.0 g		

This medium is based on Anacker Ordal (AO) agar (also known as *Cytophaga* agar) supplemented with galactose, glucose, rhamnose and skimmed milk. The size and number of colonies of *F. psychrophilum* isolated is increased compared to AO medium. Broth can be prepared without the addition of agar.

***Flexibacter polymorphus* media** (Lewin, 1974)

Reagent	Amount	Preparation of media	Description of growth characteristics
NaCl	20.0 g	Dissolve reagents in distilled water. Adjust pH to 7.5 and autoclave at 121°C for 20 min.	The organism is maintained in 5 ml of the broth medium and subcultured twice weekly. Filaments several hundred microns in length and 1.5 μm wide are produced. They have rounded ends, are flexuous, unbranched and cylindrical. The growing filaments may be peach coloured. Cultures in late growth phase may show a refractile granule at each end, which is more easily seen after cell lysis.
$MgSO_4.7H_2O$	5.0 g		
KCl	1.0 g		
$CaCl_2.2H_2O$	1.0 g		
Fe (soluble salt)	0.5 mg		
B (soluble salt)	0.5 mg		
Mn (soluble salt)	0.5 mg		
Co (soluble salt)	0.01 mg		Cobalamin is essential for growth.
Cu (soluble salt)	0.01 mg		Agar can be added to the medium for plate media.
Mo (soluble salt)	0.01 mg		
Zn (soluble salt)	0.01 mg		The organism also grows on agar medium prepared with seawater and yeast extract (10 mg/ml). Colonies may be peach coloured with a filamentous margin.
Distilled water	1000 ml		

Flexibacter **maintenance medium** (Lewin and Lounsbery, 1969; Lewin, 1974)

Reagent	Amount	Preparation of media	Description of growth characteristics
Tryptone (Difco)	1.0 g	Add all reagents to filtered seawater and adjust pH to 7.5. Dispense media into tubes. Autoclave at 121°C for 20 min. Optimum cobalamin is 0.3 µg/l.	The organisms are maintained in 5 ml of the broth medium, incubated between 22 and 32°C, and subcultured twice weekly.
Casamino acids (Difco)	1.0 g		
Monosodium glutamate	5.0 g		
Sodium glycerophosphate	0.1 g		The Flexibacteria may show a range of pigmented filaments ranging from pink, orange or yellow colouration. On solid medium they may produce fimbriate margins.
Cobalamin	1.0 ug		
Fe (soluble salt)	0.5 mg		
B (soluble salt)	0.5 mg		
Mn (soluble salt)	0.5 mg		Agar can be added to the medium for plate media.
Co (soluble salt)	0.01 mg		
Cu (soluble salt)	0.01 mg		A semi-solid medium can be produced using 0.3% agar, and viable filaments can be maintained for up to 1 month when grown in this medium at room temperature.
Mo (soluble salt)	0.01 mg		
Zn (soluble salt)	0.01 mg		
Filtered seawater	1000 ml		

Glycerol Lab Lemco broth (Used for storage of cultures at −80°C)

Reagent	Amount	Preparation of media	Description of growth characteristics
Lab Lemco broth (Oxoid)	0.64 g	Add all reagents together. Pipetting of glycerol may be facilitated by pre-warming. Dispense into 2 ml volumes into Bijou bottles. Autoclave at 121°C for 15 min. Store media at 4°C.	Used as −80°C storage medium. Using a sterile cotton-tipped swab, scrape bacterial growth from an agar plate and inoculate into 1 ml of Glycerol Lab Lemco medium in a Nunc tube. Use an inoculum of approximately tube 5 McFarland opacity. Place tube in cryobox in −80°C freezer. This medium is also suitable for liquid nitrogen storage.
Glycerol	20 ml		
Distilled water	80 ml		

Helicobacter **selective media**. Use Skirrow's medium (also known as VPT media; Skirrow, 1977)

Reagent	Amount	Preparation of media	Description of growth characteristics
Blood agar base No. 2 (Oxoid)	20 g	The agar is dissolved in water and autoclaved at 121°C for 15 min. After cooling to 50°C add 15 ml of sterile equine blood. Reconstitute 1 vial of supplement using 2 ml of sterile distilled water and add to cooled media. Mix thoroughly. Pour media into plates to an approximate depth of 3 mm. Store plates at 4°C.	Other bases such as Columbia agar base (Oxoid), or Brucella medium base (Oxoid), or tryptone soy agar may be used to prepare this medium. However, it is suggested that blood agar base No. 2 is more nutritionally rich and also that trimethoprim is more active in this base.
Distilled water	500 ml		
Sterile equine blood	15 ml		To reduce contamination from other organisms, the sample can be filtered through a 0.65 µm filter and the filtrate cultured to the plate (Butzler *et al.*, 1973).
Oxoid Campylobacter supplement (Skirrow)	2 ml		The plates are incubated at 37°C in a microaerophilic atmosphere of N_2, H_2, CO_2 (80:10:10) for 2–4 weeks. Gas generating packs are available from commercial suppliers such as MGC Anaero Pak™ Campylo from Mitsubishi Gas Chemical Company.
			Helicobacter species will grow as pin-point colonies and also appear as a thin spreading film across the plate.

This medium was developed by M.B. Skirrow (1977) for the isolation of *Campylobacter* species. It is also suitable for the isolation of *Helicobacter* species and is widely quoted in many references. It is also available as ready-made media from a number of media producers. The antibiotics are often supplied as a supplement and, depending on the originators of the medium, the supplements will be named accordingly. For example, Oxoid produces 'Campylobacter selective supplement (Skirrow) for additions of vancomycin, polymyxin and trimethoprim, and Campylobacter selective supplement (Blaser-Wang) for additions of vancomycin, polymyxin, trimethoprim, amphotericin B, and cephalothin'. The additive containing the amphotericin B is recommended when fungal contamination is suspected. The original concentrations described by Skirrow (1977) are vancomycin (10 mg/l), polymyxin B (2.5 IU/ml), and trimethoprim (5 mg/l). A series of papers describing the isolation and identification of *Helicobacter cetorum* from dolphins and whales (Harper *et al.*, 2000, 2002a,b) describes the use of TVP and CVA media from Remel. These are, in essence, Skirrow's media and Blaser-Wang media, respectively.

Hsu-Shotts agar – HS (For *Flavobacterium columnare, F. psychrophilum*; Bullock *et al.*, 1986)

Reagent	Amount	Preparation of media	Description of growth characteristics
Tryptone	2 g	Add all reagents to distilled water, except neomycin sulphate. Autoclave at 121°C for 15 min. Cool to 50°C. Add filter-sterilized neomycin sulphate, mix well and pour into Petri dishes.	This is a semi-selective medium for the isolation of *Flavobacterium columnare*. At 48 h characteristic yellow, spreading colonies are seen.
Yeast extract	0.5 g		
Gelatin	3 g		
Agar	15 g		
Distilled water	1000 ml		
Neomycin sulphate	4.0 µg/ml	Prepare solution and filter-sterilize.	

Note: HSM. Addition of 18.7 g/l sea salts (Sigma) to make a 50% seawater concentration (17.5%° salinity), and 200 IU/ml polymyxin B is suitable for isolation of *Tenacibaculum* (*Flexibacter*) *maritimum* (Chen *et al.*, 1995).

Inositol horse serum (Suspension medium for freeze-drying organisms)

Reagent	Amount	Preparation of media	Description of growth characteristics
Inositol	5 g	Dissolve inositol in horse serum. Filter through a 0.45 µm filter followed by further filtration through a 0.22 µm filter for sterilization. Check sterility. Dispense into McCartney bottles and store at 4°C.	Place 1 ml of sterile inositol horse serum into a freeze-dried vial (Wheaton serum bottle). Using a sterile cotton swab or loop emulsify a heavy suspension of bacteria in the medium. Snap-freeze in liquid nitrogen and follow instructions for freeze-drying relevant to the equipment being used.
Horse serum	100 ml		

KDM2. For growth and isolation of *Renibacterium salmoninarum* (Evelyn, 1977)

Reagent	Amount	Preparation of media	Description of growth characteristics
Tryptone	1 g	Add all reagents to distilled water and adjust pH to 6.5–6.8 (with NaOH). Dispense into tubes and autoclave at 121°C for 20 min.	Incubate plates at 15°C for up to 2 months. Initial growth is visible between 2 to 8 weeks. Colonies are pin-point to 2 mm in size. Old colonies may have a granular or crystalline appearance. To assist in prevention of overgrowth by fast growing colonies, examine plates every few days and aseptically remove these colonies.
Yeast extract	0.05 g		
L-cysteine (chlorhydrate)	0.1 g		
Agar	1.5 g		
Distilled water	100 ml	Store at 4°C for 1 month.	
Fetal calf serum	5–10%	For use, heat tubes. Cool to 50°C and add calf serum. Pour into plates.	

KDMC. For *Renibacterium salmoninarum* (Daly and Stevenson, 1985)

Reagent	Amount	Preparation of media	Description of growth characteristics
KDM2 medium		As above.	Addition of charcoal acts as a detoxifying agent.
Activated charcoal (Difco)	0.1%	For use heat tubes. Cool to 50°C and add charcoal. Pour into plates.	

Marine salt agar with blood (MSA-B) (Used as a general-purpose medium for isolation of organisms from the marine environment)

Reagent	Amount	Preparation of media	Description of growth characteristics
Tryptone soy agar (Oxoid)	20.0 g	Dissolve TSA and NaCl in water and autoclave at 121°C for 15 min. Cool in waterbath to 50°C.	TSA with added blood has the advantage of providing differentiation between haemolytic and non-haemolytic *Vibrio* spp. Some *Vibrio* species will be haemolytic on blood agar and non-haemolytic on MSA-B despite the latter medium providing better growth for most marine *Vibrio* species.
NaCl	7.5 g (= 2% w/v final conc)	Aseptically add 15 ml blood to cooled agar, mix well and pour into Petri dishes to depth of approximately 3 mm. Store at 4°C.	
Distilled water	500 ml		
Sterile equine blood	15.0 ml		

Marine 2216 agar (Difco) (ZoBell, 1941)

Reagent	Amount	Preparation of media	Description of growth characteristics
Bacto marine agar 2216 (Difco – complete medium)	55.1 g	Add powder to distilled water and boil to dissolve. Autoclave at 121°C for 15 min. pH 7.6 ± 0.2.	Plates are a light amber colour and slightly opalescent.
Distilled water	1000 ml		

MA 2216 can be purchased from Difco as a broth or agar. Composition of MA 2216: 5.0 g Bacto peptone; 1.0 g Bacto yeast extract; 0.1 g Fe(III) citrate; 19.45 g NaCl; 5.9 g $MgCl_2$ (dried); 3.24 g $NaSO_4$; 1.80 g $CaCl_2$; 0.55 g KCl; 0.16 g Na_2CO_3; 0.08 g KBr; 34.0 mg $SrCl_2$; 22.0 mg H_3BO_3; 4.0 mg Na-silicate; 2.4 mg NaF; 1.6 mg $(NH_4)NO_3$; 8.0 mg Na_2HPO_4; 15.0 g agar; 1000 ml distilled water.

Medium K (Used for isolation of *Chryseobacterium* (*Flavobacterium*) *scophthalmum*; Mudarris *et al.*, 1994)

Reagent	Amount	Preparation of media	Description of growth characteristics
Yeast extract (Oxoid)	1 g	Add all ingredients to seawater and adjust to pH 7.2. Autoclave at 121°C for 15 minutes.	Orange-pigmented colonies are shiny, smooth, round, raised entire, 5–6 mm at 48 h, 25°C.
Beef extract (Oxoid)	5 g		
Casein (Oxoid)	6 g		
Tryptone (Oxoid)	2 g		On primary culture, isolates may show slight gliding motility, but lose this after storage.
Anhydrous $CaCl_2$	1 g		
Agar (Oxoid No. 1)	15 g		
750 ml seawater (aged 30 days)			

Middlebrook 7H10-ADC medium (Used for isolation of *Mycobacterium abscessus*; Teska *et al.*, 1997)

Reagent	Amount	Preparation of media	Description of growth characteristics
NH_4SO_4	0.05 g	Add chemical reagents to distilled water. Autoclave at 121°C for 15 min. Cool to 50°C. Aseptically add sterile bovine albumin, nalidixic acid and cycloheximide, pyridoxine and biotin.	Incubate at 25°C for 14–28 days. Growth occurs in 7 days. Young colonies may exhibit a light blue-green tint. With age, colonies are off-white to tan in colour.
KH_2PO_4	0.15 g		
Na_2HPO_4	0.15 g		
Sodium citrate	0.04 g		
$MgSO_4$	0.0025 g		
$CaCl_2$	0.00005 g	Media can be prepared as broths or add 1.5 g of agar for preparation of plates.	
$ZnSO_4$	0.0001 g		
$CuSO_4$	0.0001 g		
L-glutamic acid	0.05 g		
Ferric ammonium citrate	0.004 g		
Pyridoxine	0.0001 g		
Biotin	0.00005 g		
Malachite green	0.025 g		
Glycerol	0.5 ml		
Nalidixic acid (35 μg/ml)	0.0035 g		
Cycloheximide (400 μg/ml)	0.04 g		
Bovine albumin V	0.5 ml		
D-glucose	0.2 g		
Catalase	0.0003%		
Distilled water	100 ml		

This medium is modified from Middlebrook's medium (Middlebrook *et al.*, 1960) by the addition of ADC (albumin, catalase and dextrose).

Mycoplasma medium – general purpose medium. Bacto Pleuropneumonia-like organism (PPLO) agar and broth with supplements (Difco). (Used for isolation of *Mycoplasmas* species. This commercial medium, with the added supplements, is based on Hayflick's medium)

Reagent	Amount	Preparation of media	Description of growth characteristics
PPLO agar	35 g	Agar plates	Mince or grind tissue and place a few loopfuls
Distilled water	700 ml	Add dehydrated media to distilled water and autoclave at 121°C for 15 min. Cool to 50–60°C. Aseptically add 300 ml Mycoplasma supplement or 300 ml of Mycoplasma supplement S. Mix well. Dispense into small (5 cm) Petri dishes. Store in plastic bags at 4°C.	into a 3 ml broth. Incubate broth at 25 and 37°C. A change in pH results in a colour change from red to yellow, and may indicate growth of *Mycoplasma*. Bacterial contamination may also change the pH of the medium. The bacterial contamination can be filtered off by passing the broth through a 0.22 μm filter. Place drops of the filtered broth to a fresh broth and to a plate.
PPLO broth	21 g	Broths	
Distilled water	700 ml	Add dehydrated media to distilled water and autoclave at 121°C for 15 min. Cool to 50–60°C. Add either supplement as for agar.	Subculture 2 drops from a broth culture to a PPLO agar plate every 3 or 4 days. Incubate plates in air and 5–10% CO_2 at 25 and 37°C. Subculturing on a Tuesday and a Friday may be a convenient time frame for the laboratory.
Phenol red			
Mycoplasma supplement	1 vial	Supplement	
Sterile distilled water	30 ml	Rehydrate vial with distilled water and swirl to dissolve. Aseptically add 1 vial per 70 ml of PPLO agar or broth.	Examine plates under a stereomicroscope for presence of typical 'fried-egg' type colonies.
Mycoplasma supplement S	1 vial	Supplement S	To differentiate *Mycoplasma* colonies from contaminating bacterial colonies, stain colonies with Dienes stain. *Mycoplasma* colonies stain blue with a distinctly dense blue centre and a lighter blue periphery. Mycoplasma colonies retain the stain for 24 h, whereas bacterial colonies decolorize after 30 min. See Dienes stain in 'Tests' section.
Sterile distilled water	30 ml	Rehydrate vial with distilled water and swirl to dissolve. Aseptically add 1 vial per 70 ml of PPLO agar or broth.	
DNA (optional)		DNA solution	
DNA (optional) (Calbiochem polymerized calf thymus DNA) 0.2% solution	1%	Prepare 0.2% solution by dissolving 0.2 g DNA in 100 ml distilled water. Autoclave at 121°C for 15 min. Add 10 ml to 1000 ml of above medium to produce a final concentration of 1%.	
Distilled water 100 ml			

Formulations

- PPLO agar = per litre contains 50 g Bacto beef heart infusion, 10 g Bacto peptone, 5 g sodium chloride, Bacto agar 14 g. The broth does not contain the agar component.
- Mycoplasma supplement = per 30 ml vial is Bacto yeast extract 0.01 g, 1.6 g desiccated horse serum.
- Mycoplasma supplement S = Per 30 ml vial contains Bacto yeast extract 0.01 g, 1.6 g of desiccated horse serum, 55,000 units penicillin, 50 mg thallium acetate.
- Bacto heart infusion broth (Difco) contains 500 g beef heart infusion, 10 g bacto-tryptone, 5 g sodium chloride.
- The addition of 0.2% DNA can be added to the above medium. DNA is recommended for primary isolation of the bovine mycoplasma *M. bovigenitalium*, and is thought to stimulate the growth of other *Mycoplasma* species as well. Therefore it is recommended as an optional media component (Freundt, 1983).

Mycoplasma medium – modified Hayflick medium (Chanock *et al.*, 1962) (Modified and used for isolation of *Mycoplasma mobile*; Kirchhoff and Rosengarten, 1984. Also used for many animal mycoplasma species)

Reagent	Amount	Preparation of media	Description of growth characteristics
Bacto PPLO broth (Difco)	16.8 g	Dissolve PPLO broth in distilled water and autoclave at 121°C for 15 min. Cool to 50°C for agar containing media, or room temperature for broths, and aseptically add other filter-sterilized ingredients. Dispense in 3 ml volumes to bijou tubes for broths, or 5 ml volumes to small Petri dishes for agar plates.	See above under Mycoplasma medium – general-purpose medium.
Horse or bovine serum	200 ml		
Yeast extract (50% w/v)	10 ml		
DNA	0.02 g		
Penicillin	2000 IU/ml		
Thallium acetate (1.25% w/v)	10 ml		
Distilled water	800 ml		

For agar plates add 1.0% w/v purified agar (Oxoid) or use PPLO agar. *Note:* Handle thallous acetate with caution. Use gloves and mask.

Mycoplasma medium (Used for isolation of *Mycoplasma* from crocodiles; Kirchhoff *et al.*, 1997)

Reagent	Amount	Preparation of media	Description of growth characteristics
Brain heart infusion broth (Oxoid)	37 g	Dissolve brain heart infusion, yeast extract and glycerol in distilled water. Glycerol can be warmed slightly to facilitate pipetting. Autoclave at 121°C for 15 min. Prepare other reagents and filter sterilize. Add each reagent to cooled autoclaved media. Dispense into 5 cm Petri dishes. Store in plastic bags at 4°C.	Incubate plates in a candle jar or similar CO_2 generating atmosphere, at 37°C.
Yeast extract (Oxoid)	2 g		
Glycerol	8 ml		
Thallium acetate (10% w/v)	2.5 ml		
Ampicillin (5% w/v)	2 ml		
NAD (1% w/v)	10 ml		
Distilled water	700 ml		

NAD, nicotinamide adenine dinucleotide; TTC, tetrazolium chloride.
Kirchhoff *et al.* (1997) also used this medium as a base to study the metabolism of glucose, arginine hydrolysis, TTC and phosphatase activity.

Nutrient agar (Used as a general isolation medium)

Reagent	Amount	Preparation of media	Description of growth characteristics
Nutrient agar (Oxoid)	14.0 g	Add agar powder to distilled water. Autoclave at 121°C for 15 min. Cool to 50°C and dispense into Petri dishes. Store at 4°C.	
Distilled water	500 ml		

Organic growth requirements

Some strains of *Vibrio*, e.g. *Vibrio ordalii* (previously *V. anguillarum* type II), *Moritella* (*Vibrio*) *marina* and some strains of *Vibro* (*Photobacterium*) *logei* need 0.05% yeast extract for growth (Baumann *et al.*, 1980). A 10× stock solution of yeast extract can be prepared and then 250 µl added to 5 ml of broth media.

Packer's plates (Selective medium for *Erysipelothrix*; Packer, 1943)

Reagent	Amount	Preparation of media	Description of growth characteristics
Columbia blood agar base (Oxoid)	20.0 g	Add agar base to water and autoclave at 121°C for 15 min.	Colonies are 0.5–1 mm at 24–48 h. It is suggested this medium is used in conjunction with Wood's
Distilled water	500 ml	Cool to 50°C and aseptically	broth. Approximately 1 g of chopped or minced
Sterile equine blood	25 ml	add sodium azide, crystal violet and blood. Mix well and pour into Petri dishes.	tissue or sample material is placed into 10 ml of Wood's broth. Incubate at 25°C and subculture from broth to BA and Packer's plates at 24 and
Crystal violet (stock = 0.25 g in 100 ml of distilled water)	2 ml	Prepare stock solutions of crystal violet and sodium azide.	48 h. Examine plates at 24 and 48 h for typical colonies. Colonies of *Erysipelothrix rhusiopathiae*
Sodium azide (NaH₃) (stock = 1 g in 100 ml of distilled water)	12.5 ml	Autoclave at 121°C for 15 min.	are 0.5–1 mm, grey-green at 48 h, with slight α-haemolysis around the colony. They resemble α-haemolytic *Streptococci*.
			See photographic section.

Pasteurella multocida selective media. NB plates (Rutter *et al.*, 1984)

Reagent	Amount	Preparation of media	Description of growth characteristics
Columbia agar base (Oxoid)	7.8 g	Add agar to distilled water and autoclave	This is a semi-selective medium.
Distilled water	200 ml	at 121°C for 15 min. Cool to 50°C and	Colonies of *Pasteurella multocida*
Sterile equine blood	12 ml	aseptically add blood and 1 ml of antibiotic stock. Mix well and pour into Petri dishes.	appear as grey non-haemolytic colonies 1–2 mm and may be
Neomycin sulphate (2.0 µg/ml)	4.0 mg	Prepare antibiotic stock by adding	slightly mucoid.
Bacitracin (3.5 µg/ml)	7.0 mg	neomycin and bacitracin to normal saline.	
Normal saline	10 ml	Add 1 ml to cooled agar and blood mixture as described above.	

Peptone yeast medium – PY (For growth of *Cytophaga*, *Flavobacterium* and *Sphingobacterium* species; Takeuchi and Yokota, 1992)

Reagent	Amount	Preparation of media	Description of growth characteristics
Peptone	1 g	Add all reagents to distilled water and pH to 7.0. Autoclave	For growth of *Pedobacter heparinus*,
Yeast extract	0.2 g	at 121°C for 15 min. Cool to 50°C and dispense into Petri	*Sphingobacterium spiritivorum*, *S.*
Glucose	0.2 g	dishes.	*multivorum* and *Flavobacterium*
NaCl	0.2 g		species. Incubate at 28°C for 2 days.
Agar	1.5 g	Can be prepared as a broth by omitting the agar.	
Distilled water	100 ml		

Poly plates (See under *Dermatophilus congolensis*)

PYS-2 Medium (For growth of *Vibrio rumoiensis*; Yumoto *et al.*, 1999)

Reagent	Amount	Preparation of media	Description of growth characteristics
Polypeptone	8 g	Add all reagents to distilled water and pH to 7.5.	Colonies of *Vibrio rumoiensis* are circular
Yeast extract	3 g	Autoclave at 121°C for 15 min. Cool to 50°C and	and colourless. Growth can occur between
NaCl	5 g	aseptically dispense into Petri dishes.	2 and 34°C with an optimum at 27–30°C.
Agar	15 g		
Distilled water	1000 ml		

R2A agar (Oxoid CM 906, or Difco) (Used for isolation of Antarctic *Flavobacterium* spp.; McCammon and Bowman, 2000)

Reagent	Amount	Preparation of media	Description of growth characteristics
Agar CM906	18.1 g	Add powder to distilled water and adjust pH to 7.2. Boil to dissolve. Sterilize by autoclaving at 121°C for 15 min. Cool to 50°C and pour into Petri dishes.	This is a nutritionally reduced medium that enhances recovery of heterotrophic bacteria from treated waters, and assists in recovery of bacteria that are stressed or chlorine-tolerant. The sodium pyruvate increases the recovery of stressed cells.
Distilled water	1000 ml		

Composition g/l of Oxoid CM 906: yeast extract (0.5), tryptone (0.25), peptone (0.75), dextrose (0.5), starch (0.50), di-potassium phosphate (0.3), magnesium sulphate (0.024), sodium pyruvate (0.3), agar (15.0).

Rogosa acetate agar (Oxoid). See under 'Tests' section

Strontium chloride B enrichment broth (Use for isolation of *Salmonella* spp. and *Edwardsiella tarda*; Iveson, 1971)

Reagent	Amount	Preparation of media	Description of growth characteristics
Bacto tryptone (Difco)	0.5 g	Add all reagents to the distilled water. Dispense into 10 ml volumes in McCartney bottles and autoclave at 121°C for 20 min. Final concentration of strontium chloride is 3.4% and pH is 5.0–5.5.	Place macerated or chopped sample (0.5 ml) or swab of sample into broth. Incubate at 37°C for 24 and 48 h. Subculture from broth on to selective media of MCA or DCA. On MCA plates the colonies of *Edwardsiella tarda* and *Salmonella* spp. appear as non-lactose fermenting colonies. On DCA the colonies appear as pale pink to clear colonies with or without a black centre (H$_2$S). *E. tarda* colonies are slightly smaller in size than *Salmonella* and the black centre takes longer to develop.
Sodium chloride	0.8 g		
Potassium dihydrogen phosphate	0.1 g		
Strontium chloride 60%	6.0 ml		
Distilled water	100 ml		

Seawater – artificial. See under artificial seawater

Serum-dextrose agar (Non-selective medium for *Brucella* species; Alton and Jones, 1967)

Reagent	Amount	Preparation of media	Description of growth characteristics
Nutrient agar	95 ml	Prepare nutrient agar and cool to 50°C. Add 5 ml of a stock solution of serum-dextrose per 95 ml of nutrient agar. Mix and pour into Petri dishes.	After 4 days growth, smooth colonies of *Brucella* are small, round, glistening and blue or blue-green in colour when examined by obliquely reflected light. Rough colonies are yellow-white in colour and have a dry granular appearance.
Serum-dextrose	5 ml to 95 ml of nutrient agar		
Stock solution of serum-dextrose	1 g of dextrose per 5 ml of serum	For a stock solution of serum-dextrose, dissolve pure dextrose (glucose) in inactivated serum, at a rate of 5 g of dextrose per 5 ml of serum. Filter-sterilize and store at 4°C or –20°C.	

This medium is available from Oxoid as Brucella medium base (Code CM169). Inactivated horse serum (5%) is added to the autoclaved base medium.

Shieh medium + Tobramycin (SM-T) (Selective medium for *Flavobacterium columnare* and *F. psychrophilum*; Decostere *et al.* 1997; Shieh, 1980)

Reagent	Amount	Preparation of media	Description of growth characteristics
Peptone (Difco)	5 g	Suspend all chemicals into 1000 ml of distilled water and pH to 7.2. Autoclave at 121°C for 15 min. Cool to 50°C and added filter-sterilized solution of tobramycin. Mix and pour media into Petri dishes.	Using a cotton-tipped swab, collect samples from skin lesions and gills of fish and inoculate the plate. Colonies of *F. columnare* appear at 2–5 days as yellow-pigmented, rhizoid colonies with a thin spreading growth. *See photographic section.*
Yeast extract (Difco)	0.5 g		
Sodium acetate	0.01 g		
BaCl$_2$ (H$_2$O)$_2$	0.01 g		
K$_2$HPO$_4$	0.1 g		
KH$_2$PO$_4$	0.05 g		Early colonies may be viewed with the aid of a stereomicroscope and a plug of the colony cut out using an unsealed Pasteur pipette prepared in the shape of a 'hockey stick'. Subculture the colony plug to a fresh plate.
MgSO$_4$7H$_2$O	0.3 g	Ensure the moisture content remains in the media by storing plates at 4°C in sealed plastic bags. Growth of *F. columnare* is optimal when the plates are fresh, but reduces as the surface moisture decreases.	
CaCl$_2$2H$_2$O	0.0067 g		
FeSO$_4$7H$_2$O	0.001 g		
NaHCO$_3$	0.05 g		Shieh medium plus tobramycin reduces the growth and colony size of *Aeromonas hydrophila* and prevents growth of *A. salmonicida*, *S. putrefaciens* and *Ps. fluorescens*.
Tobramycin	0.5 μg		
Noble agar (Difco)	10 g		
Distilled water (pH 7.2)	1000 ml		

Note: The original method states 1 μg/ml of tobramycin; however, 0.5 μg/ml is recommended for Australian strains of *F. columnare* as growth may be inhibited or reduced at 1 μg/ml of tobramycin (Dr Annette Thomas, Department of Primary Industries, Queensland, 2000, personal communication).

SKDM. (Selective medium for *Renibacterium salmoninarum*, Austin *et al.*, 1983)

Reagent	Amount	Preparation of media	Description of growth characteristics
Tryptone	1.0 g	Add all reagents to distilled water and adjust pH to 6.8. Autoclave at 121°C for 15 min, then cool to 50°C.	Inoculate plates with infectious material and incubate in a humid atmosphere at 15°C for 12 weeks. Examine plates regularly for the presence of pin-point colonies. Maximum size of colonies is 2 mm. Colonies are white or creamy, shiny, smooth, raised and round. Old colonies may have a granular or crystalline appearance.
Yeast extract	0.05 g		
Cycloheximide	0.005 g		
Agar	1.0 g		
Distilled water	100 ml		
Fetal calf serum	10.0 g	Filter-sterilize L-cysteine hydrochloride, D-cycloserine, polymyxin B sulphate and oxolinic acid. Add these plus sterile fetal calf serum to autoclaved media. Pour into plates.	
L-cysteine hydrochloride	0.1 g		
D-cycloserine	0.00125 g		
Polymyxin B sulphate	0.0025 g		
Oxolinic acid	0.00025 g		

Shewanella marine agar (SMA) (Used for isolation of *Shewanella woodyi*; Makemson *et al.*, 1997)

Reagent	Amount	Preparation of media	Description of growth characteristics
Bacto peptone (Difco)	5.0 g	Add all reagents to distilled water. Autoclave at 121°C for 15 min.	A medium suitable for growth of luminous marine bacteria, especially Shewanella.
Bacto yeast extract (Difco)	1.0 g		
Bacto agar (Difco)	15.0 g		
1x sea salts (see below)	200 ml		
Distilled water	1000 ml		

5× sea salts stock: 2.58 M NaCl, 0.125 M MgCl$_2$, 0.125 M MgSO$_4$ 0.1 M KCl, distilled water 1000 ml, pH to 7.5.

Siem selective medium (Selective medium for *Aerococcus viridans*; Stewart, 1972; Gjerde, 1984)

Reagent	Amount	Preparation of media	Description of growth characteristics
Glucose	6.5 g	Add powders to distilled water and dissolve. Adjust pH to 7.4. Autoclave.	Growth of Gram-positive cocci is promoted with a change of medium from purple to yellow because of acid production. Incubate at 25°C for 5 days. Subculture yellow-coloured broth to BA.
Yeast extract	4.5 g		
Tryptone	15.0 g		
NaCl	6.4 g		
Phenyl ethyl alcohol	2.5 g		
Bromocresol purple	0.008 g		
Distilled water	1000 ml		

Skirrow's medium See under Helicobacter selective medium.

SWT (A seawater-based complex medium for growth of marine organisms; Nealson, 1978; Boettcher *et al.*, 1999)

Reagent	Amount	Preparation of media	Description of growth characteristics
Tryptone	0.5 g	Add reagents to distilled water and autoclave at 121°C for 15 min. Cool to 50°C and aseptically add filter-sterilized seawater. Pour into Petri dishes.	Used for cultivation of *Vibrio fischeri*, *Roseobacter* spp., *Stappia* spp.
Yeast extract	0.3 g		
Glycerol	0.3 g		
Seawater	70 ml		
Distilled water	30 ml		
Agar	1.2 g		

TCBS cholera medium (TCBS) (Semi-selective for growth of *Vibrio* spp.; Kobayashi *et al.*, 1963; Nicholls *et al.*, 1976)

Reagent	Amount	Preparation of media	Description of growth characteristics
TCBS cholera medium (Oxoid)	44.0 g	Add powder to distilled water. Autoclave at 121°C for 15 min. Cool to 50°C and dispense into Petri dishes. Store at 4°C.	This selective medium will grow the majority of *Vibrio* species. Some strains, however, such as *Vibrio* (*Listonella*) *ordalii* do not grow on this medium. *Aeromonas* spp. and *Pseudomonas* will grow weakly as small colonies approximately 1 mm at 24–48 h.
Distilled water	500 ml		
			Vibrio species that ferment sucrose appear as yellow colonies whereas sucrose-negative isolates appear as green colonies.

Notes:

• Some *Vibrio* species grow poorly on TCBS and a few strains do not grow at all on this medium. Some brands of TCBS are more inhibitory than others. Eiken and Oxoid brands supported the growth of a greater number of *Vibrio* species than BBL or Difco brands, in particular for *V. cholerae* and non-cholera *Vibrio* spp. Also, it is recommended that each batch of medium is tested for growth of *Vibrio* spp. (Nicholls *et al.*, 1976).

- TCBS detects a lower number of *Vibrio* colonies than MSA-B when performing TBCs on water. Colonies that grow on MSA-B and not on TCBS on primary culture often grow on TCBS when subcultured to that medium, and identify as *Vibrio* spp. biochemically.
- After prolonged incubation, sucrose-fermenting colonies that initially appeared yellow may turn green once the sucrose in the medium has been used up or the pH of the medium changes.
- Formulation: yeast extract (5 g), peptone (10 g), sodium thiosulphate (10 g), sodium chloride (10 g), ox bile (8 g), sucrose (20 g), NaCl (10 g), ferric chloride (1 g), bromothymol blue (0.04 g), thymol blue (0.04 g), agar (1.4 g), water (1000 ml) to pH 8.6.

Tryptone soya agar (TSA) (Used as a general-purpose isolation medium)

Reagent	Amount	Preparation of media	Description of growth characteristics
TSA (Oxoid)	20.0 g	Dissolve TSA in water and autoclave at 121°C for 15 min. Cool in waterbath to 50°C and pour into Petri dishes to depth of approximately 3 mm. Store at 4°C.	Suitable as a general-purpose medium. Add NaCl (7.5 g) to give a final concentration of 2% to isolate bacteria from marine sources.
Distilled water	500 ml		

Tryptone yeast extract salt medium (TYES) (Used for the growth of *Flavobacterium columnare*; Triyanto and Wakabayashi, 1999)

Reagent	Amount	Preparation of media	Description of growth characteristics
Tryptone	0.4 g	Dissolve reagents in water and autoclave at 121°C for 15 min. Cool in waterbath to 50°C and pour into Petri dishes to depth of approximately 3 mm. Store at 4°C.	Used for the growth of *F. columnare*. Incubate plates at 25°C for 24 h. Colonies appear as yellow-pigmented, rhizoid, mucoid, or honeycomb-shaped colonies.
Yeast extract	0.04 g		
MgSO$_4$.7H$_2$O	0.05 g		
CaCl$_2$.10H$_2$O	0.05 g		
Distilled water	100 ml		

Tryptone yeast extract glucose agar (TYG) (Used for growth of *F. psychrophilum, F. columnare, T. maritimum,* Cipriano *et al.,* 1996)

Reagent	Amount	Preparation of media	Description of growth characteristics
Tryptone	0.2 g	Add all reagents, except neomycin, to distilled water. Autoclave at 121°C for 15 min. Cool to 50°C, and aseptically add filter-sterilized neomycin. Pour into Petri dishes.	Colonies are yellow with a thin spreading margin, which indicates the ability of the organism to exhibit gliding motility. Optimal temperature for *F. psychrophilum* is 14–20°C, for *F. columnare* 22–30°C, and 15–34°C for *T. maritimum*. Plates are examined daily for up to 7 days. Examination for colonies using a stereomicroscope may assist in early detection of characteristic colonies.
Yeast extract	0.05 g		
Glucose	0.3 g		
Agar	1.5 g		
Distilled water	100 ml		
Neomycin sulphate	0.4 mg		

Addition of 10 IU/ml of polymyxin B sulphate may assist as a selective medium for *F. columnare* (Shamsudin and Plumb, 1996).

TYG-M. Addition of 4 µg/ml neomycin sulphate, 200 IU/ml polymyxin B and 18.7 g/l ASW can be used for isolation of *Tenacibaculum (Flexibacter) maritimum* (Chen *et al.,* 1995).

VVM Selective medium for *Vibrio vulnificus* (Cerdà-Cuéllar *et al.*, 2001)

Reagent	Amount	Preparation of media	Description of growth characteristics
D-cellobiose	15.0 g	Dissolve all reagents in distilled water and bring to boil. Cool to 50°C and adjust pH to 8.5 using 5 M NaOH. (This media does not require autoclaving.)	VVM plates are violet-blue in colour. *Vibrio vulnificus* is seen as a bright yellow colony with a yellow halo of diffusion due to the fermentation of cellobiose. Other cellobiose fermenting *Vibrio* spp. such as *V. campbellii*, *V. harveyi* and *V. navarrensis* will grow on VVM agar. Some strains of *V. aestuarianus*, *V. alginolyticus*, *L. anguillarum* may also grow on this medium.
NaCl	10.0 g		
Yeast extract	4.0 g		
MgCl$_2$.6H$_2$O	4.0 g		
KCl	4.0 g		
Cresol red	40.0 mg		
Bromothymol blue	40.0 mg		
Polymyxin B	10^5 U/l		
Colistin methanesulfonate	10^5 U/l		
Agar	15.0 g		
Distilled water	1000 ml		

A modification of this medium, termed VVMc, has the same reagents except the polymyxin B, and recovery rates are virtually the same (Cerdà-Cuéllar *et al.*, 2001).

VAM (Presumptive differentiation medium for *Listonella* (*Vibrio*) *anguillarum*; Alsina *et al.*, 1994)

Reagent	Amount	Preparation of media	Description of growth characteristics
Sorbitol	15 g	Dissolve all reagents (except ampicillin) in distilled water, and boil. Cool to 50°C. Adjust pH to 8.6 with 5 M NaOH. Aseptically add ampicillin. Pour into Petri dishes. Store at 15°C. Plates are viable for 3 weeks. After this time the ampicillin begins to lose activity.	The uninoculated medium is a violet-blue colour. After 48 h incubation at 25°C, *L. anguillarum* colonies are flat, round and bright yellow with a diffuse yellow halo due to the fermentation of sorbitol. *V. fluvialis*, *V. harveyi* and *V. metschnikovii* are able to grow as yellow colonies on VAM, and therefore must be differentiated biochemically from *L. anguillarum*. *V. alginolyticus* grows as a non-swarming blue colony.
Yeast extract	4 g		
Bile salts	5 g		
NaCl	35 g		
Ampicillin	10 mg		
Cresol red	40 mg		
Bromothymol blue	40 mg		
Agar	15 g		
Distilled water	1000 ml		

Wood's broth (Selective broth for detection of *Erysipelothrix rhusiopathiae*; Wood, 1965)

Reagent	Amount	Preparation of media	Description of growth characteristics
Bacto tryptose (Difco)	7.5 g	Add tryptose, Lab Lemco and NaCl to distilled water, pH to 7.5 and autoclave at 121°C for 15 min. Cool to room temperature.	Place swab of material or chopped pieces of tissue into medium. Incubate at 37°C for 48 h. Subculture to BA at 24 and 48 h.
Lab Lemco powder (Oxoid)	1.5 g		
NaCl	2.5 g		
Distilled water	500 ml		
Sterile horse serum (not inactivated)	25 ml	Add antibiotics to 10 ml of sterile distilled water. Aseptically add horse serum and antibiotic solution to cooled base. Dispense 10 ml volumes to McCartney bottles and store at 4°C.	Colonies of *Erysipelothrix* are 0.5 mm to 1 mm, grey-green at 48 h, with slight α-haemolysis around the colony. They resemble α-haemolytic *Streptococci*. *See photographic section.*
Kanamycin	200 mg		
Neomycin	25 mg		
Vancomycin	12.5 mg		

Note: Wood and Packer (1972) modified this medium by replacing the distilled water with 0.1 M phosphate buffer (12.02 g Na$_2$HPO$_4$, 2.09 g of KH$_2$PO$_4$ and 1000 ml distilled water).

Yersinia selective agar (*Yersinia ruckeri* does not grow on this medium)

Reagent	Amount	Preparation of media	Description of growth characteristics
Yersinia selective agar base (Oxoid)	29.0 g	Add Yersinia selective base to distilled water and autoclave at 121°C for 15 min. Cool to 50°C.	Yersinia species grow as pink colonies with a darker 'bulls eye' centre at 24–48 h.
Distilled water	500 ml		
Oxoid Yersinia Selective Supplement	2 ml	Reconstitute 1 vial of supplement by aseptically adding 1 ml distilled water and 1 ml ethanol. Mix to dissolve and aseptically add to base. Pour into Petri dishes.	

7.2 Biochemical Test Media

Aesculin

Reagent	Amount	Preparation of media	Description of test
Aesculin	0.2 g	Dissolve all chemicals, except aesculin, by boiling. Cool and add aesculin while stirring. Dispense 5 ml of media into tubes. Autoclave at 121°C for 15 min. Store tubes in the dark.	Bacterial cells are inoculated into aesculin broth and incubated at appropriate temperature for 24–48 h. Development of a black colour is positive. Because some bacteria, especially some *Vibrio*, cause blackening of the medium due to melanin production, true hydrolysis of aesculin must be tested for loss of fluorescence at 354 nm longwave UV light (MacFaddin, 1980; Choopun *et al.*, 2002). Fluorescence indicates a negative reaction.
Ferric citrate	0.1 g		
Peptone water (Oxoid)	3.0 g		
Distilled water	200 ml		

Arginine dihydrolase (Møller). See under Decarboxylases

Arginine dihydrolase (Thornley). See under Decarboxylases

***Brucella*: Assay of metabolic activity** (Alton and Jones, 1967; Jahans *et al.*, 1997)

Reagent	Amount	Preparation of media	Description of test
L-alanine	1.25 g/l	Dissolve various metabolic substrates in sterile PBS, pH 7.2 at a concentration of 1.25 g/l. Sterilize each solution by filtration through 0.22 μm filter membrane. Store at 4°C. Prepare MTT solution in PBS at 1.0 g/l. Sterilize by filtration through a 0.22 μm filter.	Substrate mediated metabolic activity is assayed by the stimulation of tetrazolium reduction. Prepare a cell suspension in PBS and adjust the concentration to 10^{10} organisms per ml. In a flat-bottomed microtitre plate, place 100 μl of each substrate into each well. Place 50 μl of cell suspension in each well and incubate at 37°C in 10% CO_2 for 18 h. After incubation, place 50 μl of MTT in each well. Incubate 1 h at room temperature and add 50 μl of formaldehyde to each well. After 2–4 h, read the optical density at 630 nm. The metabolic index for each substrate is expressed as a ratio of the OD with substrate against the OD without substrate.
L-asparagine	1.25 g/l		
L-glutamic acid	1.25 g/l		
L-arginine	1.25 g/l		
DL-ornithine	1.25 g/l		
L-lysine	1.25 g/l		
D-galactose	1.25 g/l		
D-ribose	1.25 g/l		
D-xylose	1.25 g/l		
Meso-erythritol	1.25 g/l		
Urocanic acid	1.25 g/l		
3-[4,5-dimethylthiazol-2-yl]-2,5-diphenyltetrazolium bromide (MTT)	1.0 g/l		
Formaldehyde	40%		

Brucella: growth on substrate media (Alton and Jones, 1967; Jahans *et al.*, 1997)

Reagent	Amount	Preparation of media	Description of test
Basic fuchsin	20 µg/ml (1/50,000)	A 0.1% stock solution of each dye is made in distilled water, and sterilized by being held in boiling water for 1 h. Stock solution should be renewed after 3 months.	Prepare a suspension of bacterial cells in 0.5 ml of sterile normal saline.
Safranin O	100 µg/ml (1/10,000)	Each dye is then added to a base media such as tryptone-soy agar or serum-dextrose agar. Alton and Jones (1967) suggest that a laboratory should determine the concentration of each dye that gives optimum results, the range being somewhere between 1:25,000 and 1:100,000 (10–40 µg of dye per ml of medium), using the FAO/WHO reference strains. The concentrations in the 'amount' column are suggested by Jahans *et al.* (1997).	The unknown culture should be inoculated on to the same plate along with the reference strains. Therefore mark off an appropriate number of areas on a plate. Using a loop of culture suspension, make five separate streaks without recharging the loop on to each specifically marked area. Also inoculate a control plate that contains basic medium only and no dye.
Thionin	20 µg/ml (1/50,000)		
		The required amount of the dye stock solution is added to the melted base medium, mixed and poured into Petri dishes and allowed to set. Label each plate with appropriate dye identification.	Incubate plates at 37°C with and without CO_2 for 4 days. Record results of 'growth' or 'no growth' in the presence of each dye.

Brucella selective media. See Farrell's medium

Carbohydrate fermentation (General-purpose medium for detection of carbohydrate fermentation; Vera, 1948, 1950)

Reagent	Amount	Preparation of media	Description of test
Cystine tryptic medium (Gibco)	7.13 g	**Basal medium preparation**	Fermentation is detected by a colour change from red to yellow.
Distilled water	250 ml	Boil distilled water and cysteine tryptic medium.	
Carbohydrate (see below)	Sterilization method	Dispense 5 ml volumes into 10 ml tubes.	
		Carbohydrate preparation	

Carbohydrate	Sterilization	1% final concentration	To prepare the carbohydrates, make a 10% solution (3 g in 30 ml). For salicin use a 4% solution (1.2 g in 30 ml of distilled water).	References may state that carbohydrates are added at 0.5% to 1.0%. However, because reversion of the reaction may occur when 0.5% is used, using a concentration of 1% ensures against depletion of the carbohydrate by the organism.
Arabinose	Filter			
Aesculin	Autoclave			
Glucose	Autoclave		**For carbohydrates that can be autoclaved**	
Inositol	Autoclave		Pipette 0.5 ml of carbohydrate solution per 5 ml of basal medium. Autoclave at 10 lb per 10 min.	
Lactose	Autoclave			
Maltose	Filter		**For filter-sterilization**	
Mannitol	Autoclave		Filter-sterilize the carbohydrate solution through a 0.22 µm filter.	
Mannose	Filter		Autoclave basal medium 5 ml in 10 ml tubes. Cool to 50°C and aseptically add the filter-sterilized carbohydrate solution, 0.5 ml to 5 ml of basal medium.	
Salicin	Autoclave			
Sorbitol	Autoclave			
Sucrose	Filter			
Trehalose	Autoclave			
Xylose	Filter			

Cystine tryptic medium contains Bacto tryptose, L-cystine, sodium chloride, sodium sulphite, agar and phenol red. With the addition of yeast extract at 0.01%, this medium is also suitable for testing carbohydrate fermentation of the *Cytophaga, Flavobacterium, Flexibacter, Tenacibaculum* group, plus certain *Vibrio* species that require yeast extract, such as *V. ordalii, Moritella marina* and some strains of *V. logei.* Baumann *et al.* (1980) adds yeast extract at 0.05% for strains of *Vibrio* and *Photobacterium* spp. that require organic growth factors.

Notes:

● In the literature, some original recipes used bromothymol blue as the pH indicator; however, it may be toxic to a number of marine bacteria, and therefore phenol red is the recommended pH indicator in carbohydrate fermentation tests (Leifson, 1963).

- Peptone-containing media may give misleading results for carbohydrate breakdown because weak acid production may be masked by peptone-breakdown products (Vera, 1950).
- Sucrose should be filter-sterilized, as it is thermolabile and will give false positive results if sterilized by autoclaving (Stanier *et al.*, 1966).
- To prevent colour change of the medium caused by CO_2 build-up in screw-capped tubes when inoculated with organism of interest, ensure that lids are loosened for about an hour before results are read. Preparing tubed media to only contain half their volume in media also helps to prevent pH reversion caused by CO_2 build-up.
- Most carbohydrates can be stored at room temperature for 2–3 weeks, with the main stocks stored at 4°C for longer term storage. However, some carbohydrates should always be stored at 4°C to prevent unwanted colour change in the medium. These are arabinose, mannitol, xylose and also the oxidative-fermentative media and ONPG media.

Alternative methods for carbohydrate fermentation for specific bacteria

Carbohydrate fermentation media for *Flavobacterium* spp.

- Broth medium containing 0.05% tryptone, 0.05% yeast extract, 0.0018% phenol red, and filter-sterilized carbohydrate at a final concentration of 0.2% (Wakabayashi *et al.*, 1986).
- AO medium was used as a basal medium for carbohydrate fermentation; however, it was unsuitable for this test, as colour changes were seen in the negative control tube (Bernardet and Grimont, 1989).

Carbohydrate fermentation media for marine bacteria (Leifson, 1963, used by Gauthier, 1976b)

Reagent	Amount	Preparation of media	Description of test
Casitone (Difco)	0.1 g	The ingredients are dissolved in half-strength seawater and the pH is adjusted with HCl to 7.5. Autoclave and re-check the pH, which should be about 8. Adjust with HCl if necessary. Filter-sterilize the carbohydrate and add aseptically to the base medium. Dispense 3 ml into 13 × 100 mm tubes.	Fermentation is indicated by a change in colour from red to yellow.
Yeast extract	0.01 g		
Ammonium sulphate	0.05 g		
Tris buffer	0.05 g		
Agar	0.3 g		
Phenol red	0.001 g		
Artificial seawater	Half strength		
Carbohydrate	1%		

Notes: Einar Leifson (1963) modified this recipe from the original medium published in 1953 (Hugh and Leifson, 1953). The pH indicator in the 1953 medium (bromothymol blue) was found to be toxic for a number of marine bacteria and therefore phenol red is recommended. This media is also used to test for oxidation/fermentation. Add individual carbohydrates to the base medium for fermentation tests. This method is similar to the general-purpose medium.

Carbohydrate fermentation media for *Pseudomonas* and *Flavobacterium* spp. (Gilardi, 1983)

Reagent	Amount	Preparation of media	Description of test
OF basal medium (Difco)	9.4 g	Add reagent in distilled water and dissolve by boiling. Dispense into 100 ml aliquots. Autoclave at 121°C for 15 min. To each 100 ml aliquot aseptically add 10 ml of a 10% carbohydrate solution. Aseptically dispense each 100 ml of respective carbohydrate into 5 ml aliquots in 10 ml tubes.	Colour change to yellow indicates a positive fermentation reaction.
Distilled water	1000 ml		
Carbohydrate 10% solution	10 ml	Filter-sterilize and add as above.	

Composition of Difco OF medium: tryptone (2.0 g), sodium chloride (5.0 g), di-potassium phosphate (0.3 g), agar (2.0 g), bromothymol blue (0.08 g), in 1000 ml distilled water, pH 6.8.

Carbohydrate fermentation media using acid from ammonium salt sugar (ASS). Used for *Sphingobacterium, Flavobacterium, Shewanella putrefaciens, Pseudomonas* spp. Used by Holmes *et al.* (1975) for testing *Pseudomonas* spp. Used by Bernardet and Grimont (1989) as an inoculation medium for API 50CH when testing *Flavobacterium*. Quoted in Cowan and Steel (1970) from the method of Smith *et al.* (1952).

Reagent	Amount	Preparation of media	Description of test
$(NH_4)_2HPO_4$	1.0 g	Add ingredients to distilled water and dissolve by boiling or steaming.	Inoculate and incubate at the appropriate temperature. Examine over 7 days. A colour change from purple to yellow is indicative of fermentation of carbohydrate.
KCl	0.2 g		
$MgSO_4.7H_2O$	0.2 g	Add indicator and autoclave at 115°C for 20 min.	
Yeast extract	0.2 g		
Agar	20 g	Allow medium to cool to 60°C and add filter-sterilized carbohydrate solution.	
Distilled water	1000 ml		
Bromocresol purple 0.2% solution	4 ml	Dispense into tubes and incline tube so that medium sets as a slope.	
Carbohydrate 10% solution	100 ml		

Notes:

- Bacteria that do not give a reliable fermentation reaction with peptone-containing medium should be tested in ASS. In general, peptone-containing media give the least positive reactions and ASS the most positive reactions (Cowan and Steel, 1993 edition).
- Reactions using this medium are not reported in this book. *Flavobacterium* spp., reactions reported in this book were the same by both the peptone-containing method (Carbohydrate general purpose medium) and ASS.

Carbohydrate fermentation media for *Vibrio* spp. (The following media were used by Baumann et al. (1971) for Vibrio spp. and Beneckea – adapted from methods of Stanier et al., 1966)

- Basal medium (BM): 50 mM tris (hydroxymethyl)aminomethane (Tris)-hydrochloride (pH 7.5), 190 mM NH_4Cl, 0.33 mM $K_2HPO_4.3H_2O$, 0.1 mM $FeSO_4.7H_2O$ and half strength seawater.
- Basal medium agar (BMA): Mix equal volumes of double-strength BM with 20 g of Ionagar (Oxoid) per litre.
- Yeast extract broth (YEB): Add 5 g/l of yeast extract (Difco) to BM.
- For Yeast extract agar (YEA) add 20 g of agar (Difco) per litre of YEB.
- Carbohydrate fermentation medium (F-2): To YEB add 100 mM Tris-hydrochloride (pH 7.5), 1 g/l of Ionagar (Oxoid), 1 g/l of sodium thioglycollate, and 10 g/l of filter-sterilized glucose.
- Carbohydrate fermentation medium (F-3): To BM add 25 mM Tris hydrochloride (pH 7.5), 0.5 g/l of yeast extract (Difco), 1 g/l of Ionagar (Oxoid), 2 ml/l of a 1.6% (w/v) alcoholic solution of bromocresol purple, and 10 g/l of filter-sterilized carbohydrate.

Carbohydrate fermentation media for *Nocardia* spp. Basal inorganic nitrogen medium (Gordon *et al.*, 1974, Quoted from Ayers *et al.*, 1919)

Reagent	Amount	Preparation of media	Description of test
$(NH_4)_2HPO_4$	1.0 g	Add all reagents except carbohydrate to distilled water and adjust pH to 7.0. Dispense 5 ml into tubes and autoclave at 121°C for 15 min.	Production of acid due to fermentation of the carbohydrate is seen as a colour change from purple to yellow. Incubate at 28°C and read from 7–28 days. With the addition of yeast extract, this medium is the same as ASS.
KCl	0.2 g		
$MgSO_4.7H_2O$	0.2 g		
Agar	15		
Distilled water	1000 ml		
Bromocresol purple 0.04% solution	15 ml	Aseptically add 0.5 ml of a 10% solution of each carbohydrate (autoclaved separately) to tubes and allow to set on a slant.	
Carbohydrates to be tested (10% solution)	0.5 ml		
Optional Yeast extract (Difco)			

Note: When testing *N. seriolae*, add 2% w/v yeast extract (Difco) to medium (Kudo *et al.*, 1988).

Carbon dioxide atmosphere (Used when incubation conditions require a carbon dioxide atmosphere)

For laboratories that do not have access to a carbon-dioxide incubator, the following methods offer a cheap alternative.

Method 1: Candle jar method. Place agar plates in an airtight container. Place a lighted candle in the jar and seal the lid. This method produces 2% CO_2 (J. Lloyd, Department of Agriculture, Western Australia, 1985, personal communication) with other reports suggesting 6–8% CO_2 (Cottral, 1978). If the lid is plastic, then protect from the heat produced by the candle by using an aluminium foil-covered device.

Method 2: 'ENO salts' method (Lloyd, 1985). 7.5 g of ENO salts per 10 l container generates 4% CO_2. Use an appropriate container that has a well-sealed lid such as a household food container or Mitsubishi Gas Chemical Company anaerobe box. Place 3 g of ENO salts (Sigma) into 20 ml of water for a container that measures $22 \times 22 \times 8$ cm or 10 g of ENO salts into 40 ml of water for a container that measures $30 \times 30 \times 14$ cm. This technique needs to be performed quickly, therefore place salts into a small piece of tissue paper. Place the water into a 50 ml urine container and place into the incubation container. Quickly add the salts to the water and seal lid. Place box at appropriate temperature. ENO is a sparkling antacid that contains per 5 g, sodium bicarbonate (2.32 g), sodium carbonate (0.5 g), citric acid anhydrous (2.18 g) and which can be purchased through Sigma chemicals.

Catalase test (from Cowan and Steel, 1970)

Smear a colony on to a glass slide. Place a drop of 30% H_2O_2 on the smear. The presence of bubbles indicates a positive reaction. When picking growth from a blood agar plate, make sure no blood-containing agar is placed on the slide, as false-positive results may occur.

Cellulose digestion (Wakabayashi *et al.*, 1989)

To a broth culture medium appropriate to the growth requirements of the organism being tested, add a piece of cellulose filter paper. Observe for disintegration of the filter paper. A characteristic of the *Cytophaga* species is that they digest cellulose, therefore when testing an organism suspected of being a *Cytophaga* species use a broth of Anacker-Ordal medium.

Coomassie Brilliant Blue agar (CBBA). Used to detect the A+ protein layer of *Aeromonas salmonicida* (Udey, 1982; Evenberg *et al.*, 1985; Cipriano and Bertolini, 1988; Markwardt *et al.*, 1989)

Reagent	Amount	Preparation of media	Description of growth characteristics
TSA	44.0 g	Add Coomassie Blue and TSA to distilled water. Autoclave at 121°C for 15 min. Cool to 50°C and pour into Petri dishes.	Coomassie Brilliant Blue is a protein-specific dye that results in production of dark blue colonies of bacteria that have the A-layer protein. A-layer positive *A. salmonicida* stain deep blue. This medium is not selective and other bacteria will produce blue-coloured colonies. However, it assists in the isolation and detection of *A. salmonicida*.
Coomassie Brilliant Blue R250 (Bio-Rad)	0.1 g		
Distilled water	1000 ml		

Citrate: Simmons method (Simmons, 1926)

Reagent	Amount	Preparation of media	Description of test
Simmons citrate (Difco)	3.63 g	Suspend media in distilled water, and bring to the boil. Dispense 3 ml of media into tubes. Autoclave at 121°C for 15 min. Lay on a slight incline to create slopes.	Development of a deep blue colour is positive. Simmons citrate (Difco) contains 1% NaCl. Citrate is tested as a sole carbon source. (Christensen's citrate does not test for citrate as the sole carbon source, as it contains other nutrients).
Distilled water	150 ml		

Congo Red. Tests for presence of extracellular galactosamine glycan in the *Flavobacteriaceae* (Johnson and Chilton (1966) who quote personal communication with E.J. Ordal for this test; McCurdy, 1969)

Reagent	Amount	Preparation of media	Description of test
Congo Red	10 mg	Weigh 10 mg of Congo Red and add to 100 ml of distilled water. Final concentration is 0.01% aqueous. Mix well, label bottle and store at room temperature.	Place 1–2 drops of Congo Red on a few isolated colonies growing on AO or Shieh's medium. A positive reaction is red-colour development of colonies. This colour lasts for a number of hours. This test detects the presence of extracellular glucans.
Distilled water	100 ml		

Congo Red agar (Used for detection of A protein layer of *A. salmonicida*; Ishiguro *et al.*, 1985) Prepare tryptic soy agar (Difco) containing 30 g/ml of Congo Red. Incubate at 20°C and examine for red-coloured colonies after 48 h.

Dienes stain (Used to differentiate *Mycoplasma* from *Bacteria*; Dienes, 1939; Hayflick, 1965).

Reagent	Amount	Preparation of media	Description of test
Methylene blue	2.5 g	Dissolve all reagents in distilled water. Store in a screw-cap bottle.	*Method I.* Place some stain on to a coverslip and allow to dry. Agar blocks containing colonies are excised and placed on to a glass slide with colonies uppermost. Place coverslip on to agar block containing colonies and press coverslip lightly so that there is good contact between the colonies and the stain.
Azure II	1.2 g		
Maltose	10 g		
Sodium carbonate (anhydrous)	0.25 g		
Distilled water	100 ml		*Method II.* Using a cotton wool swab directly apply the stain to the *Mycoplasma* colonies.
			Mycoplasma colonies stain blue with a distinctly dense blue centre and a lighter blue periphery. *Mycoplasma* colonies retain the stain for 24 h, whereas bacterial colonies decolorize after 30 min.

Some methods contain 0.2 g of benzoic acid.

Digitonin (Used to differentiate *Mycoplasma* species from *Acholeplasma* species)

Reagent	Amount	Preparation of media	Description of test
Digitonin (1.5% w/v)	1.5 g	Dissolve digitonin in ethanol. Warm to 37°C to dissolve.	Plate suspect *Mycoplasma* colony to a 5 cm Mycoplasma agar plate. This will either be from a broth or agar plate. When transferring colonies from an agar plate, use a sterile spatula or open-ended 'hockey stick' (made from a bent Pasteur pipette), to select a plug of agar containing a colony. Invert this agar plug, colony side down, on to a fresh agar plate and rub over the surface. If inoculating the plate from a broth culture, pipette the broth culture on to the plate and then remove the excess. Allow the plate to dry for a few hours. Place the digitonin disc in the middle of the plate, and incubate at appropriate temperature under CO_2 for 3–4 days.
Ethanol	100 ml	Place 20 µl of stock solution on to a 6 mm sterile filter paper disc (Oxoid). Dry at 37°C for 1 h.	
			Mycoplasmas are sensitive to digitonin and so a 4 mm zone of inhibition of growth is seen from the edge of the disc (zone of 14 mm diameter).
			Acholeplasma are resistant to digitonin and a zone of less than 1 mm is seen from the edge of the disc.

DNase (Used by West and Colwell (1984) and reported from all sources used in this manual)

Reagent	Amount	Preparation of edia	Description of test
DNase test agar (Oxoid)	19.50 g	Suspend agar in water in 1000 ml. Autoclave at 121°C for 15 min then cool to 50°C in waterbath. Mix well and pour media into Petri dishes.	Spot-inoculate or streak as a single line, a loop of bacterial culture on to the plate. Incubate plates for 2–7 days at 24°C. Flood plate with 1% HCl to precipitate the DNA. A positive reaction is seen as a clear zone around the bacterial streak. *See photographic section.*
Distilled water	500 ml		

See under HCl for preparation of 1M HCl.

Decarboxylases and arginine dihydrolase (Used in Cowan and Steel, 1970, West and Colwell, 1984)

Reagent	Amount	Preparation of media	Description of test
Decarboxylase base Møller (Difco) (Møller, 1955)	2.1 g	In separate bottles, dissolve each individual amino acid in the 200 ml of distilled water and decarboxylase base. Dispense 5 ml into 10 ml tubes. Autoclave at 121°C for 15 min.	A tube of basal medium without amino acid is always inoculated in parallel with the test media. Inoculate with heavy inoculum. Overlay all tubes with paraffin oil, and incubate for 7–14 days for some strains. Most strains can be read at 48 h.
Distilled water	200 ml		
L-arginine HCl	2.0 g		
L-ornithine monohydrochloride	2.0 g		Check the bottom of tube and the cloudiness of the tube for signs of sufficient inoculum and growth.
L-lysine monohydrochloride	2.0 g		
			For salt-requiring strains, add 0.5 ml of 20% NaCl to tubes before inoculation with organism.
Control – use decarboxylase base and distilled water only			See under Paraffin oil for sterilization procedures for the oil.

Arginine dihydrolase (ADH) (Method of Thornley, 1960. Recommended for marine bacteria)

Reagent	Amount	Preparation of media	Description of test
Bacto-Peptone (Difco)	0.1 g	Add reagents to distilled water and pH to 6.8. Dispense 5 ml into 10 ml tubes. Autoclave at 121°C for 15 min.	A tube of basal medium without amino acid is always inoculated in parallel with the test media. Inoculate with heavy inoculum. Overlay all tubes with paraffin oil, and incubate for 7–14 days for some strains. Most strains can be read at 48 h.
NaCl	0.5 g		
K$_2$HPO$_4$	0.03 g	*Note:* the original pH described by Thornley was 7.2, however pH 6.8 is recommended by West and Colwell (1984).	
Agar	0.3 g		
Phenol red	0.001 g		Check the bottom of tube and the cloudiness of the tube for signs of sufficient inoculum and growth.
Arginine HCl	1.0 g		
Distilled water			

Note: Some *Vibrio* species are negative in Møller's method for ADH, but positive by the method of Thornley. These are; *Vibrio mediterranei, V. mytili, V. orientalis, Vibrio splendidus* biogroup I and some strains of *V. tubiashii* (Macián *et al.*, 1996). The glucose in Møller's medium appears to inhibit the reaction due to catabolite repression of the inducible ADH system (Macián *et al.*, 1996). Baumann *et al.* (1971) found that *Photobacterium* spp. can produce alkaline products in Thornley's medium, yet none possesses a constitutive ADH system when tested with more sensitive analytical methods (West and Colwell, 1984). Different brands of peptone were found to give different results and the Difco bacto-peptone was recommended (Thornley, 1960).

Thornley's medium was modified by Baumann and Baumann (1981) by using basal medium without the Tris-HCl. See under Alternative media for carbohydrate fermentation, media for *Vibrio* species, for this recipe.

Furunculosis agar (For detection of pigment production from *Aeromonas salmonicida*; Bernoth and Artz, 1989)

Reagent	Amount	Preparation of media	Description of test
Bacto-tryptone (Difco)	10 g	Add all ingredients to distilled water. Adjust pH to 7.3, and autoclave at 121°C for 15 min. Cool to 50°C and pour into Petri dishes.	Pigment production is best detected by incubating at 15–20°C for up to 7 days. At 25°C less pigment production is seen.
Yeast extract (Difco)	5 g		
L-tyrosine (Merck)	1 g		
NaCl (Merck)	2.5 g		Pigment is seen as a diffuse brown coloration around each colony.
Agar (Oxoid L11)	15 g		
Distilled water	1000 ml		

Note: Furunculosis agar is recommended as a superior medium for detection of pigment production, as a greater number of strains show pigment on FA than on TSA, NA or BHIA (Hirvelä-Koski *et al.*, 1994, Hänninen and Hirvelä-Koski, 1997). However, FA is not the best medium for primary isolation of *Aeromonas salmonicida* (Bernoth and Artz, 1989) and BA is recommended. (It may be possible that by adding blood to FA, the medium would be a suitable primary isolation medium.)

Gelatinase (0% AND 3% NaCl) (Smith and Goodner, 1958. Used by West and Colwell, 1984)

Reagent	Amount 0% NaCl	Amount 3% NaCl	Preparation of media	Description of test
Difco bacto Peptone	1.0 g	1.0 g	Add reagents to distilled water and autoclave at 121°C for 15 min. Cool to 50°C.	Spot inoculate the gelatin plate on both the 0% NaCl and 3% NaCl sides with approximately equal amounts of bacterial growth. Incubate at an appropriate temperature 24–48 h.
Difco yeast extract	0.25 g	0.25 g		
Oxoid gelatin	3.75 g	3.75 g		
NaCl	–	7.50 g	Use split Petri dishes, and label plates 0% and 3% per side of each plate. Pour appropriate media into each half of each plate.	Production of gelatinase is seen as either a cloudy or clear zone around the area of bacterial growth. The plate should be held up to the light and read against a darkish background. Chilling the plates may provide a sharper contrast between zones and the unaffected portions of the media. Flooding the plate with ammonium sulphate may assist with definition of zones of clearing.
Oxoid agar No 1	3.75 g	3.75 g		
Distilled water	250 ml	250 ml	Store at 4°C.	

This plate method of Smith and Goodner (1958) is used as it detects a change in the composition of gelatin rather than its liquefaction and thus is a very sensitive method. Either use a split plate with one side 0% NaCl and the other side containing 3% NaCl. This allows the salt requirement of the organism to be detected at the same time. Alternatively, use small (5 cm) Petri dishes for each NaCl concentration.

Glucose yeast extract agar (GYEA) (For colonial morphology and survival at 50°C of *Nocardia* species; Gordon *et al.*, 1974)

Reagent	Amount	Preparation of media	Description of test
Yeast extract	10 g	Add reagents to water and pH to 6.8. Autoclave at 121°C for 15 min. May be poured into plates, or prepared as tubed slants.	Plate media is used for observation of colonial morphology.
Glucose	10 g		
Agar	15 g		Tubed slants are used for testing survival at 50°C.
Tap water	1000 ml		

Haemolysis
Record from growth on BA or MSA-B within 7 days. For some *Vibrio* species, haemolysis may be more pronounced on BA than on MSA-B even though their preferred growth medium is MSA-B.

HCl (1N) (For detection of hydrolysis of DNA in DNase medium)

Reagent	Amount	Preparation of media	Description of test
Concentrated HCl, 32%	9.85 ml	Add 9.85 ml of 32% **acid to** 80 ml **water**, then make up to 100 ml.	To a 24 or 48 h growth on a DNase plate flood with 1 M HCl. Wait 1 min and observe over a black tile for zone of clearing around bacterial growth.
Distilled water	80 ml		
			See photographic section.

Safety note: Always add acid to water, not vice versa. Other concentrations of HCl are available. To prepare from 35.4%, add 8.9 ml to 80 ml of water, and then make up to 100 ml. To prepare from 37%, add 8.5 ml of acid to 80 ml of water, and then make up to 100 ml.

Hippurate hydrolysis (Hwang and Ederer, 1975)

Reagent	Amount	Preparation of media	Description of test
Hippuric acid sodium salt	0.15 g	Dissolve hippuric acid in water. Dispense 0.4 ml per sterile, 5 ml yellow-capped plastic tube. Store at −20°C.	Inoculate medium with a large loopful of bacterial growth from culture plate. Incubate at appropriate temperature for 2.5 h. Add 200 µl of ninhydrin reagent and incubate a further 10 min. Development of deep purple colour is positive. (Original reference suggests a purple colour as deep as the colour of Crystal violet used in the Gram stain.)
Distilled water	15 ml		

Reagent: Ninhydrin.

Ninhydrin reagent (Used for detection of hippurate hydrolysis; Hwang and Ederer, 1975)

Reagent	Amount	Preparation of media	Description of test
Ninhydrin	0.35 g	Dissolve ninhydrin in the acetone/butanol mixture. Dispense into 5 ml tubes and store in the dark.	To a mixture of organisms incubated in hippurate solution for 2.5 h, add 200 µl of ninhydrin reagent. Incubate 10 min and observe formation of purple layer.
Acetone	5 ml		
Butanol	5 ml		

Hydrogen sulphide production

Many of the references for the biochemical reactions use a method whereby a lead acetate strip (H_2S indicator) is suspended over a tube of growth medium containing cysteine (sulphur source). This is a very sensitive method. However, the preparation of the lead acetate strips is hazardous, (for method see Cowan and Steel, 1970). Alternative methods are available but may not have the same sensitivity. The test can be performed in an API 20E or a triple sugar iron tube. A hydrogen sulphide Biostrip (catalogue number TM343) is available from MedVet Science. When using these media be aware of the growth requirements of the test organism and the sensitivity of the test.

Indole (Cowan and Steel method 2, 1970; Colwell and West, 1984; MacFaddin, 1980)

Reagent	Amount	Preparation of media	Description of test
Tryptone broth (Difco)	2.5 g	Combine reagents and pH to 7.5. Dispense 5 ml per tube and autoclave at 121°C for 15 min.	Use a heavy inoculum and incubate 48 h at 25 or 37°C depending on growth requirements of the bacterium. To read, add 6–7 drops of Kovács reagent and shake tube. A positive result is development of a cherry red colour in the upper reagent layer on top of the broth medium. No colour development indicates a negative result.
NaCl	1.25 g		
Distilled water	250 ml		
			For organisms that require salt, add 0.5 ml of 20% NaCl to 5 ml of medium. Even though a salt-requiring organism may show growth in the tube at 48 h, if NaCl is not present a false-negative reaction can still occur. *See photographic section.*

Results between API 20E and tryptone broth may differ. Add NaCl to a final concentration of 2% for marine isolates (i.e. add 500 µl of 20% NaCl to 5 ml of test medium).

Indoxyl acetate hydrolysis (Mills and Gherna, 1987)

Indoxyl acetate discs may be prepared by making a 10% (wt/vol) solution of indoxyl acetate in acetone. Add 50 µl to a blank disc, 0.64 cm in diameter. Discs are available from Oxoid. Allow discs to dry and store at 4°C in an amber-coloured bottle. Shelf life of discs is approximately 6 months. Discs are also available commercially from companies such as Remel.

Test: Bacterial growth from an agar plate is scraped on to an indoxyl acetate disc and a drop of distilled water is added. A positive result is the development of a dark blue colour within 5–10 min. Alternatively, colonies from an agar plate can be emulsified in 0.3 ml of distilled water. Add an indoxyl acetate disc. A positive result is the development of a dark blue colour within 10–15 min (Mills and Gherna, 1987).

Kovács indole reagent (Kovács, 1928; Cowan and Steel, 1970; MacFaddin, 1980)

Reagent	Amount	Preparation of media	Description of test
p-dimethylaminobenzaldehyde	5 g	Handle p-dimethylaminobenzaldehyde and HCl with caution.	See under 'Indole'.
Iso-amyl alcohol	75 ml	Dissolve p-dimethylaminobenzaldehyde by warming in a waterbath at 50°C. Only leave long enough to dissolve, otherwise it turns pink then dark brown. Cool and slowly add the HCl to the aldehyde–alcohol mixture. The reagent should be light yellow to light brown in colour. Store reagent in a brown glass-stoppered bottle at 4°C.	
Concentrated hydrochloric acid	25 ml		

KOH (Tests for presence of Flexirubin pigments; Reichenbach *et al.*, 1974, 1981)

Reagent	Amount	Preparation of media	Description of test
KOH	20%		Place 1–2 drops of KOH on to a few isolated young colonies on AO medium. Detection of a brown or red pigment is positive. Detects flexirubin pigment (Reichenbach *et al.*, 1981). Some reports indicate a purple colour as positive (Mudarris and Austin 1989).

Note: Pigment production may be influenced by factors in the growth medium such as yeast extract, which promotes pigment production through stimulated growth. Also, pH can affect pigment, where a lower pH may reduce pigment production (Reichenbach *et al.*, 1974). (Reichenbach states 5% KOH in methanol.)

MacConkey agar

Reagent	Amount	Preparation of media	Description of test
MacConkey agar (Oxoid)	26.0 g	Suspend the agar in the distilled water and autoclave at 121°C for 20 min. Cool to 50°C and pour into Petri dishes. Store plates at 4°C.	Organisms that ferment lactose are seen as dark pink colonies. Non-lactose fermenting colonies are seen as yellow-clear colonies. *Salmonella* spp. and *Edwardsiella tarda* will appear as non-lactose fermenting colonies.
Distilled water	500 ml		

Marine oxidative-fermentative medium (MOF) (Leifson, 1963). See under Oxidative fermentative marine.

McFarland (nephelometer) standards preparation

McFarland Tube No.	Sulfuric acid 1% aqueous solution (ml)	Barium chloride, 1% aqueous solution (ml)	Corresponding density of bacteria – 10^6	International Units (IU) of opacity
1	9.9	0.1	300	3
5	9.5	0.5	1500	15

Method: Mix the 1% sulphuric acid and the 1% barium chloride according to the table. Use clean, clear glass tubes. Plug or cap the tubes and seal with parafilm. Store tubes upright.
Use: When comparing bacterial density against the tubes, ensure that the bacterial suspension is in a similar sized glass tube to the standard. It may be best to prepare the standards in a tube or bottle that the bacterial suspensions will be normally prepared in. Sensitivity to vibriostatic agent (0/129 discs), prepare lawn inoculum at McFarland tube 1.
API 20E, API 20NE, API 50CH. Inoculate with inoculum prepared to turbidity of McFarland tube 1.
API ZYM. Inoculate with inoculum prepared to turbidity of McFarland tube 5.

(Taken from Difco manual, referenced in *Gradwohl's Clinical Laboratory Methods and Diagnosis*. A.C. Sonnenwirth and L. Jarett (eds). C.V. Mosby Company, 1980, p.1363.)

Motility – Hanging drop method (Recommended in West and Colwell, 1984)

Place a drop of suspension from an early stationary phase broth culture on to a coverslip. Place small dollops of vaseline or plasticine at the corners. Invert coverslip and place on to a glass slide so that the drop of culture hangs free of the glass slide. Examine under phase contrast, or bright-field with the substage condenser aperture closed down to give greater contrast. Weakly motile strains may not be detected in semi-solid motility medium. For some organisms, motility is temperature dependent; therefore, ensure that growth medium is incubated at the appropriate temperature.

MRVP test medium (Clark and Lubs, 1915; Cowan and Steel, 1970; used by West and Colwell, 1984)

Reagent	Amount	Preparation of media	Description of test
MRVP medium (Oxoid)	3.75 g	Dissolve reagent in distilled water and autoclave at 121°C for 15 min. Dispense 5 ml per tube.	Add NaCl for marine organisms. Must incubate for 2–3 days before testing.
Distilled water	250 ml		VP reaction: Red coloration may develop up to 18 h after the addition of reagents.

For Voges-Proskauer test, see under VP.

Methyl Red reagent

Reagent	Amount	Preparation of media	Description of test
Methyl Red	40 mg	Add methyl red reagent to ethanol and heat to 50°C in a waterbath to assist dissolution. Make up to 100 ml total with distilled water.	After incubation at optimal temperature for at least 2 days, add 3–4 drops of Methyl Red reagent. Persistence of a red colour indicates a positive result.
Ethanol 95%	40 ml		
		More ethanol can be added if a precipitate forms.	*See photographic section.*

Mycoplasma characterization tests (Aluotto *et al.*, 1970 with modifications)

Reagent	Amount	Preparation of media	Description of test
Heart infusion broth – stock		Add dehydrated media to distilled water. Using 5M NaOH adjust pH to 7.6.	Inoculate a test and substrate control tube with 1 ml of a 24 h culture grown in broth.
Heart infusion broth	25 g		
Distilled water	1000 ml		
Basal medium		Prepare yeast extract stock, and phenol red stocks.	Read tests daily for up to 2 weeks. A pH drop of 0.5 unit or more in the glucose tube compared with the appropriate substrate control tube is a positive reaction.
Heart infusion broth (Difco) stock	74 ml	To 74 ml of HIB add horse serum, yeast extract, phenol red and test substrate. Using 5 M NaOH of 5 M HCl, adjust to appropriate pH depending on test substrate.	
Horse serum (heated 56°C for 30 min)	10 ml		
Yeast extract (Oxoid) stock solution of 10% w/v. Filter-sterilize.	5 ml		A rise of 0.5 pH unit in the arginine or urea tubes compared with the appropriate substrate tubes indicates a positive reaction.
Phenol red (0.5% w/v) autoclaved	1 ml	Prepare substrate control tubes, that is, basal medium without test substrate.	The pH values are read by comparison with a set of standards ranging in pH from 5.6 to 8.4.
Test substrate		Filter sterilize and dispense in 5 ml amounts to sterile bijou or other screw-capped bottles.	

Glucose (10% w/v)	10 ml	pH 7.6	
Arginine (0.2%)	10 ml	pH 7.0	
Urea (10% w/v)	10 ml	pH 7.0	
OF test		Dispense 1 ml of sterile medium into 15 × 45 mm tubes.	Inoculate duplicate test and substrate control tubes with 0.5 ml of an overnight broth culture. Overlay one test and one control tube with sterile paraffin oil. Incubate at appropriate temperature and examine daily for up to 2 weeks for pH change compared with pH standards.
Basal medium	90 ml		
Glucose (10% w/v)	10 ml, pH 7.6		
			Fermentative organisms produce acid in both the aerobic and anaerobic tubes, whereas oxidative organisms produce acid in the aerobic tube only.
Tetrazolium (TTC) reduction		Aseptically add reagents to HIA stock and pour into 5 cm Petri dishes.	Inoculate duplicate plates with agar plugs taken from dense growth on culture plates. Agar plugs can be excised using a scalpel blade, alcohol flame-sterilized spatula, or open-ended 'hockey stick' made from a bent Pasteur pipette. Slide the inverted agar plug over the surface of the TTC plates.
HIA stock	74 ml		
Horse serum	20 ml	TTC can also be tested in broth rather than on agar plates.	
Yeast extract stock	5 ml		
TTC (2% w/v)	1 ml		
			Incubate one plate aerobically and the other anaerobically for up to 2 weeks.
			Development of a pink-red colour in 3–4 days is positive.
Phosphatase		Aseptically add reagents to HIA stock and pour into 5 cm Petri dishes.	Inoculate 3 plates with a drop from a 24 h broth culture. Incubate plates at the appropriate temperature. On days 3, 7 and 14, flood plate with 5 M NaOH. A positive reaction is the appearance of a red colour.
HIA	74 ml		
Horse serum	20 ml	Phosphatase can also be tested in broth rather than on agar plates.	
Yeast extract stock	5 ml		
Phenolphthalein diphosphate (sodium salt) (1% w/v)	1 ml		
Film and Spots 1		Aseptically remove egg yolks and homogenize with an equal amount of sterile distilled water. Add the homogenate to HIA at a final concentration of 10%. Pour into 5 mm Petri dishes.	Inoculate the egg yolk medium with test organism, and incubate in CO_2 at 37°C for up to 14 days. Examine macroscopically using reflected light. An iridescent or pearly film seen over areas of heavy growth indicates a positive reaction.
HIA	90 ml		
Egg yolk	10 ml		
Film and Spots 2		An alternative method is to use 20% horse serum in the medium.	Inoculate plates with test organism and incubate in CO_2 at 37°C for up to 14 days. The development of small spots in the medium indicates a positive reaction.
HIA	80 ml		
Horse serum	20 ml		
Catalase			Flood a plate containing 24–48 h growth with 30% hydrogen peroxide. Production of bubbles indicates a positive reaction.
HIA plate			
See also Digitonin and Dienes stain			

Note: This is the original medium proposed by Aluotto *et al.* (1970). Bacto heart infusion broth (Difco) contains 500 g beef heart infusion, 10 g bacto-tryptone, 5 g sodium chloride. An alternative is to use Difco PPLO broth with Mycoplasma supplement, to which the phenol red and test substrate is added. Some Mycoplasmas may be inhibited by 1% arginine, as originally proposed by Aluotto *et al.* (1970), therefore 0.2% is recommended. TTC = 2, 3, 5-triphenyltetrazolium chloride.

pH standards: Add phenol red to basal medium and dispense 3-ml lots into 5 ml tubes. Adjust pH in tubes to give a pH range of 5.6–8.4.

Control organisms:

Glucose positive = *M. bovirhinis* ATCC 19884, and negative = *M. arthritidis* ATCC 19611.

Arginine hydrolysis positive = *M. arthritidis* ATCC 19611, and negative is *M. bovirhinis* ATCC 19884.

Urea positive = T-strain mycoplasma, and negative = *M. arthritidis* ATCC 19611.

OF test organisms: Oxidative = *M. pneumoniae* ATCC 15531, Fermentative = *M. bovirhinis* ATCC 19884.

TTC control organisms: Positive aerobic and anaerobic = *M. bovirhinis* ATCC 19884. Negative for both is *M. arthritidis* ATCC 19611.

Phosphatase control organisms: Positive = *M. arthritidis* ATCC 19611. Negative = *M. bovirhinis* ATCC 19884.

Film and spots control organisms: Positive = *M. gallinarum* ATCC 19708. Negative = *M. arthritidis* ATCC 19611.

Sterol requirement: Inoculate into serum-free basal medium, and basal medium with 5 mg bovine serum albumin per ml, 0.01 mg of palmitic acid per ml, and different concentrations of cholesterol (1.0, 5.0, 10.0, 20 mg/ml). Growth is determined by sedimenting the mycoplasma and testing the medium for protein content by the biuret method. Paper disc inhibition method: Use a 1.5% w/v solution of digitonin (Sigma), and 5, 10, 20% w/v solutions of sodium polyanetholesulphonate (Koch-Light Labs, UK). The widths of the zones of growth inhibition are measured. Zones of inhibition are 5–10 mm wide.

Ninhydrin reagent Used for detection of hippurate hydrolysis (from Barrow and Feltham, 1993). See under Hippurate hydrolysis.

Nitrate broth (Quoted in Crosby, 1967; Cowan and Steel, 1970; West and Colwell, 1984)

Reagent	Amount	Preparation of media	Description of test
KNO₃	0.25 g	Add reagents to distilled water and autoclave at 121°C for 15 min. Dispense 5 ml into tubes.	Inoculate media with organisms and incubate at the appropriate temperature.
Nutrient broth (Oxoid)	3.25 g		
Distilled water	250 ml		For organisms that require salt, add 0.5 ml of 20% NaCl to 5 ml of test media.

Reagent	Amount	Preparation of media	Description of test
Nitrate reagent A		Add sulphanilic acid to distilled water, then dissolve in the acetic acid.	After incubation for 24 or 48 h, add 5 drops of nitrate reagent A, and 5 drops of nitrate reagent B to nitrate broth. Appearance of a red colour is indicative of a positive reaction.
Sulphanilic acid	1.28 g		
Distilled water	110 ml		
Acetic acid	50 ml		
			Tubes that do not show a red colour should be tested with zinc dust (match head amount). Development of a red colour indicates a true negative result, whereas no colour indicates that nitrate is absent, indicating that it was reduced by the organism to nitrite, which in turn was reduced – therefore, a positive result.
Nitrate reagent B		Handle with caution and use mask and gloves.	
Dimethyl-α-naphthylamine	0.96 ml		
Distilled water	110 ml	Add dimethyl-α-naphthylamine to distilled water. Then add acetic acid and dissolve by heating to 50°C in a waterbath.	
Acetic acid	50 ml		

Nocardia **spp.: growth medium**. See Glucose yeast extract agar

0/129. See under Vibrio discs for description of test.

0/129 = 2:4-diamino-6:7-diisopropyl pteridine (0/129) phosphate. Discs are available from Oxoid and other companies. Available at concentrations of 10 µg and 150 µg.

ONPG – *o*-nitrophenyl-β-D-galactosidase (Cowan and Steel (1970) using the method of Lowe, 1962)

Reagent	Amount	Preparation of media	Description of test
Oxoid peptone water	0.9 g	Dissolve the peptone water in distilled water and autoclave at 121°C for 20 min.	Inoculate a tube of ONPG broth and incubate at the appropriate temperature for 24–48 h.
Distilled water	60 ml		
ONPG	0.15 g	Dissolve the ONPG in the phosphate solution at pH 7.5 and filter-sterilize. Aseptically add to the peptone water, and dispense 2.5 ml into sterile tubes.	A yellow colour indicates a positive result and indicates the presence of the enzyme β-galactosidase.
Na₂HPO₄	0.035 g		
Distilled water	25 ml		
		Store at 4°C and keep protected from the light.	

ONPG discs are available from Oxoid. Discs are recommended when testing *Flavobacteriaceae*.

Oxidase test (Kovács, 1956. Used in Cowan and Steel, 1970; West and Colwell, 1984)

Reagent	Amount	Preparation of media	Description of test
Tetramethyl-*p*-phenylenediamine	1% aqueous solution	Prepare a 1% solution of oxidase reagent in water. Use immediately. The oxidase reagent must be stored in a stoppered dark glass bottle, protected from the light and stored in the fridge. It should not be used if it has become deep blue in colour.	Place a piece of filter paper in an empty Petri dish. Dampen with freshly prepared oxidase reagent. Using a platinum loop or wooden orange stick or toothpick, smear a streak of bacterial growth across the paper. Appearance of a deep purple colour in 10–30 sec indicates a positive reaction. A purple colour after 2 min may be a false positive. Discard once the filter paper becomes a blue colour. Do not test for oxidase reaction from media that contains carbohydrate such as TCBS and MCA (Jones, 1981).

Note: Commercial oxidase strips are also available and are recommended, as they offer a standardized test method.

Oxidative-fermentative (O-F) media (Media of Hugh and Leifson, 1953. For marine organisms, use marine oxidative fermentative medium of Leifson, 1963)

Reagent	Amount	Preparation of media	Description of test
Difco bacto tryptone	0.8 g	Mix all ingredients, except horse serum, in distilled water and boil for 1 min to dissolve. Cool slightly and pH to 7.1 using 10 M NaOH. Dispense 5 ml to tubes and autoclave at 121°C for 15 min. Cool to 50°C and aseptically add 2 drops of sterile horse serum per tube. Store at 4°C.	Fermentative organisms produce an acid reaction (yellow colour) in both the sealed and open tubes. Oxidative organisms produce acid at the surface of the open tube. Growth is seen at the surface of the medium with little or no growth at the bottom of the open tube, and no growth in the sealed tube. Organisms that fail to either ferment or oxidize glucose may produce an alkaline reaction (purple) at the surface of the open tube.
NaCl	2.0 g		
K₂HPO₄	0.12 g		
Glucose	4.0 g		
Oxoid agar No 1	0.8 g		
Indicator (see below)	24 ml		
Sterile horse serum not inactivated	2 drops per 5 ml tube		
Distilled water	400 ml		

Notes from Hugh and Leifson (1953): Metabolism of carbohydrate involves two different mechanisms. One mechanism is called fermentation and occurs without oxygen and is therefore an anaerobic process. The other mechanism is called oxidation and occurs in the presence of oxygen and is therefore an aerobic process. The OF medium from Oxoid is based on Hugh and Leifson (1953).

OF indicator

Reagent	Amount	Preparation of media	Description of test
Cresol red	0.15 g	Dissolve the bromothymol blue in water, 0.3 ml of a 1% solution is added to each 100 ml of medium.	
Bromothymol blue	0.10 g		
NaOH	0.20 g		
Distilled water	500 ml		

Marine oxidative fermentative medium for marine organisms (MOF) (Used for carbohydrate metabolism for marine organisms; Leifson, 1963)

Reagent	Amount	Preparation of media	Description of test
Casitone (Difco)	0.1 g	Dissolve ingredients in distilled water and adjust pH to 7.5. Autoclave at 121°C for 15 min. Autoclave artificial seawater separately and add to ingredients in distilled water.	Inoculate 2 tubes of medium with organism and overlay the medium in one tube with a layer of sterile paraffin oil to approximately 1 cm depth, or 0.5 ml. Incubate tubes at appropriate temperature.
Yeast extract	0.01 g		
Ammonium sulphate	0.05 g		
Tris buffer	0.05 ml		
Agar	0.3 g	Aseptically add 10 ml filter-sterilized glucose per 100 ml. Aseptically dispense 5 ml to 10 ml tubes.	Organisms that ferment carbohydrate acidify the medium in both tubes, whereas aerobic organisms acidify the medium in the 'open' tube only.
Distilled water	50 ml		
Artificial seawater	50 ml		
Phenol red (0.1% stock)	1 ml	Phenol red: Used at a 0.001% final concentration. (Prepare a 0.1% solution and use 1 ml of this per 100 ml of medium).	For fermenting organisms record 'F' on results sheet. For aerobes or oxidizing organisms, record 'O' on results sheet.
Glucose	1.0%	Prepare a 10% solution and filter-sterilize.	

Note: The indicator bromothymol blue, which is used in the conventional OF medium, may be toxic to some marine organisms.

Paraffin oil (Used for overlay in decarboxylase tests and OF test)

Reagent	Amount	Preparation of media	Description of test
Paraffin oil	As required	Dispense into 100 ml bottles or any appropriate-sized bottle. Hot-air sterilize at 160°C for 60 min. Do not autoclave as the oil goes cloudy.	Used as overlay in ADH, LDC, ODC and OF tests. For ease of use, oil can be dispensed into a 1 l Schott bottle and capped with a Socorex 2 ml dispenser unit. Dispense 0.5 ml into 10 ml media tubes.

Physiological saline. *See* saline

Rogosa agar (RA) (Oxoid manual) (Selective medium for *Lactobacilli*; Rogosa *et al.*, 1951)

Reagent	Amount	Preparation of media	Description of test
Rogosa agar (Oxoid)	82 g	Suspend Rogosa agar in distilled water and bring to the boil to dissolve. Add glacial acetic acid and mix thoroughly. Heat to 90–100°C for 2–3 min with frequent agitation. Distribute into sterile tubes or Petri dishes. Do not autoclave.	May assist to differentiate *Lactobacilli* from other Gram-positive bacteria such as *Carnobacterium*, *Arcanobacterium* and *Vagococcus* species.
Distilled water	1000 ml		
Glacial acetic acid	1.32 ml		

Ingredients (g/l) tryptone (10.0), yeast extract (5.0), glucose (20.0), sorbitan mono-oleate (1.0), potassium dihydrogen phosphate (6.0), ammonium citrate (2.0), sodium acetate (25.0), magnesium sulphate (0.575), manganese sulphate (0.12), ferric sulphate (0.034), agar (20.0).

Saline: physiological saline

Reagent	Amount	Preparation of media	Description of test
NaCl	0.85 g	Add salt to distilled water. Dispense 10 ml into McCartney bottles. Autoclave at 121°C for 15 min.	May be used as an inoculum for commercial kit tests and 'Biochem Set'.
Distilled water	100 ml		

Salt – 20% stock solution

Reagent	Amount	Preparation of media	Description of test
NaCl	100 g	Add NaCl to distilled water. Autoclave at 121°C for 15 min.	For ease of use, 20% NaCl can be dispensed into a 1 l Schott bottle and capped with a Socorex 2 ml dispenser unit with 0.5 ml graduations. Dispense 0.5 ml per 5 ml of media to produce final NaCl concentration of 2% in test.
Distilled water	400 ml		

In most cases, the carbohydrate media tubes do not require NaCl in addition to the 2% NaCl inoculum fluid. However, liquid tube media such as aesculin, ADC, ODC, LDC, the decarboxylase control, MRVP, indole (TW) and nitrate do require that 500 µl of a 20% NaCl solution is added to the 5 ml of media.

Salt tolerance
0%, 3% NaCl. See Gelatin/ NaCl split plate. Using a wire loop, spot inoculate equal amounts of the bacterium to the 0% NaCl side and the 3% NaCl side. After 1–2 days incubation at the appropriate temperature, examine for areas of clearing or opacity. *See photographic section.*

10% NaCl. For tolerance to 10% salt, dispense equal volumes of TSB and 20% NaCl stock solution into a sterile bijou bottle or 10 ml tube. Add bacteria to turbidity of 0.5 or tube 1 McFarland standard. Incubate at appropriate temperature for 24–48 h. Look for obvious increase in growth of organism as seen by cloudiness in the medium.

Survival at 50°C (Used for *Nocardia* species; Gordon *et al.*, 1974)
Inoculate slants of glucose yeast extract agar, and incubate in a 50°C waterbath for 8 h. Remove from waterbath, cool quickly and incubate at 28°C for 3 weeks. Examine tubes for growth.

Tryptophane deaminase (TDA) reagent (For use in the API 20E kit. Can be purchased commercially)

Reagent	Amount	Preparation of media	Description of test
Ferric chloride	3.4 g	Dissolve the ferric chloride in 90 ml of water, and then make up to 100 ml.	Used in the API 20E kit. A positive is a brown colour. *Proteus* strains tend to be positive.
Distilled water	100 ml		

Triple sugar iron agar (TSI)

Reagent	Amount	Preparation of media	Description of test
Triple sugar iron agar (Oxoid)	9.75 g	Dissolve reagent in distilled water and pH to 7.4. Dispense in 5 ml aliquots to 10 ml tubes. Autoclave at 121°C for 15 min. Allow to cool and solidify in a slanted position to give deep butts.	Inoculate the tube using a straight wire. Stab into medium, and then zig-zag the slope. H_2S producers will turn the medium black. Fermentative organisms will acidify the TSI butt (yellow), whereas non-fermentative organisms will grow on the slant only and show either no pH change or an alkaline reaction and, rarely, an acidic reaction.
Distilled water	150 ml		For some organisms this test method may not be as sensitive as using a lead acetate paper strip.
			A hydrogen sulphide Biostrip (catalogue number TM343) is available from MedVet Science.

Urea (Christensen, 1946)

Part A

Reagent	Amount	Preparation of media	Description of test
Oxoid agar No 1	3.75 g	Add agar to distilled water and autoclave at 121°C for 15 min. Cool to 50°C.	Add 1% NaCl for marine organisms.
Distilled water	225 ml		

Part B

Reagent	Amount	Preparation of media	Description of test
BBL brand urea agar base	8.7 g	Add reagents to distilled water and filter-sterilize through a 0.22 µm Millipore filter.	Inoculate the slant heavily over the entire surface with a loopful of bacteria harvested from an agar medium. Bright pink indicates a positive reaction.
Distilled water	30 ml	Add Part A to 25 ml of Part B. Aseptically dispense 5 ml volumes per tube. Allow to solidify in a sloped position.	

Note: Urea slopes should be prepared with a deep butt and a short slant (Gilardi, 1983).

Vibrio discs

0/129 = 2:4-diamino-6:7-diisopropyl pteridine (0/129) phosphate, vibriostatic reagent (µg). The *Vibrio* spp. (including *Listonella* spp., *Moritella* spp. and *Photobacterium* spp.) are sensitive to this compound, commonly called vibriostatic agent (Shewan *et al.*, 1954). This test assists in the differentiation of *Vibrio* species from other Gram-negative rods, particularly *Aeromonas* species, which are resistant to vibriostatic agent. Almost all *Vibrio* species are sensitive to 0/129 at 150 µg and some are sensitive to 10 µg. However, *Vibrio cholerae* 0139 has developed resistance to 0129 at 150 µg concentration (Albert *et al.*, 1993; Islam *et al.*, 1994).

Treat this test in the same manner as 'sensitivity testing' used to test an organism as sensitive or resistant to an antimicrobial reagent. Prepare an inoculum in normal saline to the opacity of McFarland tube 1, a suspension that is just visible to the eye. Use a sterile cotton-tipped swab to lawn inoculate the plate. Use BA for freshwater organisms and MSA-B for marine organisms. Place the two discs on to the agar surface so that they are at least 4 cm apart. Invert the plate, and incubate at the appropriate temperature for 24 h. If there is insufficient growth incubate a further 24 h; however, normally the zones are recorded at 24 h as further growth, especially when testing *Vibrio* species that swarm, may show a false resistant result. Record zone sizes as 'sensitive' (S) or 'resistant' (R). For the 0/129 500 µg disc, a zone of 9 mm is susceptible for *Vibrio* species (Bernardet and Grimont, 1989). A zone size of 22 mm is considered sensitive for *Photobacterium damselae* ssp. *damselae* (Love *et al.* 1981).

The discs are purchased from Oxoid or Rosco Diagnostics. The Oxoid codes for the two concentrations are DD14 for 0129 10 µg, and DD 15 for 150 µg.

Voges-Proskauer reaction See MRVP test (Clark and Lubs, 1915; Voges and Proskauer, 1898, as reported in Cowan and Steel, 1970. Used by West and Colwell, 1984)

Use for the detection of acetoin from glucose fermentation. The incubation time and temperature rather than the method influence the production of acetoin.

MRVP test medium (The commercial medium is based on Clark and Lubs medium, 1915)

Reagent	Amount	Preparation of media	Description of test
MRVP medium (Oxoid)	3.75 g	Dissolve reagent in distilled water and autoclave at 121°C for 15 min. Dispense 5 ml per tube.	Add NaCl for marine organisms. Must incubate for 2–3 days before testing.
Distilled water	250 ml		Add VP reagent A, and VP reagent B. Red coloration may develop up to 18 h after the addition of reagents.

MR = methyl red; VP = Voges-Proskauer.

VP test reagents (Barritt, 1936)

Reagent	Amount	Preparation of media	Description of test
Reagent A		Dissolve α-naphthol in ethanol. Store at 4°C.	After incubation of MRVP medium for 48 h at appropriate temperature, place 1 ml into a test tube. Add 0.6 ml of reagent A, and 0.2 ml of reagent B. Examine for pink colour up to 4 h at room temperature.
α-naphthol	5.0 g		
Absolute ethanol	100 ml		
Reagent B		Weigh KOH. Make up to 100 ml final volume with distilled water. Store at 4°C.	
KOH	40.0 g		
Distilled water	Add to final 100 ml		Tests for acetylmethylcarbinol (acetoin).

Note: VP reagent I and II in the API 20E kit are also suitable. Transfer 250 µl of medium to a microfuge tube. Add 150 µl and 50 µl of reagent I and II respectively, shaking after each addition. Read after 10–20 min.

Further Reading and Other Information Sources

Books on Fish Diseases

Austin, B. and Austin, D.A. (1999) *Bacterial Fish Pathogens: Disease of Farmed and Wild Fish*. 3rd revised edn. Praxis Publishing, Chichester, UK.

Woo, P.T.K. and Bruno, D.W. (eds) (1999) *Fish Diseases and Disorders*. Vol. 3: *Viral, Bacterial and Fungal Infections*. CAB International, Wallingford, UK.

Bergey's Manual of Systematic Bacteriology, Vol I and Vol II. (1984) Holt, J.G. (ed.) Lippincott Williams and Wilkins, Baltimore, Maryland.

Diagnostic Manual for Aquatic Animal Diseases, 3rd edn (2000) Office International Des Epizooties (OIE), 12 rue de Prony, F-75017, Paris, France.

Plumb, J.A. (1999) *Health Maintenance and Principal Microbial Diseases of Cultured Fishes*. Iowa State University Press.

The fourth edition of the OIE *Diagnostic Manual for Aquatic Animal Diseases* will be available in July 2003. These manuals are also available on line. http://www.oie.int/eng/normes/fmanual/A_summry.htm

Books on Biochemical Identification Tests

Cowan, S. and Steel, K. (1970) *Manual for the Identification of Medical Bacteria*. Cambridge University Press, Cambridge.

Barrow, G.I. and Feltham, R.K.A. (1993) *Cowan and Steel's Manual for the Identification of Medical Bacteria*, 3rd edn. Cambridge University Press, Cambridge.

MacFaddin, J.F. (1980) *Biochemical Tests for Identification of Medical Bacteria*, 2nd edn. Williams and Wilkins, Baltimore, Maryland.

MacFaddin, J.F. (2000) *Biochemical Tests for Identification of Medical Bacteria*, 3rd edn. Williams and Wilkins, Baltimore, Maryland.

Other Biochemical Identification Schemes

Alsina, M. and Blanch, A. (1994) A set of keys for biochemical identification of environmental *Vibrio* species. *Journal of Applied Bacteriology*, 76, 79–85.

Alsina, M. and Blanch, A. (1994) Improvement and update of a set of keys for biochemical identification of *Vibrio* species. *Journal of Applied Bacteriology*, 77, 719–721.

Carson, J., Wagner, T., Wilson, T. and Donachie, L. (2001) Miniaturised tests for computer-assisted identification of motile *Aeromonas* species with an improved probability matrix. *Journal of Applied Microbiology*, 90, 190–200.

Schmidtke, L.M. and Carson, J. (1994) Characteristics of *Vagococcus salmoninarum* isolated from diseased salmonid fish. *Journal of Applied Bacteriology*, 77, 229–236.

Journals

Applied and Environmental Microbiology   http://aem.asm.org/
Aquaculture   http://www.elsevier.com/locate/aquaculture
Bulletin of the European Association of
   Fish Pathologists
Current Microbiology   http://link.springer.de/link/service/journals/00284/
Diseases of Aquatic Organisms   http://www.int-res.com/journals/dao/
Fish Pathology
International Journal of Systematic and
   Evolutionary Microbiology   http://ijs.sgmjournals.org/
Journal of Applied Ichthyology   http://www.blackwell-synergy.com/Journals/issuelist.asp?journal=jai
Journal of Applied Microbiology   http://www.blackwell-synergy.com/Journals/issuelist.asp?journal=jam
Journal of Aquatic Animal Health
Journal of Clinical Microbiology   http://jcm.asm.org/
Journal of Fish Diseases
Veterinary Microbiology   http://www.elsevier.nl/locate/vetmic

Culture Collections

Ecole Nationale Vétérinaire de Toulouse, 23 chemin des Capelles, F-31076 Toulouse cedex 03, France. http://www.bacterio.cict.fr/collections.html

ACAM: Australian Collection of Antarctic Microorganisms, Antarctic CRC, University of Tasmania, Hobart, Australia.

AHLDA: Animal Health Laboratories, Department of Agriculture. 3 Baron-Hay Court, South Perth, Western Australia 6151.

ATCC: American Type Culture Collection. Corporate: ATCC, 10801 University Boulevard, Manassas, VA 20110–2209, USA. Products & Services Orders: ATCC, PO Box 1549, Manassas, VA 20108–1549, USA. http://www.atcc.org/

CCUG: Culture Collection, University of Göteborg, Department of Clinical Bacteriology, Institute of Clinical Bacteriology, Immunology, and Virology, Guldhedsgatn 10A s-413, 46 Göteborg, Sweden.

CDC: Center for Disease Control, 1600 Clifton Rd, Atlanta, Georgia 30333, USA.

CECT: Coleccion Espanola de Cultivos Tipo, Universidad de Valencia, Burjassot, Spain.

CIP: Collection de l'Institut Pasteur, Institut Pasteur, 28 Rue du Docteur Roux, 75724 Paris Cedex 15, France.

CNCM: Collection Nationale de Culture de Microorganismes, Institut Pasteur, Paris, France.

DSMZ (DSM): Deutsche Sammlung von Mikroorganismen und Zellkulturen GmbH, Mascheroder Weg 1B, D-38124, Braunschweig, Germany. http://www.dsmz.de/dsmzhome.htm

IAM: Institute of Molecular and Cellular Biosciences (formerly Institute of Applied Microbiology, Culture Collection – IAMCC), The University of Tokyo, Yayoi, Bunkyo-Ku, Tokyo, Japan.

KMM: Collection of Marine Microorganisms, Pacific Institute of Bioorganic Chemistry, Vladivostok, Russia.

NCFB: National Collection of Food Bacteria (previously named NCDO). Transferred from the IFR (Institute of Food Research), Reading, to National Collections of Industrial, Food and Marine Bacteria, 23 Machar Drive, Aberdeen AB24 3RY, UK.

NCIMB: National Collection of Industrial and Marine Bacteria, National Collections of Industrial, Food and Marine Bacteria, 23 Machar Drive, Aberdeen AB24 3RY, UK. http://www.ncimb.co.uk/ncimb.htm

NCTC: National Collection of Type Cultures, Central Public Health Laboratory, Colindale Ave., London NW9 5HT, UK. www.phls.co.uk

RVAU: Royal Veterinary and Agricultural University, Copenhagen, Denmark.

UB: University of Barcelona, Barcelona, Spain.

Bacterial Names/Taxonomy/Nomenclature

Bacterial nomenclature is continually being updated at the following websites:

DSMZ (DSM): Deutsche Sammlung von Mikroorganismen und Zellkulturen GmbH, Mascheroder Weg 1B, D-38124, Braunschweig, Germany. http://www.dsmz.de/dsmzhome.htm

List of Bacterial Names with Standing in Nomenclature: http://www.bacterio.cict.fr/
NCBI: http://www.ncbi.nlm.nih.gov/Taxonomy/taxonomyhome.html/

Fish Disease Web Sites

http://www.fishdisease.net This website is for aquatic animal health professionals and contains information on
   leaflets on different diseases, an image library on parasites, notification of conferences, jobs and contacts.
http://www.fishbase.org/home.htm or http://www.fishbase.org/search.html This website contains information on
   fishes that may be useful to fisheries managers, scientists and others. One of its features is the latest scientific
   name of the different fishes.

Appendix

Common Name and Scientific Name of Aquatic Animals

Common name	Scientific name
Abalone	*Haliotis discus hannai*
Adriatic sturgeon	*Acipenser naccarii*
African cichlid	*Nimbochromis venustus*
Amazon freshwater dolphin	*Inia geoffrensis*
Amberjack	*Seriola dumerili*
American alligator	*Alligator mississippiensis*
American crayfish	*Orconectes limosus, Pacifastacus leniusculus, Procambarus clarkii*
American eel	*Anguilla rostrata*
American plaice	*Hippoglossoides platessoides*
Antarctic fur seal	*Arctocephalus gazella*
Arctic char	*Salvelinus alpinus* L.
Atlantic bottlenose dolphin	*Tursiops truncatus*
Atlantic cod	*Gadus morhua* L.
Atlantic croaker	*Micropogon undulatus*
Atlantic menhaden	*Brevoortia tyrannus* Latrobe
Atlantic salmon	*Salmo salar* L.
Atlantic walrus	*Odobenus rosmarus rosmarus*
Atlantic white-sided dolphin	*Lagenorhynchus acutus*
Australian oyster	*Saccostrea commercialis*
Ayu	*Plecoglossus altivelis*
Balloon molly	*Poecilia* spp.
Banana prawn	*Penaeus merguiensis*
Barramundi	*Lates calcarifer* (Bloch)
Beluga whale	*Delphinapterus leucas*
Bighead carp	*Aristichthys nobilis*
Black acara	*Cichlasoma bimaculatum*
Black mullet	*Mugil cephalus*
Black scraper	*Novodon modestus*
Black skirted tetra	*Hyphessobrycon* spp.
Blenny	*Zoarces viviparus*
Blue fish	*Pomatomus saltatrix*
Blue mackerel	*Scomber australasicus*
Blue manna crab	*Portunus pelagicus*
Blue shrimp	*P. (Litopenaeus) stylirostris*

Continued

281

Common name	Scientific name
Bottlenose whale	*Hyperodoon ampullatus*
Boney bream	*Nematolosa come* (Richardson)
Borneo mullet	*Liza macrolepis*
Bottle-nosed dolphin	*Tursiops truncatus*
Bowhead whale	*Balaena mysticetus*
Bream	*Abramis brama*
Brine shrimp	*Artemia*
Brook salmon	*Salvelinus fontinalis*
Brook trout	*Salvelinus fontinalis* (Mitchill)
Brown bullhead	*Ictalurus nebulosus* (Lesueur)
Brown shark	*Carcharhinus plumbeus*
Brown-spotted grouper	*Epinephelus tauvina, E. coioides*
Brown trout	*Salmo trutta* m. *fario, Salmo trutta* m. *lacustris* L.
Burnett salmon	*Polydactylus sheridani* (Macleay)
Californian sea lion	*Zalophus californianus*
Canadian shrimp	*Lismata amboiens*
Carp	*Cyprinus carpio* L.
Catfish	*Clarius batrachus* L.
Caucasian carp	*Carassius carassius*
Chanchito	*Chichlasoma facetum* (Jenyns)
Channel catfish	*Ictalurus punctatus* (Rafinesque)
Chinook salmon	*Oncorhynchus tschawytscha*
Chub	*Leuciscus cephalis*
Chum salmon	*Oncorhynchus keta* (Walbaum)
Cichlid	*Oreochromis mossambicus*
Clam	*Tapes philippinarum*
Cod	*Gadus morhua*
Coho salmon	*Oncorhynchus kisutch*
Common carp	*Cyprinus carpio* L.
Common dolphin	*Delphinus delphis*
Common seal	*Phoca vitulina*
Common snook	*Centropomus undecimalis*
Common wolf-fish	*Anarhichas lupus*
Coral prawn	*Metapenaeopsis* spp.
Crucian carp	*Carassius carassius*
Cultured flounder	*Paralichthys olivaceus*
Cutthroat trout	*Salmo clarki*
Cuttle fish	*Sepia officinalis*
Dab	*Limanda limanda*
Dace	*Leuciscus leuciscus* L.
Damselfish	*Chromis punctipinnis*
Damselfish	*Pomacentridae, Amphiprion clarkii* (Bennett), *Amblyglyphidodon curacao* (Bloch)
Danio	*Danio devario*
Discus fish	*Symphysodon discus, S. aequifasciatus*
Dolphin	*Tursiops truncatus, T. gephyreus*
Dolphin fish	*Coryphaena hippurus* L.
Eastern freshwater cod	*Maccullochella ikei*
Eastern mosquitofish	*Gambusia holbrooki*
Eastern painted turtle	*Chrysemys picta picta*
Eel	*Anguilla japonica, A. reinhardtii*
Elephant seal	*Mirounga leonina*
European crayfish	*Astacus leptodactylus, A. pachypus, A. torrentium, A. astacus, Austropotamobius pallipes*
European eel	*Anguilla anguilla*
European sea bass	*Dicentrarchus labrax* Serranidae

Common name	Scientific name
Fairy shrimp	*Branchipus schaefferi* (Fisher),*Chirocephalus diaphanus* (Prévost), *Streptocephalus torvicornis* (Waga)
False killer whale	*Pseudorca crassidens*
Farmed mussel	*Perna perna*
Fathead minnow	*Pimephales promelas*
Fighting fish	*Betta splendens*
Firemouth cichlid	*Cichlasoma meeki*
Flat-tailed mullet	*Liza dussumieri* (Valenciennes)
Flounder	*Paralichthys olivaceus, P. flesus*
Flounder	*Platichthys flesus*
Four bearded rockling	*Enchelyopus cimbrius* L.
Freshwater cod (Australian native)	*Maccullochella* spp.
Freshwater dolphin	*Inia geoffrensis*
Freshwater prawn	*Macrobranchium rosenbergii*
Fur seal	*Arctocephalus australis*
Gilthead sea bream	*Sparus auratus*
Golden shiner	*Notemigonus crysoleucas* (Mitchell)
Goldfish	*Carassius auratus* L.
Goldsinny wrasse	*Ctenolabrus rupestris*
Gourami (three-spot)	*Trichogaster trichopterus*
Grass carp	*Ctenopharyngodon idella*
Grayling	*Thymallus thymallus* L.
Greater weever	*Trachinus draco*
Green knife fish	*Eigemannia virescens*
Green moray eel	*Gymnothorax funebris*
Green sturgeon	*Acipenser medirostris*
Greenback flounder	*Rhombosolea tapirina* Gunther
Grey seal	*Halichoerus grypus*
Grouper	*Epinephelus guaz, E. coioides*
Guppy	*Poecilia reticulata* (Peters), *Lebistes reticulatus*
Haddock	*Melanogrammus aeglefinus* L.
Halibut	*Hippoglossus hippoglossus* L.
Harbour porpoise	*Phocoena phocoena*
Harbour seal	*Phoca vitulina*
Harp seal	*Phoca groenlandica*
Herring	*Arripis georgianus*
Hooded seal	*Cystophora cristata*
Horse mackerel	*Trachurus trachurus*
Iberian toothcarp	*Aphanius iberus*
Japanese abalone	*Sulculus diversicolor supratexta*
Japanese eel	*Anguilla japonica*
Japanese flounder	*Paralichthys olivaceus*
Japanese medaka	*Oryzias latipes*
Jewel tetra	*Hyphessobrycon callistus* (Boulenger)
Johnston crocodiles (freshwater)	*Crocodylus johnstoni*
Killer whale	*Orcinus orca*
King prawn	*Penaeus latisulcatus*
Knife fish	*Gymnotus carapo*
Lake trout	*Salmo trutta* m. *lacustris, Salvelinus namaycush* Walbaum
Largemouth bass	*Micropterus salmoides*
Lemon shark	*Negaprion brevirostris*
Little penguin	*Eudyptula minor*
Living dace	*Tribolodon hakonensis* Gunther
Loach	*Misgurnus anguillicaudatus* Cantor
Lobster	*Homarus gammarus* L.

Continued

Common name	Scientific name
Local mussel	*Mutilus edulis*
Long-tom	*Tylosurus macleayanus* (Ogilby)
Mackerel	*Scomber scombrus*
Manila clam	*Tapes philippinarum, T. decussatus, Ruditapes philippinarum*
Masu salmon	*Oncorhynchus masou*
Menhaden	*Brevoortia patronus*
Minke whale	*Balaenoptera acutorostrata*
Minnow	*Phoxinus phoxinus* L.
Molly	*Poecilia velifera* (Regan)
Mud crab	*Scylla serrata*
Mullet	*Mugil cephalus*
Murray cod	*Maccullochella peeli*
Mussel (Far-eastern)	*Crenomytilus grayanus* and *Patinopecten yessoensis*
Mussel	*Protothaca jedoensis, Mytilus edulis, Mytilus galloprovincialis*
Narwahl whale	*Monodon monocerus*
Neon tetra	*Paracheirodon innesi, Hyphessobrycon innesi*
New Zealand fur seal	*Arctocephalus forsteri*
New Zealand mussel	*Perna canaliculus*
New Zealand sea lion	*Phocarctos hookeri*
Nile tilapia	*Oreochromis niloticus*
North-east Atlantic mackerel	*Scomber scombrus*
Northern elephant seal	*Mirounga angustirostris*
Northern fur seal	*Callorhinus ursinus*
Northern pike	*Esox lucius* L.
Northern right whale dolphin	*Lissodelphis borealis*
Nurse shark	*Orectolobus ornatus*
Octopus	*Octopus vulgaris, O. joubini*
One-spot bream	*Diplodus sargus*
Ornamental fish	*Pterophyllum scalare*
Oscar	*Astronotus ocellatus*
Oscar	*Apistogramma ocellatus*
Otter	*Lutra lutra*
Oyster	*Ostrea edulis*
Oyster (eastern)	*Crassostreae virginica*
Pacific herring	*Clupea harengus pallasi*
Pacific oyster	*Crassostrea gigas*
Pacific salmon	*Oncorhynchus* spp.
Pacific staghorn sculpin	*Leptocottus armatus*
Pacific white shrimp	*Penaeus vannamei*
Pacific white-sided dolphin	*Lagenorhynchus obliquidens*
Paradise fish	*Macropodus opercularis* (L.)
Pejerrey	*Odonthestes bonariensis*
Penguins	*Aptenodytes patagonica, Eudyptes crestatus, Pyoscelis papua, Spheniscus demersus, Spheniscus humboldti*
Perch	*Perca fluviatilis*
Pike	*Esox lucius*
Pilchard	*Sardinops neopilchardus*
Pilot whale	*Globicephala scammoni*
Pink salmon	*Oncorhynchus gorbuscha*
Pink snapper	*Chrysophrys unicolor*
Pinkfish	*Lagodon rhomboides*
Pirarucu	*Arapaima gigas* Cuvier
Plaice	*Pleuronectes platessa*
Pompanos	*Trachinotus carolinus* L.
Rabbitfish	*Siganus rivulatus* (Forsskål)
Rainbow trout	*Oncorhynchus mykiss* (Walbaum)

Common name	Scientific name
Rainbow and steelhead trout	*Salmo gairdneri*
Red abalone	*Haliotis rufescens*
Red algae	*Jainia* spp.
Red claw crayfish	*Cherax quadricarinatus*
Red drum, Redfish	*Sciaenops ocellatus*
Red-eared slider turtle	*Chrysemys scripta elegans*
Red sea bream	*Pagrus major*
Red swamp crawfish	*Procambarus clarkii*
Redtail catfish	*Phractocephalus hemiliopterus*
Ringed seal	*Phoca hispida*
Risso dolphin	*Grampus griseus*
Roach	*Rutilus rutilus* L.
Rohu	*Labeo rohita*
Rosy barb	*Puntius conchonius*
Rudd	*Scardinius erythrophthalmus*
Sablefish	*Anoplopoma fimbria* (Pallas)
Salmon	*Oncorhynchus kisutch* (Walbaum)
Saltwater crocodile	*Crocodylus porosus*
Sand eel	*Ammodytes lancea* (Cuvier), *Hyperoplus lanceolatus* (Lesauvege)
Sand lance	*Ammodytes personatus* Girard
Sand whiting	*Sillago ciliata* Cuvier
Saratoga	*Scleropages leichardii*
Sardine	*Sardinops melanostictus, Sardinops sagnax*
Scallop	*Pecten maximus, Argopecten purpuratus*
Scaly mackerel fish	*Amblygaster postera*
Sea bass	*Dicentrarchus labrax*
Sea bream	*Pagrus major, Evynnis japonicus, Sparus aurata, Acanthopagrus latus*
Sea catfish	*Arius felis*
Sea horse	*Hippocampus angustus, H. barbouri, H. whitei, H. kuda*
Sea lion	*Otaria flavescens*
Sea mullet	*Mugil cephalus* L.
Sea trout	*Salmo trutta* m. *trutta* L.
Sea turtle	*Chelonia mydas*
Sea-urchin	*Paracentrotus lividu*
Sepiolid squid	*Euprymna scolopes*
Shotted halibut	*Eopsetta grigorjewi*
Shubunkin	*Carassius* spp.
Siamese fighting fish	*Betta splendens* Regan
Signal crayfish	*Pacifastacus leniusculus*
Silver black porgy	*Acanthopagrus cuvieri*
Silver bream	*Blicca bjoerkna*
Silver bream	*Acanthopagrus butcheri, A. australis* (Owen)
Silver carp	*Hypophthalmichthys molitrix* Valenciennes
Silver mullet	*Mugil curema* (Valenciennes), *Mugil cephalus* (L.)
Silver molly	*Poecilia* spp.
Silver perch	*Bidyanus bidyanus* (Mitchell)
Silver trout	*Cynoscion nothus*
Small abalone	*Haliotis diversicolor supertexta*
Smallmouth bass	*Micropterus dolomieui*
Snakehead fish	*Channa striatus*
Snakehead fish	*Ophicephalus punctatus, O. striatus*
Snub-nose garfish	*Arrhamphus sclerolepsis* (Gunther)

Continued

Common name	Scientific name
Sockeye salmon	*Oncorhynchus nerka* (Walbaum)
Softshell clam	*Mya arenaria*
Sole	*Solea solea*
South African abalone	*Haliotis midae*
South American side-necked turtle	*Podocnemis unifelis*
Southern elephant seal	*Mirounga leonina*
Sowerby's beaked whale	*Mesoploden bidens*
Spanish mackerel	*Scomber japonicus*
Spanner crab	*Ranina ranina*
Spiny soft-shelled turtle	*Trionyx spinifer*
Spot	*Leiostomus xanthurus*
Spotted dolphin	*Stenella plagiodon*
Spotted moray eel	*Gymnothorax moringa*
Spotted wolf-fish	*Anarhichas minor*
Squid	*Loligo pealei*
Squid	*Sepiola*
Squid	*Teuthoidea* species
Starfish	*Asterias rubens*
Stingray	*Dasyatis pastinaca*
Striped bass	*Morone saxatilis* (Walbaum), *M. chrysops*
Striped dolphin	*Stenella coeruleoalba*
Striped mullet	*Mugil cephalus*
Striped-neck musk turtle	*Sternotherus minor peltifer*
Sturgeon	*Acipenser naccarii*
Sunfish	*Mola mola*
Tasmanian lobster	*Jasus novaehollandiae*
Tilapia	*Oreochromis niloticus, O. aurus*
Tilapia	*Tilapia nilotica, Tilapia aurea, Tilapia mosambica*
Tilapia	*Sarotherodon aureus* (Steindachner)
Tom cod	*Gadus microgadus*
Tropical shrimp	*Stenopus hispidus*
Trout cod	*Maccullochella macquariensis*
Turbot	*Scophthalmus maximus* L.
Turtle	*Dermochelys coriacea*
Turtle	*Pseudemis scripta*
Viviparous blenny	*Zoarces viviparus*
Weddell seal	*Leptonychotes weddellii*
Western rock lobster	*Panulirus cygnus*
White catfish	*Ictalurus catus* L.
White clawed crayfish	*Austropotamobius pallipes*
White leg shrimp	*P. (Litopenaeus) vannamei*
White perch	*Roccus americanus, Morone americanus* (Gremlin)
White whale	*Delphinapterus leucas*
Whitefish	*Coregonus* spp.
Whitespotted rabbitfish	*Siganus canaliculatus*
Whiting	*Merlangius merlangus*
Wolf-fish	*Anarchichas lupus* L.
Wrasse	*Labrus berggylta*
Yellow bass	*Morone mississippiensis*
Yellow perch	*Perca flavescens*
Yellowtail	*Seriola quinqueradiata, S. lalandi*
Yellowfin bream	*Acanthopagrus australis* (Owen)
Zebra danio	*Brachydanio rerio*

This list was taken from the references used in this manual.

Glossary of Terms

α or αH	(referring to greening of agar) Alpha haemolysis
A	Alkaline reaction
AAHRL	Australian Animal Health Reference Laboratory
ACAM	Australian Collection of Antarctic Microorganisms
ADH	Arginine dihydrolase
Aes	Aesculin
AFB	Acid-fast bacteria
AHL	Animal Health Laboratories
AHLDA	Animal Health Laboratories Department of Agriculture Culture Collection
Ala	L-alanine
Amp	Ampicillin disc 10 μg
ANA	Plate media for anaerobic bacteria
AO	Anacker Ordal agar for *Flavobacteria*
AO-M	Anacker Ordal agar with added NaCl for growth of marine *Flavobacteria*
API 50CH	API identification system from bioMérieux. Carbohydrate tests
API 20E	API identification system from bioMérieux. Fermentation and enzyme tests
API 20NE	API identification system. Utilization tests
API Rapid ID32 Strep	API system for *Streptococci* and other Gram-positives
API 20 Strep	API identification system for *Streptococci* and other Gram-positive organisms
API ZYM	API identification system. Enzyme tests
Arab	L-arabinose
Arg	Arginine
Asp	L-asparagine
ASW	Artificial sea water
AT	Annealing temperature
ATCC	American Type Culture Collection, Rockville, Maryland, USA
β	Beta (refers to clear zone or β haemolysis)
BA	Blood agar
BGD	Bacterial gill disease
βH	β-haemolysis
BHA	Bacterial haemorrhagic ascites
BHIA	Brain heart infusion agar
BKD	Bacterial kidney disease
bp	Base pairs (in kilo-bases)
BRD	Brown ring disease
Brucella agar	Used for isolation of *Brucella* species

C	Number of cycles in the PCR reaction
Ca	Calcium
Cat	Catalase
CBBA	Coomassie brilliant blue agar
CCA	Cellobiose Colistin agar for isolation of *Vibrio vulnificus*
CCRC	Culture Centre for Research and Collection
CCUG	Culture Collection of the University of Göteborg, Department of Clinical Bacteriology, Göteborg, Sweden
CDC	Centres for Disease Control and Prevention, Atlanta, Georgia, USA
CFPA	Selective media for isolation of *Bordetella bronchiseptica*
CFU	colony forming units
Cit	Citrate
CNCM	Collection Nationale de Culture de Microorganismes (Institut Pasteur, Paris, France)
CO_2	Carbon dioxide
CPC	Cellobiose Polymyxin B Colistin agar for isolation of *Vibrio vulnificus*
Cr	Coccoid rods
CR	Congo Red
CSF	Cerebrospinal fluid
cv	Curved rods
d	Days
DCA	Desoxycholate-citrate agar
Dmso	Dimethyl sulphoxide
DNase	Test for the detection of hydrolysis of DNA
DSM	Deutsche Sammlung von Mikroorganismen und Zellkulturen GmbH, Braunschweig, Germany
EIM	*Edwardsiella ictaluri* medium
EM	Electron microscopy
ERM	Enteric red mouth
ESC	Enteric septicaemia of catfish
Ery	*Meso*-erythritol
F	Fermentative (facultative anaerobe)
FA	Furunculosis agar for *Aeromonas salmonicida*
FAO	Food and Agriculture Organization of the United Nations
FINE	Flounder necrotizing enteritis
FM	Farrell's medium for *Brucella* species
FPM	*Flavobacterium psychrophilum* medium
G	Gelatin
G	Green-coloured colony on TCBS plate
Glid	Gliding motility
Gal	D-galactose
Glu	Glucose
Glut	Glutamic acid
Gm	Gram stain reaction (blue colour = positive, red = negative)
GUD	Goldfish ulcer disease
h	Hours
H_2S	Hydrogen sulphide
HCl	Hydrochloric acid
HG	(DNA) hybridization group
Hip	Hippurate
HS	Hsu-Shotts agar for *Flavobacterium columnare*
HSM	For the isolation of *Tenacibaculum maritimum*
I	Inert reaction (in OF test)
ID	Identification
IGS	Intergenic spacer
Ind	Indole

Inos	Inositol
ISP No. 2	Yeast malt extract agar (Difco)
JOD	Juvenile oyster disease
KCl	Potassium chloride
KDM2	Medium for growth of *Renibacterium salmoninarum*
KDMC	Medium for growth of *Renibacterium salmoninarum*
Kf	Cephalothin 30 µg disc
KOH	Potassium hydroxide. 20%
KUMA	Kumamoto Prefectural Institute of Public Health (culture collection prefix)
Lac	Lactose
LDC	Lysine decarboxylase
LJM	Lowenstein-Jensen medium for *Mycobacteria* species
LMG	Laboratorium Microbiologie Gent Culture Collection, Universiteit Gent, Belgium
LPS	Lipopolysaccharide
Lys	Lysine
MA 2216	Medium for growth of marine organisms. Commercially available
MAF	Modified acid fast stain
Malt	Maltose
Man	Mannitol
Man An	Fermentation of mannitol under anaerobic conditions
Mano	Mannose
MCA	MacConkey agar
Mg	Magnesium
Middlebrook's media	For the isolation of *Mycobacterium* species
min	Minute
MOF	Marine oxidative-fermentative medium
Mot	Motility
MR	Methyl Red
MRVP	Methyl Red Voges-Proskauer
MSA-B	Marine salt agar – blood medium for growth of marine organisms
N	Negative
NA	Nutrient agar
NaCl	Salt
NaCl 0/3	Plate media containing either 0% NaCl or 3% NaCl concentration
NB	Nutrient broth
NCFB	National Collection of Food Bacteria, Agricultural and Food Research Council (AFRC) Institute of Food Research, Reading Laboratory, Reading, Berkshire, UK (formerly National Collection of Dairy Organisms)
NCIM	National Collection of Industrial Microorganisms, National Chemical Laboratory, India
NCIMB	National Collection of Industrial and Marine Bacteria, Aberdeen, Scotland, UK
NCTC	National Collection of Type Cultures, Central Public Health Laboratory, London, UK
ND	Not done
Neg	Negative
NG	No growth
NH	Non-haemolytic
Nit	Nitrate
NK	Not known
NLF	Non-lactose fermenting
nmol	Nanomoles
NVI	National Veterinary Institute, Oslo, Norway
nm	Nanometre
nt	Not tested

O	Oxidative (aerobic metabolism)
OD	Optical density
ODC	Ornithine decarboxylase
OF	Oxidative-fermentative test
ONPG	*o*-Nitrophenyl β-D-galactopyranoside
Orn	Ornithine
Ox	Oxidase
0%–3%	Split plate containing gelatin, and concentration of salt at 0% and 3%
0129	Vibriostatic agent – 2:4-diamino- 6:7-diisopropyl pteridine phosphate
Packer's plates	Selective media for the isolation of *Erysipelothrix* species
PBS	Phosphate buffered saline
Pig	Pigment production
pmol	Picomoles
Poly plates	Media for the isolation of *Dermatophilus* species
Pos	Positive
PS	Partially sensitive to vibriostatic disc
PY	Peptone yeast medium for *Pedobacter* and *Sphingobacterium* species
PYR	L-Pyrrolidonyl-β-naphthylamide
PYS-2	Peptone yeast medium 2
R	Resistant
R2A	Medium for isolation of marine *Flavobacterium* and other organisms
RAA	Rogosa acetate agar
RBC	Red blood cells
Rib	D-ribose
S	Sensitive
SS	Media for the isolation of *Salmonella* and *Shigella*
SAB	Sabouraud's medium for fungi
Sal	Salicin
Shieh medium	For the isolation of *Flavobacterium columnare*
Siem agar	Selective medium for isolation of *Aerococcus viridans*
SKDM	Selective medium for *Renibacterium salmoninarum*
Skirrow's medium	For the isolation of *Helicobacter* species
Sor	Sorbitol
Suc	Sucrose
SW	Swarming growth on agar plate
SWT	Seawater-based complex medium
THA	Todd-Hewitt agar
TB Lab	Tuberculosis laboratory
TCBS	Thiosulphate–citrate–bile salts–sucrose agar
TE	Tris EDTA buffer
Temp	Temperature
Tm	Melting temperature
Tre	Trehalose
TSA	Tryptone soy agar
TSA-B	Tryptone soy agar with added blood
TSA+NaCl	Tryptone soy agar with added salt (2% final concentration)
TSB	Tryptic soy broth
TSI	Triple sugar iron
TYG	Tryptone yeast extract glucose agar
TYG-M	Tryptone yeast extract glucose agar with added NaCl for growth of *T. maritimum*
Uro	Urocanic acid
V	Variable reaction in literature
VAM	*Vibrio anguillarum* medium for isolation of *Listonella anguillarum*
VP	Voges-Proskauer test

VPT	Vancomycin, Polymyxin, Trimethoprim – Skirrow's medium for the isolation of *Helicobacter* species
Vs	Variable or slow reaction
UK	United Kingdom
USA	United States of America
UV	Ultraviolet light
VVM	*Vibrio vulnificus* medium
W	Weak
WHO	World Health Organization
Wood's Broth	Selective broth for the isolation of *Erysipelothrix* species
XLD	Xylose lysine desoxycholate agar for isolation of *Salmonella* species
Xyl	D-xylose
Y	Yellow-coloured colony on TCBS plate
YSA	*Yersinia* selective agar
ZN	Ziehl–Neilson strain for Mycobacteria
–	Negative reaction
–cr	Gram-negative curved rod
–α	Negative result, but may show α-haemolysis after a week
+	Positive reaction
+gb	Gram positive cocco-bacilli
+g$_+$	Glucose fermentation positive, gas produced
+g.	Glucose fermentation positive, no gas produced
+rt	Positive at room temperature (25°C) but negative at 37°C
+s	Slow positive reaction. May take 2–4 days for reaction to occur
+sr	Gram-positive, short rod
+w	Weak positive reaction

References

1 Abbott, S.L., Cheung, W.K.W., Kroske-Bystrom, S., Malekzadeh, T. and Janda, J.M. (1992) Identification of *Aeromonas* strains to the genospecies level in the clinical laboratory. *Journal of Clinical Microbiology* 30, 1262–1266.

2 Abbott, S.L., Seli, L.S., Catino, M. Jr., Hartley, M.A. and Janda, J.M. (1998) Misidentification of unusual *Aeromonas* species as members of the genus *Vibrio*: a continuing problem. *Journal of Clinical Microbiology* 36, 1103–1104.

3 Abu-Samra, M.T. and Walton, G.S. (1977) Modified techniques for the isolation of *Dermatophilus* spp. from infected material. *Sabouraudia* 15, 23–27.

4 Acuna, M.T., Diaz, G., Bolanos, H., Barquero, C., Sanchez, O., Sanchez, L.M., Mora, G., Chaves, A. and Campos, E. (1999) Sources of *Vibrio mimicus* contamination of turtle eggs. *Applied and Environmental Microbiology* 65, 336–338.

5 Aguirre, A.A., Balazs, G.H., Zimmerman, B. and Spraker, T. (1994) Evaluation of Hawaiian green turtles (*Chelonia mydas*) for potential pathogens associated with fibropapillomas. *Journal of Wildlife Diseases* 30, 8–15.

6 Ahmet, Z., Stanier, P., Harvey, D. and Holt, D. (1999) New PCR primers for the sensitive detection and specific identification of Group B β-haemolytic streptococci in cerebrospinal fluid. *Molecular and Cellular Probes* 13, 349–357.

7 Aiso, K., Simidu, V. and Hasuo, K. (1968) Microflora in the digestive tract of inshore fish in Japan. *Journal of General Microbiology* 52, 361–364.

8 Akagawa, M. and Yamasato, K. (1989) Synonymy of *Alcaligenes aquamarinus*, *Alcaligenes faecalis* subsp. *homari,* and *Deleya aesta*: *Deleya aquamarina* comb. nov. as the type species of the genus *Deleya*. *International Journal of Systematic Bacteriology* 39, 462–466.

9 Albert, M., Ansaruzzaman, M., Bardhan, P., Faruque, A., Faruque, S., Islam, M., Mahalanabis, D., Sack, R., Salam, M., Siddique, A., Yunus, M. and Zaman, K. (1993) Large epidemic of cholera-like disease in Bangladesh caused by *Vibrio cholerae* 0139 synonym Bengal. *Lancet* 342, 387–390.

10 Alcaide, E., Amaro, C., Todolí, R. and Oltra, R. (1999) Isolation and characterization of *Vibrio parahaemolyticus* causing infection in Iberian toothcarp *Aphanius iberus*. *Diseases of Aquatic Organisms* 35, 77–80.

11 Alcaide, E., Gil-Sanz, C., Sanjuán, E., Esteve, D., Amaro, C. and Silveira, L. (2001) *Vibrio harveyi* causes disease in seahorse, *Hippocampus* sp. *Journal of Fish Diseases* 24, 211–313.

12 Aleksic, S., Steigerwalt, A., Bockemühl, J., Huntley-Carter, G. and Brenner, D. (1987) *Yersinia rohdei* sp. nov. isolated from human and dog faeces and surface water. *International Journal of Systematic Bacteriology* 37, 327–332.

13 Ali, A., Carnahan, A., Altwegg, M., Lüthy-Hottenstein, J. and Joseph, S. (1996) *Aeromonas bestiarum* sp. nov. (formerly genomospecies DNA group 2 *A. hydrophila*), a new species isolated from non-human sources. *Medical Microbiology Letters* 5, 156–165.

14 Allam, B., Paillard, C., Howard, A. and Pennec, M.L. (2000) Isolation of the pathogen *Vibrio tapetis* and defense parameters in brown ring diseased Manila clams *Ruditapes philippinarum* cultivated in England. *Diseases of Aquatic Organisms* 41, 105–113.

15 Allen, D., Austin, B. and Colwell, R. (1983) *Aeromonas media*, a new species isolated from river water. *International Journal of Systematic Bacteriology* 33, 599–604.

16 Alsina, M. and Blanch, A.R. (1994) A set of keys for biochemical identification of environmental *Vibrio*

species. *Journal of Applied Bacteriology* 76, 79–85.

17 Alsina, M. and Blanch, A.R. (1994) Improvement and update of a set of keys for biochemical identification of *Vibrio* species. *Journal of Applied Bacteriology* 77, 719–721.

18 Alsina, M., Martínez-Picado, J., Jofre, J. and Blanch, A. (1994) A medium for presumptive identification of *Vibrio anguillarum*. *Applied and Environmental Microbiology* 60, 1681–1683.

19 Altmann, K., Marshall, M., Nicholson, S., Hanna, P. and Gudkovs, N. (1992) Glucose repression of pigment production in atypical isolates of *Aeromonas salmonicida* responsible for goldfish ulcer disease. *Microbios* 72, 215–220.

20 Alton, G.G. and Jones, L.M. (1967) *Laboratory Techniques in Brucellosis*. World Health Organization, Geneva.

21 Altwegg, M., Steigerwalt, A.G., Altwegg-Bissig, R., Lüthy-Hottenstein, J. and Brenner, D.J. (1990) Biochemical identification of *Aeromonas* geno-species isolated from humans. *Journal of Clinical Microbiology* 28, 258–264.

22 Aluotto, B., Wittler, R., Williams, C. and Faber, J. (1970) Standardized bacteriologic techniques for the characterization of *Mycoplasma* species. *International Journal of Systematic Bacteriology* 20, 35–58.

23 Alvarez, J.D., Austin, B., Alvarez, A.M. and Reyes, H. (1998) *Vibrio harveyi*: a pathogen of penaeid shrimps and fish in Venezuela. *Journal of Fish Diseases* 21, 313–316.

24 Amann, R., Binder, B., Olson, R., Chisholm, S., Devereux, R. and Stahl, D. (1990) Combination of 16S rRNA-targeted oligonucleotide probes with flow cytometry for analyzing mixed microbial populations. *Applied and Environmental Microbiology* 56, 1919–1925.

25 Amaro, C. and Biosca, E. (1996) *Vibrio vulnificus* biotype 2, pathogenic for eels, is also an opportunistic infection for humans. *Applied and Environmental Microbiology* 62, 1454–1457.

26 Amaro, C., Hor, L.-I., Marco-Noales, E., Bosque, T., Fouz, B. and Alcaide, E. (1999) Isolation of *Vibrio vulnificus* serovar E from aquatic habitats in Taiwan. *Applied and Environmental Microbiology* 65, 1352–1355.

27 Anacker, R.L. and Ordal, E.J. (1955) Study of the bacteriophage infecting the myxobacterium *Chondrococcus columnaris*. *Journal of Bacteriology* 70, 738–741.

28 Anacker, R.L. and Ordal, E.J. (1959) Studies on the myxobacterium *Chondrococcus columnaris*. I. Serological typing. *Journal of Bacteriology* 78, 25–32.

29 Angka, S.L., Lam, T.J. and Sin, Y.M. (1995) Some virulence characteristics of *Aeromonas hydrophila*

in walking catfish (*Clarias gariepinus*). *Aquaculture* 130, 103–112.

30 Anguiano-Beltrán, C., Searcy-Bernal, R. and Lizárraga-Partida, M. (1998) Pathogenic effects of *Vibrio alginolyticus* on larvae and postlarvae of the red abalone *Haliotis rufescens*. *Diseases of Aquatic Organisms* 33, 119–122.

31 Angulo, L., Lopez, J., Vicente, J. and Saborido, A. (1994) Haemorrhagic areas in the mouth of farmed turbot, *Scophthalmus maximus* (L.). *Journal of Fish Diseases* 17, 163–169.

32 Aoki, T., Park, C.-I., Yamashita, H. and Hirono, I. (2000) Species-specific polymerase chain reaction primers for *Lactococcus garvieae*. *Journal of Fish Diseases* 23, 1–6.

33 Arias, C., Verdonck, L., Swings, J., Aznar, R. and Garay, E. (1997) A polyphasic approach to study the intraspecific diversity amongst *Vibrio vulnificus* isolates. *Systematic and Applied Microbiology* 20, 622–633.

34 Arias, C., Aznar, R., Pujalte, M. and Garay, E. (1998) A comparison of strategies for the detection and recovery of *Vibrio vulnificus* from marine samples of the Western Mediterranean coast. *Systematic and Applied Microbiology* 21, 128–134.

35 Aronson, J.D. (1926) Spontaneous tuberculosis in salt water fish. *Journal of Infectious Diseases* 39, 314–320.

36 Ashburner, L.D. (1977) Mycobacteria in hatchery-confined chinook salmon (*Oncorhynchus tshawytscha* Walbaum). *Journal of Fish Biology* 10, 523–528.

37 Ashdown, L.R. (1979a) An improved screening technique for isolation of *Pseudomonas pseudomallei* from clinical specimens. *Pathology* 11, 293–297.

38 Ashdown, L.R. (1979b) Identification of *Pseudomonas pseudomallei* in the clinical laboratory. *Journal of Clinical Pathology* 32, 500–504.

39 Auling, G., Reh, M., Lee, C.M. and Schlegel, H.G. (1978) *Pseudomonas pseudoflava*, a new species of hydrogen-oxidising bacteria: its differentiation from *Pseudomonas flava* and other yellow-pigmented, Gram-negative, hydrogen-oxidising species. *International Journal of Systematic Bacteriology* 28, 82–95.

40 Austin, B. (1993) Recovery of 'atypical' isolates of *Aeromonas salmonicida*, which grow at 37°C, from ulcerated non-salmonids in England. *Journal of Fish Diseases* 16, 165–168.

41 Austin, B. and Austin, D.A. (1999) *Bacterial Fish Pathogens: Diseases of Farmed and Wild Fish*. Praxis Publishing, Chichester, UK.

42 Austin, B. and Stobie, M. (1992a) Recovery of *Serratia plymuthica* and presumptive *Pseudomonas pseudoalcaligenes* from skin lesions in rainbow trout, *Oncorhynchus mykiss* (Walbaum),

otherwise infected with enteric redmouth. *Journal of Fish Diseases* 15, 541–543.

43 Austin, B. and Stobie, M. (1992b) Recovery of *Micrococcus luteus* and presumptive *Planococcus* sp. from moribund fish during an outbreak of rainbow trout, *Oncorhynchus mykiss* (Walbaum), fry syndrome in England. *Journal of Fish Diseases* 15, 203–206.

44 Austin, B., Zachary, A. and Colwell, R. (1978) Recognition of *Beneckea natriegens* (Payne *et al.*) Baumann *et al.* as a member of the genus *Vibrio*, as previously proposed by Webb and Payne. *International Journal of Systematic Bacteriology* 28, 315–317.

45 Austin, B., Rodgers, C.J., Forns, J.M. and Colwell, R.R. (1981) *Alcaligenes faecalis* subsp *homari* subsp. nov., a new group of bacteria isolated from moribund lobsters. *International Journal of Systematic Bacteriology* 31, 72–76.

46 Austin, B., Embley, T. and Goodfellow, M. (1983) Selective isolation of *Renibacterium salmoninarum*. *FEMS Microbiology Letters* 17, 111–114.

47 Austin, D., McIntosh, D. and Austin, B. (1989) Taxonomy of fish associated *Aeromonas* spp., with the description of *Aeromonas salmonicida* subsp. *smithia* subsp. nov. *Systematic and Applied Microbiology,* 11, 277–290.

48 Austin, B., Gonzalez, C., Stobie, M., Curry, J. and McLoughlin, M. (1992) Recovery of *Janthinobacterium lividum* from diseased rainbow trout, *Oncorhynchus mykiss* (Walbaum), in Northern Ireland and Scotland. *Journal of Fish Diseases* 15, 357–359.

49 Austin, B., Austin, D.A., Blanch, A.R., Cerdà, M., Grimont, F., Grimont, P.A.D., Jofre, J., Koblavi, S., Larsen, J.L., Pedersen, K., Tiainen, T., Verdonck, L. and Swings, J. (1997) A comparison of methods for the typing of fish-pathogenic *Vibrio* spp. *Systematic and Applied Microbiology* 20, 89–101.

50 Austin, B., Austin, D., Dalsgaard, I., Gudmundsdóttir, B., Høie, S., Thornton, J., Larsen, J., O'Hici, B. and Powell, R. (1998) Characterization of atypical *Aeromonas salmonicida* by different methods. *Systematic and Applied Microbiology* 21, 50–64.

51 Aydin, S., Çelebi, S. and Akyurt, I. (1997) Clinical, haematological and pathological investigations of *Escherichia vulneris* in rainbow trout (*Oncorhynchus mykiss*). *Fish Pathology* 32, 29–34.

52 Ayers, S., Rupp, P. and Johnson, W. Jr. (1919) A study of the alkali-forming bacteria found in milk. *Bulletin No. 782* US Department of Agriculture.

53 Backman, S., Ferguson, H.W., Prescott, J.F. and Wilcock, B.P. (1990) Progressive panophthalmitis in chinook salmon, *Oncorhynchus tshawytscha* (Walbaum): a case report. *Journal of Fish Diseases* 13, 345–353.

54 Bader, J., Shoemaker, C. and Klesius, P. (2003) Rapid detection of columnaris disease in channel catfish (*Ictalurus punctatus*) with a new species-specific 16-S rRNA gene-based PCR primer for *Flavobacterium columnare*. *Journal of Microbiological Methods* 52, 209–220.

55 Baharaeen, S. and Vishniac, H. (1982) *Cryptococcus lupi* sp. nov., an Antarctic Basidioblastomycete. *International Journal of Systematic Bacteriology* 32, 229–232.

56 Bakopoulos, V., Adams, A. and Richards, R. (1995) Some biochemical properties and antibiotic sensitivities of *Pasteurella piscicida* isolated in Greece and comparison with strains from Japan, France and Italy. *Journal of Fish Diseases* 18, 1–7.

57 Bakopoulos, V., Peric, Z., Rodger, H., Adams, A. and Richards, R. (1997) First report of fish pasteurellosis from Malta. *Journal of Aquatic Animal Health* 9, 26–33.

58 Balebona, M.C., Zorrilla, I., Moriñigo, M. and Borrego, J. (1998) Survey of bacterial pathologies affecting farmed gilt-head bream (*Sparus aurata* L.) in southwestern Spain from 1990–1996. *Aquaculture* 166, 19–35.

59 Banin, E., Israely, T., Kushmaro, A., Loya, Y., Orr, E. and Rosenberg, E. (2000) Penetration of the coral-bleaching bacterium *Vibrio shiloi* into *Oculina patagonica*. *Applied and Environmental Microbiology* 66, 3031–3036.

60 Baptista, T., Romalde, J. and Toranzo, A. (1996) First occurrence of Pasteurellosis in Portugal affecting cultured gilthead seabream (*Sparus aurata*). *Bulletin of the European Association of Fish Pathologists* 16, 92–95.

61 Barbeyron, T., L'Haridon, S., Corre, E., Kloareg, B. and Potin, P. (2001) *Zobellia galactanovorans* gen. nov., sp. nov., a marine species of *Flavobacteriaceae* isolated from a red alga, and classification of [*Cytophaga*] *uliginosa* (ZoBell and Upham 1944) Reichenbach 1989 as *Zobellia uliginosa* gen. nov., comb. nov. *International Journal of Systematic and Evolutionary Microbiology* 51, 985–997.

62 Barritt, M.M. (1936) The intensification of the Voges-Proskauer reaction by the addition of α-naphthol. *Journal of Pathology and Bacteriology* 42, 441–454.

63 Barrow and Feltham (1993) *Cowan and Steel's Manual for the Identification of Medical Bacteria*. Cambridge University Press, Cambridge.

64 Baumann, P. and Baumann, L. (1981) In: Starr, M., Stolp, H., Trüper, H., Balows, A. and Schlegel, H. (eds) *The Prokaryotes*, Vol. II. Springer, Berlin, pp. 1302–1331.

65 Baumann, P., Baumann, L. and Mandel, M. (1971) Taxonomy of marine bacteria: the genus *Beneckea*. *Journal of Bacteriology* 107, 268–294.

66 Baumann, L., Baumann, P., Mandel, M. and Allen, R. (1972) Taxonomy of aerobic marine Eubacteria. *Journal of Bacteriology* 110, 402–429.

67 Baumann, P., Baumann, L., Bang, S. and Woolkalis, M. (1980) Reevaluation of the taxonomy of *Vibrio*, *Beneckea*, and *Photobacterium*: abolition of the genus *Beneckea*. *Current Microbiology* 4, 127–132.

68 Baumann, L., Bowditch, R. and Baumann, P. (1983a) Description of *Deleya* gen. nov. created to accommodate the marine species *Alcaligenes aestus*, *A. pacificus*, *A. cupidus*, *A. venustus*, and *Pseudomonas marina*. *International Journal of Systematic Bacteriology* 33, 793–802.

69 Baumann, P., Bowditch, R., Baumann, L. and Beaman, B. (1983b) Taxonomy of marine *Pseudomonas* species: *P. stanieri* sp. nov., *P. perfectomarina* sp. nov., nom. rev.; *P. nautica*; and *P. doudoroffii*. *International Journal of Systematic Bacteriology* 33, 857–865.

70 Baumann, P., Furniss, A.L. and Lee, J.V. (1984) In: Krieg, N.R. and Holt, J.G. (eds) *Bergey's Manual of Systematic Bacteriology*, Vol. 1. Williams & Wilkins, Baltimore, pp. 518–538.

71 Baya, A., Lupiani, B., Hetrick, F., Roberson, B., Lukacovic, R., May, E. and Poukish, C. (1990a) Association of *Streptococcus* sp. with fish mortalities in the Chesapeake Bay and its tributaries. *Journal of Fish Diseases* 13, 251–253.

72 Baya, A., Toranzo, A., Núñez, S., Barja, J.L. and Hetrick, F.M. (1990b) Association of a *Moraxella* sp. and a reo-like virus with mortalities of striped bass, *Morone saxatilis*. In: Perkins, F. and Cheng, T. (eds) *Pathology in Marine Science*. Proceedings of the Third International Colloquium on Pathology in Marine Aquaculture held in Gloucester Point, Virginia, 2–6 October, 1988, pp. 91–99.

73 Baya, A., Toranzo, A., Lupiani, B., Li, T., Roberson, B. and Hetrick, F. (1991) Biochemical and serological characterization of *Carnobacterium* spp. isolated from farmed and natural populations of striped bass and catfish. *Applied and Environmental Microbiology* 57, 3114–3120.

74 Baya, A., Li, T., Lupiani, B. and Hetrick, F. (1992a) *Bacillus cereus*, a pathogen for striped bass. In: *Eastern Fish Health and American Fisheries Society Fish Health Section Workshop. 16–19 June 1992*. Auburn University, Auburn, p. 67.

75 Baya, A., Lupiani, B., Bandín, I., Hetrick, F., Figueras, A., Carnahan, A., May, E. and Toranzo, A. (1992b) Phenotypic and pathobiological properties of *Corynebacterium aquaticum* isolated from diseased striped bass. *Diseases of Aquatic Organisms* 14, 115–126.

76 Baya, A., Toranzo, A., Lupiani, B., Santos, Y. and Hetrick, F. (1992c) *Serratia marcescens*: a potential pathogen for fish. *Journal of Fish Diseases* 15, 15–26.

77 Bein, S. (1954) A study of certain chromogenic bacteria isolated from 'red tide' water with a description of a new species. *Bulletin of Marine Science of the Gulf and Caribbean* 4, 110–119.

78 Bej, A., Patterson, D., Brasher, C., Vickery, M., Jones, D. and Kaysner, C. (1999) Detection of total and hemolysin-producing *Vibrio parahaemolyticus* in shellfish using multiplex PCR amplification of *tl*, *tdh* and *trh*. *Journal of Microbiological Methods* 36, 215–225.

79 Bejerano, Y., Sarig, S., Horne, M. and Roberts, R. (1979) Mass mortalities in silver carp *Hypophthalmichthys molitrix* (Valenciennes) associated with bacterial infection following handling. *Journal of Fish Diseases* 2, 49–56.

80 Beji, A., Mergaert, J., Gavini, F., Izard, D., Kersters, K., Leclerc, H. and de Ley, J. (1988) Subjective synonymy of *Erwinia herbicola*, *Erwinia milletiae*, and *Enterobacter agglomerans* and redefinition of the Taxon by genotypic and phenotypic data. *International Journal of Systematic Bacteriology* 38, 77–88.

81 Benediktsdóttir, E., Helgason, S. and Sigurjónsdóttir, H. (1998) *Vibrio* spp. isolated from salmonids with shallow skin lesions and reared at low temperature. *Journal of Fish Diseases* 21, 19–28.

82 Benediktsdóttir, E., Verdonck, L., Spröer, C., Helgason, S. and Swings, J. (2000) Characterization of *Vibrio viscosus* and *Vibrio wodanis* isolated at different geographical locations: a proposal for reclassification of *Vibrio viscosus* as *Moritella viscosa* comb. nov. *International Journal of Systematic and Evolutionary Microbiology* 50, 479–488.

83 Ben-Haim, Y. and Rosenberg, E. (2002) A novel *Vibrio* sp. pathogen of the coral *Pocillopora damicornis*. *Marine Biology* 141, 47–55.

84 Ben-Haim, Y., Thompson, F., Thompson, C., Cnockaert, M., Hoste, B., Swings, J. and Rosenberg, E. (2003) *Vibrio coralliilyticus* sp. nov., a temperature-dependent pathogen of the coral *Pocillopora damicornis*. *International Journal of Systematic and Evolutionary Microbiology* 53, 309–315.

85 Bercovier, H., Steigerwalt, A., Guiyoule, A., Huntley-Carter, G. and Brenner, D.J. (1984) *Yersinia aldovae* (formerly *Yersinia enterocolitica*-like group X2): a new species of *Enterobacteriacea* isolated from aquatic ecosystems. *International Journal of Systematic Bacteriology* 34, 166–172.

86 Bercovier, H., Ursing, J., Brenner, D., Steigerwalt, A., Fanning, G., Carter, G. and Mollaret, H. (1980) *Yersinia kristensenii*: a new species of *Enterobacteriaceae* composed of sucrose-negative strains (formerly called atypical *Yersinia enterocolitica* or *Yersinia enterocolitica*-like. *Current Microbiology* 4, 219–224.

87 Bergman, S., Selig, M., Collins, M.D., Farrow, J.A.E., Baron, E.J., Dickersin, G.R. and Ruoff, K.L. (1995) 'Streptococcus milleri' strains displaying a gliding type of motility. International Journal of Systematic Bacteriology 45, 235–239.

88 Bernadet, J.F. (1989) 'Flexibacter columnaris': first description in France and comparison with bacterial strains from other origins. Diseases of Aquatic Organisms 6, 37–44.

89 Bernadet, J.-F. and Grimont, P. (1989) Deoxyribonucleic acid relatedness and phenotypic characterization of Flexibacter columnaris sp. nov., nom. rev., Flexibacter psychrophilus sp. nov., nom. rev., and Flexibacter maritimus Wakabayashi, Hikida, and Masumura 1986. International Journal of Systematic Bacteriology 39, 346–354.

90 Bernadet, J.F., Campbell, A.C. and Buswell, J.A. (1990) Flexibacter maritimus is the agent of 'black patch necrosis' in Dover sole in Scotland. Diseases of Aquatic Organisms 8, 233–237.

91 Bernadet, J.-F., Kerouault, B. and Michel, C. (1994) Comparative study on Flexibacter maritimus strains isolated from farmed sea bass (Dicentrarchus labrax) in France. Fish Pathology 29, 105–111.

92 Bernadet, J.-F., Segers, P., Vancanneyt, M., Berthe, F., Kersters, K. and Vandamme, P. (1996) Cutting a Gordian knot: emended classification and description of the Genus Flavobacterium, emended description of the Family Flavobacteriaceae, and proposal of Flavobacterium hydatis nom. nov. (Basonym, Cytophaga aquatilis Strohl and Tait 1978). International Journal of Systematic Bacteriology 46, 128–148.

93 Bernadet, J.-F., Nakagawa, Y. and Holmes, B. (2002) Proposed minimal standards for describing new taxa of the family Flavobacteriaceae and emended description of the family. International Journal of Systematic and Evolutionary Microbiology 52, 1049–1070.

94 Bernoth, E.-M. (1990) Autoagglutination, growth on tryptone-soy-Coomassie-agar, outer membrane protein patterns and virulence of Aeromonas salmonicida strains. Journal of Fish Diseases 13, 145–155.

95 Bernoth, E.-M. and Artz, G. (1989) Presence of Aeromonas salmonicida in fish tissue may be overlooked by sole reliance on furunculosis-agar. Bulletin of the European Association of Fish Pathologists 9, 5–6.

96 Berthe, F., Michel, C. and Bernadet, J.-F. (1995) Identification of Pseudomonas anguilliseptica isolated from several fish species in France. Diseases of Aquatic Organisms 21, 151–155.

97 Biosca, E., Esteve, C., Garay, E. and Amaro, C. (1993) Evaluation of the API 20E system for identification and discrimination of Vibrio vulnificus biotypes 1 and 2. Journal of Fish Diseases 16, 79–82.

98 Biosca, E.G., Amaro, C., Larsen, J.L. and Pedersen, K. (1997) Phenotypic and genotypic characterization of Vibrio vulnificus: proposal for the substitution of the subspecific taxon biotype for serovar. Applied and Environmental Microbiology 63, 1460–1466.

99 Birkbeck, T.H., Billcliffe, B., Laidler, A. and Cox, D.I. (2000) The relationship between Aeromonas sp. NCIMB 2263, a causative agent of skin lesions in Atlantic salmon, Vibrio marinus (Moritella marina) and Vibrio viscosus. Journal of Fish Diseases 23, 281–283.

100 Birkbeck, T.H., Laidler, L.A., Grant, A.N. and Cox, D.I. (2002) Pasteurella skyensis sp. nov., isolated from Atlantic salmon (Salmo salar L.). International Journal of Systematic and Evolutionary Microbiology 52, 699–704.

101 Bisharat, N., Agmon, V., Finkelstein, R., Raz, R., Ben-Dror, G., Lerner, L., Soboh, S., Colodner, R., Cameron, D., Wykstra, D., Swerdlow, D. and Farmer, J. III. (1999) Clinical, epidemiological, and microbiological features of Vibrio vulnificus biogroup 3 causing outbreaks of wound infection and bacteraemia in Israel. Lancet 354, 1421–1424.

102 Blackall, P. and Doheny, C. (1987) Isolation and characterisation of Bordetella avium and related species and an evaluation of their role in respiratory disease in poultry. Australian Veterinary Journal 64, 235–238.

103 Blanco, M., Gibello, A., Vela, A., Moreno, M., Domínguez, L. and Fernández-Garayzábal, J. (2002) PCR detection and PFGE DNA macrorestriction analyses of clinical isolates of Pseudomonas anguilliseptica from winter disease outbreaks in sea bream Sparus aurata. Diseases of Aquatic Organisms 50, 19–27.

104 Boettcher, K., Barber, B. and Singer, J. (1999) Use of antibacterial agents to elucidate the etiology of juvenile oyster disease (JOD) in Crassostrea virginica and numerical dominance of an α-proteobacterium in JOD-affected animals. Applied and Environmental Microbiology 65, 2534–2539.

105 Boettcher, K., Barber, B. and Singer, J. (2000) Additional evidence that juvenile oyster disease is caused by a member of the Roseobacter group and colonization of nonaffected animals by Stappia stellulata-like strains. Applied and Environmental Microbiology 66, 3924–3930.

106 Boettcher, K. and Ruby, E. (1990) Depressed light emission by symbiont Vibrio fischeri of the sepiolid squid Euprymna scolopes. Journal of Bacteriology 172, 3701–3706.

107 Boomker, J., Henton, M., Naudé, T. and Hunchzermeyer, F. (1984) Furunculosis in rainbow trout (Salmo gairdneri) raised in sea water. Onderstepoort Journal of Veterinary Research 51, 91–94.

108 Borrego, J.J., Castro, D., Luque, A., Paillard, C., Maes, P., Garcia, M.T. and Ventosa, A. (1996) *Vibrio tapetis* sp. nov., the causative agent of the brown ring disease affecting cultured clams. *International Journal of Systematic Bacteriology* 46, 480–484.

109 Borucinska, J. and Frasca, S. Jr. (2002) Naturally occurring lesions and micro-organisms in two species of free-living sharks: the spiny dogfish, *Squalus acanthias* L., and the smooth dogfish, *Mustelus canis* (Mitchill), from the north-western Atlantic. *Journal of Fish Diseases* 25, 287–298.

110 Bouvet, P. and Grimont, P. (1986) Taxonomy of the genus *Acinetobacter* with recognition of *Acinetobacter baumannii* sp. nov., *Acinetobacter haemolyticus* sp. nov., *Acinetobacter johnsonii* sp. nov., and *Acinetobacter junii* sp. nov. and emended description of *Acinetobacter calcoaceticus* and *Acinetobacter lwoffii*. *International Journal of Systematic Bacteriology* 36, 228–240.

111 Bowenkamp, K., Frasca, S. Jr., Draghi, A. II, Tsongalis, G., Koerting, C., Hinckley, L., Guise, S.D., Montali, R., Goertz, C., Aubin, D.S. and Dunn, J. (2001) *Mycobacterium marinum* dermatitis and panniculitis with chronic pleuritis in a captive white whale (*Delphinapterus leucas*) with aortic rupture. *Journal of Veterinary Investigation* 13, 524–530.

112 Bowman, J., McCammon, S., Nichols, D., Skerratt, J., Rea, S., Nichols, P. and McMeekin, T. (1997) *Shewanella gelidimarina* sp. nov. and *Shewanella frigidimarina* sp. nov., novel Antarctic species with the ability to produce eicosapentaenoic acid (20:5ω3) and grow anaerobically by dissimilatory Fe (III) reduction. *International Journal of Systematic Bacteriology* 47, 1040–1047.

113 Bowman, J.P. and Nichols, D.S. (2002) *Aequorivita* gen. nov., a member of the family *Flavobacteriaceae* isolated from terrestrial and marine Antarctic habitats. *International Journal of Systematic and Evolutionary Microbiology* 52, 1533–1541.

114 Bowser, P., Rosemark, R. and Reiner, C. (1981) A preliminary report of vibriosis in cultured American lobsters, *Homarus americanus*. *Journal of Invertebrate Pathology* 37, 80–85.

115 Bozal, N., Tudela, E., Rosselló-Mora, R., Lalucat, J. and Guinea, J. (1997) *Pseudoalteromonas antarctica* sp. nov., isolated from an Antarctic coastal environment. *International Journal of Systematic Bacteriology* 47, 345–351.

116 Bragg, R.R., Huchzermeyer, H.F. and Hanisch, M.A. (1990) *Mycobacterium fortuitum* isolated from three species of fish in South Africa. *Onderstepoort Journal of Veterinary Research* 57, 101–102.

117 Bransden, M.P., Carson, J., Munday, B.L., Handlinger, J.H., Carter, C.G. and Nowak, B.F. (2000) Nocardiosis in tank-reared Atlantic salmon, *Salmo salar* L. *Journal of Fish Diseases* 23, 83–85.

118 Brasher, C., DePaola, A., Jones, D. and Bej, A. (1998) Detection of microbial pathogens in shellfish with multiplex PCR. *Current Microbiology* 37, 101–107.

119 Brauns, L., Hudson, M. and Oliver, J. (1991) Use of the polymerase chain reaction in detection of culturable and non-culturable *Vibrio vulnificus* cells. *Applied and Environmental Microbiology* 57, 2651–2655.

120 Brayton, P.R., Bode, R.B., Colwell, R.R., MacDonell, M.T., Hall, H.L., Grimes, D.J., West, P.A. and Bryant, T.N. (1986) *Vibrio cincinnatiensis* sp. nov., a new human pathogen. *Journal of Clinical Microbiology* 23, 104–108.

121 Brenner, D., Steigerwalt, A., Falcao, D., Weaver, R. and Fanning, G.R. (1976) Characterization of *Yersinia enterocolitica* and *Yersinia pseudotuberculosis* by deoxyribonucleic acid hybridization and biochemical reactions. *International Journal of Systematic Bacteriology* 26, 180–194.

122 Brenner, D.J., Farmer, J.J. III, Fanning, G.R., Steigerwalt, A.G., Klykken, P., Wathen, H.G., Hickman, F.W. and Ewing, W.H. (1978) Deoxyribonucleic acid relatedness of *Proteus* and *Providencia* species. *International Journal of Systematic Bacteriology* 28, 269–282.

123 Brenner, D.J., Hickman-Brenner, F.W., Lee, J.V., Steigerwalt, A.G., Fanning, G.R., Hollis, D.G., Farmer, J.J. III, Weaver, R.E., Joseph, S.W. and Seidler, R.J. (1983) *Vibrio furnissii* (formerly aerogenic biogroup *Vibrio fluvialis*), a new species isolated from human feces and the environment. *Journal of Clinical Microbiology* 18, 816–824.

124 Brenner, D.J., Müller, H.E., Steigerwalt, A.G., Whitney, A.M., O'Hara, C.M. and Kämpfer, P. (1998) Two new *Rahnella* genomospecies that cannot be phenotypically differentiated from *Rahnella aquatilis*. *International Journal of Systematic Bacteriology* 48, 141–149.

125 Brew, S.D., Perrett, L.L., Stack, J.A., MacMillan, A.P. and Staunton, N.J. (1999) Human exposure to *Brucella* recovered from a sea mammal. *Veterinary Record* 144, 483.

126 Bricker, B., Ewalt, D., MacMillan, A., Foster, G. and Brew, S. (2000) Molecular characterization of *Brucella* strains isolated from marine mammals. *Journal of Clinical Microbiology* 38, 1258–1262.

127 Bromage, E.S., Thomas, A. and Owens, L. (1999) *Streptococcus iniae*, a bacterial infection in barramundi *Lates calcarifer*. *Diseases of Aquatic Organisms* 36, 177–181.

128 Brown, D.R., Farley, J.M., Zacher, L.A., Carlton, J.M.-R., Clippinger, T.L., Tully, J.G. and Brown, M.B. (2001) *Mycoplasma alligatoris* sp. nov. from American alligators. *International*

Journal of Systematic and Evolutionary Microbiology 51, 419–424.

129 Brown, D.R., Nogueira, M.F., Schoeb, T.R., Vliet, K.A., Bennett, R.A., Pye, G.W. and Jacobson, E.R. (2001) Pathology of experimental mycoplasmosis in American alligators. *Journal of Wildlife Diseases* 37, 671–679.

130 Brown, G., Sutcliffe, I. and Cummings, S. (2001) Reclassification of [*Pseudomonas*] *doudoroffii* (Baumann *et al.* 1983) into the genus *Oceanomonas* gen. nov., as *Oceanomonas doudoroffii* comb. nov., and description of a phenol-degrading bacterium from estuarine water as *Oceanomonas baumannii* sp. nov. *International Journal of Systematic and Evolutionary Microbiology* 51, 67–72.

131 Brown, L., Iwama, G., Evelyn, T., Nelson, W. and Levine, R. (1994) Use of the polymerase chain reaction (PCR) to detect DNA from *Renibacterium salmoninarum* within individual salmonid eggs. *Diseases of Aquatic Organisms* 18, 165–171.

132 Bruno, D., Griffiths, J., Petrie, J. and Hastings, T. (1998a) *Vibrio viscosus* in farmed Atlantic salmon *Salmo salar* in Scotland: field and experimental observations. *Diseases of Aquatic Organisms* 34, 161–166.

133 Bruno, D., Griffiths, J., Mitchell, C., Wood, B., Fletcher, Z., Drobniewski, F. and Hastings, T. (1998b) Pathology attributed to *Mycobacterium chelonae* infection among farmed and laboratory-infected Atlantic salmon *Salmo salar*. *Diseases of Aquatic Organisms* 33, 101–109.

134 Buck, J., Meyers, S. and Leifson, E. (1963) *Pseudomonas (Flavobacterium) piscicida* Bein comb. nov. *Journal of Bacteriology* 4, 1125–1126.

135 Buller, N.B. (2003) Unpublished.

136 Bullock, G.L., Hsu, T.C. and Shotts, E.B. Jr (1986) Columnaris disease of fishes. *U.S. Fish and Wildlife Service Fish Disease Leaflet* 72.

137 Bullock, G.L., Stuckey, H.M. and Shotts, E.B. Jr (1978) Enteric redmouth bacterium: comparison of isolates from different geographic areas. *Journal of Fish Diseases* 1, 351–356.

138 Butzler, J.P., Dekeyser, P., Detrain, M. and Dehaen, F. (1973) Related vibrio in stools. *Journal of Pediatrics* 82, 493–495.

139 Cai, Y., Benno, Y., Nakase, T. and Oh, T.-K. (1998) Specific probiotic characterization of *Weissella hellenica* DS-12 isolated from flounder intestine. *Journal of General and Applied Microbiology* 44, 311–316.

140 Candan, A., Kucker, M. and Karatas, S. (1996) Pasteurellosis in cultured sea bass (*Dicentrarchus labrax*) in Turkey. *Bulletin of the European Association of Fish Pathologists* 16, 150–153.

141 Cann, D.C. and Taylor, L.Y. (1982) A outbreak of botulism in rainbow trout, *Salmo gairdneri* Richardson, farmed in Britain. *Journal of Fish Diseases* 5, 393–399.

142 Carnahan, A., Chakraborty, T., Fanning, G., Verma, D., Ali, A., Janda, M. and Joseph, S. (1991) *Aeromonas trota* sp. nov., an ampicillin-susceptible species isolated from clinical specimens. *Journal of Clinical Microbiology* 29, 1206–1210.

143 Carnahan, A., Fanning, G.R. and Joseph, S.W. (1991) *Aeromonas janadaei* (formerly genospecies DNA group 9 *A. sobria*), a new sucrose-negative species isolated from clinical specimens. *Journal of Clinical Microbiology* 29, 560–564.

144 Carson, J. and Handlinger, J. (1988) Virulence of the aetiological agent of goldfish ulcer disease in Atlantic salmon, *Salmo salar* L. *Journal of Fish Diseases* 11, 471–479.

145 Carson, J., Schmidtke, L.M. and Munday, B.L. (1993) *Cytophaga johnsonae*: a putative skin pathogen of juvenile farmed barramundi, *Lates calcarifer* Bloch. *Journal of Fish Diseases* 16, 209–218.

146 Castro, D., Martínez-Manzanares, E., Luque, A., Fouz, B., Moriñigo, M., Borrego, J. and Toranzo, A. (1992) Characterization of strains related to brown ring disease outbreaks in southwestern Spain. *Diseases of Aquatic Animals* 14, 229–236.

147 Cepeda, C., García-Márquez, S. and Santos, Y. (2003) Detection of *Flexibacter maritimus* in fish tissue using nested PCR amplification. *Journal of Fish Diseases* 26, 65–70.

148 Cerdà-Cuéllar, M., Permin, L., Larsen, J. and Blanch, A. (2001) Comparison of selective medium for the detection of *Vibrio vulnificus* in environmental samples. *Journal of Applied Microbiology* 91, 322–327.

149 Cerdà-Cuéllar, M., Ramon, A., Rosselló-Mora, R.A., Lalucat, J., Jofre, J. and Blanch, A. (1997) *Vibrio scophthalmi* sp. nov., a new species from Turbot (*Scophthalmus maximus*). *International Journal of Systematic Bacteriology* 47, 58–61.

150 Chan, K., Baumann, L., Garza, M. and Baumann, P. (1978) Two new species of *Alteromonas*: *Alteromonas espejiana* and *Alteromonas undina*. *International Journal of Systematic Bacteriology* 28, 217–222.

151 Chang, C., Jeong, J., Shin, J., Lee, E. and Son, H. (1999) *Rahnella aquatilis* sepsis in an immuno-competent adult. *Journal of Clinical Microbiology* 37, 4161–4162.

152 Chanock, R., Hayflick, L. and Barile, M. (1962) Growth on artificial medium of an agent associated with atypical pneumonia and its identification as a pleuropneumonia-like organism. *Proceedings of the National Academy of Sciences USA* 48, 41–48.

153 Chapman, P., Cipriano, R. and Teska, J. (1991) Isolation and phenotypic characterization of an oxidase-negative *Aeromonas salmonicida* causing furunculosis in coho salmon (*Oncorhynchus kisutch*). *Journal of Wildlife Diseases* 27, 61–67.

154 Chen, M.F., Henry-Ford, D. and Groff, J.M. (1995) Isolation and characterization of *Flexibacter maritimus* from marine fishes of California. *Journal of Aquatic Animal Health* 7, 318–326.

155 Chen, S.-C., Lee, J.-L., Lai, C.-C., Gu, Y.-W., Wang, C.-T., Chang, H.-Y. and Tsai, K.-H. (2000) Nocardiosis in sea bass, *Lateolabrax japonicus*, in Taiwan. *Journal of Fish Diseases* 23, 299–307.

156 Chen, S.-C., Lin, Y.-D., Liaw, L.-L. and Wang, P.-C. (2001) *Lactococcus garvieae* infection in the giant freshwater prawn *Macrobranchium rosenbergii* confirmed by polymerase chain reaction and 16S rDNA sequencing. *Diseases of Aquatic Organisms* 45, 45–52.

157 Chen, S.-C., Liaw, L.-L., Su, H.-Y., Ko, S.-C., Wu, C.-Y., Chaung, H.-C., Tsai, Y.-H., Yang, K.-L., Chen, Y.-C., Chen, T.-H., Lin, G.-R., Cheng, S.-Y., Lin, Y.-D., Lee, J.-L., Lai, C.-C., Weng, Y.-J. and Chu, S.-Y. (2002) *Lactococcus garvieae*, a cause of disease in grey mullet, *Mugil cephalus* L., in Taiwan. *Journal of Fish Diseases* 25, 727–732.

158 Choopun, N., Louis, V., Huq, A. and Colwell, R. (2002) Simple procedure for rapid identification of *Vibrio cholerae* from the aquatic environment. *Applied and Environmental Microbiology* 68, 995–998.

159 Chopra, A., Houston, C., Peterson, J. and Jin, G.-F. (1993) Cloning, expression, and sequence analysis of a cytolytic enterotoxin gene from *Aeromonas hydrophila*. *Canadian Journal of Microbiology* 39, 513–523.

160 Chow, K.H., Ng, T.K., Yuen, K.Y. and Yam, W.C. (2001) Detection of RTX toxin gene in *Vibrio cholerae* by PCR. *Journal of Clinical Microbiology* 39, 2594–2597.

161 Chowdhury, M.A.R., Yamanaka, H., Miyoshi, S., Aziz, K.M.S. and Shinoda, S. (1989) Ecology of *Vibrio mimicus* in aquatic environments. *Applied and Environmental Microbiology* 55, 2073–2078.

162 Christensen, P. (1980) *Flexibacter canadensis* sp. nov. *International Journal of Systematic Bacteriology* 30, 429–432.

163 Christensen, P. (1980) Description and taxonomic status of *Cytophaga heparina* (Payza and Korn) comb. nov. (Basionym: *Flavobacterium heparinum* Payza and Korn 1956). *International Journal of Systematic Bacteriology* 30, 473–475.

164 Christensen, W.B. (1946) Urea decomposition as a means of differentiating *Proteus* and paracolon cultures from each other and from *Salmonella* and *Shigella* types. *Journal of Bacteriology* 52, 461–466.

165 Chun, J., Seong, C.-N., Bae, K., Lee, K.-J., Kang, S.-O., Goodfellow, M. and Hah, Y. (1998) *Nocardia flavorosea* sp. nov. *International Journal of Systematic Bacteriology* 48, 901–905.

166 Cipriano, R. and Bertolini, J. (1988) Selection for virulence in the fish pathogen *Aeromonas salmonicida*, using coomassie brilliant blue agar. *Journal of Wildlife Diseases* 24, 672–678.

167 Cipriano, R., Schill, W., Pyle, S. and Horner, R. (1986) An epizootic in chinook salmon (*Oncorhynchus tshawytscha*) caused by a sorbitol-positive serovar 2 strain of *Yersinia ruckeri*. *Journal of Wildlife Diseases* 22, 488–492.

168 Cipriano, R., Schill, W.B., Teska, J.D. and Ford, L.A. (1996) Epizootiological study of bacterial cold-water disease in Pacific Salmon and further characterization of the etiological agent, *Flexibacter psychrophila*. *Journal of Aquatic Animal Health* 8, 28–36.

169 Clark, W., Hollis, D., Weaver, R. and Riley, P. (1984) *Identification of Unusual Pathogenic Gram-negative Aerobic and Facultatively Anaerobic Bacteria*. US Department of Health and Human Services, Public Health Service, Centres for Disease Control, Atlanta.

170 Clark, W.M. and Lubs, H.A. (1915) The differentiation of bacteria of the colon-aerogenes family by the use of indicators. *Journal of Infectious Diseases* 17, 161–173.

171 Clavareau, C., Wellemans, V., Walravens, K., Tryland, M., Verger, J.-M., Grayon, M., Cloeckaert, A., Letesson, J.-J. and Godfroid, J. (1998) Phenotypic and molecular characterization of a *Brucella* strain isolated from a minke whale (*Balaenoptera acutorostrata*). *Microbiology* 144, 3267–3273.

172 Cloeckaert, A., Verger, J.-M., Grayon, M., Paquet, J.-Y., Garin-Bastuji, B., Foster, G. and Godfroid, J. (2001) Classification of *Brucella* spp. isolated from marine mammals by DNA polymorphism at the *omp2* locus. *Microbes and Infection* 3, 729–738.

173 Coleman, S., Melanson, D., Biosca, E. and Oliver, J. (1996) Detection of *Vibrio vulnificus* biotypes 1 and 2 in eels and oysters by PCR amplification. *Applied and Environmental Microbiology* 62, 1368–1382.

174 Collins, M., Farrow, J., Phillips, B. and Kandler, O. (1983) *Streptococcus garvieae* sp. nov. and *Streptococcus plantarum* sp. nov. *Journal of General Microbiology* 129, 3427–3431.

175 Collins, M.D., Farrow, J.A.E., Katic, V. and Kandler, O. (1984) Taxonomic studies on *Streptococci* of serological groups E, P, U and V: description of *Streptococcus porcinus* sp. nov. *Systematic and Applied Microbiology* 5, 402–413.

176 Collins, M.D., Farrow, J.A.E., Phillips, B.A., Ferusu, S. and Jones, D. (1987) Classification of *Lactobacillus divergens*, *Lactococcus piscicola*, and some catalase-negative, asporogenous, rod-shaped bacteria from poultry in a new genus, *Carnobacterium*. *International Journal of Systematic Bacteriology* 37, 310–316.

177 Collins, M.D., Ash, C., Farrow, J.A.E., Wallbanks, S. and Williams, A.M. (1989) 16S Ribosomal

ribonucleic acid sequence analysis of lactococci and related taxa. Description of *Vagococcus fluvialis* gen. nov., sp. nov. *Journal of Applied Bacteriology* 67, 453–460.

178 Collins, M.D., Martinez-Murcia, A.J. and Cai, J. (1993) *Aeromonas enteropelogenes* and *Aeromonas ichthiosmia* are identical to *Aeromonas trota* and *Aeromonas veronii*, respectively, as revealed by small-subunit rRNA sequence analysis. *International Journal of Systematic Bacteriology* 43, 855–856.

179 Collins, M.D. and Lawson, P.A. (2000) The genus *Abiotrophia* (Kawamura *et al.*) is not monophyletic; proposal of *Granulicatella* gen. nov., *Granulicatella adiacens* comb. nov., *Granulicatella elegans* comb. nov., and *Granulicatella balaenopterae*. *International Journal of Systematic and Evolutionary Microbiology* 50, 365–369.

180 Collins, M., Hoyles, L., Hutson, R., Foster, G. and Falsen, E. (2001) *Corynebacterium testudinoris* sp. nov., from a tortoise, and *Corynebacterium felinum* sp. nov., from a Scottish wild cat. *International Journal of Systematic and Evolutionary Microbiology* 51, 1349–1352.

181 Collins, M., Hutson, R., Foster, G., Falsen, E. and Weiss, N. (2002a) *Isobaculum melis* gen. nov., sp., nov., a *Carnobacterium*-like organism isolated from the intestine of a badger. *International Journal of Systematic and Evolutionary Microbiology* 52, 207–210.

182 Collins, M. D., Hoyles, L., Foster, G., Falsen, E. and Weiss, N. (2002b) *Arthrobacter nasiphocae* sp. nov., from the common seal (*Phoca vitulina*). *International Journal of Systematic and Evolutionary Microbiology* 52, 569–571.

183 Colorni, A., Diamant, A., Eldar, A., Kvitt, H. and Zlotkin, A. (2002) *Streptococcus iniae* infections in Red Sea cage-cultured and wild fishes. *Diseases of Aquatic Organisms* 49, 165–170.

184 Coquet, L., Cosette, P., Quillet, L., Petit, F., Junter, G.-A. and Jouenne, T. (2002) Occurrence and phenotypic characterization of *Yersinia ruckeri* strains with biofilm-forming capacity in a rainbow trout farm. *Applied and Environmental Microbiology* 68, 470–475.

185 Corbel, M. and Morgan, W.B. (1975) Proposal for minimal standards for descriptions of new species and biotypes of the genus *Brucella*. *International Journal of Systematic Bacteriology* 25, 83–89.

186 Cornick, J., Morrison, C., Zwicker, B. and Shum, G. (1984) Atypical *Aeromonas salmonicida* infection in Atlantic cod, *Gadus morhua* L. *Journal of Fish Diseases* 7, 495–499.

187 Costa, R., Mermoud, I., Koblavi, S., Morlet, B., Haffner, P., Berthe, F., Legroumellec, M. and Grimont, P. (1998) Isolation and characterization of bacteria associated with a *Penaeus stylirostris*

disease (Syndrome 93) in New Caledonia. *Aquaculture* 164, 297–309.

188 Cottew, G.S. (1983) Recovery and identification of caprine and ovine mycoplasmas In: Razin, S. and Tully, J. (eds) *Methods in Mycoplasmology*, Vol. II. Academic Press, London, pp. 91–104.

189 Cottral, G.E. (ed.) (1978) *Manual of Standardized Methods for Veterinary Microbiology*. Cornell University Press, Ithaca, New York, pp 675.

190 Cousins, D.V. and Lloyd, J.M. (1988) Rapid identification of *Haemophilus somnus*, *Histophilus ovis* and *Actinobacillus seminis* using the API ZYM system. *Veterinary Microbiology* 17, 75–81.

191 Cowan, S.T. and Steel, K.J. (1970) *Manual for the Identification of Medical Bacteria*. Cambridge University Press, Cambridge.

192 Coyne, V. and Al-Harthi, L. (1992) Induction of melanin biosynthesis in *Vibrio cholerae*. *Applied and Environmental Microbiology* 58, 2861–2865.

193 Crosby, N.T. (1967) The determination of nitrite in water using Cleve's acid 1-naphthylamine-7-sulphonic acid. *Proceedings of the Society for Water Treatment and Examination* 16, 51.

194 Crumlish, M., Dung, T.T., Turnbull, J.F., Ngoc, N.T.N. and Ferguson, H.W. (2002) Identification of *Edwardsiella ictaluri* from diseased freshwater catfish, *Pangasius hypophthalmus* (Sauvage), cultured in the Mekong Delta, Vietnam. *Journal of Fish Diseases* 25, 733–736.

195 Cruz, J., Saraiva, A., Eiras, J., Branco, R. and Sousa, J. (1986) An outbreak of *Pleisiomonas shigelloides* in farmed rainbow trout, *Salmo gairdneri* Richardson, in Portugal. *Bulletin of the European Association of Fish Pathologists* 6, 20–22.

196 Dakin, W., Howell, D., Sutton, R., O'Keefe, M. and Thomas, P. (1974) Gastroenteritis due to non-agglutinable (non-cholera) vibrios. *Medical Journal of Australia* 2, 487–490.

197 Dalsgaard, I. and Paulsen, H. (1986) Atypical *Aeromonas salmonicida* isolated from diseased sand-eels, *Ammodytes lancea* (Cuvier) and *Hyperoplus lanceolatus* (Lesauvege). *Journal of Fish Diseases* 9, 361–364.

198 Dalsgaard, I., Jurgens, O. and Mortensen, A. (1988) *Vibrio salmonicida* isolated from farmed Atlantic salmon in the Faroe Islands. *Bulletin of the European Association of Fish Pathologists* 8, 53–54.

199 Dalsgaard, A., Dalsgaard, I., Høi, L. and Larsen, J.L. (1996) Comparison of a commercial biochemical kit and an oligonucleotide probe for identification of environmental isolates of *Vibrio vulnificus*. *Letters in Applied Microbiology* 22, 184–188.

200 Dalsgaard, I., Gudmundsdóttir, B., Helgason, S., Høie, S., Thoresen, O., Wichardt, U. and Wiklund, T. (1998) Identification of atypical *Aeromonas*

salmonicida: inter-laboratory evaluation and harmonization of methods. *Journal of Applied Microbiology* 84, 999–1006.

201 Dalsgaard, I., Høi, L., Siebeling, R.J. and Dalsgaard, A. (1999) Indole-positive *Vibrio vulnificus* isolated from disease outbreaks on a Danish eel farm. *Diseases of Aquatic Organisms* 35, 187–194.

202 Daly, J. and Stevenson, R. (1985) Charcoal agar, a new growth medium for the fish disease bacterium *Renibacterium salmoninarum*. *Applied and Environmental Microbiology* 50, 868–871.

203 Danley, M.L., Goodwin, A.E. and Killian, H.S. (1999) Epizootics in farm-raised channel catfish, *Ictalurus punctatus* (Rafinesque), caused by the enteric redmouth bacterium *Yersinia ruckeri*. *Journal of Fish Diseases* 22, 451–456.

204 Daoust, P.-Y., Larson, B. and Johnson, G. (1989) Mycobacteriosis in yellow perch (*Perca flavescens*) from two lakes in Alberta. *Journal of Wildlife Diseases* 25, 31–37.

205 Daskalov, H., Stobie, M. and Austin, B. (1998) *Klebsiella pneumoniae*: a pathogen of rainbow trout (*Oncorhynchus mykiss*, Walbaum). *Bulletin of the European Association of Fish Pathologists* 18, 26–28.

206 Daskalov, H., Austin, D. and Austin, B. (1999) An improved growth medium for *Flavobacterium psychrophilum*. *Letters in Applied Microbiology* 28, 297–299.

207 Davies, R. (1990) O-serotyping of *Yersinia ruckeri* with special emphasis on European isolates. *Veterinary Microbiology* 22, 299–307.

208 Davies, R. L. (1991) Clonal analysis of *Yersinia ruckeri* based on biotypes, serotypes and outer membrane protein-types. *Journal of Fish Diseases* 14, 221–228.

209 Davies, R. L. and Frerichs, G. N. (1989) Morphological and biochemical differences among isolates of *Yersinia ruckeri* obtained from wide geographical areas. *Journal of Fish Diseases* 12, 357–365.

210 Davis, B.R., Fanning, G.R., Madden, J.M., Steigerwalt, A.G., Bradford, H.B., Smith, H.L. and Brenner, D.J. (1981) Characterisation of biochemically atypical *Vibrio cholerae* strains and designation of a new pathogenic species, *Vibrio mimicus*. *Journal of Clinical Microbiology* 14, 631–639.

211 Davis, H.S. (1922) A new bacterial disease of freshwater fishes. *Bulletin of the U.S. Bureau of Fisheries, Washington, DC* 38, 261–280.

212 Davis, J. and Sizemore, R. (1982) Incidence of *Vibrio* species associated with Blue Crabs (*Callinectes sapidus*) collected from Galveston Bay, Texas. *Applied and Environmental Microbiology* 43, 1092–1097.

213 Decostere, A., Haesebrouck, F. and Devriese, L. (1997) Shieh medium supplemented with tobramycin for selective isolation of *Flavobacterium columnare* (*Flexibacter columnaris*) from diseased fish. *Journal of Clinical Microbiology* 35, 322–324.

214 Decostere, A., Haesebrouck, F. and Devriese, L. A. (1998) Characterization of four *Flavobacterium columnare* (*Flexibacter columnaris*) strains isolated from tropical fish. *Veterinary Microbiology* 62, 35–45.

215 DeLong, E., Wickham, G. and Pace, N. (1989) Phylogenetic strains: ribosomal RNA-based probes for the identification of single cells. *Science* 243, 1360–1363.

216 Denner, E.B.M., Vybiral, D., Fischer, U.R., Velimirov, B. and Busse, H.-J. (2002) *Vibrio calviensis* sp. nov., a halophilic, facultatively oligotrophic 0.2 μm-filterable marine bacterium. *International Journal of Systematic and Evolutionary Microbiology* 52, 549–553.

217 Desolme, B. and Bernardet, J.-F. (1996) Freeze-drying of *Flavobacterium columnare*, *Flavobacterium psychrophilum* and *Flexibacter maritimus*. *Diseases of Aquatic Organisms* 27, 77–80.

218 Diamant, A., Banet, A., Ucko, M., Colorni, A., Knibb, W. and Kvitt, H. (2000) Mycobacteriosis in wild rabbitfish *Siganus rivulatus* associated with cage farming in the Gulf of Eilat, Red Sea. *Diseases of Aquatic Organisms* 39, 211–219.

219 Dienes, L. (1939) 'L' organism of Klieneberger and *Streptobacillus moniliformis*. *Journal of Infectious Diseases* 65, 24–42.

220 Dierckens, K.R., Vandenberghe, J., Beladjal, L., Huys, G., Mertens, J. and Swings, J. (1998) *Aeromonas hydrophila* causes 'black disease' in fairy shrimps (Anostraca; Crustacea). *Journal of Fish Diseases* 21, 113–119.

221 Diggles, B., Carson, J., Hine, P., Hickman, R. and Tait, M. (2000) *Vibrio* species associated with mortalities in hatchery-reared turbot (*Colistium nudipinnis*) and brill (*C. guntheri*) in New Zealand. *Aquaculture* 183, 1–12.

222 DiSalvo, L., Blecka, J. and Zebal, R. (1978) *Vibrio anguillarum* and larval mortality in a Californian coastal shellfish hatchery. *Applied and Environmental Microbiology* 35, 219–221.

223 Dodson, S.V., Maurer, J.J. and Shotts, E.B. (1999) Biochemical and molecular typing of *Streptococcus iniae* isolated from fish and human cases. *Journal of Fish Diseases* 22, 331–336.

224 Doménech, A., Fernández-Garayzábal, J.F., Pascual, C., García, J.A., Cutúli, M.T., Moreno, M.A., Collins, M.D. and Domínguez, L. (1996) Streptococcosis in cultured turbot, *Scophthalmus maximus* (L.), associated with *Streptococcus parauberis*. *Journal of Fish Diseases* 19, 33–38.

225 Doménech, A., Fernández-Garayzábal, J.F., García, J.A., Cutuli, M.T., Blanco, M., Gibello, A., Moreno, M.A. and Domínguez, L. (1999) Association of *Pseudomonas anguilliseptica* infection with

'winter disease' in sea bream, *Sparus aurata* L. *Journal of Fish Diseases* 22, 69–71.

226 Donlon, J., McGettigan, S., O'Brien, P. and Carra, P.O. (1983) Re-appraisal of the nature of the pigment produced by *Aeromonas salmonicida*. *FEMS Microbiology Letters* 19, 285–290.

227 Doukas, V., Athanassopoulou, F., Karagouni, E. and Dotsika, E. (1998) *Aeromonas hydrophila* infection in cultured sea bass, *Dicentrarchus labrax* L., and *Puntazzo puntazzo* Cuvier from the Aegean Sea. *Journal of Fish Diseases* 21, 317–320.

228 Drancourt, M., Bollet, C., Carta, A. and Rousselier, P. (2001) Phylogenetic analysis of *Klebsiella* species delineate *Klebsiella* and *Raoultella* gen. nov., with description of *Raoultella ornithinolytica* comb. nov., *Raoultella terrigena* comb. nov. and *Raoultella planticola* comb. nov. *International Journal of Systematic and Evolutionary Microbiology* 51, 925–932.

229 Dunbar, S. and Clarridge, J. III (2000) Potential errors in recognition of *Erysipelothrix rhusiopathiae*. *Journal of Clinical Microbiology* 38, 1302–1304.

230 Eaves, L. and Ketterer, P. (1994) Mortalities in red claw crayfish *Cherax quadricarinatus* associated with systemic *Vibrio mimicus* infection. *Diseases of Aquatic Organisms* 19, 233–237.

231 Egan, S., Holmstrom, C. and Kjelleberg, S. (2001) *Pseudoalteromonas ulvae* sp. nov., a bacterium with antifouling activities isolated from the surface of a marine alga. *International Journal of Systematic and Evolutionary Microbiology* 51, 1499–1504.

232 Egidius, E., Wiik, R., Andersen, K., Hoff, K.A. and Hjeltnes, B. (1986) *Vibrio salmonicida* sp. nov., a new fish pathogen. *International Journal of Systematic Bacteriology* 36, 518–520.

233 Eldar, A., Bejerano, Y. and Bercovier, H. (1994) *Streptococcus shiloi* and *Streptococcus difficile*: two new Streptococcal species causing a meningoencephalitis in fish. *Current Microbiology* 28, 139–143.

234 Eldar, A., Bejerano, Y., Livoff, A., Horovitcz, A. and Bercovier, H. (1995a) Experimental streptococcal menigo-encephalitis in cultured fish. *Veterinary Microbiology* 43, 33–40.

235 Eldar, A., Frelier, P.F., Assenta, L., Varner, P.W., Lawhon, S. and Bercovier, H. (1995b) *Streptococcus shiloi*, the name for an agent causing septicemic infection in fish, is a junior synonym of *Streptococcus iniae*. *International Journal of Systematic Bacteriology* 45, 840–842.

236 Eldar, A., Ghittino, C., Asanta, L., Bozzetta, E., Goria, M., Prearo, M. and Bercovier, H. (1996) *Enterococcus seriolicida* is a junior synonym of *Lactococcus garvieae*, a causative agent of septicaemia and meningoencephalitis in fish. *Current Microbiology* 32, 85–88.

237 Eldar, A., Gloria, M., Ghittino, C., Zlotkin, A. and Bercovier, H. (1999) Biodiversity of *Lactococcus garvieae* strains isolated from fish in Europe, Asia, and Australia. *Applied and Environmental Microbiology* 65, 1005–1008.

238 Elliott, J. and Facklam, R. (1996) Antimicrobial susceptibilities of *Lactococcus lactis* and *Lactococcus garvieae* and a proposed method to discriminate between them. *Journal of Clinical Microbiology* 34, 1296–1298.

239 Enger, Ø., Nygaard, H., Solberg, M., Schei, G., Nielsen, J. and Dundas, I. (1987) Characterization of *Alteromonas denitrificans* sp. nov. *International Journal of Systematic Bacteriology* 37, 416–421.

240 Esteve, C., Amaro, C., Biosca, E. and Garay, E. (1995) Biochemical and toxigenic properties of *Vibrio furnissii* isolated from a European eel farm. *Aquaculture* 132, 81–90.

241 Esteve, C., Gutiérrez, M. C. and Ventosa, A. (1995) *Aeromonas encheleia* sp. nov., isolated from European eels. *International Journal of Systematic Bacteriology* 45, 462–466.

242 Evans, J.J., Klesius, P.H., Gilbert, P.M., Shoemaker, C.A., Sarawi, M.A.A., Landsberg, J., Duremdez, R., Marzouk, A.A. and Zenki, S.A. (2002) Characterization of β-haemolytic Group B *Streptococcus agalactiae* in cultured seabream, *Sparus auratus* L., and wild mullet, *Liza klunzingeri* (Day), in Kuwait. *Journal of Fish Diseases* 25, 505–513.

243 Evelyn, T. (1977) An improved growth medium for the kidney disease bacterium and some notes on using the medium. *Bulletin de L'Office International des Epizooties* 87, 511–513.

244 Evelyn, T. and Prosperi-Porta, L. (1989) Inconsistent performance of KDM2, a culture medium for the kidney disease bacterium *Renibacterium salmoninarum*, due to variation in the composition of its peptone ingredient. *Diseases of Aquatic Organisms* 7, 227–229.

245 Evelyn, T., Prosperi-Porta, L. and Ketcheson, J. (1990) Two new techniques for obtaining consistent results when growing *Renibacterium salmoninarum* on KDM2 culture medium. *Diseases of Aquatic Organisms* 9, 209–212.

246 Evenberg, D., Versluis, R. and Lugtenberg, B. (1985) Biochemical and immunological characterization of the cell surface of the fish pathogenic bacterium *Aeromonas salmonicida*. *Biochimica et Biophysica Acta* 815, 233–244.

247 Ewalt, D.R., Payeur, J.B., Martin, B.M., Cummins, D.R. and Miller, W.G. (1994) Characteristics of a *Brucella* species from a bottlenose dolphin (*Tursiops truncatus*). *Journal of Veterinary Diagnostic Investigation* 6, 448–452.

248 Ewing, W. and Davies, B. (1972) Biochemical characterization of *Citrobacter diversus* (Burkey) Werkman and Gillen and designation of the

neotype strain. *International Journal of Systematic Bacteriology* 22, 12–18.

249 Ewing, W.H. and Fife, M.A. (1972) *Enterobacter agglomerans* (Beijerinck) comb. nov. (the Herbicola-Lathyri bacteria). *International Journal of Systematic Bacteriology* 22, 4–11.

250 Ewing, W.H., Ross, A.J., Brenner, D.J. and Fanning, G.R. (1978) *Yersinia ruckeri* sp. nov., the redmouth (RM) bacterium. *International Journal of Systematic Bacteriology* 28, 37–44.

251 Farkas, J. (1985) Filamentous *Flavobacterium* sp. isolated from fish with gill disease in cold water. *Aquaculture* 44, 1–10.

252 Farmer, J. III and McWhorter, A. (1984) Genus *Edwardsiella*. In: Krieg, N.R. and Holt, J.G. (eds) *Bergey's Manual of Systematic Bacteriology*, Vol. I. Williams & Wilkins, Baltimore, Maryland, pp. 486–491.

253 Farrell, I.D. (1974) The development of a new selective medium for the isolation of *Brucella abortus* from contaminated sources. *Research in Veterinary Science* 16, 280–286.

254 Farto, R., Montes, M., Pérez, M., Nieto, T., Larsen, J. and Pedersen, K. (1999) Characterization by numerical taxonomy and ribotyping of *Vibrio splendidus* biovar I and *Vibrio scophthalmi* strains associated with turbot cultures. *Journal of Applied Microbiology* 86, 796–804.

255 Fearrington, E., Rand, C., Mewborn, A. and Wilkerson, J. (1974) Non-cholera vibrio septicemia and meningoencephalitis. *Annals of Internal Medicine*, 81, 401.

256 Ferragut, C., Izard, D., Gavini, F., Kersters, K., Ley, J.D. and Leclerc, H. (1983) *Klebsiella trevisanii*: a new species from water and soil. *International Journal of Systematic Bacteriology* 33, 133–142.

257 Fidopiastis, P.M., Boletzky, S.V. and Ruby, E.G. (1998) A new niche for *Vibrio logei*, the predominant light organ symbiont of squids in the genus *Sepiola*. *Journal of Bacteriology* 180, 59–64.

258 Fields, P., Popovic, T., Wachsmuth, K. and Olsvik, O. (1992) Use of Polymerase Chain Reaction for detection of toxigenic *Vibrio cholerae* 01 strains from the Latin American cholera epidemic. *Journal of Clinical Microbiology* 30, 2118–2121.

259 Filler, G., Ehrich, J., Strauch, E. and Beutin, L. (2000) Acute renal failure in an infant associated with cytotoxic *Aeromonas sobria* isolated from patient's stool and from aquarium water as suspected source of infection. *Journal of Clinical Microbiology* 38, 469–470.

260 Foo, J.T.W., Ho, B. and Lam, T.L. (1985) Mass mortality in *Siganus canaliculatus* due to streptococcal infection. *Aquaculture* 49, 185–195.

261 Forbes, L., Nielsen, O., Measures, L. and Ewalt, D. (2000) Brucellosis in ringed seals and harp seals from Canada. *Journal of Wildlife Diseases* 36, 595–598.

262 Foster, G., Jahans, K.L., Reid, R.J. and Ross, H.M. (1996a) Isolation of *Brucella* species from cetaceans, seals and an otter. *Veterinary Record* 138, 583–586.

263 Foster, G., Ross, H.M., Malnick, H., Willems, A., Garcia, P., Reid, R.J. and Collins, M.D. (1996b) *Actinobacillus delphinicola* sp. nov., a new member of the family *Pasteurellaceae* Pohl (1979) 1981 isolated from sea mammals. *International Journal of Systematic Bacteriology* 46, 648–652.

264 Foster, G., Ross, H.M., Hutson, R.A. and Collins, M.D. (1997) *Staphylococcus lutrae* sp. nov., a new coagulase-positive species isolated from otters. *International Journal of Systematic Bacteriology* 47, 724–726.

265 Foster, G., Ross, H.M., Patterson, I.A.P., Hutson, R.A. and Collins, M.D. (1998) *Actinobacillus scotiae* sp. nov., a new member of the family *Pasteurellaceae* Pohl (1979) 1981 isolated from porpoises (*Phocoena phocoena*). *International Journal of Systematic Bacteriology* 48, 929–933.

266 Foster, G., Ross, H., Malnick, H., Willems, A., Hutson, R., Reid, R. and Collins, M. (2000) *Phocoenobacter uteri* gen. nov., sp. nov., a new member of the family *Pasteurellaceae* Pohl (1979) 1981 isolated from a harbour porpoise (*Phocoena phocoena*). *International Journal of Systematic and Evolutionary Microbiology* 50, 135–139.

267 Foster, G., MacMillan, A.P., Godfroid, J., Howie, F., Ross, H.M., Cloeckaert, A., Reid, R.J., Brew, S. and Patterson, I.A.P. (2002) A review of *Brucella* sp. infection of sea mammals with particular emphasis on isolates from Scotland. *Veterinary Microbiology* 90, 563–580.

268 Fouz, B., Larsen, J., Nielsen, B., Barja, J. and Toranzo, A. (1992) Characterization of *Vibrio damsela* strains isolated from turbot *Scophthalmus maximus* in Spain. *Diseases of Aquatic Organisms* 12, 155–166.

269 Freundt, E.A. (1983) Culture media for classic mycoplasmas. In: Razin, S. and Tully, J. (eds) *Methods in Mycoplasmology*, Vol. I. Academic Press, London, pp. 128–139.

270 Friedman, C., Beaman, B., Chun, J., Goodfellow, M., Gee, A. and Hedrick, R. (1998) *Nocardia crassostreae* sp nov., the causal agent of nocardiosis in Pacific oysters. *International Journal of Systematic Bacteriology* 48, 237–246.

271 Fuhrmann, H., Böhm, K.H. and Schlotfeldt, H.-J. (1984) On the importance of enteric bacteria in the bacteriology of freshwater fish. *Bulletin of the European Association of Fish Pathologists* 4, 42–46.

272 Fujino, T., Sakazaki, R. and Tamura, K. (1974) Designation of the type strain of *Vibrio parahaemolyticus* and description of 200 strains of the species. *International Journal of Systematic Bacteriology* 24, 447–449.

273 Fukuda, Y., Matsuoka, S., Mizuno, Y. and Narita, K. (1996) *Pasteurella piscicida* infection in cultured juvenile Japanese flounder. *Fish Pathology* 31, 33–38.

274 Funke, G., Ramos, C., Fernández-Garayzábal, J., Weiss, N. and Collins, M. (1995) Description of human-derived Centers for Disease Control Coryneform Group 2 bacteria as *Actinomyces bernardiae* sp. nov. *International Journal of Systematic Bacteriology* 45, 57–60.

275 Furniss, A.L. and Donovan, T.J. (1974) The isolation and identification of *Vibrio cholerae*. *Journal of Clinical Pathology* 27, 764–766.

276 Furniss, A.L., Lee, J.V. and Donovan, T.J. (1977) Group F, a new vibrio? *Lancet* ii, 565–566.

277 Furniss, A.L., Lee, J.V. and Donovan, T.J. (1978) The Vibrios. *Public Health Laboratory Science. Monograph Series No. 11.* London. pp. 3–57.

278 Gales, N., Wallace, G. and Dickson, J. (1985) Pulmonary Cryptococcosis in a striped dolphin (*Stenella coeruleoalba*). *Journal of Wildlife Diseases* 21, 443–446.

279 Garcia, M.T., Ventosa, A., Ruiz-Berraquero, F. and Kocur, M. (1987) Taxonomic study and amended description of *Vibrio costicola*. *International Journal of Systematic Bacteriology* 37, 251–256.

280 Garland, C.D., Nash, G.V., Sumner, C.E. and McMeekin, T.A. (1983) Bacterial pathogens of oyster larvae (*Crassostrea gigas*) in a Tasmanian hatchery. *Australian Journal of Marine and Freshwater Research* 34, 483–487.

281 Gatesoupe, F.J., Lambert, C. and Nicolas, J.L. (1999) Pathogenicity of *Vibrio splendidus* strains associated with turbot larvae, *Scophthalmus maximus*. *Journal of Applied Microbiology* 87, 757–763.

282 Gauger, E. and Gómez-Chiarri, M. (2002) 16S ribosomal DNA sequencing confirms the synonymy of *Vibrio harveyi* and *V. carchariae*. *Diseases of Aquatic Organisms* 52, 39–46.

283 Gauthier, M.J. (1976a) *Alteromonas rubra* sp. nov., a new marine antibiotic-producing bacterium. *International Journal of Systematic Bacteriology* 26, 459–466.

284 Gauthier, M.J. (1976b) Morphological, physiological, and biochemical characteristics of some violet-pigmented bacteria isolated from seawater. *Canadian Journal of Microbiology* 22, 138–149.

285 Gauthier, M.J. (1977) *Alteromonas citrea*, a new Gram-negative, yellow-pigmented species from seawater. *International Journal of Systematic Bacteriology* 27, 349–354.

286 Gauthier, M. and Breittmayer, V. (1979) A new antibiotic-producing bacterium from seawater: *Alteromonas aurantia* sp. nov. *International Journal of Systematic Bacteriology* 29, 366–372.

287 Gauthier, M.J. (1982) Validation of the name *Alteromonas luteoviolacea*. *International Journal of Systematic Bacteriology* 32, 82–86.

288 Gauthier, G., Gauthier, M. and Christen, R. (1995a) Phylogenetic analysis of the genera *Alteromonas*, *Shewanella*, and *Moritella* using genes coding for small-subunit rRNA sequences and division of the genus *Alteromonas* into two genera, *Alteromonas* (emended) and *Pseudo-alteromonas* gen. nov., and proposal of twelve new species combinations. *International Journal of Systematic Bacteriology* 45, 755–761.

289 Gauthier, G., Lafay, B., Ruimy, R., Breittmayer, V., Nicolas, J.L., Gauthier, M. and Christen, R. (1995b) Small-subunit rRNA sequences and whole DNA relatedness concur for the reassignment of *Pasteurella piscicida* (Snieszko *et al.*) Janssen and Surgalla to the genus *Photobacterium* as *Photobacterium damsela* subsp. *piscicida* comb. nov. *International Journal of Systematic Bacteriology* 45, 139–144.

290 Gavini, F., Ferragut, C., Izard, D., Trinel, P., Leclerc, H., Lefebvre, B. and Mossel, D. (1979) *Serratia fonticola*, a new species from water. *International Journal of Systematic Bacteriology* 29, 92–101.

291 Gavini, F., Mergaert, J., Beji, A., Mielcarek, C., Izard, D., Kersters, K. and Ley, J.D. (1989) Transfer of *Enterobacter agglomerans* (Beijerinck 1888) Ewing and Fife 1972 to *Pantoea* gen. nov. as *Pantoea agglomerans* comb. nov. and description of *Pantoea dispersa* sp. nov. *International Journal of Systematic Bacteriology* 39, 337–345.

292 Geraci, J., Sauer, R. and Medway, W. (1966) Erysipelas in Dolphins. *American Journal of Veterinary Research* 27, 597–606.

293 Gibello, A., Blanco, M.M., Moreno, M.A., Cutúli, M.T., Doménech, A., Domínguez, L. and Fernández-Garayzábal, J.F. (1999) Development of a PCR assay for detection of *Yersinia ruckeri* in tissues of inoculated and naturally infected trout. *Applied and Environmental Microbiology* 65, 346–350.

294 Gibson, L.F., Woodworth, J. and George, A.M. (1998) Probiotic activity of *Aeromonas media* on the Pacific oyster, *Crassostrea gigas*, when challenged with *Vibrio tubiashii*. *Aquaculture* 169, 111–120.

295 Giebel, J., Meier, J., Binder, A., Flossdorf, J., Poveda, J., Schmidt, R. and Kirchhoff, H. (1991) *Mycoplasma phocarhinis* sp. nov. and *Mycoplasma phocacerebrale* sp. nov., two new species from harbour seals (*Phoca vitulina* L.). *International Journal of Systematic Bacteriology* 41, 39–44.

296 Gil, P., Vivas, J., Gallardo, C.S. and Rodriguez, L.A. (2000) First isolation of *Staphylococcus warneri*, from diseased rainbow trout,

Oncorhynchus mykiss (Walbaum), in Northwest Spain. *Journal of Fish Diseases* 23, 295–298.

297 Gilardi, G. (1983) *Identification of Glucose-nonfermenting Gram-negative Rods.* American Society for Microbiology, Washington, DC.

298 Gilmartin, W., Vainik, P. and Neill, V. (1979) Salmonellae in feral pinnipeds off the southern California coast. *Journal of Wildlife Diseases* 15, 511–514.

299 Gjerde, J. (1984) Occurrence and characterization of *Aerococcus viridans* from lobsters, *Homarus gammarus* L., dying in captivity. *Journal of Fish Diseases* 7, 355–362.

300 Glazebrook, J.S. and Campbell, R.S.F. (1990a) A survey of the diseases of marine turtles in northern Australia. I. Farmed turtles. *Diseases of Aquatic Organisms* 9, 83–95.

301 Glazebrook, J.S. and Campbell, R.S.F. (1990b) A survey of the diseases of marine turtles in northern Australia. II. Oceanarium-reared and wild turtles. *Diseases of Aquatic Organisms* 9, 97–104.

302 Goldenberger, D., Perschil, I., Ritzler, M. and Altwegg, M. (1995) A simple 'universal' DNA extraction procedure using SDS and proteinase K is compatible with direct PCR amplification. *PCR Methods and Applications* 4, 368–370.

303 Goldman, C., Loureiro, J., Quse, V., Corach, D., Calderon, E., Caro, R., Boccio, J., Heredia, S., Carlo, M.D. and Zubillaga, M. (2002) Evidence of *Helicobacter* sp. in dental plaque of captive dolphins (*Tursiops gephyreus*). *Journal of Wildlife Diseases* 38, 644–648.

304 Gomez-Gil, B., Tron-Mayén, L., Roque, A., Turnbull, J., Inglis, V. and Guerra-Flores, A. (1998) Species of *Vibrio* isolated from hepatopancreas, haemolymph and digestive tract of a population of healthy juvenile *Penaeus vannamei*. *Aquaculture* 163, 1–9.

305 Gomez-Gil, B., Thompson, F., Thompson, C. and Swings, J. (2003a) *Vibrio rotiferianus* sp. nov., isolated from cultures of the rotifer *Brachionus plicatilis*. *International Journal of Systematic and Evolutionary Microbiology* 53, 239–243.

306 Gomez-Gil, B., Thompson, F., Thompson, C. and Swings, J. (2003b) *Vibrio pacinii* sp. nov., from cultured aquatic organisms. *International Journal of Systematic and Evolutionary Microbiology* 53, 1569–1573.

307 Goodwin, AE., Roy, J.S. Jr., Grizzle, J.M. and Goldspy, M.T. Jr. (1994) *Bacillus mycoides*: a bacterial pathogen of channel catfish *Ictalurus punctatus*. *Diseases of Aquatic Organisms* 18, 173–179.

308 Gordon, M.A. (1976) Characterization of *Dermatophilus congolensis*: its affinities with the *Actinomycetales* and differentiation from *Geodermatophilusi*. In: Lloyd, D.H. and Sellers,

K.C. (eds) *Dermatophilus Infection in Animals and Man.* Academic Press, London, pp. 187–201.

309 Gordon, R., Barnett, D., Handerhan, J. and Pang, C.H.-N. (1974) *Nocardia coeliaca*, *Nocardia autotrophica*, and the Nocardin strain. *International Journal of Systematic Bacteriology* 24, 54–63.

310 Graevenitz, A., Bowman, J., Notaro, C.D. and Ritzler, M. (2000) Human infection with *Halomonas venusta* following fish bite. *Journal of Clinical Microbiology* 38, 3123–3124.

311 Grandis, S. de, Krell, P., Flett, D. and Stevenson, R. (1988) Deoxyribonucleic acid relatedness of serovars of *Yersinia ruckeri*, the enteric redmouth bacterium. *International Journal of Systematic Bacteriology* 38, 49–55.

312 Greenwood, A. and Taylor, D. (1978) Clostridial myositis in marine mammals. *Veterinary Record* 103, 54–55.

313 Greipsson, S. and Priest, F. (1983) Numerical taxonomy of *Hafnia alvei*. *International Journal of Systematic Bacteriology* 33, 470–475.

314 Grimes, D., Stemmler, J., Hada, H., May, E., Maneval, D., Hetrick, F., Jones, R., Stoskopf, M. and Colwell, R. (1984) *Vibrio* species associated with mortality of sharks held in captivity. *Microbial Ecology* 10, 271–282.

315 Grimes, D., Gruber, S. and May, E. (1985) Experimental infection of lemon sharks, *Negaprion brevirostris* (Poey), with *Vibrio* species. *Journal of Fish Diseases* 8, 173–180.

316 Grimes, D. J., Jacobs, D., Swartz, D., Brayton, P. and Colwell, R. (1993) Numerical taxonomy of gram-negative, oxidase-positive rods from carcharhinid sharks. *International Journal of Systematic Bacteriology* 43, 88–98.

317 Grimont, P., Grimont, F., Richard, C. and Sakazaki, R. (1980) *Edwardsiella hoshinae*, a new species of *Enterobacteriaceae*. *Current Microbiology* 4, 347–351.

318 Gudmundsdóttir, B.K., Hastings, T.S. and Ellis, A.E. (1990) Isolation of a new toxic protease from a strain of *Aeromonas salmonicida* subspecies *achromogenes*. *Diseases of Aquatic Organisms* 9, 199–208.

319 Guerinot, M.L., West, P.A., Lee, J.V. and Colwell, R.R. (1982) *Vibrio diazotrophicus* sp. nov., a marine nitrogen-fixing bacterium. *International Journal of Systematic Bacteriology* 32, 350–357.

320 Gunn, B. and Colwell, R. (1983) Numerical taxonomy of Staphylococci isolated from the marine environment. *International Journal of Systematic Bacteriology* 33, 751–759.

321 Hada, H.S., West, P.A., Lee, J.V., Stemmler, J. and Colwell, R.R. (1984) *Vibrio tubiashii* sp. nov., a pathogen of bivalve mollusks. *International Journal of Systematic Bacteriology* 34, 1–4.

322 Hahnel, G.B. and Gould, R.W. (1982) Effects of temperature on biochemical reactions and drug

resistance of virulent and avirulent *Aeromonas salmonicida. Journal of Fish Diseases* 5, 329–337.

323 Hänninen, M.-L. and Hirvelä-Koski, V. (1997) Molecular and phenotypic methods for the characterization of atypical *Aeromonas salmonicida. Veterinary Microbiology* 56, 147–158.

324 Hansen, G., Bergh, O., Michaelsen, J. and Knappskog, D. (1992) *Flexibacter ovolyticus* sp. nov., a pathogen of eggs and larvae of Atlantic halibut, *Hippoglossus hippoglossus* L. *International Journal of Systematic Bacteriology* 42, 451–458.

325 Hansen, G.H., Raa, J. and Olafsen, J.A. (1990) Isolation of *Enterobacter agglomerans* from dolphin fish, *Coryphaena hippurus* L. *Journal of Fish Diseases* 13, 93–96.

326 Hao, M.V. and Komagata, K. (1985) A new species of *Planococcus, P. kocurii* isolated from fish, frozen foods, and fish curing brine. *Journal of General and Applied Microbiology* 31, 441–455.

327 Harper, C., Dangler, C., Xu, S., Feng, Y., Shen, Z., Sheppard, B., Stamper, A., Dewhirst, F., Paster, B. and Fox, J. (2000) Isolation and characterization of a *Helicobacter* sp. from the gastric mucosa of dolphins, *Lagenorhynchus acutus* and *Delphinus delphis. Applied and Environmental Microbiology* 66, 4751–4757.

328 Harper, C.G., Feng, Y., Xu, S., Taylor, N., Kinsel, M., Dewhirst, F., Paster, B., Greenwell, M., Levine, G., Rogers, A. and Fox, J. (2002a) *Helicobacter cetorum* sp. nov., a urease-positive *Helicobacter* species isolated from dolphins and whales. *Journal of Clinical Microbiology* 40, 4536–4543.

329 Harper, C., Xu, S., Feng, Y., Dunn, J.L., Taylor, N., Dewhirst, F. and Fox, J. (2002b) Identification of novel *Helicobacter* spp. from a Beluga whale. *Applied and Environmental Microbiology* 68, 2040–2043.

330 Håstein, T. and Holt, G. (1972) The occurrence of vibrio disease in wild Norwegian fish. *Journal of Fish Biology* 4, 33–37.

331 Håstein, T., Saltveit, S. and Roberts, R. (1978) Mass mortality among minnows *Phoxinus phoxinus* (L.) in Lake Tveitevatn, Norway, due to an aberrant strain of *Aeromonas salmonicida. Journal of Fish Diseases* 1, 241–249.

332 Hatai, K., Egusa, S. and Chikahata, H. (1975) *Pseudomonas chlororaphis* as a fish pathogen. *Bulletin of the Japanese Society of Scientific Fisheries* 41, 1203.

333 Hawke, J., Plakas, S., Minton, R., McPhearson, R., Snider, T. and Guarino, A. (1987) Fish Pasteurellosis of cultured striped bass (*Morone saxatilis*) in coastal Alabama. *Aquaculture* 65, 193–204.

334 Hawke, J.P., McWhorter, A.C., Steigerwalt, A.G. and Brenner, D.J. (1981) *Edwardsiella ictaluri* sp. nov., the causative agent of enteric septicaemia of catfish. *International Journal of Systematic Bacteriology* 31, 396–400.

335 Hayflick, L. (1965) Tissue cultures and mycoplasmas. *Texas Reports on Biology and Medicine* 23, Supplement 1, 285–303.

336 Hebert, A.M. and Vreeland, R.H. (1987) Phenotypic comparison of halotolerant bacteria: *Halomonas halodurans* sp. nov., rev., comb. nov. *International Journal of Systematic Bacteriology* 37, 347–350.

337 Heckert, R., Elankumaran, S., Milani, A. and Baya, A. (2001) Detection of a new *Mycobacterium* species in wild striped bass in the Chesapeake Bay. *Journal of Clinical Microbiology* 39, 710–715.

338 Hedlund, B. and Staley, J. (2001) *Vibrio cyclotrophicus* sp. nov., a polycyclic aromatic hydrocarbon (PAH)-degrading marine bacterium. *International Journal of Systematic and Evolutionary Microbiology* 51, 61–66.

339 Hedrick, R., McDowell, T. and Groff, J. (1987) Mycobacteriosis in cultured striped bass from California. *Journal of Wildlife Diseases* 23, 391–395.

340 Hendrie, M., Hodgkiss, W. and Shewan, J. (1970) The identification, taxonomy and classification of luminous bacteria. *Journal of General Microbiology* 64, 151–169.

341 Hendrie, M., Hodgkiss, W. and Shewan, J. (1971a) Proposal that the species *Vibrio anguillarum* Bergman 1909, *Vibrio piscium* David 1927, and *Vibrio ichthyodermis* (Wells and ZoBell) Shewan, Hobbs, and Hodgkiss 1960 be combined as a single species, *Vibrio anguillarum. International Journal of Systematic Bacteriology* 21, 64–68.

342 Hendrie, M., Hodgkiss, W. and Shewan, J. (1971b) Proposal that *Vibrio marinus* (Russell 1891) Ford 1927 be amalgamated with *Vibrio fischeri* (Beijerinck 1889) Lehmann and Neumann 1896. *International Journal of Systematic Bacteriology* 21, 217–221.

343 Henley, M.W. and Lewis, D.H. (1976) Anaerobic bacteria associated with epizootics in grey mullet (*Mugil cephalus*) and redfish (*Sciaenops ocellata*) along the Texas gulf coast. *Journal of Wildlife Diseases* 12, 448–453.

344 Henrichsen, J. (1972) Bacterial surface translocation: a survey and classification. *Bacteriological Reviews* 36, 478–503.

345 Herbst, L., Costa, S., Weiss, L., Johnson, L., Bartell, J., Davis, R., Walsh, M. and Levi, M. (2001) Granulomatous skin lesions in Moray eels caused by a novel *Mycobacterium* species related to *Mycobacterium triplex. Infection and Immunity* 69, 4639–4646.

346 Hickman, F.W., Farmer, J.J. III., Hollis, D.G., Fanning, G.R., Steigerwalt, A.G., Weaver, R.E. and Brenner, D.J. (1982) Identification of *Vibrio hollisae* sp. nov. from patients with diarrhea. *Journal of Clinical Microbiology* 15, 395–401.

347 Hickman-Brenner, F.W., MacDonald, K.L., Steigerwalt, A.G., Fanning, G.R., Brenner, D.J.

and Farmer, J.J. III (1987) *Aeromonas veronii*, a new ornithine decarboxylase-positive species that may cause diarrhea. *Journal of Clinical Microbiology* 25, 900–906.

348 Hickman-Brenner, F.W., Fanning, G.R., Arduino, M.J., Brenner, D.J. and Farmer, J.J. III (1988) *Aeromonas schubertii*, a new mannitol-negative species found in human clinical specimens. *Journal of Clinical Microbiology* 26, 1561–1564.

349 Hicks, C., Kinoshita, R. and Ladds, P. (2000) Pathology of melioidosis in captive marine mammals. *Australian Veterinary Journal* 78, 193–195.

350 Hill, A. (1985) *Mycoplasma testudinis*, a new species isolated from a Tortoise. *International Journal of Systematic Bacteriology* 35, 489–492.

351 Hirono, I., Masuda, T. and Aoki, T. (1996) Cloning and detection of the hemolysin gene of *Vibrio anguillarum*. *Microbial Pathogenesis* 21, 173–182.

352 Hirvelä-Koski, V., Koski, P. and Niiranen, H. (1994) Biochemical properties and drug resistance of *Aeromonas salmonicida* in Finland. *Diseases of Aquatic Organisms* 20, 191–196.

353 Hiu, S.F., Holt, R.A., Sriranganathan, N., Seidler, R.J. and Fryer, J.L. (1984) *Lactobacillus piscicola*, a new species from salmonid fish. *International Journal of Systematic Bacteriology* 34, 393–400.

354 Hogardt, M., Trebesius, K., Geiger, A., Hornef, M., Rosenecker, J. and Heesemann, J. (2000) Specific and rapid detection by fluorescent in situ hybridization of bacteria in clinical samples obtained from cystic fibrosis patients. *Journal of Clinical Microbiology* 38, 818–825.

355 Høi, L., Dalsgaard, I. and Dalsgaard, A. (1998a) Improved isolation of *Vibrio vulnificus* from seawater and sediment with Cellobiose-Colistin Agar. *Applied and Environmental Microbiology* 64, 1721–1724.

356 Høi, L., Dalsgaard, I., DePaola, A., Siebeling, R.J. and Dalsgaard, A. (1998b) Heterogeneity among isolates of *Vibrio vulnificus* recovered from eels (*Anguilla anguilla*) in Denmark. *Applied and Environmental Microbiology* 64, 4676–4682.

357 Høie, S., Heum, M. and Thoresen, O. (1997) Evaluation of a polymerase chain reaction-based assay for the detection of *Aeromonas salmonicida* ss *salmonicida* in Atlantic salmon *Salmo salar*. *Diseases of Aquatic Organisms* 30, 27–35.

358 Hollis, D., Weaver, R., Baker, C. and Thornsberry, C. (1976) Halophilic *Vibrio* species isolated from blood cultures. *Journal of Clinical Microbiology* 3, 425–431.

359 Holmes, B. (1986) Identification and distribution of *Pseudomonas stutzeri* in clinical material. *Journal of Applied Bacteriology* 60, 401–411.

360 Holmes, B., Lapage, S.P. and Malnick, H. (1975) Strains of *Pseudomonas putrefaciens* from clinical material. *Journal of Clinical Pathology* 28, 149–155.

361 Holmes, B., Owen, R., Evans, A., Malnick, H. and Willcox, W. (1977a) *Pseudomonas paucimobilis*, a new species isolated from human clinical specimens, the hospital environment, and other sources. *International Journal of Systematic Bacteriology* 27, 133–146.

362 Holmes, B., Snell, J. and Lapage, S. (1977b) Revised description, from clinical isolates, of *Flavobacterium odoratum* Stutzer and Kwaschnina 1929, and designation of the neotype strain. *International Journal of Systematic Bacteriology* 27, 330–336.

363 Holmes, B., Snell, J. and Lapage, S. (1978) Revised description, from clinical strains, of *Flavobacterium breve* (Lustig) Bergey et al. 1923 and proposal of the neotype strain. *International Journal of Systematic Bacteriology* 28, 201–208.

364 Holmes, B., Owen, R.J. and Weaver, R.E. (1981) *Flavobacterium multivorum*, a new species isolated from human clinical specimens and previously known as group IIk, biotype 2. *International Journal of Systematic Bacteriology* 31, 21–34.

365 Holmes, B., Owen, R.J. and Hollis, D.G. (1982) *Flavobacterium spiritivorum*, a new species isolated from human clinical specimens. *International Journal of Systematic Bacteriology* 32, 157–165.

366 Holmes, B., Owen, R.J., Steigerwalt, A.G. and Brenner, D.J. (1984) *Flavobacterium gleum*, a new species found in human clinical specimens. *International Journal of Systematic Bacteriology* 34, 21–25.

367 Hommez, J., Devriese, L. and Castryck, F. (1983) Improved media for the isolation of *Bordetella bronchiseptica*. In: Pedersen, K. and Nielsen, N. (eds) *Atrophic Rhinitis in Pigs*. Commission of the European Communities, Luxembourg, pp. 98–104.

368 Hoyles, L., Foster, G., Falsen, E., Thomson, L. and Collins, M. (2001) *Facklamia miroungae* sp. nov., from a juvenile southern elephant seal (*Mirounga leonina*). *International Journal of Systematic and Evolutionary Microbiology* 51, 1401–1403.

369 Hoyles, L., Lawson, P., Foster, G., Falsen, E., Ohlén, M., Grainger, J. and Collins, M. (2000) *Vagococcus fessus* sp. nov., isolated from a seal and harbour porpoise. *International Journal of Systematic and Evolutionary Microbiology* 50, 1151–1154.

370 Hoyles, L., Pascual, C., Falsen, E., Foster, G., Grainger, J. and Collins, M. (2001) *Actinomyces marimammalium* sp. nov., from marine mammals. *International Journal of Systematic and Evolutionary Microbiology* 51, 151–156.

371 Huang, C.-Y., Garcia, J.-L., Patel, B.K.C., Cayot, J.-L., Baresi, L. and Mah, R.A. (2000) *Salinivibrio costicola* subsp *vallismortis* subsp nov., a halotolerant facultative anaerobe from Death Valley,

and an emended description of *Salinivibrio costicola*. *International Journal of Systematic and Evolutionary Microbiology* 50, 615–622.

372 Hugh, R. and Leifson, E. (1953) The taxonomic significance of fermentative versus oxidative metabolism of carbohydrates by various Gram negative bacteria. *Journal of Bacteriology* 66, 24–66.

373 Hughes, K., Jr, Duncan, R., Jr and Smith, S. (2002) Renomegaly associated with a mycobacterial infection in summer flounder *Paralichthys dentatus*. *Fish Pathology* 37, 83–86.

374 Humphrey, J.D., Lancaster, C., Gudkovs, N. and McDonald, W. (1986) Exotic bacterial pathogens *Edwardsiella tarda* and *Edwardsiella ictaluri* from imported ornamental fish *Betta splendens* and *Puntius conchonius*, respectively: isolation and quarantine significance. *Australian Veterinary Journal* 63, 369–371.

375 Humphrey, J., Lancaster, C., Gudkovs, N. and Copland, J. (1987) The disease status of Australian salmonids: bacteria and bacterial diseases. *Journal of Fish Diseases* 10, 403–410.

376 Humphry, D., George, A., Black, G. and Cummings, S. (2001) *Flavobacterium frigidarium* sp. nov., an aerobic, psychrophilic, xylanolytic and laminarinolytic bacterium from Antarctica. *International Journal of Systematic and Evolutionary Microbiology* 51, 1235–1243.

377 Huq, A., Alam, M., Parveen, S. and Colwell, R. (1992) Occurrence of resistance to vibriostatic compound 0/129 in *Vibrio cholerae* 01 isolated from clinical and environmental samples in Bangladesh. *Journal of Clinical Microbiology* 30, 219–221.

378 Huq, M.I., Alam, A.K.M.J., Brenner, D.J. and Morris, G.K. (1980) Isolation of *Vibrio*-like group, EF-6, from patients with diarrhea. *Journal of Clinical Microbiology* 11, 621–624.

379 Huys, G., Kämpfer, P., Altwegg, M., Coopman, R., Janssen, P., Gillis, M. and Kersters, K. (1997a) Inclusion of Aeromonas DNA hybridization group 11 in *Aeromonas encheleia* and extended description of the species *Aeromonas eucrenophila* and *A. encheleia*. *International Journal of Systematic Bacteriology* 47, 1157–1164.

380 Huys, G., Kämpfer, P., Altwegg, M., Kersters, I., Lamb, A., Coopman, R., Lüthy-Hottenstein, J., Vancanneyt, M., Janssen, P. and Kersters, K. (1997b) *Aeromonas popoffii* sp. nov., a mesophilic bacterium isolated from drinking water production plants and reservoirs. *International Journal of Systematic Bacteriology* 47, 1165–1171.

381 Huys, G., Kämpfer, P. and Swings, J. (2001) New DNA-DNA hybridization and phenotypic data on the species *Aeromonas ichthiosmia* and *Aeromonas allosaccharophila*: *A. ichthiosmia* Schubert *et al.* 1990 is a later synonym of *A. veronii*

Hickman-Brenner *et al.* 1987. *Systematic and Applied Microbiology* 24, 177–182.

382 Huys, G., Denys, R. and Swings, J. (2002a) DNA-DNA reassociation and phenotypic data indicate synonymy between *Aeromonas enteropelogenes* Schubert *et al.* 1990 and *Aeromonas trota* Carnahan *et al.* 1991. *International Journal of Systematic and Evolutionary Microbiology* 52, 1969–1972.

383 Huys, G., Kämpfer, P., Albert, M.J., Kühn, I., Denys, R. and Swings, J. (2002b) *Aeromonas hydrophila* subsp. *dhakensis* subsp. nov., isolated from children with diarrhoea in Bangladesh, and extended description of *Aeromonas hydrophila* subsp. *hydrophila* (Chester 1901) Stanier 1943 (Approved Lists 1980). *International Journal of Systematic and Evolutionary Microbiology* 52, 705–712.

384 Hwang, M.-N. and Ederer, G. (1975) Rapid hippurate hydrolysis method for presumptive identification of Group B streptococci. *Journal of Clinical Microbiology* 1, 114–115.

385 Iida, T., Sakata, C., Kawatsu, H. and Fukuda, Y. (1997) Atypical *Aeromonas salmonicida* infection in cultured marine fish. *Fish Pathology* 32, 65–66.

386 Innis, M.A. and Gelfand, D.H. (1990) Optimisation of PCRs. In: Innis, M.A., Gelfand, D.H., Sninsky, J.J. and White, T.J. (eds) *PCR Protocols: a Guide to Methods and Applications*. Academic Press, San Diego, pp. 3–12.

387 Ishiguro, E., Ainsworth, T., Trust, T. and Kay, W. (1985) Congo Red agar, a differential medium for *Aeromonas salmonicida*, detects the presence of the cell surface protein array involved in virulence. *Journal of Bacteriology* 164, 1233–1237.

388 Ishimaru, K., Akagawa-Matsushita, M. and Muroga, K. (1995) *Vibrio penaeicida* sp. nov., a pathogen of Kuruma Prawns (*Penaeus japonicus*). *International Journal of Systematic Bacteriology* 45, 134–138.

389 Ishimaru, K., Akagawa-Matsushita, M. and Muroga, K. (1996) *Vibrio ichthyoenteri* sp. nov., a pathogen of Japanese Flounder (*Paralichthys olivaceus*) larvae. *International Journal of Systematic Bacteriology* 46, 155–159.

390 Ishimaru, K. and Muroga, K. (1997) Taxonomic re-examination of two pathogenic *Vibrio* species isolated from milkfish and swimming crab. *Fish Pathology* 32, 59–64.

391 Isik, K., Chun, J., Hah, Y. and Goodfellow, M. (1999) *Nocardia salmonicida* nom. rev., a fish pathogen. *International Journal of Systematic Bacteriology* 49, 833–837.

392 Islam, M., Hasan, M., Miah, M., Yunus, M., Zaman, K. and Albert, M. (1994) Isolation of *Vibrio cholerae* 0139 synonym Bengal from the aquatic environment in Bangladesh: implications for

disease transmission. *Applied and Environmental Microbiology* 60, 1684–1686.

393 Itoh, H., Kuwata, G., Tateyama, S., Yamashita, K., Inoue, T., Kataoka, H., Ido, A., Ogata, K., Takasaki, M., Inoue, S., Tsubouchi, H. and Koono, M. (1999) *Aeromonas sobria* infection with severe soft tissue damage and segmental necrotizing gastroenteritis in a patient with alcoholic liver cirrhosis. *Pathology International* 49, 541–546.

394 Ivanova, E., Chun, J., Romanenko, L., Matte, M., Mikhailov, V., Frolova, G., Huq, A. and Colwell, R. (2000) Reclassification of *Alteromonas distincta* Romanenko *et al.* 1995 as *Pseudoalteromonas distincta* comb. nov. *International Journal of Systematic and Evolutionary Microbiology* 50, 141–144.

395 Ivanova, E., Kiprianova, E., Mikailov, V., Levanova, G., Garagulya, A., Gorshkova, N. and Yumoto, N. (1996) Characterisation and identification of marine *Alteromonas nigrifaciens* strains and emendation of the description. *International Journal of Systematic Bacteriology* 46, 223–228.

396 Ivanova, E., Kiprianova, E., Mikhailov, V., Levanova, G., Garagulya, A., Gorshkova, N., Vysotskii, M., Nicolau, D., Yumoto, N., Taguchi, T. and Yoshikawa, S. (1998) Phenotypic diversity of *Pseudoalteromonas citrea* from different marine habitats and emendation of the description. *International Journal of Systematic and Evolutionary Microbiology* 48, 247–256.

397 Ivanova, E.P., Sawabe, T., Gorshkova, N.M., Svetashev, V.I., Mikhailov, V.V., Nicolau, D.V. and Christen, R. (2001) *Shewanella japonica* sp. nov. *International Journal of Systematic and Evolutionary Microbiology* 51, 1027–1033.

398 Ivanova, E.P., Shevchenko, L.S., Sawabe, T., Lysenko, A.M., Svetashev, V.I., Gorshkova, N.M., Satomi, M., Christen, R. and Mikhailov, V.V. (2002) *Pseudoalteromonas maricaloris* sp. nov., isolated from an Australian sponge, and reclassification of [*Pseudoalteromonas aurantia*] NCIMB 2033 as *Pseudoalteromonas flavipulchra* sp. nov. *International Journal of Systematic and Evolutionary Microbiology* 52, 263–271.

399 Iveson, J. (1971) Strontium chloride B and E.E. enrichment broth media for the isolation of *Edwardsiella*, *Salmonella* and *Arizona* species from tiger snakes. *Journal of Hygiene* 69, 323–330.

400 Iwamoto, Y., Suzuki, Y., Kurita, A., Watanabe, Y., Shimizu, T., Ohgami, H. and Yanagihara, Y. (1995a) *Vibrio trachuri* sp. nov., a new species isolated from diseased Japanese horse mackerel. *Microbiology and Immunology* 39, 831–837.

401 Iwamoto, Y., Suzuki, Y., Kurita, A., Watanabe, Y., Shimizu, T., Ohgami, H. and Yanagihara, Y. (1995b) Rapid and sensitive PCR detection of *Vibrio trachuri* pathogenic to Japanese Horse

Mackerel (*Trachurus japonicus*). *Microbiology and Immunology* 39, 1003–1006.

402 Izard, D., Ferragut, C., Gavini, F., Kersters, K., Ley, J. D. and Leclerc, H. (1981) *Klebsiella terrigena*, a new species from soil and water. *International Journal of Systematic Bacteriology* 31, 116–127.

403 Izumikawa, K. and Ueki, N. (1997) Atypical *Aeromonas salmonicida* infection in cultured Schlegel's black rockfish. *Fish Pathology* 32, 67–68.

404 Jahans, K.L., Foster, G. and Broughton, E.S. (1997) The characterisation of *Brucella* strains isolated from marine mammals. *Veterinary Microbiology* 57, 373–382.

405 Janda, J.M. and Abbott, S.L. (2002) Bacterial identification for publication: when is enough enough? *Journal of Clinical Microbiology* 40, 1887–1891.

406 Janda, J.M., Abbott, S.L., Khashe, S., Kellogg, G.H. and Shimada, T. (1996) Further studies on biochemical characteristics and serologic properties of the genus *Aeromonas*. *Journal of Clinical Microbiology* 34, 1930–1933.

407 Jansen, G., Mooibroek, M., Idema, J., Harmsen, H., Welling, G. and Degener, J. (2000) Rapid identification of bacteria in blood cultures by using fluorescently labeled oligonucleotide probes. *Journal of Clinical Microbiology* 38, 814–817.

408 Jasmin, A. and Baucom, J. (1967) *Erysipelothrix insidiosa* infections in the Caiman (*Caiman crocodilus*) and the American crocodile (*Crocodilus acutus*). *American Journal of Veterinary Clinical Pathology* 1, 173–177.

409 Jensen, M., Tebo, B., Baumann, P., Mandel, M. and Nealson, K. (1980) Characterization of *Alteromonas hanedai* (sp. nov.), a nonfermentative luminous species of marine origin. *Current Microbiology* 3, 311–315.

410 Jiravanichpaisal, P., Miyazaki, T. and Limsuwan, C. (1994) Histopathology, biochemistry, and pathogenicity of *Vibrio harveyi* infecting black tiger prawn *Penaeus monodon*. *Journal of Aquatic Animal Health* 6, 27–35.

411 Jöborn, A., Olsson, J., Westerdahl, A., Conway, P. and Kjelleberg, S. (1997) Colonization in the fish intestinal tract and production of inhibitory substances in intestinal mucus and faecal extracts by *Carnobacterium* sp. strain K1. *Journal of Fish Diseases* 20, 383–392.

412 Jöborn, A., Dorsch, M., Olsson, J., Westerdahl, A. and Kjelleberg, S. (1999) *Carnobacterium inhibens* sp. nov., isolated from the intestine of Atlantic salmon (*Salmo salar*). *International Journal of Systematic Bacteriology* 49, 1891–1898.

413 Johnson, J.L. and Chilton, W.S. (1966) Galactosamine glycan of *Chondrococcus columnaris*. *Science* 152, 1247–1248.

414 Johnson, R., Colwell, R., Sakazaki, R. and Tamura, K. (1975) Numerical taxonomy study of

the *Enterobacteriaceae*. *International Journal of Systematic Bacteriology* 25, 12–37.

415 Jones, A. (1981) Effect of carbohydrate content of culture media on Kovac's oxidase test, with particular reference to *Vibrio* spp. *Medical Laboratory Sciences*, 38, 133–137.

416 Jones, M.W. and Cox, D.I. (1999) Clinical disease in seafarmed Atlantic salmon (*Salmo salar*) associated with a member of the family Pasteurellaceae – a case history. *Bulletin of the European Association of Fish Pathologists* 19, 75–78.

417 Joseph, S., Colwell, R. and Kaper, J. (1983) *Vibrio parahaemolyticus* and related halophilic vibrios. *Critical Reviews in Microbiology* 10, 77–124.

418 Kalina, G.P., Antonov, A.S., Turova, T.P. and Grafova, T.I. (1984) *Allomonas enterica* gen. nov., sp. nov.: deoxyribonucleic acid homology between *Allomonas* and some other members of the *Vibrionaceae*. *International Journal of Systematic Bacteriology* 34, 150–154.

419 Kaminski, G. and Suter, I. (1976) Human infection with *Dermatophilus congolensis*. *Medical Journal of Australia* 1, 443–447.

420 Kämpfer, P. and Altwegg, M. (1992) Numerical classification and identification of *Aeromonas* genospecies. *Journal of Applied Bacteriology* 72, 341–351.

421 Kanamoto, T., Sato, S. and Inoue, M. (2000) Genetic heterogeneities and phenotypic characteristics of strains of the genus *Abiotrophia* and proposal of *Abiotrophia para-adiacens* sp. nov. *Journal of Clinical Microbiology* 38, 492–498.

422 Kaneko, K.-I. and Hashimoto, N. (1982) Five biovars of *Yersinia enterocolitica* delineated by numerical taxonomy. *International Journal of Systematic Bacteriology* 32, 275–287.

423 Kapperud, G., Bergan, T. and Lassen, J. (1981) Numerical taxonomy of *Yersinia enterocolitica* and *Yersinia enterocolitica*-like bacteria. *International Journal of Systematic Bacteriology* 31, 401–419.

424 Kariya, T., Kubota, S., Nakamura, Y. and Kira, K. (1968) Nocardial infection in cultured yellowtails (*Seriola quinqueradiata* and *S. purpurascens*). I. Bacteriological study. *Fish Pathology* 3, 16–23.

425 Karunasagar, I., Karunasagar, I. and Pai, P. (1992) Systemic *Citrobacter freundii* infection in common carp, *Cyrinus carpio* L., fingerlings. *Journal of Fish Diseases* 15, 95–98.

426 Kasornchandra, J., Rogers, W. and Plumb, J. (1987) *Edwardsiella ictaluri* from walking catfish, *Clarias batrachus* L., in Thailand. *Journal of Fish Diseases* 10, 137–138.

427 Kaznowski, A. (1998) Identification of *Aeromonas* strains of different origin to the genomic species level. *Journal of Applied Microbiology* 84, 423–430.

428 Kent, M. (1982) Characteristics and identification of *Pasteurella* and *Vibrio* species pathogenic to fishes using API 20 E (Analytabs Products) Multitube test strips. *Canadian Journal of Fisheries and Aquatic Sciences* 39, 1725–1729.

429 Ketterer, P.J. and Eaves, L.E. (1992) Deaths in captive eels (*Anguilla reinhardtii*) due to *Photobacterium* (*Vibrio*) *damsela*. *Australian Veterinary Journal* 69, 203–204.

430 Keyes, M.C., Crews, F.W. and Ross, A.J. (1968) *Pasteurella multocida* isolated from a Californian Sea Lion (*Zalophus californianus*). *Journal of American Veterinary Medical Association* 153, 803–804.

431 Khan, A.A. and Cerniglia, C.E. (1997) Rapid and sensitive method for the detection of *Aeromonas caviae* and *Aeromonas trota* by polymerase chain reaction. *Letters in Applied Microbiology* 24, 233–239.

432 Khan, A.A., Nawaz, M.S., Khan, S.A. and Cerniglia, C.E. (1999) Identification of *Aeromonas trota* (hybridization group 13) by amplification of the aerolysin gene using polymerase chain reaction. *Molecular and Cellular Probes* 13, 93–98.

433 Khashe, S. and Janda, J.M. (1998) Biochemical and pathogenic properties of *Sherwanella alga* and *Sherwanella putrefaciens*. *Journal of Clinical Microbiology* 36, 783–787.

434 Kiiyukia, C., Nakajima, A., Nakai, T., Muroga, K., Kawakami, H. and Hashimoto, H. (1992) *Vibrio cholerae* non-01 isolated from Ayu fish (*Plecoglossus altivelis*) in Japan. *Applied and Environmental Microbiology* 58, 3078–3082.

435 Kim, J.-H., Lee, J.-K., Yoo, H.-S., Shin, N.-R., Shin, N.-S., Lee, K.-H. and Kim, D.-Y. (2002) Endocarditis associated with *Escherichia coli* in a sea lion (*Zalophus californianus*). *Journal of Veterinary Diagnostic Investigation* 14, 260–262.

436 Kim, Y., Okuda, J., Matsumoto, C., Takahashi, N., Hashimoto, S. and Nishibuchi, M. (1999) Identification of *Vibrio parahaemolyticus* strains at the species level by PCR targeted to the *toxR* gene. *Journal of Clinical Microbiology* 37, 1173–1177.

437 Kimura, B., Hokimoto, S., Takahashi, H. and Fujii, T. (2000) *Photobacterium histaminum* Okuzumi *et al.* 1994 is a later subjective synonym of *Photobacterium damselae* subsp *damselae* (Love *et al.* 1981) Smith *et al.* 1991. *International Journal of Systematic and Evolutionary Microbiology* 50, 1339–1342.

438 Kimura, T. (1969) A new subspecies of *Aeromonas salmonicida* as an etiological agent of furunculosis on 'sakuramasu' (*Oncorhynchus masou*) and pink salmon (*O. gorbuscha*) rearing for maturity. Part 1. On the morphological and physiological properties. *Fish Pathology* 3, 34–44.

439 Kirchhoff, H. and Rosengarten, R. (1984) Isolation of a motile mycoplasma from fish. *Journal of General Microbiology* 130, 2439–2445.

440 Kirchhoff, H., Beyene, P., Fischer, M., Flossdorf, J., Heitmann, J., Khattab, B., Lopatta, D., Rosengarten, R., Seidel, G. and Yousef, C. (1987) *Mycoplasma mobile* sp. nov., a new species from fish. *International Journal of Systematic Bacteriology* 37, 192–197.

441 Kirchhoff, H., Mohan, K., Schmidt, R., Runge, M., Brown, D.R., Brown, M.B., Foggin, C.M., Muvavarirwa, P., Lehmann, H. and Flossdorf, J. (1997) *Mycoplasma crocodyli* sp. nov., a new species from crocodiles. *International Journal of Systematic Bacteriology* 47, 742–746.

442 Kitao, T., Aoki, T. and Sakoh, R. (1981) Epizootic caused by β-haemolytic Streptococcus species in cultured freshwater fish. *Fish Pathology* 15, 301–307.

443 Klein, B., Kleingeld, D. and Bohm, K. (1993) First isolations of *Pleisiomonas shigelloides* from samples of cultured fish in Germany. *Bulletin of the European Association of Fish Pathologists* 13, 70–72.

444 Kloos, W.E. and Schleifer, K.H. (1975) Isolation and characterization of Staphylococci from human skin. II. Descriptions of four new species: *Staphylococcus warneri, Staphylococcus capitus, Staphylococcus hominus*, and *Staphylococcus simulans*. *International Journal of Systematic Bacteriology* 25, 62–79.

445 Kobayashi, T., Enomoto, S., Sakazaki, R. and Kuwahara, S. (1963) A new selective medium for pathogenic vibrios TCBS Agar (modified Nakanishi's Agar). *Japanese Journal of Bacteriology* 18, 387–391.

446 Koch, C., Schumann, P. and Stackebrandt, E. (1995) Reclassification of *Micrococcus agilis* (Ali-Cohen 1889) to the genus *Arthrobacter* as *Arthrobacter agilis* comb. nov. and emendation of the genus *Arthrobacter. International Journal of Systematic Bacteriology* 45, 837–839.

447 Kodama, H., Nakanishi, Y., Yamamoto, F., Mikami, T., Izawa, H., Imagawa, T., Hashimoto, Y. and Kudo, N. (1987) *Salmonella arizonae* isolated from a pirarucu, *Arapaima gigas* Cuvier, with septicaemia. *Journal of Fish Diseases* 10, 509–512.

448 Kong, R., Lee, S., Law, T., Law, S. and Wu, R. (2002) Rapid detection of six types of bacterial pathogen in marine waters by multiplex PCR. *Water Research* 36, 2802–2812.

449 Königsson, M., Pettersson, B. and Johansson, K.-E. (2001) Phylogeny of the seal mycoplasmas *Mycoplasma phocae* corrig., *Mycoplasma phocicerebrale* corrig. and *Mycoplasma phocirhinis* corrig. based on sequence analysis of 16S rDNA. *International Journal of Systematic and Evolutionary Microbiology* 51, 1389–1393.

450 Koppang, E.O., Fjølstad, M., Melgård, B., Vigerust, M. and Sørum, H. (2000) Non-pigmented-producing isolates of *Aeromonas salmonicida*
subspecies *salmonicida*: isolation, identification, transmission and pathogenicity in Atlantic salmon, *Salmo salar* L. *Journal of Fish Diseases* 23, 39–48.

451 Kovács, N. (1928) Eine vereinfachte Methode zum Nachweis der Indolbildung durch Bakterien. *Zeitschrift fur Immunitatsforschung-Immunobiology* 44, 311–315.

452 Kozińska, A., Figueras, M., Chacon, M. and Soler, L. (2002) Phenotypic characteristics and pathogenicity of *Aeromonas* genomospecies isolated from common carp (*Cyprinus carpio* L.). *Journal of Applied Microbiology* 93, 1034–1041.

453 Kraxberger-Beatty, T., McGarey, D., Grier, H. and Lim, D. (1990) *Vibrio harveyi*, an opportunistic pathogen of common snook, *Centropomus undecimalis* (Bloch), held in captivity. *Journal of Fish Diseases* 13, 557–560.

454 Krovacek, K., Huang, K., Sternberg, S. and Svenson, S.B. (1998) *Aeromonas hydrophila* septicaemia in a grey seal (*Halichoerus grypus*) from the Baltic Sea: a case study. *Comparative Immunology, Microbiology and Infectious Diseases* 21, 43–49.

455 Kudo, T., Hatai, K. and Seino, A. (1988) *Nocardia seriolae* sp. nov., causing Nocardiosis of cultured fish. *International Journal of Systematic Bacteriology* 38, 173–178.

456 Kuijper, E.J., Steigerwalt, A.G., Schoenmakers, B.S.C.I.M., Peeters, M.F., Zanen, H.C. and Brenner, D.J. (1989) Phenotypic characterisation and DNA relatedness in human fecal isolates of *Aeromonas* spp. *Journal of Clinical Microbiology* 27, 132–138.

457 Kurup, P. and Schmitt, J. (1973) Numerical taxonomy of *Nocardia. Canadian Journal of Microbiology* 19, 1035–1048.

458 Kushmaro, A., Banin, E., Loya, Y., Stackebrandt, E. and Rosenberg, E. (2001) *Vibrio shiloi* sp. nov., the causative agent of bleaching of the coral *Oculina patagonica. International Journal of Systematic and Evolutionary Microbiology* 51, 1383–1388.

459 Kusuda, R. and Yamaoka, M. (1972) Etiological studies on bacterial pseudotuberculosis in cultured yellowtail with *Pasteurella piscicida* as the causative agent. I. On the morphological and biochemical properties. *Bulletin of the Japanese Society of Scientific Fisheries* 38, 1325–1332.

460 Kusuda, R., Toyoshima, T. and Nishioka, J. (1974) Characteristics of a pathogenic *Pseudomonas* isolated from cultured crimson sea breams. *Fish Pathology* 9, 71–78.

461 Kusuda, R. and Toyoshima, T. (1976) Characteristics of a pathogenic *Pseudomonas* isolated from cultured yellowtail. *Fish Pathology* 11, 133–139.

462 Kusuda, R., Kawakami, K. and Kawai, K. (1987) A fish-pathogenic *Mycobacterium* sp. isolated from

an epizootic of cultured yellowtail. *Nippon Suisan Gakkaishi* 53, 1797–1904.

463 Kusuda, R., Yokoyama, J. and Kawai, K. (1986) Bacteriological study on cause of mass mortalities in cultured black sea bream fry. *Bulletin of the Japanese Society of Scientific Fisheries* 52, 1745–1751.

464 Kusuda, R., Kawai, K., Salati, F., Banner, C.R. and Fryer, J.L. (1991) *Enterococcus seriolicida* sp. nov., a fish pathogen. *International Journal of Systematic Bacteriology* 41, 406–409.

465 Kusuda, R., Dohata, N., Fukuda, Y. and Kawai, K. (1995) *Pseudomonas anguilliseptica* infection of Striped Jack. *Fish Pathology* 30, 121–122.

466 Lacoste, A., Jalabert, F., Malham, S., Cueff, A., Gélébart, F., Cordevant, C., Lange, M. and Poulet, S. (2001a) A *Vibrio splendidus* strain is associated with summer mortality of juvenile oysters *Crassostrea gigas* in the Bay of Morlaix (North Brittany, France). *Diseases of Aquatic Organisms* 46, 139–145.

467 Lacoste, A., Jalabert, F., Malham, S., Cueff, A. and Poulet, S. (2001b) Stress and stress-induced neuroendocrine changes increase the susceptibility of juvenile oysters (*Crassostrea gigas*) to *Vibrio splendidus*. *Applied and Environmental Microbiology* 67, 2304–2309.

468 Laidler, L., Treasurer, J., Grant, A. and Cox, D. (1999) Atypical *Aeromonas salmonicida* infection in wrasse (Labridae) used as cleaner fish of farmed Atlantic salmon, *Salmo salar* L., in Scotland. *Journal of Fish Diseases* 22, 209–213.

469 Lallier, R. and Higgins, R. (1988) Biochemical and toxigenic characteristics of *Aeromonas* spp. isolated from diseased mammals, moribund and healthy fish. *Veterinary Microbiology* 18, 63–71.

470 Lambert, C., Nicolas, J. L., Cilia, V. and Corre, S. (1998) *Vibrio pectenicida* sp. nov., a pathogen of scallops (*Pecten Maximus*) larvae. *International Journal of Systematic Bacteriology* 48, 481–487.

471 Lane, D. (1991) 16S/23S rRNA sequencing. In: Stackebrandt, E. and Goodfellow, M. (eds) *Nucleic Acid Techniques in Bacterial Systematics*. John Wiley & Sons, Chichester, pp. 115–147.

472 Lane, D., Pace, B., Olsen, G., Stahl, D., Sogin, M. and Pace, N. (1985) Rapid determination of 16S ribosomal RNA sequences for phylogenetic analyses. *Proceedings of the National Academy of Sciences USA* 82, 6955–6959.

473 Langdon, J. (1988) Fish diseases: refresher course for veterinarians. In: *Post-Graduate Committee in Veterinary Science, Proceedings No 106*. University of Sydney, 225–259.

474 Lansdell, W., Dixon, B., Smith, N. and Benjamin, L. (1993) Isolation of several *Mycobacterium* species from fish. *Journal of Aquatic Animal Health* 5, 73–76.

475 Larsen, J. and Pedersen, K. (1996) Atypical *Aeromonas salmonicida* isolated from diseased Turbot (*Scophthalmus maximus* L.). *Acta Veterinaria Scandinavica* 37, 139–146.

476 Laurent, F., Provost, F. and Boiron, P. (1999) Rapid identification of clinically relevant *Nocardia* species to genus level by 16S rRNA gene PCR. *Journal of Clinical Microbiology* 37, 99–102.

477 Lawson, P.A., Foster, G., Falsen, E., Ohlén, M. and Collins, M.D. (1999a) *Vagococcus lutrae* sp. nov., isolated from the common otter (*Lutra lutra*). *International Journal of Systematic Bacteriology* 49, 1251–1254.

478 Lawson, P.A., Foster, G., Falsen, E., Sjøden, B. and Collins, M.D. (1999b) *Abiotrophia balaenopterae* sp. nov., isolated from the minke whale (*Balaenoptera acutorostrata*). *International Journal of Systematic Bacteriology* 49, 503–506.

479 Lawson, P., Foster, G., Falsen, E., Ohlén, M. and Collins, M. (2000) *Atopobacter phocae* gen. nov., sp. nov., a novel bacterium isolated from common seals. *International Journal of Systematic and Evolutionary Microbiology* 50, 1755–1760.

480 Lawson, P.A., Falsen, E., Foster, G., Eriksson, E., Weiss, N. and Collins, M. (2001) *Arcanobacterium pluranimalium* sp. nov., isolated from porpoise and deer. *International Journal of Systematic and Evolutionary Microbiology* 51, 55–59.

481 Lee, C.-Y., Pan, S.-F. and Chen, C.-H. (1995) Sequence of a cloned pR72H fragment and its use for detection of *Vibrio parahaemolyticus* in shellfish with the PCR. *Applied and Environmental Microbiology* 61, 1311–1317.

482 Lee, J., Kim, J.S., Nahm, C.H., Choi, J.W., Kim, J., Pai, S.H., Moon, K.H., Lee, K. and Chong, Y. (1999) Two cases of *Chromobacterium violaceum* infection after injury in a subtropical region. *Journal of Clinical Microbiology* 37, 2068–2070.

483 Lee, J.V., Donovan, T.J. and Furniss, A.L. (1978) Characterization, taxonomy, and emended description of *Vibrio metschnikovii*. *International Journal of Systematic Bacteriology* 28, 99–111.

484 Lee, J.V., Hendrie, M.S. and Shewan, J.M. (1979) Identification of *Aeromonas, Vibrio* and related organisms In: Skinner, F.A. and Lovelock, D.W. (eds) *Identification Methods for Microbiologists*. The Society of Applied Bacteriology. Technical Series No. 14. Academic Press, London and New York, pp. 152–166.

485 Lee, J.V., Shread, P., Furniss, A.L. and Bryant, T. (1981) Taxonomy and description of *Vibrio fluvialis* sp. nov. (synonym Group F Vibrios, Group EF6). *Journal of Applied Bacteriology* 50, 73–94.

486 Lee, K.-H. and Ruby, E.G. (1995) Symbiotic role of the viable but nonculturable state of *Vibrio fischeri* in Hawaiian coastal seawater. *Applied and Environmental Microbiology* 61, 278–283.

487 Lee, S., Wang, H., Law, S., Wu, R. and Kong, R. (2002) Analysis of the 16S-23S rDNA intergenic spacers (IGSs) of marine vibrios for species-specific signature DNA sequences. *Marine Pollution Bulletin* 44, 412–420.

488 Lee, S.E., Kim, S.Y., Kim, S.J., Kim, H.S., Shin, J.H., Choi, S.H., Chung, S.S. and Rhee, J.H. (1998) Direct identification of *Vibrio vulnificus* in clinical specimens by nested PCR. *Journal of Clinical Microbiology* 36, 2887–2892.

489 Leifson, H. (1963) Determination of carbohydrate metabolism of marine bacteria. *Journal of Bacteriology* 85, 1183–1184.

490 LeJeune, J. and Rurangirwa, F. (2000) Polymerase chain reaction for definitive identification of *Yersinia ruckeri*. *Journal of Veterinary Investigation* 12, 558–561.

491 Leon, G., Maulen, N., Figueroa, J., Villaneuva, J., Rodriguez, C., Vera, M. and Krauskopf, M. (1994) A PCR-based assay for the identification of the fish pathogen *Renibacterium salmoninarum*. *FEMS Microbiology Letters* 115, 131–136.

492 Leonardo, M.R., Moser, D.P., Barbieri, E., Brantner, C.A., MacGregor, B.J., Paster, B.J., Stackebrandt, E. and Nealson, K.H. (1999) *Shewanella pealeana* sp. nov., a member of the microbial community associated with the accessory nidamental gland of the squid *Loligo pealei*. *International Journal of Systematic Bacteriology* 49, 1341–1351.

493 Lewin, R. and Lounsbery, D. (1969) Isolation, cultivation and characterization of *Flexibacteria*. *Journal of General Microbiology* 58, 145–170.

494 Lewin, R.A. (1974) *Flexibacter polymorphus*, a new marine species. *Journal of General Microbiology* 82, 393–403.

495 Lightner, D.V. and Redman, R.M. (1998) Shrimp diseases and current diagnostic methods. *Aquaculture* 164, 201–220.

496 Lincoln, S.P., Fermor, T.R. and Tindall, B.J. (1999) *Janthinobacterium agaricidamnosum* sp. nov., a soft rod pathogen of *Agaricus bisporus*. *International Journal of Systematic Bacteriology* 49, 1577–1589.

497 Lio-Po, G.D., Albright, L.J., Michel, C. and Leaño, E.M. (1998) Experimental induction of lesions in snakeheads (*Ophicephalus striatus*) and catfish (*Clarias batrachus*) with *Aeromonas hydrophila*, *Aquaspirillum* sp., *Pseudomonas* sp. and *Streptococcus* sp. *Journal of Applied Ichthyology* 14, 75–79.

498 Liston, J. (1957) The occurrence and distribution of bacterial types on flatfish. *Journal of General Microbiology* 16, 205.

499 Liu, P.-C., Chen, Y.-C., Huang, C.-Y. and Lee, K.-K. (2000) Virulence of *Vibrio parahaemolyticus* isolated from cultured small abalone, *Haliotis diversicolor supertexta*, with withering syndrome. *Letters in Applied Microbiology* 31, 433–437.

500 Llewellyn, L.C. (1980) A bacterium with similarities to the redmouth bacterium and *Serratia liquefaciens* (Grimes and Hennerty) causing mortalities in hatchery reared salmonids in Australia. *Journal of Fish Diseases* 3, 29–39.

501 Lloyd, J. (1985) *Estimation of Amount of Carbondioxide Produced by Two Incubation Methods*. Department of Agriculture, Western Australia.

502 Logan, N. (1989) Numerical taxonomy of violet-pigmented, gram-negative bacteria and description of *Iodobacter fluviatile* gen. nov., comb. nov. *International Journal of Systematic Bacteriology* 39, 450–456.

503 Lönnström, L., Wiklund, T. and Bylund, G. (1994) *Pseudomonas anguilliseptica* isolated from Baltic herring *Clupea harengus membras* with eye lesions. *Diseases of Aquatic Organisms* 18, 143–147.

504 Love, M., Teebken-Fisher, D., Hose, J.E., Farmer, J.J. III, Hickman, F.W. and Fanning, G.R. (1981) *Vibrio damsela*, a marine bacterium, causes skin ulcers on the Damselfish *Chromis punctipinnis*. *Science* 214, 1139–1140.

505 Lowe, G.H. (1962) The rapid detection of lactose fermentation in paracolon organisms by the demonstration of β-D-galactosidase. *Journal of Medical Laboratory Technology* 19, 21–25.

506 Lunder, T., Sørum, H., Holstad, G., Steigerwalt, A., Mowinckel, P. and Brenner, D. (2000) Phenotypic and genotypic characterization of *Vibrio viscosus* sp. nov. and *Vibrio wodanis* sp. nov. isolated from Atlantic Salmon (*Salmo salar*) with 'winter ulcer'. *International Journal of Systematic and Evolutionary Microbiology* 50, 427–450.

507 Lupiani, B., Baya, A.M., Magariños, B., Romalde, J.L., Li, T., Roberson, B.S., Hetrick, F.M. and Toranzo, A.E. (1993) *Vibrio mimicus* and *Vibrio cholerae* non-01 isolated from wild and hatchery-reared fish. *Gyobyo Kenkyu* 28, 15–26.

508 MacDonell, M.T. and Colwell, R.R. (1985) Phylogeny of the Vibrionaceae, and recommendation for two new genera, *Listonella* and *Shewanella*. *Systematic and Applied Microbiology* 6, 171–182.

509 MacDonell, M.T., Singleton, F.L. and Hood, M.A. (1982) Diluent composition for use of API 20 E in characterising marine and estuarine bacteria. *Applied and Environmental Microbiology* 44, 423–427.

510 MacFaddin, J.F. (1980) *Biochemical Tests for Identification of Medical Bacteria*. Williams & Wilkins, Baltimore, Maryland.

511 Macián, M., Garay, E. and Pujalte, M. (1996) The arginine dihydrolase (ADH) system in the identification of some marine *Vibrio* species. *Systematic and Applied Microbiology* 19, 451–456.

512 Macián, M.C., Ludwig, W., Schleifer, K.-H., Garay, E. and Pujalte, M. (2000) *Vibrio pelagius*: differences of the Type strain deposited at various culture collections. *Systematic and Applied Microbiology* 23, 373–375.

513 Macián, M.C., Ludwig, W., Aznar, R., Grimont, P.A.D., Schleifer, K.H., Garay, E. and Pujalte, M.J. (2001a) *Vibrio lentus* sp. nov., isolated from Mediterranean oysters. *International Journal of Systematic and Evolutionary Microbiology* 51, 1449–1456.

514 Macián, M.C., Ludwig, W., Schleifer, K., Pujalte, M. and Garay, E. (2001b) *Vibrio agarivorans* sp. nov., a novel agarolytic marine bacterium. *International Journal of Systematic and Evolutionary Microbiology* 51, 2031–2036.

515 MacKenzie, K. (1988) Presumptive mycobacteriosis in North-east Atlantic mackerel, *Scomber scombrus*. *Journal of Fish Biology* 32, 263–275.

516 MacKnight, K., Chow, D., See, B. and Vedros, N. (1990) Melioidosis in a macaroni penguin *Eudyptes chrysolophus*. *Diseases of Aquatic Organisms* 9, 105–107.

517 MacLeod, R.A. (1968) On the role of inorganic ions in the physiology of marine bacteria. *Advances in Microbiology of the Sea* 1, 95–126.

518 Magariños, B., Romalde, J., Bandín, I., Fouz, B. and Toranzo, A. (1992) Phenotypic, antigenic, and molecular characterization of *Pasteurella piscicida* strains isolated from fish. *Applied and Environmental Microbiology* 58, 3316–3322.

519 Maher, M., Palmer, R., Gannon, F. and Smith, T. (1995) Relationship of a novel bacterial fish pathogen to *Streptobacillus moniliformis* and the Fusobacteria group, based on 16S ribosomal RNA analysis. *Systematic and Applied Microbiology* 18, 79–84.

520 Mainster, M.E., Lynd, F.T., Cragg, P.C. and Karger, J. (1973) Treatment of multiple cases of *Pasteurella multocida* and staphylococcal pneumonia in *Alligator mississippiensis* on a herd basis. In: *Annual Proceedings of the American Association of Zoo Veterinarians*, Houston, pp. 34–36.

521 Makemson, J., Fulayfil, N., Landry, W., Ert, L.V., Wimpee, C., Widder, E. and Case, J. (1997) *Shewanella woodyi* sp. nov., an exclusively respiratory luminous bacterium isolated from the Alboran Sea. *International Journal of Systematic Bacteriology* 47, 1034–1039.

522 Manefield, M., Harris, L., Rice, S., Nys, R. de and Kjelleberg, S. (2000) Inhibition of luminescence and virulence in the black tiger prawn (*Penaeus monodon*) pathogen *Vibrio harveyi* by intercellular signal antagonists. *Applied and Environmental Microbiology* 66, 2079–2084.

523 Marchesi, J., Sato, T., Weightman, A., Martin, T., Fry, J., Hiom, S. and Wade, W. (1998) Design and evaluation of useful bacterium-specific PCR primers that amplify genes coding for bacterial 16S rRNA. *Applied and Environmental Microbiology* 64, 795–799.

524 Markwardt, N., Gocha, Y. and Klontz, G. (1989) A new application for Coomassie brilliant blue agar; detection of *Aeromonas salmonicida* in clinical samples. *Diseases of Aquatic Organisms* 6, 231–233.

525 Marshall, B. and Warren, J. (1984) Unidentified curved bacilli in the stomach of patients with gastritis and peptic ulceration. *Lancet* i, 1311–1315.

526 Martinez-Murcia, A. (1999) Phylogenetic positions of *Aeromonas encheleia*, *Aeromonas popoffii*, Aeromonas DNA hybridization Group 11 and Aeromonas hybridization Group 501. *International Journal of Systematic Bacteriology* 49, 1403–1408.

527 Martinez-Murcia, A., Esteve, C., Garay, E. and Collins, M. (1992) *Aeromonas allosaccharophila* sp. nov., a new mesophilic member of the genus *Aeromonas*. *FEMS Microbiology Letters* 91, 199–206.

528 Massad, G. and Oliver, J.D. (1987) New selective and differential medium for *Vibrio cholerae* and *Vibrio vulnificus*. *Applied and Environmental Microbiology* 53, 2262–2264.

529 Masters, A., Ellis, T., Carson, J., Sutherland, S. and Gregory, A. (1995) *Dermatophilus chelonae* sp. nov., isolated from Chelonids in Australia. *International Journal of Systematic Bacteriology* 45, 50–56.

530 Mauel, M., Miller, D., Frazier, K. and Hines II, M. (2002) Bacterial pathogens isolated from cultured bullfrogs (*Rana castesbeiana*). *Journal of Veterinary Diagnostic Investigation* 14, 431–433.

531 Mawdesley-Thomas, L.E. (1969) Furunculosis in goldfish. *Journal of Fish Biology* 1, 19–23.

532 McCammon, S., Innes, B., Bowman, J., Franzmann, P., Dobson, S., Holloway, P., Skerratt, J., Nichols, P. and Rankin, L. (1998) *Flavobacterium hibernum* sp. nov., a lactose-utilizing bacterium from a freshwater Antarctic lake. *International Journal of Systematic Bacteriology* 48, 1405–1412.

533 McCammon, S.A. and Bowman, J.P. (2000) Taxonomy of Antarctic *Flavobacterium* species: description of *Flavobacterium gillisiae* sp. nov., *Flavobacterium tegetincola* sp. nov., and *Flavobacterium xanthum* sp. nov., nom. rev. and reclassification of [*Flavobacterium*] *salegens* as *Salegentibacter salegens* gen. nov., com. nov. *International Journal of Systematic and Evolutionary Microbiology* 50, 1055–1063.

534 McCarthy, D.H. (1975) Fish furunculosis caused by *Aeromonas salmonicida* var. *achromogenes*. *Wildlife Diseases* 11, 489–493.

535 McCarthy, D.H. (1977) The identification and significance of atypical strains of *Aeromonas*

salmonicida. Bulletin of the International Office of Epizootics 87, 459–463.

536 McCarthy, D. and Johnson, K. (1982) A serotypic survey and cross-protection test of North American field isolates of *Yersinia ruckeri. Journal of Fish Diseases* 5, 323–328.

537 McCurdy, H.D. (1969) Study on the taxonomy of Myxobacterales. I. Record of Canadian isolates and survey of methods. *Canadian Journal of Microbiology* 15, 1453–1461.

538 McIntosh, S. and Austin, B. (1990) Recovery of an extremely proteolytic form of *Serratia liquefaciens* as a pathogen of Atlantic salmon, *Salmo salar*, in Scotland. *Journal of Fish Biology* 36, 765–772.

539 McVicar, A. and White, P. (1979) Fin and skin necrosis of cultivated Dover sole *Solea solea* (L.). *Journal of Fish Diseases* 2, 557–562.

540 Mendes, E., Queiroz, D., Dewhirst, F., Paster, B., Moura, S. and Fox, J. (1996) *Helicobacter trogontum* sp. nov., isolated from the rat intestine. *International Journal of Systematic Bacteriology* 46, 916–921.

541 Michel, C., Bernardet, J.-F. and Dinand, D. (1992) Phenotypic and genotypic studies of *Pseudomonas anguilliseptica* strains isolated from farmed European eels (*Anguilla anguilla*) in France. *Gyobyo Kenkyu* 27, 229–232.

542 Michel, C., Nougayrède, P., Eldar, A., Sochon, E. and Kinkelin, P. de. (1997) *Vagococcus salmoninarum*, a bacterium of pathological significance in rainbow trout *Oncorhynchus mykiss* farming. *Diseases of Aquatic Organisms* 30, 199–208.

543 Michel, C., Messiaen, S. and Bernardet, J.-F. (2002) Muscle infections in imported neon tetra, *Paracheirodon innesi* Myers: limited occurrence of microsporidia and predominance of severe forms of columnaris disease caused by an Asian genomovar of *Flavobacterium columnare. Journal of Fish Diseases* 25, 253–263.

544 Middlebrook, G., Cohn, M.L., Dye, W.E., Russell, W.F. and Levy, D. (1960) Microbiologic procedures of value in tuberculosis. *Acta Tuberculosea Scandinavica* 38, 66–81.

545 Mills, C. and Gherna, R. (1987) Hydrolysis of indoxyl acetate by Campylobacter species. *Journal of Clinical Microbiology* 25, 1560–1561.

546 Miriam, A., Griffiths, S., Lovely, J. and Lynch, W. (1997) PCR and Probe-PCR assays to monitor broodstock Atlantic Salmon (*Salmo salar* L.) ovarian fluid and kidney tissue for presence of DNA of the fish pathogen *Renibacterium salmoninarum. Journal of Clinical Microbiology* 35, 1322–1326.

547 Mitchell, A. and Goodwin, A. (2000) The isolation of *Edwardsiella ictaluri* with a limited tolerance for aerobic growth from Channel Catfish. *Journal of Aquatic Animal Health* 12, 297–300.

548 Miyashita, T. (1984) *Pseudomonas fluorescens* and *Edwardsiella tarda* isolated from diseased tilapia. *Fish Pathology* 19, 45–50.

549 Miyata, M., Inglis, V. and Aoki, T. (1996) Rapid identification of *Aeromonas salmonicida* subspecies *salmonicida* by polymerase chain reaction. *Aquaculture* 141, 13–24.

550 Miyazaki, T., Kubota, S., Kaige, N. and Miyashita, T. (1984) A histopathological study of Streptococcal disease in Tilapia. *Fish Pathology* 19, 167–172.

551 Mohney, L., Poulos, B., Brooker, J., Cage, G. and Lightner, D. (1998) Isolation and identification of *Mycobacterium peregrinum* from the Pacific White Shrimp *Penaeus vannamei. Journal of Aquatic Animal Health* 10, 83–88.

552 Molitoris, E., Marii, M.A., Joseph, S.W., Krichevsky, M.I., Fanning, G.R., Last, G., El-Mishad, A.M., Batawi, Y.A.E. and Colwell, R.R. (1989) Numerical taxonomy and deoxyribonucleic acid relatedness of environmental and clinical *Vibrio* species isolated in Indonesia. *International Journal of Systematic Bacteriology* 39, 442–449.

553 Møller, V. (1955) Simplified test for some amino acid decarboxylases and for the arginine dihydrolase system. *Acta Pathologica et Microbiologica Scandinavica* 36, 158–172.

554 Moreno, E., Cloeckaert, A. and Moriyón, I. (2002) *Brucella* evolution and taxonomy. *Veterinary Microbiology* 90, 209–227.

555 Morris, J.G.J., Wilson, R., Hollis, D., Weaver, R., Miller, H., Tacket, C., Hickman, F. and Blake, P. (1982) Illness caused by *Vibrio damsela* and *Vibrio hollisae. Lancet* June 5, 1294–1297.

556 Mudarris, M. and Austin, B. (1989) Systemic disease in turbot *Scophthalmus maximus* caused by a previously unrecognized *Cytophaga*-like bacterium. *Diseases of Aquatic Organisms* 6, 161–166.

557 Mudarris, M., Austin, B., Segers, P., Vancanneyt, M., Hoste, B. and Bernardet, J. F. (1994) *Flavobacterium scophthalmum* sp. nov., a pathogen of turbot (*Scophthalmus maximus* L.). *International Journal of Systematic Bacteriology* 44, 447–453.

558 Müller, H., Fanning, G.R. and Brenner, D.J. (1995) Isolation of *Serratia fonticola* from Mollusks. *Systematic and Applied Microbiology* 18, 279–284.

559 Müller, H.E. (1983) *Providencia friedericiana*, a new species isolated from Penguins. *International Journal of Systematic Bacteriology* 33, 709–715.

560 Mullis, K.B. and Faloona, F.A. (1987) Specific synthesis of DNA *in vitro* via a polymerase-catalyzed chain reaction. In: Wu, R. (ed.) *Methods in Enzymology: Recombinant DNA*, Vol. 155. Academic Press, New York, pp. 335–350.

561 Muroga, K., Yamanoi, H., Hironaka, Y., Yamamoto, S., Tatani, M., Jo, Y., Takahashi, S. and Hanada, H. (1984) Detection of *Vibrio anguillarum* from wild fingerlings of ayu

Plecoglossus altivelis. Bulletin of the Japanese Society of Scientific Fisheries 50, 591–596.

562 Mutters, R., Ihm, P., Pohl, S., Frederiksen, W. and Mannheim, W. (1985) Reclassification of the genus Pasteurella Trevisan 1887 on the basis of deoxyribonucleic acid homology, with proposals for the new species Pasteurella dagmatis, Pasteurella canis, Pasteurella stomatis, Pasteurella anatis, and Pasteurella langaa. International Journal of Systematic Bacteriology 35, 309–322.

563 Myhr, E., Larsen, J., Lillehaug, A., Gudding, R., Heum, M. and Håstein, T. (1991) Characterization of Vibrio anguillarum and closely related species isolated from farmed fish in Norway. Applied and Environmental Microbiology 57, 2750–2757.

564 Nagai, T. and Iida, Y. (2002) Occurrence of bacterial kidney disease in cultured Ayu. Fish Pathology 37, 77–81.

565 Nair, G.B. and Holmes, B. (1999) International Committee on the Systematic Bacteriology Subcommittee on the Taxonomy of Vibrionaceae. International Journal of Systematic Bacteriology 49, 1945–1947.

566 Nakagawa, Y., Sakane, T. and Yokota, A. (1996) Emendation of the genus Planococcus and transfer of Flavobacterium okeanokoites Zobell and Upham 1944 to the genus Planococcus as Planococcus okeanokoites comb. nov. International Journal of Systematic Bacteriology 46, 866–870.

567 Nakai, T., Fujiie, N., Muroga, K., Arimoto, M., Mizuta, Y. and Matsuoka, S. (1992) Pasteurella piscicida infection in hatchery-reared juvenile striped jack. Gyobyo Kenkyu 27, 103–108.

568 Nakai, T., Hanada, H. and Muroga, K. (1985) First records of Pseudomonas anguilliseptica infection in cultured ayu, Plecoglossus altivelis. Fish Pathology 20, 481–484.

569 Nakajima, K., Muroga, K. and Hancock, R.E.W. (1983) Comparison of fatty acid, protein, and serological properties distinguishing outer membranes of Pseudomonas anguilliseptica strains from those of fish pathogens and other pseudomonads. International Journal of Systematic Bacteriology 33, 1–8.

570 Nakatsugawa, T. (1983) A streptococcal disease of cultured flounder. Fish Pathology 17, 281–285.

571 Nealson, K.H. (1978) Isolation, identification, and manipulation of luminous bacteria. Methods in Enzymology 57, 153–156.

572 Nelson, E.J. and Ghiorse, W.C. (1999) Isolation and identification of Pseudoalteromonas piscicida strain Cura-d associated with diseased damselfish (Pomacentridae) eggs. Journal of Fish Diseases 22, 253–260.

573 Nesterenko, O.A., Nogina, T.M., Kasumova, S.A., Kvasnikov, E.I. and Batrakov, S.G. (1982) Rhodococcus luteus nom. nov. and Rhodococcus maris nom. nov. International Journal of Systematic Bacteriology 32, 1–14.

574 Nguyen, H. and Kanai, K. (1999) Selective agars for the isolation of Streptococcus iniae from Japanese flounder, Paralichthys olivaceus, and its cultural environment. Journal of Applied Microbiology 86, 769–776.

575 Nicholls, K.M., Lee, J.V. and Donovan, T.J. (1976) An evaluation of commercial thiosulphate citrate bile salt sucrose agar (TCBS). Journal of Applied Bacteriology 41, 265–269.

576 Nicolas, J.L., Basuyaux, O., Mazurié, J. and Thébault, A. (2002) Vibrio carchariae, a pathogen of the abalone Haliotis tuberculata. Diseases of Aquatic Organisms 50, 35–43.

577 Nicols, D.S., Hart, P., Nicols, P.D. and McMeekin, T.A. (1996) Enrichment of the rotifer Brachionus plicatilis fed an Antarctic bacterium containing polyunsaturated fatty acids. Aquaculture 147, 115–125.

578 Nielsen, M.E., Høi, L., Schmidt, A., Qian, D., Shimada, T., Shen, J. and Larsen, J. (2001) Is Aeromonas hydrophila the dominant motile Aeromonas species that causes disease outbreaks in aquaculture production in the Zhejiang Province in China? Diseases of Aquatic Organisms 46, 23–29.

579 Nieto, T., López, L., Santos, Y., Núñez, S. and Toranzo, A. (1990) Isolation of Serratia plymuthica as an opportunistic pathogen in rainbow trout, Salmo gairdneri Richardson. Journal of Fish Diseases 13, 175–177.

580 Nishibuchi, M., Doke, S., Toizumi, S., Umeda, T., Yoh, M. and Miwatani, T. (1988) Isolation from a coastal fish of Vibrio hollisae capable of producing a hemolysin similar to the thermostable direct hemolysin of Vibrio parahaemolyticus. Applied and Environmental Microbiology 54, 2144–2146.

581 Nishimori, E., Hasegawa, O., Numata, T. and Wakabayashi, H. (1998) Vibrio carchariae causes mass mortalities in Japanese abalone, Sulculus diversicolor supratexta. Fish Pathology 33, 495–502.

582 Nishimori, E., Kita-Tsukamoto, K. and Wakabayashi, H. (2000) Pseudomonas plecoglossicida sp. nov., the causative agent of bacterial haemorrhagic ascites in ayu, Plecoglossus altivelis. International Journal of Systematic and Evolutionary Microbiology 50, 83–89.

583 Nishimura, Y., Kinpara, M. and Iizuka, H. (1989) Mesophilobacter marinus gen. nov., sp. nov.: an aerobic coccobacillus isolated from seawater. International Journal of Systematic Bacteriology 39, 378–381.

584 Noga, E. and Berkhoff, H. (1990) Pathological and microbiological features of Aeromonas salmonicida infection in the American eel (Anguilla rostrata). Fish Pathology 25, 127–132.

585 Nogi, Y., Kato, C. and Horikoshi, K. (1998) *Moritella japonica* sp. nov., a novel barophilic bacterium isolated from a Japan Trench sediment. *Journal of General and Applied Microbiology* 44, 289–295.

586 Nogi, Y., Masui, N. and Kato, C. (1998) *Photobacterium profundum* sp. nov., a new, moderately barophilic species isolated from a deep-sea sediment. *Extremophiles* 2, 1–7.

587 Novoa, B., Luque, A., Castro, D., Borrego, J. and Figueras, A. (1998) Characterization and infectivity of four bacterial strains isolated from Brown Ring Disease-affected clams. *Journal of Invertebrate Pathology* 71, 34–41.

588 Nozue, H., Hayashi, T., Hashimoto, Y., Ezaki, T., Hamasaki, K., Ohwada, K. and Terawaki, Y. (1992) Isolation and characterisation of *Shewanella alga* from human clinical specimens and emendation of the description of *S. alga* Simidu *et al.*, 1990, 335. *International Journal of Systematic Bacteriology* 42, 628–634.

589 Oakey, H., Gibson, L. and George, A. (1999) DNA probes specific for *Aeromonas hydrophila* (HG1). *Journal of Applied Microbiology* 86, 187–193.

590 Obendorf, D.L., Carson, J. and McManus, T.J. (1987) *Vibrio damselae* infection in a stranded leatherback turtle (*Dermochelys coriacea*). *Journal of Wildlife Diseases* 23, 666–668.

591 Odile, M., Bouvet, M., Grimont, P., Richard, C., Aldova, E., Hausner, O. and Gabrhelova, M. (1985) *Budvicia aquatica* gen. nov., sp. nov.: a hydrogen sulfide-producing member of the Enterobacteriaceae. *International Journal of Systematic Bacteriology* 35, 60–64.

592 OIE (2000a) *World Animal Health in 2000: Reports on the Animal Health Status and Disease Control Methods and Tables on Incidence of List A Diseases.* Office International des Epizooties, Paris, pp. 33–37.

593 OIE (2000b) Enteric septicaemia of catfish. In: *Diagnostic Manual for Aquatic Animal Diseases.* Office International Des Epizooties, Paris, pp. 105–111.

594 Okuda, J., Nakai, T., Chang, P., Oh, T., Nishino, T., Koitabashi, T. and Nishibuchi, M. (2001) The *toxR* gene of *Vibrio* (*Listonella*) *anguillarum* controls expression of the major outer membrane proteins but not virulence in the natural host model. *Infection and Immunity* 69, 6091–6101.

595 Okuzumi, M., Hiraishi, A., Kobayashi, T. and Fujii, T. (1994) *Photobacterium histaminum* sp. nov., a histamine-producing marine bacterium. *International Journal of Systematic Bacteriology* 44, 631–636.

596 Olafsen, J., Mikkelsen, H., Giaever, H. and Hansen, G. (1993) Indigenous bacteria in haemolymph and tissues of marine bivalves at low temperatures. *Applied and Environmental Microbiology* 59, 1848–1854.

597 Oliver, J., Warner, R. and Cleland, D. (1983) Distribution of *Vibrio vulnificus* and other lactose-fermenting Vibrios in the marine environment. *Applied and Environmental Microbiology* 45, 985–998.

598 Olivier, G. (1990) Virulence of *Aeromonas salmonicida*: lack of relationship with phenotypic characteristics. *Journal of Aquatic Animal Health* 2, 119–127.

599 Onarheim, A., Wiik, R., Burghardt, J. and Stackebrandt, E. (1994) Characterisation and identification of two *Vibrio* species indigenous to the intestine of fish in cold sea water; description of *Vibrio iliopiscarius* sp. nov. *Systematic and Applied Microbiology* 17, 370–379.

600 Osorio, C.R., Barja, J.L., Hutson, R.A. and Collins, M.D. (1999) *Arthrobacter rhombi* sp. nov., isolated from Greenland halibut (*Reinhardtius hippoglossoides*). *International Journal of Systematic Bacteriology* 49, 1217–1220.

601 Osorio, C.R., Toranzo, A.E., Romalde, J.L. and Barja, J.L. (2000) Multiplex PCR assay for *ureC* and 16S rRNA genes clearly discriminates between both subspecies of *Photobacterium damselae*. *Diseases of Aquatic Organisms* 40, 177–183.

602 Ostland, V.E., Ferguson, H.W. and Stevenson, R.M.W. (1989) Case report: bacterial gill disease in goldfish *Carassius auratus*. *Diseases of Aquatic Organisms* 6, 179–184.

603 Ostland, V.E., Lumsden, J.S., MacPhee, D.D. and Ferguson, H.W. (1994) Characteristics of *Flavobacterium branchiophilum*, the cause of salmonid Bacterial Gill Disease in Ontario. *Journal of Aquatic Animal Health* 6, 13–26.

604 Ostland, V.E., Byrne, P.J., Lumsden, J.S., MacPhee, D.D., Derksen, J.A., Haulena, M., Skar, K., Myhr, E. and Ferguson, H.W. (1999a) Atypical bacterial gill disease: a new form of bacterial gill disease affecting intensively reared salmonids. *Journal of Fish Diseases* 22, 351–358.

605 Ostland, V.E., LaTrace, C., Morrison, D. and Ferguson, H. (1999b) *Flexibacter maritimus* associated with a bacterial stomatitis in Atlantic salmon smolts reared in net-pens in British Columbia. *Journal of Aquatic Animal Health* 11, 35–44.

606 Otis, V.S. and Behler, J.L. (1973) The occurrence of Salmonellae and *Edwardsiella* in the turtles of the New York Zoological Park. *Journal of Wildlife Diseases* 9, 4–6.

607 Packer, R.A. (1943) The use of sodium azide (NaN$_3$) and crystal violet in a selective medium for Streptococci and *Erysipelothrix rhusiopathiae*. *Journal of Bacteriology* 46, 343–349.

608 Padgitt, P.J. and Moshier, S.E. (1987) *Mycobacterium poriferae* sp. nov. a scotochromogenic, rapidly growing species isolated from a marine

sponge. *International Journal of Systematic Bacteriology* 37, 186–191.

609 Paillard, C. and Maes, P. (1994) Brown ring disease in the Manila clam *Ruditapes philippinarum*: establishment of a classification system. *Diseases of Aquatic Organisms* 19, 137–146.

610 Paillard, C., Maes, P. and Oubella, R. (1994) Brown ring disease in clams. *Annual Review of Fish Diseases* 4, 219–240.

611 Palmer, R., Drinan, E. and Murphy, T. (1994) A previously unknown disease of farmed Atlantic salmon: pathology and establishment of bacterial aetiology. *Diseases of Aquatic Organisms* 19, 7–14.

612 Palmgren, H., McCafferty, D., Aspan, A., Broman, T., Sellin, M., Wollin, R., Bergstrom, S. and Olsen, B. (2000) Salmonella in sub-Antarctic: low heterogeneity in Salmonella serotypes in South Georgian seals and birds. *Epidemiology and Infection* 125, 257–262.

613 Pascual, C., Foster, G., Alvarez, N. and Collins, M.D. (1998) *Corynebacterium phocae* sp. nov., isolated from the common seal (*Phoca vitulina*). *International Journal of Systematic Bacteriology* 48, 601–604.

614 Pascuale, V., Baloda, S., Dumontet, S. and Krovacek, K. (1994) An outbreak of *Aeromonas hydrophila* infection in turtles (*Pseudemis scripta*). *Applied and Environmental Microbiology* 60, 1678–1680.

615 Pavan, M.E., Abbott, S.L., Zorzopulos, J. and Janda, J.M. (2000) *Aeromonas salmonicida* subsp. *pectinolytica* subsp. nov., a new pectinase-positive subspecies isolated from a heavily polluted river. *International Journal of Systematic and Evolutionary Microbiology* 50, 119–1124.

616 Pazos, F., Santos, Y., Macías, A.R., Núñez, S. and Toranzo, A.E. (1996) Evaluation of media for the successful culture of *Flexibacter maritimus*. *Journal of Fish Diseases* 19, 193–197.

617 Pedersen, K., Kofod, H., Dalsgaard, I. and Larsen, J. (1994) Isolation of oxidase-negative *Aeromonas salmonicida* from diseased turbot *Scophthalmus maximus*. *Diseases of Aquatic Organisms* 18, 149–154.

618 Pedersen, K., Dalsgaard, I. and Larsen, J.L. (1997) *Vibrio damsela* associated with diseased fish in Denmark. *Applied and Environmental Microbiology* 63, 3711–3715.

619 Pedersen, K., Verdonck, L., Austin, B., Austin, D.A., Blanch, A.R., Grimont, P.A.D., Jofre, J., Koblavi, S., Larsen, J.L., Tianinen, T., Vigneulle, M. and Swings, J. (1998) Taxonomic evidence that *Vibrio carchariae* Grimes *et al.* 1985 is a junior synonym of *Vibrio harveyi* (Johnson and Shunk 1936) Baumann *et al.* 1981. *International Journal of Systematic Bacteriology* 48, 749–758.

620 Pedersen, K., Austin, B., Austin, D. and Larsen, J. (1999) Vibrios associated with mortality in cultured

Plaice *Pleuronectes platessa* fry. *Acta Veterinaria Scandinavica.* 40, 263–270.

621 Perera, R., Johnson, S., Collins, M. and Lewis, D. (1994) *Streptococcus iniae* associated with mortality of *Tilapia nilotica* and *T. aurea* hybrids. *Journal of Aquatic Animal Health* 6, 335–340.

622 Petrie, J., Bruno, D.W. and Hastings, T.S. (1996) Isolation of *Yersinia ruckeri* from wild, Atlantic salmon, *Salmo salar* L., in Scotland. *Bulletin of the European Association of Fish Pathologists* 16, 83–84.

623 Pickett, M. and Pedersen, M. (1970) Characterization of saccharolytic nonfermentative bacteria associated with man. *Canadian Journal of Microbiology* 16, 351–362.

624 Pidiyar, V., Kaznowski, A., Narayan, N.B., Patole, M. and Shouche, Y.S. (2002) *Aeromonas culicicola* sp. nov., from the midgut of *Culex quinquefasciatus*. *International Journal of Systematic and Evolutionary Microbiology* 52, 1723–1728.

625 Pier, G.B. and Madin, S.H. (1976) *Streptococcus iniae* sp nov., a beta-haemolytic streptococcus isolated from an Amazon freshwater dolphin, *Inia geoffrensis*. *International Journal of Systematic Bacteriology* 26, 545–553.

626 Pier, G., Madin, S. and Nakeeb, S. (1978) Isolation and characterization of a second isolate of *Streptococcus iniae*. *International Journal of Systematic Bacteriology* 28, 311–314.

627 Plumb, J. and Sanchez, D. (1983) Susceptibility of five species of fish to *Edwardsiella ictaluri*. *Journal of Fish Diseases* 6, 261–266.

628 Pollard, D., Johnson, W., Lior, H., Tyler, S. and Rozee, K. (1990) Detection of the Aerolysin gene in *Aeromonas hydrophila* by the polymerase chain reaction. *Journal of Clinical Microbiology* 28, 2477–2481.

629 Pot, B., Devriese, L.A., Hommez, J., Miry, C., Vandemeulebroecke, K., Kersters, K. and Haesebrouck, F. (1994) Characterization and identification of *Vagococcus fluvialis* strains isolated from domestic animals. *Journal of Applied Bacteriology* 77, 362–369.

630 Pu, Z., Dobos, M., Limsowtin, G. and Powell, I. (2002) Integrated polymerase chain reaction-based procedures for the detection and identification of species and subspecies of the Gram-positive bacterial genus *Lactococcus*. *Journal of Applied Microbiology* 93, 353–361.

631 Pujalte, M.-J. and Garay, E. (1986) Proposal of *Vibrio mediterranei* sp. nov.: a new marine member of the genus *Vibrio*. *International Journal of Systematic Bacteriology* 36, 278–281.

632 Pujalte, M.-J., Ortigosa, M., Urdaci, M.-C., Garay, E. and Grimont, P. (1993) *Vibrio mytili* sp. nov., from mussels. *International Journal of Systematic Bacteriology* 43, 358–362.

633 Puttinaowarat, S., Thompson, K.D., Kolk, A. and Adams, A. (2002) Identification of *Mycobacterium* spp. isolated from snakehead, *Channa striata* (Fowler), and Siamese fighting fish, *Betta splendens* (Regan), using polymerase chain reaction-reverse cross blot hybridization (PCR-RCBH). *Journal of Fish Diseases* 25, 235–243.

634 Pychynski, T., Malanowska, T. and Kozlowski, M. (1981) Bacterial flora in branchionecrosis of carp (particularly *Bacillus cereus* and *Bacillus subtilis*). *Medycyna Weterynaryjna* 37, 742–743.

635 Raguenes, G., Christen, R., Guezennec, J., Pignet, P. and Barbier, G. (1997) *Vibrio diabolicus* sp. nov., a new polysaccharide-secreting organism isolated from a deep-sea hydrothermal vent polychaete annelid, *Alvinella pompejana*. *International Journal of Systematic Bacteriology* 47, 989–995.

636 Ramos, C.P., Foster, G. and Collins, M.D. (1997) Phylogenetic analysis of the genus *Actinomyces* based on 16S rRNA gene sequences: description of *Arcanobacterium phocae* sp. nov., *Arcanobacterium bernardiae* com. nov., and *Arcanobacterium pyogenes* com. nov. *International Journal of Systematic Bacteriology* 47, 46–53.

637 Rasheed, V., Limsuwan, C. and Plumb, J. (1985) Histopathology of bullminnows, *Fundulus grandis* Baird & Girard, infected with a non-haemolytic group B *Streptococcus* sp. *Journal of Fish Diseases* 8, 65–74.

638 Ravelo, C., Magariños, B., Romalde, J. and Toranzo, A. (2001) Conventional versus miniaturized systems for the phenotypic characterization of *Lactococcus garvieae* strains. *Bulletin of the European Association of Fish Pathologists* 21, 136–144.

639 Reddacliff, G.L., Hornitzky, M., Carson, J., Petersen, R. and Zelski, R. (1993) Mortalities of goldfish, *Carassius auratus* (L.), associated with *Vibrio cholerae* (non-01) infection. *Journal of Fish Diseases* 16, 517–520.

640 Reddacliff, G.L., Hornitzky, M. and Whittington, R.J. (1996) *Edwardsiella tarda* septicaemia in rainbow trout (*Oncorhynchus mykiss*). *Australian Veterinary Journal* 73, 30.

641 Reddy, C., Cornell, C. and Fraga, A. (1982) Transfer of *Corynebacterium pyogenes* (Glage) Eberson to the genus Actinomyces as *Actinomyces pyogenes* (Glage) comb. nov. *International Journal of Systematic Bacteriology* 32, 419–429.

642 Register, K., Sacco, R. and Foster, G. (2000) Ribotyping and restriction endonuclease analysis reveal a novel clone of *Bordetella bronchiseptica* in seals. *Journal of Veterinary Investigation* 12, 535–540.

643 Reichelt, J. and Baumann, P. (1975) *Photobacterium mandapamensis* Hendrie *et al.*, a later subjective synonym of *Photobacterium leiognathi* Boisvert *et al.* *International Journal of Systematic Bacteriology* 25, 208–209.

644 Reichenbach, H., Kleinig, H. and Achenbach, H. (1974) The pigments of *Flexibacter elegans*: novel and chemosystematically useful compounds. *Archives of Microbiology* 101, 131–144.

645 Reichenbach, H., Kohl, W. and Achenbach, H. (1981) The flexirubin-type pigments, chemosystematically useful compounds. In: Reichenbach, H. and Weeks, O.B. (eds) *Proceedings of the International Symposium on Yellow Pigmented Gram-Negative Bacteria of the Flavobacterium-Cytophaga Group*. Verlag Chemie, Deerfield Beach, Florida.

646 Reichenbach-Klinke, H. and Elkan, E. (1966) *The Principal Diseases of Lower Vertebrates*. Academic Press, New York.

647 Reid, G.A. and Gordon, E. (1999) Phylogeny of marine and freshwater *Shewanella*: reclassification of *Shewanella putrefaciens* NCIMB 400 as *Shewanella frigidimarina*. *International Journal of Systematic Bacteriology* 49, 189–191.

648 Ringø, E. and Gatesoupe, F.-J. (1998) Lactic acid bacteria in fish: a review. *Aquaculture* 160, 177–203.

649 Ringø, E., Seppola, M., Berg, A., Olsen, R.E., Schillinger, U. and Holzapfel, W. (2002) Characterization of *Carnobacterium divergens* strain 6251 isolated from intestine of Arctic charr (*Salvelinus alpinus* L.). *Systematic and Applied Microbiology* 25, 120–129.

650 Riquelme, C., Toranzo, A., Barja, J., Vergara, N. and Araya, R. (1996) Association of *Aeromonas hydrophila* and *Vibrio alginolyticus* with larval mortalities of scallop (*Argopecten purpuratus*). *Journal of Invertebrate Pathology* 67, 213–218.

651 Robert-Pillot, A., Guenole, A. and Fournier, J.-M. (2002) Usefulness of R72H PCR assay for differentiation between *Vibrio parahaemolyticus* and *Vibrio alginolyticus* species: validation by DNA–DNA hybridization. *FEMS Microbiology Letters* 215, 1–6.

652 Rodriguez, L.A., Gallardo, C.S., Acosta, F., Nieto, T.P., Acosta, B. and Real, F. (1998) *Hafnia alvei* as an opportunistic pathogen causing mortality in brown trout, *Salmo trutta* L. *Journal of Fish Diseases* 21, 365–369.

653 Roggenkamp, A., Abele-Horn, M., Trebesius, K.-H., Tretter, U., Autenrieth, I.B. and Heesemann, J. (1998) *Abiotrophia elegans* sp. nov., a possible pathogen in patients with culture-negative endocarditis. *Journal of Clinical Microbiology* 36, 100–104.

654 Rogosa, M.J., Mitchell, J.A. and Wiseman, R.F. (1951) A selective medium for the isolation and enumeration of oral and fecal lactobacilli. *Journal of Bacteriology* 62, 132–133.

655 Romalde, J., Magariños, B., Fouz, B., Bandín, I., Núñez, S. and Toranzo, A. (1995) Evaluation of BIONOR Mono-kits for rapid detection of bacterial

fish pathogens. *Diseases of Aquatic Organisms* 21, 25–34.

656 Ross, A.J. and Brancato, F. (1959) *Mycobacterium fortuitum* Cruz from the tropical fish *Hyphessobrycon innesi*. *Journal of Bacteriology* 78, 392–395.

657 Ross, A.J., Rucker, R.R. and Ewing, W.H. (1966) Description of a bacterium associated with redmouth disease of rainbow trout (*Salmo gairdneri*). *Canadian Journal of Microbiology* 12, 763–770.

658 Ross, H.M., Foster, G., Reid, R.J., Jahans, K.L. and MacMillan, A.P. (1994) *Brucella* species infection in sea-mammals. *Veterinary Record* 134, 359.

659 Ruger, H.J. and Tan, T.L. (1983) Separation of *Alcaligenes denitrificans* sp. nov., nom. rev. from *Alcaligenes faecalis* on the basis of DNA base composition, DNA homology, and nitrate reduction. *International Journal of Systematic Bacteriology* 33, 85–99.

660 Ruhnke, H.L. and Madoff, S. (1992) *Mycoplasma phocidae* sp. nov., isolated from harbour seals (*Phoca vitulina* L.). *International Journal of Systematic Bacteriology* 42, 211–214.

661 Ruimy, R., Riegel, P., Carlotti, A., Boiron, P., Bernardin, G., Monteil, H., Wallace, R.J. Jr and Christen, R. (1996) *Nocardia pseudobrasiliensis* sp. nov., a new species of Nocardia which groups bacterial strains previously identified as *Nocardia brasiliensis* and associated with invasive diseases. *International Journal of Systematic Bacteriology* 46, 259–264.

662 Ruiz-Ponte, C., Cilia, V., Lambert, C. and Nicolas, J.L. (1998) *Roseobacter gallaeciensis* sp. nov., a new marine bacterium isolated from rearings and collectors of the scallop *Pecten maximus*. *International Journal of Systematic Bacteriology* 48, 537–542.

663 Rutter, J. (1981) Quantitative observations on *Bordetella bronchiseptica* infection in atrophic rhinitis of pigs. *Veterinary Record* 108, 451–454.

664 Rutter, J.M., Taylor, R.J., Crighton, W.G., Robertson, I.B. and Benson, J.A. (1984) Epidemiological study of *Pasteurella multocida* and *Bordetella bronchiseptica* in atrophic rhinitis. *Veterinary Record* 115, 615–619.

665 Saeed, M.O., Almoudi, M.M. and Al-Harbi, A.H. (1987) A *Pseudomonas* associated with disease in cultured rabbitfish *Siganus rivulatus* in the Red Sea. *Diseases of Aquatic Organisms* 3, 177–180.

666 Saeed, M.O., Alamoudi, M.M. and Al-Harbi, A.H. (1990) Histopathology of *Pseudomonas putrefaciens* associated with disease in cultured rabbitfish, *Siganus rivulatus* (Forskal). *Journal of Fish Diseases* 13, 417–422.

667 Saiki, R. (1989) The design and optimisation of the PCR. In: Erlich, H.A. (ed.) *PCR Technology: Principles and Applications for DNA Amplification*. Stockton Press, New York, pp. 7–16.

668 Saiki, R., Scharf, S., Faloona, F., Mullis, K., Horn, G., Erlich, H. and Arnheim, N. (1985) Enzymatic amplification of β-globulin genomic sequences and restriction site analysis for diagnosis of sickle cell anemia. *Science* 230, 1350–1354.

669 Salati, F., Tassi, P. and Bronzi, P. (1996) Isolation of an *Enterococcus*-like bacterium from diseased Adriatic sturgeon, *Acipenser naccarii*, farmed in Italy. *Bulletin of the European Association of Fish Pathologists* 16, 96–100.

670 Sambrook, J., Fritsch, E. and Maniatis, T. (1989) *Molecular Cloning: a Laboratory Manual*. Cold Spring Harbor Laboratory Press, Cold Spring Harbor, New York.

671 Sanders, J.E. and Fryer, J.L. (1980) *Renibacterium salmoninarum* gen. nov., sp. nov., the causative agent of bacterial kidney disease in salmonid fishes. *International Journal of Systematic Bacteriology* 30, 496–502.

672 Santacana, J.A., Conroy, D.A., Mujica, M.E., Marín, C. and López, N.D. (1982) Acid-fast bacterial infection and its control in three-spot gouramis, *Trichogaster trichopterus*. *Journal of Fish Diseases* 5, 545–547.

673 Santos, N. de, Vale, A. de, Sousa, M. and Silva, M. (2002) Mycobacterial infection in farmed turbot *Scophthalmus maximus*. *Diseases of Aquatic Organisms* 52, 87–91.

674 Santos, Y., Romalde, J.L., Bandín, I., Magariños, B., Núñez, S., Barja, J.L. and Toranzo, A.E. (1993) Usefulness of the API-20E system for the identification of bacterial fish pathogens. *Aquaculture* 116, 111–120.

675 Sato, N., Yamane, N. and Kawamura, T. (1982) Systemic *Citrobacter freundii* infection among sunfish *Mola mola* in Matsushima Aquarium. *Bulletin of the Japanese Society of Scientific Fisheries* 48, 1551–1557.

676 Saulnier, D., Avarre, J.C., Moullac, G.L., Ansquer, D., Levy, P. and Vonau, V. (2000) Rapid and sensitive PCR detection of *Vibrio penaeicida*, the putative etiological agent of Syndrome 93 in New Caledonia. *Diseases of Aquatic Organisms* 40, 109–115.

677 Sawabe, T., Makino, H., Tatsumi, M., Nakano, K., Tajima, K., Iqbal, M.M., Yumoto, I., Ezura, Y. and Christen, R. (1998a) *Pseudoalteromonas bacteriolytica* sp. nov., a marine bacterium that is the causative agent of red spot disease of *Laminaria japonica*. *International Journal of Systematic Bacteriology* 48, 769–774.

678 Sawabe, T., Sugimura, I., Ohtsuka, M., Nakano, K., Tajima, K., Ezura, Y. and Christen, R. (1998b) *Vibrio halioticoli* sp. nov., a non-motile alginolytic marine bacterium isolated from the gut of the abalone *Haliotis discus hannai*. *International Journal of Systematic Bacteriology* 48, 573–580.

679 Sawabe, T., Tanaka, R., Iqbae, M.M., Tajima, K., Ezura, Y., Ivanova, E.P. and Christen, R. (2000) Assignment of *Alteromonas elyakovii* KMM 162ᵀ and five strains isolated from spot-wounded fronds of *Laminaria japonica* to *Pseudoalteromonas elyakovii* comb. nov. and the extended description of the species. *International Journal of Systematic and Evolutionary Microbiology* 50, 265–271.

680 Schiewe, M.H., Trust, T.J. and Crosa, J.H. (1981) *Vibrio ordalii* sp. nov.: causative agent of vibriosis in fish. *Current Microbiology* 6, 343–348.

681 Schleifer, K.H. and Kloos, W.E. (1975) Isolation and characterization of Staphylococci from human skin. I. Amended descriptions of *Staphylococcus epidermidis* and *Staphylococcus saprophyticus* and descriptions of three new species: *Staphylococcus cohnii*, *Staphylococcus haemolyticus*, and *Staphylococcus xylosus*. *International Journal of Systematic Bacteriology* 25, 50–61.

682 Schmidtke, L.M. and Carson, J. (1994) Characteristics of *Vagococcus salmoninarum* isolated from diseased salmonid fish. *Journal of Applied Bacteriology* 77, 229–236.

683 Schmidtke, L.M. and Carson, J. (1995) Characteristics of *Flexibacter psychrophilus* isolated from Atlantic salmon in Australia. *Diseases of Aquatic Organisms* 21, 157–161.

684 Schubert, R. (1971) Status of the names *Aeromonas* and *Aerobacter liquefaciens* Beijerinck and designation of the neotype strain for *Aeromonas hydrophila* Stanier. *International Journal of Systematic Bacteriology* 21, 87–90.

685 Segers, P., Vancanneyt, M., Pot, B., Torck, U., Hoste, B., Dewettinck, D., Falsen, E., Kersters, K. and Vos, P. de. (1994) Classification of *Pseudomonas diminuta* Leifson and Hugh 1954 and *Pseudomonas vesicularis* Busing, Doll, and Freytag 1953 in *Brevundimonas* gen. nov. as *Brevundimonas diminuta* comb. nov., and *Brevundimonas vesicularis* comb. nov., respectively. *International Journal of Systematic Bacteriology* 44, 499–510.

686 Seibold, H.R. and Neal, J.E. (1956) *Erysipelothrix septicaemia* in the porpoise. *Journal of American Veterinary Medical Association* 128, 537–539.

687 Seidler, R., Allen, D., Colwell, R., Joseph, S. and Daily, O. (1980) Biochemical characteristics and virulence of environmental Group F bacteria isolated in the United States. *Applied and Environmental Microbiology* 40, 715–720.

688 Shah, K. and Tyagy, B. (1986) An eye disease in silver carp, *Hypophthalmichthys molitrix*, held in tropical ponds, associated with the bacterium *Staphylococcus aureus*. *Aquaculture* 55, 1–4.

689 Shamsudin, M.N. and Plumb, J.A. (1996) Morphological, biochemical, and physiological characterization of *Flexibacter columnaris* isolates from four species of fish. *Journal of Aquatic Animal Health* 8, 335–339.

690 Sherwan, J., Hodgkiss, W. and Listor, J. (1954) A method for the rapid identification of certain non-pigmented asporogenous. bacilli *Nature (London)* 63, 208–209.

691 Shieh, H.S. (1980) Studies on the nutrition of a fish pathogen, *Flexibacter columnaris*. *Microbios Letters* 13, 129–133.

692 Shieh, W.Y., Chen, A.L. and Chiu, H.H. (2000) *Vibrio aerogenes* sp. nov., a facultatively anaerobic marine bacterium that ferments glucose with gas production. *International Journal of Systematic and Evolutionary Microbiology* 50, 321–329.

693 Shiose, J., Wakabayashi, H., Tominaga, M. and Egusa, S. (1974) A report on a disease of cultured carp due to a capsulated *Pseudomonas*. *Fish Pathology* 9, 79–83.

694 Shotts, E.B. and Waltman, W.D. II. (1990) A medium for the selective isolation of *Edwardsiella ictaluri*. *Journal of Wildlife Diseases* 26, 214–218.

695 Shotts, E.B.J., Talkington, F.D., Elliott, D.G. and McCarthy, D.H. (1980) Aetiology of an ulcerative disease in goldfish, *Carassius auratus* (L.): characterization of the causative agent. *Journal of Fish Diseases* 3, 181–186.

696 Simidu, U. and Hasuo, K. (1968) Salt dependency of the bacterial flora of marine fish. *Journal of General Microbiology* 52, 347–354.

697 Simmons, J.S. (1926) A culture medium for differentiating organisms of typhoid-colon aerogenes groups and for isolation of certain fungi. *Journal of Infectious Diseases* 39, 201–214.

698 Simon, G. and Oppenheimer, C. (1968) Bacterial changes in sea water samples, due to storage and volume. *Zeitschrift für Allgemeine Mikrobiologie* 8, 209–214.

699 Simmons, G.C., Sullivan, N.D. and Green, P.E. (1972) Dermatophilosis in a lizard (*Amphibolurus barbatus*). *Australian Veterinary Journal* 48, 465–466.

700 Skaar, I., Gaustad, P., Tønjum, T., Holm, B. and Stenwig, H. (1994) *Streptococcus phocae* sp. nov., a new species isolated from clinical specimens from seals. *International Journal of Systematic Bacteriology* 44, 646–650.

701 Skirrow, M.B. (1977) Campylobacter enteritis: a 'new' disease. *British Medical Journal* 2, 9–11.

702 Smith, H.J. and Goodner, K. (1958) Detection of bacterial gelatinases by gelatin-agar plate methods. *Journal of Bacteriology* 76, 662–665.

703 Smith, I.M. and Baskerville, A.J. (1979) A selective medium facilitating the isolation of and recognition of *Bordetella bronchiseptica* in pigs. *Research in Veterinary Science* 27, 187–192.

704 Smith, N., Gordon, R. and Clark, F. (1952) Aerobic sporeforming bacteria. In: *Monograph No. 16*, US Department of Agriculture, Washington, DC.

705 Smith, S.K., Sutton, D.C., Fuerst, J.A. and Reichelt, J.L. (1991) Evaluation of the genus

Listonella and reassignment of *Listonella damsela* (Love *et al.*) MacDonell and Colwell to the genus Photobacterium as *Photobacterium damsela* comb.nov. with an emended description, *International Journal of Systematic Bacteriology* 41, 529–534.

706 Snieszko, S.F. (1981) Bacterial gill disease of freshwater fishes. *US Fish and Wildlife Service Fish Disease Leaflet 62*, pp. 1–11.

707 Snieszko, S.F., Bullock, G., Dunbar, C. and Pettijohn, L. (1964) Nocardial infection in hatchery-reared fingerling rainbow trout (*Salmo gairdneri*). *Journal of Bacteriology* 88, 1809–1810.

708 Snieszko, S.F., Bullock, G.L., Hollis, E. and Boone, J.G. (1964) *Pasteurella* sp. from an epizootic of white perch (*Roccus americanus*) in Chesapeake Bay tidewater areas. *Journal of Bacteriology* 88, 1814–1815.

709 Snipes, K.P. and Biberstein, E.L. (1982) *Pasteurella testudinis* sp. nov.: a parasite of desert tortoises (*Gopherus agassizi*). *International Journal of Systematic Bacteriology* 32, 201–210.

710 Soffientino, B., Gwaltney, T., Nelson, D.R., Specker, J.L., Mauel, M. and Gómez-Chiarri, M. (1999) Infectious necrotizing enteritis and mortality caused by *Vibrio carchariae* in summer flounder *Paralichthys dentatus* during intensive culture. *Diseases of Aquatic Organisms* 38, 201–210.

711 Sonnenwirth, A. (1970) Bacteremia with and without meningitis due to *Yersinia enterocolitica*, *Edwardsiella tarda*, *Comamonas terrigena* and *Pseudomonas maltophilia*. *Annals of the New York Academy of Sciences* 174, 488–502.

712 Sørensen, U. and Larsen, J. (1986) Serotyping of *Vibrio anguillarum*. *Applied and Environmental Microbiology* 51, 593–597.

713 Stackebrandt, E. and Kandler, O. (1979) Taxonomy of the genus *Cellulomonas*, based on phenotypic characters and deoxyribonucleic acid-deoxyribonucleic acid homology, and proposal of seven neotype strains. *International Journal of Systematic Bacteriology* 29, 272–282.

714 Stanier, R.Y., Palleroni, N.J. and Doudoroff, M. (1966) The aerobic Pseudomonads: a taxonomic study. *Journal of General Microbiology* 43, 159–271.

715 Starliper, C. (2001) Isolation of *Serratia liquefaciens* as a pathogen of Arctic char, *Salvelinus alpinus* (L.). *Journal of Fish Diseases* 24, 53–56.

716 Starliper, C., Shotts, E. and Brown, J. (1992) Isolation of *Carnobacterium piscicola* and an unidentified Gram-positive bacillus from sexually mature and post-spawning rainbow trout *Oncorhynchus mykiss*. *Diseases of Aquatic Organisms* 13, 181–187.

717 Stevenson, R.M.W. and Daly, J.G. (1982) Biochemical and serological characteristics of Ontario

isolates of *Yersinia ruckeri*. *Canadian Journal of Fisheries and Aquatic Sciences* 39, 870–876.

718 Stevenson, R. and Airdrie, D. (1984) Serological variation among *Yersinia ruckeri* strains. *Journal of Fish Diseases* 7, 247–254.

719 Stewart, J.E. (1972) The detection of *Gaffkya homari*, the bacterium pathogenic to lobsters (Genus *Homarus*). *Fisheries Research Board of Canada. New Series* 43, 1–5.

720 Strohl, W. and Tait, L. (1978) *Cytophaga aquatilis* sp. nov., a facultative anaerobe isolated from the gills of freshwater fish. *International Journal of Systematic Bacteriology* 28, 293–303.

721 Sugumar, G., Nakai, T., Hirata, Y., Matsubara, D. and Muroga, K. (1998) *Vibrio splendidus* biovar II as the causative agent of bacillary necrosis of Japanese oyster *Crassostrea gigas* larvae. *Diseases of Aquatic Organisms* 33, 111–118.

722 Sukenda and Wakabayashi, H. (2000) Tissue distribution of *Pseudomonas plecoglossicida* in experimentally infected Ayu *Plecoglossus altivelis* studied by real-time quantitative PCR. *Fish Pathology* 35, 223–228.

723 Sung, H.-H., Hwang, S.-F. and Tasi, F.-M. (2000) Response of giant freshwater prawn (*Macrobranchium rosenbergii*) to challenge by 2 strains of *Aeromonas* spp. *Journal of Invertebrate Pathology* 76, 278–284.

724 Suzuki, M. and Giovannoni, S. (1996) Bias caused by template annealing in the amplification of mixtures of 16S rRNA genes by PCR. *Applied and Environmental Microbiology* 62, 625–630.

725 Suzuki, M., Nakagawa, Y., Harayama, S. and Yamamoto, S. (2001) Phylogenetic analysis and taxonomic study of marine Cytophaga-like bacteria: proposal for *Tenacibaculum* gen. nov. with *Tenacibaculum maritimum* comb. nov. and *Tenacibaculum ovolyticum* comb. nov., and description of *Tenacibaculum mesophilum* sp. nov. and *Tenacibaculum amylolyticum* sp. nov. *International Journal of Systematic and Evolutionary Microbiology* 51, 1639–1652.

726 Sweeney, J.C. and Ridgway, S.H. (1975) Common diseases of small cetaceans. *Journal of the American Veterinary Medical Association* 167, 533–540.

727 Swenshon, M., Lammler, C. and Siebert, U. (1998) Identification and molecular characterization of beta-haemolytic Streptococci isolated from harbour porpoises (*Phocoena phocoena*) of the North and Baltic seas. *Journal of Clinical Microbiology* 36, 1902–1906.

728 Takeuchi, M. and Yokota, A. (1992) Proposals of *Sphingobacterium faecium* sp. nov., *Sphingobacterium piscium* sp. nov., *Sphingobacterium heparinum* comb. nov., *Sphingobacterium thalpophilum* com. nov. and two genospecies of the genus *Sphingobacterium*, and synonymy of

Flavobacterium yabuuchiae and *Sphingobacterium spiritivorum*. *Journal of General and Applied Microbiology* 38, 465–482.

729 Talaat, A., Reimschuessel, R. and Trucksis, M. (1997) Identification of mycobacteria infecting fish to the species level using polymerase chain reaction and restriction enzyme analysis. *Veterinary Microbiology* 58, 229–237.

730 Taylor, P. and Winton, J. (2002) Optimization of nested polymerase chain reaction assays for identification of *Aeromonas salmonicida, Yersinia ruckeri*, and *Flavobacterium psychrophilum*. *Journal of Aquatic Animal Health* 14, 216–224.

731 Teixeira, L., Merquior, V.L., Vianni, M. de C., Carvalho, M. de G., Fracalanzza, S., Steigerwalt, A., Brenner, D. and Facklam, R. (1996) Phenotypic and genotypic characterization of atypical *Lactococcus garvieae* strains isolated from water buffalos with subclinical mastitis and confirmation of *L. garvieae* as a senior subjective synonym of *Enterococcus seriolicida*. *International Journal of Systematic Bacteriology* 46, 664–668.

732 Teixeira, L.M., Carvalho, M. de G.S., Merquior, V.L.C., Steigerwalt, A.G., Brenner, D.J. and Facklam, R.R. (1997) Phenotypic and genotypic characterization of *Vagococcus fluvialis*, including strains isolated from human sources. *Journal of Clinical Microbiology* 35, 2778–2781.

733 Temprano, A., Yugueros, J., Hernanz, C., Sanchez, M., Berzal, B., Luengo, J. and Naharro, G. (2001) Rapid identification of *Yersinia ruckeri* by PCR amplification of *yrul-yruR* quorum sensing. *Journal of Fish Diseases* 24, 253–261.

734 Tendencia, E. (2002) *Vibrio harveyi* isolated from cage-cultured seabass *Lates calcarifer* Bloch in the Philippines. *Aquaculture Research* 33, 455–458.

735 Tendencia, E.A. (2002) *Vibrio harveyi* isolated from cage-cultured sea bass *Lates calcarifer* Bloch in the Philippines. *Aquaculture Research* 33, 455–458.

736 Teska, J., Twerdok, L., Beaman, J., Curry, M. and Finch, R. (1997) Isolation of *Mycobacterium abscessus* from Japanese Medaka. *Journal of Aquatic Animal Health* 9, 234–238.

737 Thoen, C. and Schliesser, T. (1984) Mycobacterial infections in cold-blooded animals. In: Kubica, G. and Wayne, L. (eds) *The Mycobacteria: a Sourcebook*, Vol. II. Marcel Dekker, New York, pp. 1297–1311.

738 Thomas, A.D., Forbes-Faulkner, J. and Parker, M. (1979) Isolation of *Pseudomonas pseudomallei* from clay layers at defined depths. *American Journal of Epidemiology* 110, 515–521.

739 Thompson, F., Hoste, B., Vandemeulebroecke, K. and Swings, J. (2001a) Genomic diversity amongst *Vibrio* isolates from different sources determined by fluorescent amplified fragment length polymorphism. *Systematic and Applied Microbiology* 24, 520–538.

740 Thompson, F., Li, Y., Gomez-Gil, B., Thompson, C., Hoste, B., Vandemeulebroecke, K., Rupp, G., Pereira, A., Bem, M.D., Sorgeloos, P. and Swings, J. (2003) *Vibrio neptunius* sp. nov., *Vibrio brasiliensis* sp. nov. and *Vibrio xuii* sp. nov., isolated from the marine aquaculture environment (bivalves, fish, rotifers and shrimps). *International Journal of Systematic and Evolutionary Microbiology* 53, 245–252.

741 Thompson, F.L., Hoste, B., Thompson, C.C., Goris, J., Gomez-Gil, B., Huys, L., Vos, P. de and Swings, J. (2002a) *Enterovibrio norvegicus* gen. nov., sp. nov., isolated from the gut of turbot (*Scophthalmus maximus*) larvae: a new member of the family Vibrionaceae. *International Journal of Systematic and Evolutionary Microbiology* 52, 2015–2022.

742 Thompson, F.L., Hoste, B., Thompson, C.C., Huys, G. and Swings, J. (2001b) The coral bleaching *Vibrio shiloi* Kushmaro et al. 2001 is a later synonym of *Vibrio mediterranei* Pujalte and Garay 1986. *Systematic and Applied Microbiology* 24, 516–519.

743 Thompson, F.L., Hoste, B., Vandemeulebroecke, K., Engelbeen, K., Denys, R. and Swings, J. (2002b) *Vibrio trachuri* Iwamoto et al. 1995 is a junior synonym of *Vibrio harveyi* (Johnson and Shunk 1936) Baumann et al. 1981. *International Journal of Systematic and Evolutionary Microbiology* 52, 973–976.

744 Thornley, M. (1960) The differentiation of pseudomonas from other gram-negative bacteria on the basis of arginine metabolism. *Journal of Applied Bacteriology* 23, 37–52.

745 Thyssen, A., Grisez, L., Houdt, R.V. and Ollevier, F. (1998) Phenotypic characterization of the marine pathogen *Photobacterium damselae* subsp. *piscicida*. *International Journal of Systematic Bacteriology* 48, 1145–1151.

746 Tison, D., Nishibuchi, M., Greenwood, J. and Seidler, R. (1982) *Vibrio vulnificus* biogroup 2: new biogroup pathogenic for eels. *Applied and Environmental Microbiology* 44, 640–646.

747 Tison, D.L. and Seidler, R.J. (1983) *Vibrio aestuarianus* : a new species from estuarine waters and shellfish. *International Journal of Systematic Bacteriology* 33, 699–702.

748 Toranzo, A.E. and Barja, J.L. (1990) A review of the taxonomy and seroepizootiology of *Vibrio anguillarum*, with special reference to aquaculture in the northwest of Spain. *Diseases of Aquatic Organisms* 9, 73–82.

749 Toranzo, A., Baya, A., Roberson, B., Barja, J., Grimes, D. and Hetrick, F. (1987) Specificity of slide agglutination test for detecting bacterial fish pathogens. *Aquaculture* 61, 81–97.

750 Toranzo, A.E., Baya, A.M., Romalde, J.L. and Hetrick, F.M. (1989) Association of *Aeromonas sobria* with mortalities of adult gizzard shad, *Dorosoma cepedianum* Lesueur. *Journal of Fish Diseases* 12, 439–448.

751 Toranzo, A.E., Barreiro, S., Casal, J.F., Figueras, A., Magariños, B. and Barja, J. (1991) Pasteurellosis in cultured gilthead seabream (*Sparus aurata*): first report in Spain. *Aquaculture* 99, 1–15.

752 Toranzo, A.E., Romalde, J.L., Núñez, S., Figueras, A. and Barja, J.L. (1993) An epizootic in farmed, market-size rainbow trout in Spain caused by a strain of *Carnobacterium piscicola* of unusual virulence. *Diseases of Aquatic Organisms* 17, 87–99.

753 Toranzo, A., Cutrín, J., Roberson, B., Núñez, S., Abell, J., Hetrick, F. and Baya, A. (1994) Comparison of the taxonomy, serology, drug resistance transfer, and virulence of *Citrobacter freundii* strains from mammals and poikilothermic hosts. *Applied and Environmental Microbiology* 60, 1789–1797.

754 Toranzo, A., Cutrín, J., Núñez, S., Romalde, J. and Barja, J. (1995) Antigenic characterization of *Enterococcus* strains pathogenic for turbot and their relationship with other Gram-positive bacteria. *Diseases of Aquatic Organisms* 21, 187–191.

755 Torrent, A., Déniz, S., Ruiz, A., Calabuig, P., Sicilia, J. and Orós, J. (2002) Esophageal diverticulum associated with *Aerococcus viridans* infection in a Loggerhead Sea Turtle (*Caretta caretta*). *Journal of Wildlife Diseases* 38, 221–223.

756 Tortoli, E., Bartoloni, A., Bozzetta, E., Burrini, C., Lacchini, C., Mantella, A., Penati, V., Simonetti, M.T. and Ghittino, C. (1996) Identification of the newly described *Mycobacterium poriferae* from tuberculous lesions of snakehead fish (*Channa striatus*). *Comparative Immunology, Microbiology and Infectious Diseases* 19, 25–29.

757 Toyama, T., Kita-Tsukamoto, K. and Wakabayashi, H. (1994) Identification of *Cytophaga psychrophila* by PCR targeted 16S ribosomal RNA. *Fish Pathology* 29, 271–275.

758 Toyama, T., Kita-Tsukamoto, K. and Wakabayashi, H. (1996) Identification of *Flexibacter maritimus*, *Flavobacterium branchiophilum* and *Cytophaga columnaris* by PCR targeted 16S ribosomal DNA. *Fish Pathology* 31, 25–31.

759 Triyanto and Wakabayashi, H. (1999) Genotypic diversity of strains of *Flavobacterium columnare* from diseased fishes. *Fish Pathology* 34, 65–71.

760 Triyanto, Kumamaru, A. and Wakabayashi, H. (1999) The use of PCR targeted 16S rDNA for identification of genomovars of *Flavobacterium columnare*. *Fish Pathology* 34, 217–218.

761 Trust, T., Khouri, A., Austen, R. and Ashburner, L. (1980) First isolation in Australia of atypical *Aeromonas salmonicida*. *FEMS Microbiology Letters* 9, 39–42.

762 Tubiash, H.S., Chanley, P.E. and Leifson, E. (1965) Bacillary necrosis, a disease of larval and juvenile bivalve mollusks. I. Etiology and epizootiology. *Journal of Bacteriology* 90, 1036–1044.

763 Udey, L.R. (1982) A differential medium for distinguishing Alr+ from Alr- phenotypes in *Aeromonas salmonicida*. In: *Proceedings of the 13th annual conference and workshop and 7th eastern fish health workshop*. International Association for Aquatic Animal Medicine, Baltimore, Maryland.

764 Udey, L., Young, E. and Sallman, B. (1977) Isolation and characterization of an anaerobic bacterium, *Eubacterium tarantellus* sp. nov., associated with striped mullet (*Mugil cephalus*) mortality in Biscayne Bay, Florida. *Journal of the Fisheries Research Board of Canada* 34, 402–409.

765 Uhland, F.C., Hélie, P. and Higgins, R. (2000) Infections of *Edwardsiella tarda* among Brook trout in Quebec. *Journal of Aquatic Animal Health* 12, 74–77.

766 Urakawa, H., Kita-Tsukamoto, K., Steven, S.E., Ohwada, K. and Colwell, R.R. (1998) A proposal to transfer *Vibrio marinus* (Russell 1891) to a new genus *Moritella* gen. nov. as *Moritella marina* comb. nov. *FEMS Microbiology Letters* 165, 373–378.

767 Urakawa, H., Kita-Tsukamoto, K. and Ohwada, K. (1999) Reassessment of the taxonomic position of *Vibrio iliopiscarius* (Onarheim *et al.* 1994) and proposal for *Photobacterium iliopiscarium* comb. nov. *International Journal of Systematic Bacteriology* 49, 257–260.

768 Urdaci, M., Marchand, M., Ageron, E., Arcos, J., Sesma, B. and Grimont, P. (1991) *Vibrio navarrensis* sp. nov. from sewerage. *International Journal of Systematic Bacteriology* 41, 290–294.

769 Urdaci, M.C., Chakroun, C. and Bernardet, J.-F. (1998) Development of a polymerase chain reaction assay for identification and detection of the fish pathogen *Flavobacterium psychrophilum*. *Research in Microbiology* 149, 519–530.

770 Ursing, J., Rosselló-Mora, R., García-Valdés, E. and Lalucat, J. (1995) Taxonomic note: a pragmatic approach to the nomenclature of phenotypically similar genomic groups. *International Journal of Systematic Bacteriology* 45, 604.

771 Valheim, M., Håstein, T., Myhr, E., Speilberg, L. and Ferguson, H.W. (2000) *Varracalbmi*: a new bacterial panophthalmitis in farmed Atlantic salmon, *Salmo salar* L. *Journal of Fish Diseases* 23, 61–70.

772 Valtonen, E.T., Rintamäki, P. and Koskivaara, M. (1992) Occurrence and pathogenicity of *Yersinia ruckeri* at fish farms in northern and central Finland. *Journal of Fish Diseases* 15, 163–171.

773 Vancanneyt, M., Segers, P., Hauben, L., Hommez, J., Devriese, L.A., Hoste, B., Vandamme, P. and Kersters, K. (1994) *Flavobacterium*

meningosepticum, a pathogen in birds. *Journal of Clinical Microbiology* 32, 2398–2403.

774 Vancanneyt, M., Segers, P., Torck, U., Hoste, B., Bernardet, J.-F., Vandamme, P. and Kersters, K. (1996) Reclassification of *Flavobacterium odoratum* (Stutzer 1929) strains to a new genus, *Myroides*, as *Myroides odoratus* comb. nov. and *Myroides odoratimimus* sp. nov. *International Journal of Systematic Bacteriology* 46, 926–932.

775 Vandamme, P., Bernardet, J.-F., Segers, P., Kersters, K. and Holmes, B. (1994) New perspectives in the classification of the Flavobacteria: description of *Chryseobacterium* gen. nov., *Bergeyella* gen. nov., and *Empedobacter* nom. rev. *International Journal of Systematic Bacteriology* 44, 827–831.

776 Vandamme, P., Devriese, L., Pot, B., Kersters, K. and Melin, P. (1997) *Streptococcus difficile* is a nonhaemolytic group B type Ib Streptococcus. *International Journal of Systematic Bacteriology* 47, 81–85.

777 Vandenberghe, J., Li, Y., Verdonck, L., Li, J., Sorgeloos, P., Xu, H. and Swings, J. (1998) Vibrios associated with *Penaeus chinensis* (Crustacea: Decapoda) larvae in Chinese shrimp hatcheries. *Aquaculture* 169, 121–132.

778 Varaldo, P., Kilpper-Bälz, R., Biavasco, F., Satta, G. and Schleifer, K.H. (1988) *Staphylococcus delphini* sp. nov., a coagulase-positive species isolated from Dolphins. *International Journal of Systematic Bacteriology* 38, 436–439.

779 Vedros, N., Quinlivan, J. and Cranford, R. (1982) Bacterial and fungal flora of wild northern fur seals (*Callorhinus ursinus*). *Journal of Wildlife Diseases* 18, 447–456.

780 Vela, A., Vázquez, J., Gibello, A., Blanco, M., Moreno, M., Liébana, P., Albendea, C., Alcalá, B., Mendez, A., Domínguez, L. and Fernández-Garayzábal, J. (2000) Phenotypic and genetic characterization of *Lactococcus garvieae* isolated in Spain from Lactococcosis outbreaks and comparison with isolates of other countries and sources. *Journal of Clinical Microbiology* 38, 3791–3795.

781 Venkateswaran, K., Dohmoto, N. and Harayama, S. (1998) Cloning and nucleotide sequence of the *gyrB* gene of *Vibrio parahaemolyticus* and its application in detection of this pathogen in shrimp. *Applied and Environmental Microbiology* 64, 681–687.

782 Venkateswaran, K., Moser, D.P., Dollhopf, M.E., Lies, D.P., Saffarini, D.A., MacGregor, B.J., Ringelberg, D.B., White, D.C., Nishijima, M., Sano, H., Burghardt, J., Stackebrandt, E. and Nealson, K.H. (1999) Polyphasic taxonomy of the genus Shewanella and description of *Shewanella oneidensis* sp. nov. *International Journal of Systematic Bacteriology* 49, 705–724.

783 Ventura, M. and Grizzle, J. (1988) Lesions associated with natural and experimental infections of *Aeromonas hydrophila* in channel catfish, *Ictalurus punctatus* (Rafinesque). *Journal of Fish Diseases* 11, 397–407.

784 Vera, H.D. (1948) A simple medium for identification and maintenance of the gonococcus and other bacteria, *Journal of Bacteriology* 55, 531–536.

785 Vera, H.D. (1950) Relation of peptones and other culture media ingredients to the accuracy of fermentation tests, *American Journal of Public Health* 40, 1267–1272.

786 Vera, P., Navas, J.I. and Fouz, B. (1991) First isolation of *Vibrio damsela* from seabream (*Sparus aurata*). *Bulletin of the European Association of Fish Pathologists* 11, 112–113.

787 Verger, J.-M., Grimont, F., Grimont, P.A.D. and Grayon, M. (1985) *Brucella*, a monospecific genus as shown by deoxyribonucleic acid hybridization. *International Journal of Systematic Bacteriology* 35, 292–295.

788 Verschuere, L., Heang, H., Criel, G., Sorgeloos, P. and Verstraete, W. (2000) Selected bacterial strains protect *Artemia* spp. from the pathogenic effects of *Vibrio proteolyticus* CW8T2. *Applied and Environmental Microbiology* 66, 1139–1146.

789 Vicente, A., Coelho, A. and Salles, C. (1997) Detection of *Vibrio cholerae* and *V. mimicus* heat-stable toxin gene sequence by PCR. *Journal of Medical Microbiology* 46, 398–402.

790 Vieira, V., Teixeira, L., Zahner, V., Momen, H., Facklam, R., Steigerwalt, A., Brenner, D. and Castro, A. (1998) Genetic relationships among the different phenotype of *Streptococcus dysgalactiae* strains. *International Journal of Systematic Bacteriology* 48, 1231–1243.

791 Vigneulle, M. and Laurencin, F.B. (1995) *Serratia liquefaciens*: a case report in turbot (*Scophthalmus maximus*) cultured in floating cages in France. *Aquaculture* 132, 121–124.

792 Vogel, B.F., Jørgensen, K., Christensen, H., Olsen, J.E. and Gram, L. (1997) Differentiation of *Shewanella putrefaciens* and *Shewanella alga* on the basis of whole-cell protein profiles, ribotyping, phenotypic characterization, and 16S rRNA gene sequence analysis. *Applied and Environmental Microbiology* 63, 2189–2199.

793 Voges, O. and Proskauer, B. (1898) *Zentralblatt für Hygiene* 28, 20–22.

794 Vos, P.D. and Trüper, H.G. (2000) Judicial Commission of the International Committee on Systematic Bacteriology IXth International (IUMS) Congress of Bacteriology and Applied Microbiology. Minutes of the meetings, 14, 15 and 18 August 1999, Sydney, Australia. *International Journal of Systematic and Evolutionary Microbiology* 50, 2239–2244.

795 Vreeland, R.H., Litchfield, C.D., Martin, E.L. and Elliot, E. (1980) *Halomonas elongata*, a new genus and species of extremely salt-tolerant bacteria. *International Journal of Systematic Bacteriology* 30, 485–495.

796 Vuddhakul, V., Nakai, T., Matsumoto, C., Oh, T., Nishino, T., Chen, C.-H., Nishibuchi, M. and Okuda, J. (2000) Analysis of *gyrB* and *toxR* gene sequences of *Vibrio hollisae* and development of *gyrB*- and *toxR*-targeted PCR methods for isolation of *V. hollisae* from the environment and its identification. *Applied and Environmental Microbiology* 66, 3506–3514.

797 Vuillaume, A., Brun, R., Chene, P., Sochon, E. and Lesel, R. (1987) First isolation of *Yersinia ruckeri* from sturgeon, *Acipenser baeri* Brandt, in south west of France, *Bulletin of the European Association of Fish Pathologists* 7, 18–19.

798 Waechter, M., Roux, F.L., Nicolas, J.-L., Marissal, E. and Berthe, F. (2002) Characterization of pathogenic bacteria of the cupped oyster *Crassostrea gigas*. *Comptes Rendus Biologies* 325, 231–238.

799 Wakabayashi, H. and Egusa, S. (1972) Characteristics of a *Pseudomonas* sp. from an epizootic of pond-cultured eels (*Anguilla japonica*). *Bulletin of the Japanese Society of Scientific Fisheries* 38, 577–587.

800 Wakabayashi, H. and Egusa, S. (1973) *Edwardsiella tarda* (*Paracolobactrum anguillimortiferum*) associated with pond-cultured eel disease. *Bulletin of the Japanese Society of Scientific Fisheries* 39, 931–936.

801 Wakabayashi, H., Hikida, M. and Masumura, K. (1986) *Flexibacter maritimus* sp. nov., a pathogen of marine fishes. *International Journal of Systematic Bacteriology* 36, 396–398.

802 Wakabayashi, H., Huh, G. and Kimura, N. (1989) *Flavobacterium branchiophila* sp. nov., a causative agent of bacterial gill disease of freshwater fishes. *International Journal of Systematic Bacteriology* 39, 213–216.

803 Wakabayashi, H., Sawada, K., Ninomiya, K. and Nishimori, E. (1996) Bacterial hemorrhagic ascites of Ayu caused by *Pseudomonas* sp. *Fish Pathology* 31, 239–240.

804 Wallace, L.J., White, F.H. and Gore, H.L. (1966) Isolation of *Edwardsiella tarda* from a seal lion and two alligators. *Journal of American Veterinary Medical Association* 149, 881–883.

805 Wallace, R.J., Brown, B., Tsukamura, M., Brown, J. and Onyi, G. (1991) Clinical and laboratory features of *Nocardia nova*. *Journal of Clinical Microbiology* 29, 2407–2411.

806 Wallach, J.D. (1977) Ulcerative shell disease in turtles: identification, prophylaxis and treatment. *International Zoo Yearbook* 17, 170–171.

807 Wallbanks, S., Martinez-Murcia, A.J., Fryer, J.L., Phillips, B.A. and Collins, M.D. (1990) 16S rRNA sequence determination for members of the genus Carnobacterium and related lactic acid bacteria and description of *Vagococcus salmoninarum* sp. nov. *Journal of Systematic Bacteriology* 40, 224–230.

808 Waltman, W.D., Shotts, E.B. and Hsu, T.C. (1986) Biochemical characteristics of *Edwardsiella ictaluri*. *Applied and Environmental Microbiology* 51, 101–104.

809 Wang, G., Tyler, K., Munro, C. and Johnson, W. (1996) Characterization of cytotoxic, hemolytic *Aeromonas caviae* clinical isolates and their identification by determining presence of a unique hemolysin gene. *Journal of Clinical Microbiology* 34, 3203–3205.

810 Wang, R.-F., Cao, W.-W. and Cerniglia, C. (1997) A universal protocol for PCR detection of 13 species of foodborne pathogen in foods. *Journal of Applied Microbiology* 83, 727–736.

811 Watson, R. (1989) The formation of primer artifacts in polymerase chain reactions. *Amplifications* 1, 5–6.

812 Wauters, G., Janssens, M., Steigerwalt, A. and Brenner, D. (1988) *Yersinia mollaretii* sp. nov. and *Yersinia bercovieri* sp. nov., formerly called *Yersinia enterocolitica* biogroups 3A and 3B. *International Journal of Systematic Bacteriology* 38, 424–429.

813 Wayne, L.G., Brenner, D.J., Colwell, R.R., Grimont, P.A.D., Kandler, O., Krichevsky, M.I., Moore, L.H., Moore, W.E.C., Murray, R.G.E., Stackebrandt, E., Starr, M.P. and Trüper, H.G. (1987) Report of the ad hoc committee on reconciliation of approaches to bacterial systematics. *International Journal of Systematic Bacteriology* 37, 463–464.

814 Weiner, R., Segall, A. and Colwell, R. (1985) Characterization of a marine bacterium associated with *Crassostrea virginica* (the Eastern Oyster). *Applied and Environmental Microbiology* 49, 83–90.

815 Weiner, R.M., Coyne, V.E., Brayton, P., West, P. and Raiken, S.F. (1988) *Alteromonas colwelliana* sp. nov., an isolate from oyster habitats. *International Journal of Systematic Bacteriology* 38, 240–244.

816 Weinstein, M., Litt, M., Kertesz, D., Wyper, P., Rose, D., Coulter, M., McGreer, A., Facklam, R., Ostach, C., Willey, B., Borczyk, A. and Low, D. (1997) Invasive infections due to a fish pathogen, *Streptococcus iniae*. *New England Journal of Medicine* 9, 589–594.

817 Weisburg, W., Barns, S., Pelletier, D. and Lane, D. (1991) 16S ribosomal DNA amplification for phylogenetic study. *Journal of Bacteriology* 173, 697–703.

818 West, P.A. and Colwell, R.R. (1984) Identification and classification of Vibrionaceae – an overview.

In: Colwell, R.R. (ed.) *Vibrios in the Environment.* John Wiley & Sons, New York, pp. 285–363.

819 West, P., Lee, J.V. and Bryant, T.N. (1983) A numerical taxonomic study of species of *Vibrio* isolated from the aquatic environment and birds in Kent, England. *Journal of Applied Bacteriology* 55, 263–282.

820 West, P.A., Brayton, P.R., Twilley, R.R., Bryant, T.N. and Colwell, R.R. (1985) Numerical taxonomy of nitrogen-fixing 'decarboxylase-negative' Vibrio species isolated from aquatic environments. *International Journal of Systematic Bacteriology* 35, 198–205.

821 West, P.A., Brayton, P.R., Bryant, T.N. and Colwell, R.R. (1986) Numerical taxonomy of Vibrios isolated from aquatic environments. *International Journal of Systematic Bacteriology* 36, 531–543.

822 Westbrook, G., O'Hara, C.M., Roman, S. and Miller, J. M. (2000) Incidence and identification of *Klebsiella planticola* in clinical isolates with emphasis on newborns. *Journal of Clinical Microbiology* 38, 1495–1497.

823 White, F.H., Simpson, C.F. and Williams, L.E. (1973) Isolation of *Edwardsiella tarda* from aquatic animal species and surface waters in Florida. *Journal of Wildlife Diseases* 9, 204–208.

824 Whittington, R. and Cullis, B. (1988) The susceptibility of salmonid fish to an atypical strain of *Aeromonas salmonicida* that infects goldfish, *Carassius auratus* (L.), in Australia. *Journal of Fish Diseases* 11, 461–470.

825 Whittington, R., Gudkovs, N., Carrigan, M., Ashburner, L. and Thurstan, S. (1987) Clinical, microbiological and epidemiological findings in recent outbreaks of goldfish ulcer disease due to atypical *Aeromonas salmonicida* in south-eastern Australia. *Journal of Fish Diseases* 10, 353–362.

826 Whittington, R.J., Djordjevic, S., Carson, J. and Callinan, R. (1995) Restriction endonuclease analysis of atypical *Aeromonas salmonicida* isolates from goldfish *Carassius auratus*, silver perch *Bidyanus bidyanus*, and greenback flounder *Rhombosolea tapirina* in Australia. *Diseases of Aquatic Organisms* 22, 185–191.

827 Wiik, R., Torsvik, V. and Egidius, E. (1986) Phenotypic and genotypic comparisons among strains of the lobster pathogen *Aerococcus viridans* and other marine *Aerococcus viridans*-like cocci. *International Journal of Systematic Bacteriology* 36, 431–434.

828 Wiklund, T. and Bylund, G. (1990) *Pseudomonas anguilliseptica* as a pathogen of salmonid fish in Finland. *Diseases of Aquatic Organisms* 8, 13–19.

829 Wiklund, T. and Bylund, G. (1993) Skin ulcer disease of flounder *Platichthys flesus* in the northern Baltic Sea. *Diseases of Aquatic Organisms* 17, 165–174.

830 Wiklund, T. and Dalsgaard, I. (1998) Occurrence and significance of atypical *Aeromonas salmonicida* in non-salmonid and salmonid fish species: a review. *Diseases of Aquatic Organisms* 32, 49–69.

831 Wiklund, T., Dalsgaard, I., Eerola, E. and Olivier, G. (1994) Characteristics of 'atypical', cytochrome oxidase-negative *Aeromonas salmonicida* isolated from ulcerated flounders (*Platichthys flesus* (L.)). *Journal of Applied Bacteriology* 76, 511–520.

832 Wiklund, T., Tabolina, I. and Bezgachina, T. (1999) Recovery of atypical *Aeromonas salmonicida* from ulcerated fish from the Baltic Sea. *ICES Journal of Marine Science* 56, 175–179.

833 Wiklund, T., Madsen, L., Bruun, M. and Dalsgaard, I. (2000) Detection of *Flavobacterium psychrophilum* from fish tissue and water samples by PCR amplification. *Journal of Applied Microbiology* 88, 299–307.

834 Willems, A., Busse, J., Goor, M., Pot, B., Falsen, E., Jantzen, E., Hoste, B., Gillis, M., Kersters, K., Auling, G. and Ley, J.D. (1989) *Hydrogenophaga*, a new genus of hydrogen-oxidising bacteria that includes *Hydrogenophaga flava* comb. nov. (formerly *Pseudomonas flava*), *Hydrogenophaga palleronii* (formerly *Pseudomonas palleronii*), *Hydrogenophaga pseudoflava* (formerly *Pseudomonas pseudoflava* and '*Pseudomonas carboxydoflava*'), and *Hydrogenophaga taeniospiralis* (formerly *Pseudomonas taeniospiralis*). *International Journal of Systematic Bacteriology* 39, 319–333.

835 Williams, A.M., Fryer, J.L. and Collins, M.D. (1990) *Lactococcus piscium* sp. nov. a new *Lactococcus* species from salmonid fish. *FEMS Microbiology Letters* 68, 109–114.

836 Willumsen, B. (1989) Birds and wild fish as potential vectors of *Yersinia ruckeri*. *Journal of Fish Diseases* 12, 275–277.

837 Wilson, B. and Holliman, A. (1994) Atypical *Aeromonas salmonicida* isolated from ulcerated chub *Leuciscus cephalis*. *Veterinary Record* 135, 185–186.

838 Wilson, K., Blitchington, R. and Greene, R. (1990) Amplification of bacterial 16S ribosomal DNA with polymerase chain reaction. *Journal of Clinical Microbiology* 28, 1942–1946.

839 Wolters, W. and Johnson, M. (1994) Enteric septicaemia resistance in blue catfish and three channel catfish strains. *Journal of Aquatic Animal Health* 6, 329–334.

840 Wong, F., Fowler, K. and Desmarchelier, P. (1995) Vibriosis due to *Vibrio mimicus* in Australian freshwater crayfish. *Journal of Aquatic Animal Health* 7, 284–291.

841 Woo, P.T.K. and Bruno, D.W. (1999) *Viral, Bacterial and Fungal Infections. Fish Diseases and Disorders.* Vol. 3. CAB International, Wallingford, UK.

842 Wood, R. (1965) A selective liquid medium utilizing antibiotics for isolation of *Erysipelothrix insidiosa*. *American Journal of Veterinary Research* 26, 1303–1308.

843 Wood, R. and Packer, R.A. (1972) Isolation of *Erysipelothrix rhusiopathiae* from soil and manure of swine-raising premises. *American Journal of Veterinary Research* 33, 1611–1620.

844 Yabuuchi, E., Kaneko, T., Yano, I., Moss, C.W. and Miyoshi, N. (1983) *Sphingobacterium* gen. nov., *Sphingobacterium spiritivorum* comb. nov., *Sphingobacterium multivorum* comb. nov., *Sphingobacterium mizutae* sp. nov.: and *Flavobacterium indologenes* sp. nov.: glucose-nonfermenting Gram-negative rods in CDC Groups IIK-2 and IIb. *International Journal of Systematic Bacteriology* 33, 580–598.

845 Yamada, Y. and Wakabayashi, H. (1999) Identification of fish-pathogenic strains belonging to the genus *Edwardsiella* by sequence analysis of *sodB*. *Fish Pathology* 34, 145–150.

846 Yang, Y., Yeh, L., Cao, Y., Baumann, L., Baumann, P., Tang, J.S. and Beaman, B. (1983) Characterization of marine luminous bacteria isolated off the coast of China and description of *Vibrio orientalis* sp. nov. *Current Microbiology* 8, 95–100.

847 Yii, K.-C., Yang, T. and Lee, K.-K. (1997) Isolation and characterization of *Vibrio carchariae*, a causative agent of gastroenteritis in the groupers, *Epinephelus coioides*. *Current Microbiology* 35, 109–115.

848 Yuasa, K., Kitancharoen, N., Kataoka, Y. and Al-Murbaty, F.A. (1999) *Streptococcus iniae*, the causative agent of mass mortality in Rabbitfish *Siganus canaliculatus* in Bahrain. *Journal of Aquatic Animal Health* 11, 87–93.

849 Yumoto, I., Kawasaki, K., Iwata, H., Matsuyama, H. and Okuyama, H. (1998) Assignment of *Vibrio* sp. strain ABE-1 to *Colwellia maris* sp. nov., a new psychrophilic bacterium. *International Journal of Systematic Bacteriology* 48, 1357–1362.

850 Yumoto, I., Iwata, H., Sawabe, T., Ueno, K., Ichise, N., Matsuyama, H., Okuyama, H. and Kawasaki, K. (1999) Characterization of a facultatively psychrophilic bacterium, *Vibrio rumoiensis* sp. nov., that exhibits high catalase activity. *Applied and Environmental Microbiology* 65, 67–72.

851 Ziemke, F., Höfle, M.G., Lalucat, J. and Rosselló-Mora, R. (1998) Reclassification of *Shewanella putrefaciens* Owen's genomic group II as *Shewanella baltica* sp. nov. *International Journal of Systematic Bacteriology* 48, 179–186.

852 Zlotkin, A., Eldar, A., Ghittino, C. and Bercovier, H. (1998a) Identification of *Lactococcus garvieae* by PCR. *Journal of Clinical Microbiology* 36, 983–985.

853 Zlotkin, A., Hershko, H. and Eldar, A. (1998b) Possible transmission of *Streptococcus iniae* from wild fish to cultured marine fish. *Applied and Environmental Microbiology* 64, 4065–4067.

854 ZoBell, C.E. (1941) Studies on marine bacteria. I. The cultural requirements of heterotrophic aerobes. *Journal of Marine Research* 4, 42–75.

855 Zorilla, I., Balebona, M.C., Moriñigo, M.A., Sarasquete, C. and Borrego, J.J. (1999) Isolation and characterisation of the causative agent of pasteurellosis, *Photobacterium damselae* spp. *piscicida*, from sole, *Solea senegalensis* (Kaup). *Journal of Fish Diseases* 22, 167–172.

Index